POLYMER SCIENCE and TECHNOLOGY

Robert O. Ebewele
Faculty of Engineering
University of Benin
Benin City, Nigeria

CRC Press
Boca Raton London New York Washington, D.C.

Library of Congress Cataloging-in-Publication Data

Ebewele, Robert Oboigbaotor.
Polymer science and technology / Robert O. Ebewele.
p. cm.
Includes bibliographical references (p. -) and index.
ISBN 0-8493-8939-9 (alk. paper)
1. Polymerization. 2. Polymers. I. Title.
TP156.P6E24 1996
668.9—dc20 95-32995
CIP

Direct all inquiries to CRC Press LLC, 2000 N.W. Corporate Blvd., Boca Raton, Florida 33431.

Visit the CRC Press Web site at www.crcpress.com

No claim to original U.S. Government works
International Standard Book Number 0-8493-8939-9
Library of Congress Card Number 95-32995
Printed in the United States of America 3 4 5 6 7 8 9 0
Printed on acid-free paper

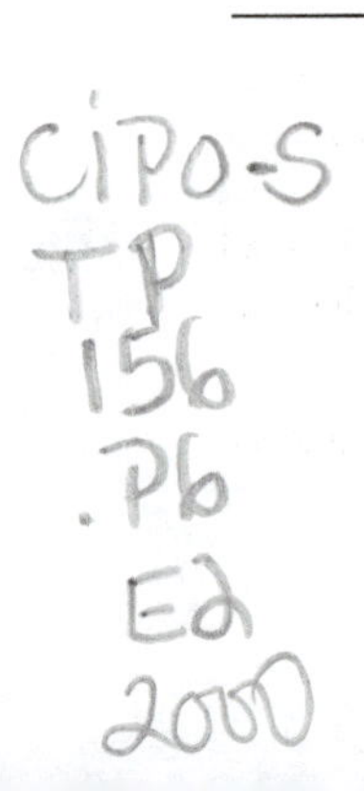

PREFACE

The book is divided into three parts. The first part covers polymer fundamentals. This includes a brief discussion of the historical development of polymers, basic definitions and concepts, and an overview of the basis for the various classifications of polymers. It also examines the requirements for polymer formation from monomers and discusses polymer structure at three levels: primary, secondary, and tertiary. The relationship between the structure of the monomers and properties of the resulting polymer is highlighted. This section continues with a discussion of polymer modification techniques. Throughout the discussion, emphasis is on the structure-property relationship and several examples are used to illustrate this concept.

The second part deals with how polymers are prepared from monomers and the transformation of polymers into useful everyday articles. It starts with a discussion of the various polymer preparation methods with emphasis on reaction mechanisms and kinetics. The control of molecular weight through appropriate manipulation of the stoichiometry of reactants and reaction conditions is consistently emphasized. This section continues with a discussion of polymer reaction engineering. Emphasis is on the selection of the appropriate polymerization process and reactor to obtain optimal polymer properties. The section terminates with a discussion of polymer additives and reinforcements and the various unit operations in polymer processing. Here again, the primary focus is on how processing conditions affect the properties of the part produced.

The third part of the book deals with the properties and applications of polymers. It starts with a discussion of polymer solution properties through the mechanical properties of polymers and concludes with an overview of the various applications of polymer materials solids. The viscoelastic nature of polymers is also treated. This section also includes a discussion of polymer fracture. The effects of various molecular and environmental factors on mechanical properties are examined.

The primary focus of the book is the ultimate property of the finished polymer product. Consequently, the emphasis throughout the book is on how various stages involved in the production of the finished product influence its properties. For example, which polymerization process will be preferable for a given monomer? Having decided on the polymerization process, which type of reactor will give optimum product properties? What is the best type of processing technique for a given polymer material? How do processing conditions affect the properties of the part produced and which polymer material is most suitable for a particular application? The book addresses the elements that must be considered to come up with appropriate answers to these types of questions. The distinguishing features of the book are intended to address certain problems associated with teaching an elementary course in polymers:

1. For a vast majority of introductory polymer courses, very frequently the instructor has to rely on several textbooks to cover the basics of polymers as none of the existing textbooks discusses the required materials satisfactorily. Most students find dealing with several textbooks in an introductory course problematic. This book attempts to remedy this problem. A deliberate effort has been made to cover most of the areas normally taught in such an introductory course. Indeed, these areas are typical of existing texts. However, the approach and depth of coverage are different. The book presents various aspects of polymer science and technology in a readily understandable way. Emphasis is on a basic, qualitative understanding of the concepts rather than rote memorization or detailed mathematical analysis. Description of experimental procedures employed in the characterization of polymers has been either completely left out or minimized. I strongly believe that this approach will appeal to those students who will be learning polymer science for the first time.
2. None of the existing texts has worked examples. It is my experience that students feel more comfortable with and generally prefer textbooks that illustrate principles being discussed with examples. I have followed this approach throughout the text. In addition, each chapter has review problems; answers are provided in a Solutions Manual. Both the worked examples and the review problems are designed to provide additional insight to the materials covered. The overall objective of this approach is to enhance the reader's understanding of the material and build his/her confidence. Emphasis throughout the book is on structure-property relationship and both the worked examples and review problems reflect this basic objective.

Robert O. Ebewele

ACKNOWLEDGMENT

In writing this book, I have had to rely on materials from various sources. These sources have been compiled as references at the end of each chapter. While I express my profound gratitude to publishers for permission to use their materials, I apologize for ideas and materials which I have inadvertently failed to acknowledge. I certainly do not lay claim to these published concepts and ideas.

The skeletal framework for this book was initiated during my student days at the University of Wisconsin, Madison and over the years, the material in the book has been constantly refined as it was being developed for use by successive generations of undergraduate and graduate students at the Ahmadu Bello University, Zaria, Nigeria. The final version of the book was written during my sabbatical leave at the Department of Chemical Engineering, University of Wisconsin, Madison, and subsequently during my leave of absence at the Forest Products Laboratory Madison, Wisconsin. I am grateful to the Ahmadu Bello University, Zaria, the University of Wisconsin, Madison and the Forest Products Laboratory, Madison for providing me unlimited access to their library materials and other facilities. Finally, I am indebted to the late Prof. J. A. Koutsky of the University of Wisconsin, Madison; Dr. George E. Myers and Mr. Bryan H. River, formerly of the Forest Products Laboratory, Madison; and a host of others for reviewing various parts of this book. Your contributions have greatly improved the quality of the book. I, however, take full responsibility for any lapses and errors that may be contained in the book.

TABLE OF CONTENTS

PART I: FUNDAMENTALS

Chapter Three
Chemical Bonding and Polymer Structure ... **59**

Chapter Four
Thermal Transitions in Polymers ... **95**

Chapter Five
Polymer Modification ... 117

PART II: POLYMER PREPARATION AND PROCESSING METHODS

Chapter Six
Condensation (Step-Reaction) Polymerization ... 161

PART III: PROPERTIES AND APPLICATIONS

Chapter Thirteen

Mechanical Properties of Polymers 343

Chapter Fourteen

Polymer Viscoelasticity 389

PART I: FUNDAMENTALS

Chapter 1

Introduction

I. HISTORICAL DEVELOPMENT

Before we go into details of the chemistry of polymers it is appropriate to briefly outline a few landmarks in the historical development of what we now know as polymers. Polymers have been with us from the beginning of time; they form the very basis (building blocks) of life. Animals, plants — all classes of living organisms — are composed of polymers. However, it was not until the middle of the 20th century that we began to understand the true nature of polymers. This understanding came with the development of plastics, which are true man-made materials that are the ultimate tribute to man's creativity and ingenuity. As we shall see in subsequent discussions, the use of polymeric materials has permeated every facet of our lives. It is hard to visualize today's world with all its luxury and comfort without man-made polymeric materials.

The plastics industry is recognized as having its beginnings in 1868 with the synthesis of cellulose nitrate. It all started with the shortage of ivory from which billiard balls were made. The manufacturer of these balls, seeking another production method, sponsored a competition. John Wesley Hyatt (in the U.S.) mixed pyroxin made from cotton (a natural polymer) and nitric acid with camphor. The result was cellulose nitrate, which he called celluloid. It is on record, however, that Alexander Parkes, seeking a better insulating material for the electrical industry, had in fact discovered that camphor was an efficient plasticizer for cellulose nitrate in 1862. Hyatt, whose independent discovery of celluloid came later, was the first to take out patents for this discovery.

Cellulose nitrate is derived from cellulose, a natural polymer. The first truly man-made plastic came 41 years later (in 1909) when Dr. Leo Hendrick Baekeland developed phenol–formaldehyde plastics (phenolics), the source of such diverse materials as electric iron and cookware handles, grinding wheels, and electrical plugs. Other polymers — cellulose acetate (toothbrushes, combs, cutlery handles, eyeglass frames); urea–formaldehyde (buttons, electrical accessories); poly(vinyl chloride) (flooring, upholstery, wire and cable insulation, shower curtains); and nylon (toothbrush bristles, stockings, surgical sutures) — followed in the 1920s.

Table 1.1 gives a list of some plastics, their year of introduction, and some of their applications. It is obvious that the pace of development of plastics, which was painfully slow up to the 1920s, picked up considerable momentum in the 1930s and the 1940s. The first generation of man-made polymers was the result of empirical activities; the main focus was on chemical composition with virtually no attention paid to structure. However, during the first half of the 20th century, extensive organic and physical developments led to the first understanding of the structural concept of polymers — long chains or a network of covalently bonded molecules. In this regard the classic work of the German chemist Hermann Staudinger on polyoxymethylene and rubber and of the American chemists W. T. Carothers on nylon stand out clearly. Staudinger first proposed the theory that polymers were composed of giant molecules, and he coined the word *macromolecule* to describe them. Carothers discovered nylon, and his fundamental research (through which nylon was actually discovered) contributed considerably to the elucidation of the nature of polymers. His classification of polymers as *condensation* or *addition* polymers persists today.

Following a better understanding of the nature of polymers, there was a phenomenal growth in the numbers of polymeric products that achieved commercial success in the period between 1925 and 1950. In the 1930s, acrylic resins (signs and glazing); polystyrene (toys, packaging and housewares industries); and melamine resins (dishware, kitchen countertops, paints) were introduced.

The search for materials to aid in the defense effort during World War II resulted in a profound impetus for research into new plastics. Polyethylene, now one of the most important plastics in the world, was developed because of the wartime need for better-quality insulating materials for such applications as radar cable. Thermosetting polyester resins (now used for boatbuilding) were developed for military use. The terpolymer acrylonitrile-butadiene-styrene (ABS), (telephone handsets, luggage,

Table 1.1 Introduction of Plastics Materials

Date	Material	Typical Use
1868	Cellulose nitrate	Eyeglass frames
1909	Phenol–formaldehyde	Telephone handsets, knobs, handles
1919	Casein	Knitting needles
1926	Alkyds	Electrical insulators
1927	Cellulose acetate	Toothbrushes, packaging
1927	Poly(vinyl chloride)	Raincoats, flooring
1929	Urea–formaldehyde	Lighting fixtures, electrical switches
1935	Ethyl cellulose	Flashlight cases
1936	Polyacrylonitrile	Brush backs, displays
1936	Poly(vinyl acetate)	Flashbulb lining, adhesives
1938	Cellulose acetate butyrate	Irrigation pipe
1938	Polystyrene	Kitchenwares, toys
1938	Nylon (polyamide)	Gears, fibers, films
1938	Poly(vinyl acetal)	Safety glass interlayer
1939	Poly(vinylidene chloride)	Auto seat covers, films, paper, coatings
1939	Melamine–formaldehyde	Tableware
1942	Polyester (cross-linkable)	Boat hulls
1942	Polyethylene (low density)	Squeezable bottles
1943	Fluoropolymers	Industrial gaskets, slip coatings
1943	Silicone	Rubber goods
1945	Cellulose propionate	Automatic pens and pencils
1947	Epoxies	Tools and jigs
1948	Acrylonitrile-butadiene-styrene copolymer	Luggage, radio and television cabinets
1949	Allylic	Electrical connectors
1954	Polyurethane	Foam cushions
1956	Acetal resin	Automotive parts
1957	Polypropylene	Safety helmets, carpet fiber
1957	Polycarbonate	Appliance parts
1959	Chlorinated polyether	Valves and fittings
1962	Phenoxy resin	Adhesives, coatings
1962	Polyallomer	Typewriter cases
1964	Ionomer resins	Skin packages, moldings
1964	Polyphenylene oxide	Battery cases, high temperature moldings
1964	Polyimide	Bearings, high temperature films and wire coatings
1964	Ethylene–vinyl acetate	Heavy gauge flexible sheeting
1965	Polybutene	Films
1965	Polysulfone	Electrical/electronic parts
1970	Thermoplastic polyester	Electrical/electronic parts
1971	Hydroxy acrylates	Contact lenses
1973	Polybutylene	Piping
1974	Aromatic polyamides	High-strength tire cord
1975	Nitrile barrier resins	Containers

safety helmets, etc.) owes its origins to research work emanating from the wartime crash program on large-scale production of synthetic rubber.

The years following World War II (1950s) witnessed great strides in the growth of established plastics and the development of new ones. The Nobel-prize-winning development of stereo-specific catalysts by Professors Karl Ziegler of Germany and Giulio Natta of Italy led to the ability of polymer chemists to "order" the molecular structure of polymers. As a consequence, a measure of control over polymer properties now exists; polymers can be tailor-made for specific purposes.

The 1950s also saw the development of two families of plastics — acetal and polycarbonates. Together with nylon, phenoxy, polyimide, poly(phenylene oxide), and polysulfone they belong to the group of plastics known as the engineering thermoplastics. They have outstanding impact strength and thermal and dimensional stability — properties that place them in direct competition with more conventional materials like metals.

The 1960s and 1970s witnessed the introduction of new plastics: thermoplastic polyesters (exterior automotive parts, bottles); high-barrier nitrile resins; and the so-called high-temperature plastics, including such materials as polyphenylene sulfide, polyether sulfone, etc. The high-temperature plastics were initially developed to meet the demands of the aerospace and aircraft industries. Today, however, they have moved into commercial areas that require their ability to operate continuously at high temperatures.

In recent years, as a result of better understanding of polymer structure–property relationships, introduction of new polymerization techniques, and availability of new and low-cost monomers, the concept of a truly tailor-made polymer has become a reality. Today, it is possible to create polymers from different elements with almost any quality desired in an end product. Some polymers are similar to existing conventional materials but with greater economic values, some represent significant improvements over existing materials, and some can only be described as unique materials with characteristics unlike any previously known to man. Polymer materials can be produced in the form of solid plastics, fibers, elastomers, or foams. They may be hard or soft or may be films, coatings, or adhesives. They can be made porous or nonporous or can melt with heat or set with heat. The possibilities are almost endless and their applications fascinating. For example, *ablation* is the word customarily used by the astronomers and astrophysicists to describe the erosion and disintegration of meteors entering the atmosphere. In this sense, long-range missiles and space vehicles reentering the atmosphere may be considered man-made meteors. Although plastic materials are generally thermally unstable, ablation of some organic polymers occurs at extremely high temperatures. Consequently, selected plastics are used to shield reentry vehicles from the severe heat generated by air friction and to protect rocket motor parts from hot exhaust gases, based on the concept known as ablation plastics. Also, there is a "plastic armor" that can stop a bullet, even shell fragments. (These are known to be compulsory attire for top government and company officials in politically troubled countries.) In addition, there are flexible plastics films that are used to wrap your favorite bread, while others are sufficiently rigid and rugged to serve as supporting members in a building.

In the years ahead, polymers will continue to grow. The growth, from all indications, will be not only from the development of new polymers, but also from the chemical and physical modification of existing ones. Besides, improved fabrication techniques will result in low-cost products. Today the challenges of recycling posed by environmental problems have led to further developments involving alloying and blending of plastics to produce a diversity of usable materials from what have hitherto been considered wastes.

II. BASIC CONCEPTS AND DEFINITIONS

The word *polymer* is derived from classical Greek ***poly*** meaning "many" and ***meres*** meaning "parts." Thus a polymer is a large molecule (macromolecule) built up by the repetition of small chemical units. To illustrate this, Equation 1.1 shows the formation of the polymer polystyrene.

$$n\,CH_2{=}CH(C_6H_5) \longrightarrow \left[-CH_2-CH(C_6H_5)- \right]_n \qquad (1.1)$$

styrene (monomer) (1) polystyrene (polymer) (2)

The styrene molecule (1) contains a double bond. Chemists have devised methods of opening this double bond so that literally thousands of styrene molecules become linked together. The resulting structure, enclosed in square brackets, is the polymer polystyrene (2). Styrene itself is referred to as a *monomer,* which is defined as any molecule that can be converted to a polymer by combining with other molecules of the same or different type. The unit in square brackets is called the *repeating unit.* Notice that the structure of the repeating unit is not exactly the same as that of the monomer even though both possess identical atoms occupying similar relative positions. The conversion of the monomer to the polymer involves a rearrangement of electrons. The residue from the monomer employed in the preparation of a polymer is referred to as the *structural unit.* In the case of polystyrene, the polymer is derived from a single monomer (styrene) and, consequently, the structural unit of the polystyrene chain is the same as its repeating unit. Other examples of polymers of this type are polyethylene, polyacrylonitrile, and polypropylene. However, some polymers are derived from the mutual reaction of two or more monomers that are chemically similar but not identical. For example, poly(hexamethylene adipamide) or nylon 6,6 (5) is made from the reaction of hexamethylenediamine (3) and adipic acid (4) (Equation 1.2).

$$H_2N-(CH_2)_6-NH_2 + HOOC-(CH_2)_4-COOH \longrightarrow H\left[\begin{matrix}H \\ | \\ N\end{matrix}-(CH_2)_6-\begin{matrix}H \\ | \\ N\end{matrix}-\begin{matrix}O \\ \| \\ C\end{matrix}-(CH_2)_4-\begin{matrix}O \\ \| \\ C\end{matrix}\right]_n OH \quad (1.2)$$

hexamethylenediamine (3) adipic acid (4) poly(hexamethylene adipamide) (5)

The repeating unit in this case consists of two structural units: $-\overset{H}{\overset{|}{N}}-(CH_2)_6-\overset{H}{\overset{|}{N}}-$, the residue from hexamethylenediamine; and $-\overset{O}{\overset{\|}{C}}-(CH_2)_4-\overset{O}{\overset{\|}{C}}-$, the residue from adipic acid. Other polymers that have repeating units with more than one structural unit include poly(ethyleneterephthalate) and proteins. As we shall see later, the constitution of a polymer is usually described in terms of its structural units.

The subscript designation, n, in Equations 1.1 and 1.2 indicates the number of repeating units strung together in the polymer chain (molecule). This is known as the *degree of polymerization (DP).* It specifies the length of the polymer molecule. Polymerization occurs by the sequential reactions of monomers, which means that a successive series of reactions occurs as the repeating units are linked together. This can proceed by the reaction of monomers to form a *dimer,* which in turn reacts with another monomer to form a *trimer* and so on. Reaction may also be between dimers, trimers, or any molecular species within the reaction mixture to form a progressively larger molecule. In either case, a series of linkages is built between the repeating units, and the resulting polymer molecule is often called a *polymer chain,* a description which emphasizes its physical similarity to the links in a chain. Low-molecular-weight polymerization products such as dimers, trimers, tetramers, etc., are referred to as *oligomers.* They generally possess undesirable thermal and mechanical properties. A high degree of polymerization is normally required for a material to develop useful properties and before it can be appropriately described as a polymer. Polystyrene, with a degree of polymerization of 7, is a viscous liquid (not of much use), whereas commercial grade polystyrene is a solid and the DP is typically in excess of 1000. It must be emphasized, however, that no clear demarcation has been established between the sizes of oligomers and polymers.

The degree of polymerization represents one way of quantifying the molecular length or size of a polymer. This can also be done by use of the term *molecular weight (MW).* By definition, MW(Polymer) = DP × MW(Repeat Unit). To illustrate this let us go back to polystyrene (2). There are eight carbon atoms and eight hydrogen atoms in the repeating unit. Thus, the molecular weight of the repeating unit is 104 ($8 \times 12 + 1 \times 8$). If, as we stated above, we are considering commercial grade polystyrene, we will be dealing with a DP of 1000. Consequently, the molecular weight of this type of polystyrene is 104,000. As we shall see later, molecular weight has a profound effect on the properties of a polymer.

Example 1.1: What is the molecular weight of polypropylene (PP), with a degree of polymerization of 3×10^4?

Solution: Structure of the repeating unit for PP

$$\left[-CH_2-\underset{\underset{CH_3}{|}}{CH}-\right] \quad \text{(Str. 1)}$$

Molecular weight of repeat unit = $(3 \times 12 + 6 \times 1) = 42$
Molecular weight of polypropylene = $3 \times 10^4 \times 42 = 1.26 \times 10^6$

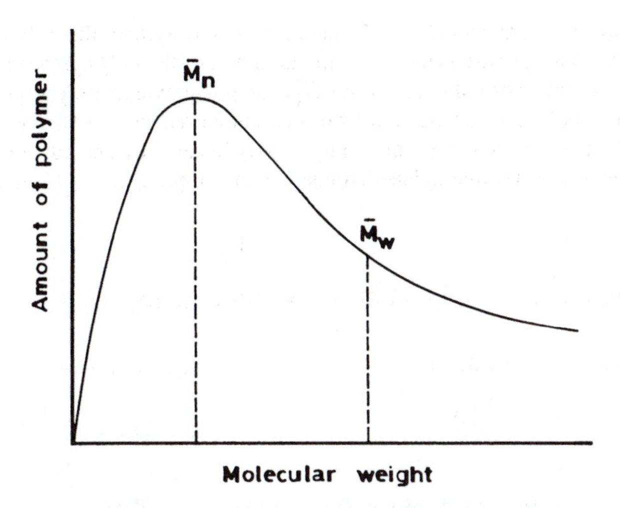

Figure 1.1 Molecular weight distribution curve.

So far, we have been discussing a single polymer molecule. However, a given polymer sample (like a piece of polystyrene from your kitchenware) is actually composed of millions of polymer molecules. For almost all synthetic polymers irrespective of the method of polymerization (formation), the length of a polymer chain is determined by purely random events. Consequently, any given polymeric sample contains a mixture of molecules having different chain lengths (except for some biological polymers like proteins, which have a single, well-defined molecular weight [monodisperse]). This means that a distribution of molecular weight exists for synthetic polymers. A typical molecular weight distribution curve for a polymer is shown in Figure 1.1.

The existence of a distribution of molecular weights in a polymer sample implies that any experimental measurement of molecular weight in the given sample gives only an average value. Two types of molecular weight averages are most commonly considered: the number-average molecular weight represented by $\overline{M}_n$, and the weight-average molecular weight $\overline{M}_w$. The number-average molecular weight is derived from measurements that, in effect, count the number of molecules in the given sample. On the other hand, the weight-average molecular weight is based on methods in which the contribution of each molecule to the observed effect depends on its size.

In addition to the information on the size of molecules given by the molecular weights $\overline{M}_w$ and $\overline{M}_n$, their ratio $\overline{M}_w/\overline{M}_n$ is an indication of just how broad the differences in the chain lengths of the constituent polymer molecules in a given sample are. That is, this ratio is a measure of polydispersity, and consequently it is often referred to as the heterogeneity index. In an ideal polymer such as a protein, all the polymer molecules are of the same size ($\overline{M}_w = \overline{M}_n$ or $\overline{M}_w/\overline{M}_n = 1$). This is not true for synthetic polymers – the numerical value of $\overline{M}_w$ is always greater than that of $\overline{M}_n$. Thus as the ratio $\overline{M}_w/\overline{M}_n$ increases, the molecular weight distribution is broader.

Example 1.2: Nylon 11 has the following structure

$$\left[-\overset{\displaystyle \text{H} \atop \displaystyle |}{\text{N}}-(\text{CH}_2)_{10}-\overset{\displaystyle \text{O} \atop \displaystyle \|}{\text{C}}- \right]_n \qquad \text{(Str. 2)}$$

If the number-average degree of polymerization, $\overline{X}_n$, for nylon is 100 and $\overline{M}_w$ = 120,000, what is its polydispersity?

Solution: We note that $\overline{X}_n$ and n(DP) define the same quantity for two slightly different entities. The degree of polymerization for a single molecule is n. But a polymer mass is composed of millions of molecules, each of which has a certain degree of polymerization. $\overline{X}_n$ is the average of these. Thus,

$$\overline{X}_n = \frac{\sum_{i=1}^{N} n_i M_r}{N}$$

where N = total number of molecules in the polymer mass
M_r = molecular weight of repeating unit
n_i = DP of molecule i.

$$\text{Now} \quad \overline{M}_n = \overline{X}_n M_r = 100\,(15 + 14 \times 10 + 28) = 18{,}300$$

$$\text{Polydispersity} = \frac{\overline{M}_w}{\overline{M}_n} = \frac{120{,}000}{18{,}300} = 6.56$$

III. CLASSIFICATION OF POLYMERS

Polymers can be classified in many different ways. The most obvious classification is based on the origin of the polymer, i.e., natural vs. synthetic. Other classifications are based on the polymer structure, polymerization mechanism, preparative techniques, or thermal behavior.

A. NATURAL VS. SYNTHETIC

Polymers may either be naturally occurring or purely synthetic. All the conversion processes occurring in our body (e.g., generation of energy from our food intake) are due to the presence of enzymes. Life itself may cease if there is a deficiency of these enzymes. Enzymes, nucleic acids, and proteins are polymers of biological origin. Their structures, which are normally very complex, were not understood until very recently. Starch — a staple food in most cultures — cellulose, and natural rubber, on the other hand, are examples of polymers of plant origin and have relatively simpler structures than those of enzymes or proteins. There are a large number of synthetic (man-made) polymers consisting of various families: fibers, elastomers, plastics, adhesives, etc. Each family itself has subgroups.

B. POLYMER STRUCTURE

1. Linear, Branched or Cross-linked, Ladder vs. Functionality

As we stated earlier, a polymer is formed when a very large number of structural units (repeating units, monomers) are made to link up by covalent bonds under appropriate conditions. Certainly even if the conditions are "right" not all simple (small) organic molecules possess the ability to form polymers. In order to understand the type of molecules that can form a polymer, let us introduce the term *functionality*. The functionality of a molecule is simply its interlinking capacity, or the number of sites it has available for bonding with other molecules under the specific polymerization conditions. A molecule may be classified as monofunctional, bifunctional, or polyfunctional depending on whether it has one, two, or greater than two sites available for linking with other molecules. For example, the extra pair of electrons in the double bond in the styrene molecules endows it with the ability to enter into the formation of two bonds. Styrene is therefore bifunctional. The presence of two condensable groups in both hexamethylenediamine ($-NH_2$) and adipic acid (–COOH) makes each of these monomers bifunctional. However, functionality as defined here differs from the conventional terminology of organic chemistry where, for example, the double bond in styrene represents a single functional group. Besides, even though the interlinking capacity of a monomer is ordinarily apparent from its structure, functionality as used in polymerization reactions is specific for a given reaction. A few examples will illustrate this.

A diamine like hexamethylenediamine has a functionality of 2 in amide-forming reactions such as that shown in Equation 1.2. However, in esterification reactions a diamine has a functionality of zero. Butadiene has the following structure:

$$\underset{1}{CH_2}=\underset{2}{CH}-\underset{3}{CH}=\underset{4}{CH_2} \qquad \text{(Str. 3)}$$

From our discussion about the polymerization of styrene, the presence of two double bonds on the structure of butadiene would be expected to prescribe a functionality of 4 for this molecule. Butadiene may indeed be tetrafunctional, but it can also have a functionality of 2 depending on the reaction conditions (Equation 1.3).

$$n\,\underset{1}{CH_2}=\underset{2}{CH}-\underset{3}{CH}=\underset{4}{CH_2} \xrightarrow[3,4]{1,2\text{ or}} \left[\begin{array}{c} -CH_2-CH- \\ | \\ CH \\ \| \\ CH_2 \end{array}\right]_n \quad (7)$$

$$\xrightarrow{1,4} \left[-CH_2-CH=CH-CH_2-\right]_n \quad (8) \qquad (1.3)$$

Since there is no way of making a distinction between the 1,2 and 3,4 double bonds, the reaction of either double bond is the same. If either of these double bonds is involved in the polymerization reaction, the residual or unreacted double bond is on the structure attached to the main chain [i.e., part of the pendant group (7)]. In 1,4 polymerization, the residual double bond shifts to the 2,3 position along the main chain. In either case, the residual double bond is inert and is generally incapable of additional polymerization under the conditions leading to the formation of the polymer. In this case, butadiene has a functionality of 2. However, under appropriate reaction conditions such as high temperature or cross-linking reactions, the residual unsaturation either on the pendant group or on the backbone can undergo additional reaction. In that case, butadiene has a total functionality of 4 even though all the reactive sites may not be activated under the same conditions. Monomers containing functional groups that react under different conditions are said to possess *latent functionality.*

Now let us consider the reaction between two monofunctional monomers such as in an esterification reaction (Equation 1.4).

$$\underset{\substack{\text{acid}\\(9)}}{R-COOH} + \underset{\substack{\text{alcohol}\\(10)}}{R'-OH} \longrightarrow \underset{\substack{\text{ester}\\(11)}}{R-\overset{\overset{\displaystyle O}{\|}}{C}-O-R'} \qquad (1.4)$$

You will observe that the reactive groups on the acid and alcohol are used up completely and that the product ester (11) is incapable of further esterification reaction. But what happens when two bifunctional molecules react? Let us use esterification once again to illustrate the principle (Equation 1.5).

$$\underset{\substack{\text{bifunctional}\\(12)}}{HOOC-R-COOH} + \underset{\substack{\text{bifunctional}\\(13)}}{HO-R'OH} \longrightarrow \underset{\substack{\text{bifunctional}\\(14)}}{HOOC-R-\overset{\overset{\displaystyle O}{\|}}{C}-O-R'-OH} \qquad (1.5)$$

The ester (14) resulting from this reaction is itself bifunctional, being terminated on either side by groups that are capable of further reaction. In other words, this process can be repeated almost indefinitely. The same argument holds for polyfunctional molecules. It is thus obvious that the generation of a polymer through the repetition of one or a few elementary units requires that the molecule(s) must be at least bifunctional.

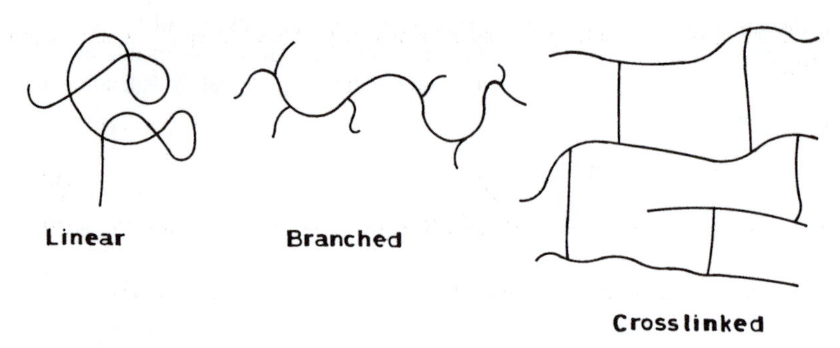

Figure 1.2 Linear, branched, and cross-linked polymers.

The structural units resulting from the reaction of monomers may in principle be linked together in any conceivable pattern. Bifunctional structural units can enter into two and only two linkages with other structural units. This means that the sequence of linkages between bifunctional units is necessarily linear. The resulting polymer is said to be *linear.* However, the reaction between polyfunctional molecules results in structural units that may be linked so as to form nonlinear structures. In some cases the side growth of each polymer chain may be terminated before the chain has a chance to link up with another chain. The resulting polymer molecules are said to be *branched.* In other cases, growing polymer chains become chemically linked to each other, resulting in a *cross-linked* system (Figure 1.2).

The formation of a cross-linked polymer is exemplified by the reaction of epoxy polymers, which have been used traditionally as adhesives and coatings and, more recently, as the most common matrix in aerospace composite materials. Epoxies exist at ordinary temperatures as low-molecular-weight viscous liquids or prepolymers. The most widely used prepolymer is diglycidyl ether of bisphenol A (DGEBA), as shown below (15):

$$\overset{\overset{O}{\diagup\ \diagdown}}{CH_2-CH}-CH_2-O-C_6H_4-\underset{CH_3}{\overset{CH_3}{\underset{|}{\overset{|}{C}}}}-C_6H_4-O-CH_2-\overset{\overset{O}{\diagup\ \diagdown}}{CH-CH_2} \qquad \text{(Str. 4)}$$

diglycidyl ether of bisphenol A (DGEBA)
(15)

The transformation of this viscous liquid into a hard, cross-linked three-dimensional molecular network involves the reaction of the prepolymer with reagents such as amines or Lewis acids. This reaction is referred to as *curing.* The curing of epoxies with a primary amine such as hexamethylene-diamine involves the reaction of the amine with the epoxide. It proceeds essentially in two steps:

1. The attack of an epoxide group by the primary amine

$$\underset{\text{1°amine 1°amine (16)}}{H_2N-R-NH_2} + \underset{\text{epoxide (17)}}{\overset{\overset{O}{\diagup\ \diagdown}}{CH_2-CH}-} \longrightarrow \underset{\text{1°amine 2°amine (18)}}{H_2N-R-\overset{\overset{H}{|}}{N}-CH_2-\overset{\overset{OH}{|}}{CH}-} \qquad (1.6)$$

2. The combination of the resulting secondary amine with a second epoxy group to form a branch point (19).

$$\underset{\text{1°amine}\quad\text{2°amine}}{H_2N-R-\overset{\overset{H}{|}}{N}-CH_2-\overset{\overset{OH}{|}}{CH}-} + \underset{\text{epoxide}}{CH_2\overset{O}{\diagup\diagdown}CH-} \longrightarrow \underset{\text{branch point (19)}}{H_2N-R-\overset{\overset{\overset{CH-OH}{|}}{CH_2}}{\underset{|}{N}}-CH_2-\overset{\overset{OH}{|}}{CH}-} \quad (1.7)$$

The presence of these branch points ultimately leads to a cross-linked infusible and insoluble polymer with structures such as (20).

$$-\overset{\overset{OH}{|}}{CH}-CH_2-\underset{\underset{\underset{|}{CH-OH}}{|}}{\underset{CH_2}{\underset{|}{N}}}-R-\overset{\overset{\overset{|}{CH-OH}}{|}}{\overset{CH_2}{\overset{|}{N}}}-CH_2-\overset{\overset{OH}{|}}{CH}- \quad \text{(Str. 5)}$$

(20)

In this reaction, the stoichiometric ratio requires one epoxy group per amine hydrogen. Consequently, an amine such as hexamethylenediamine has a functionality of 4. Recall, however, that in the reaction of hexamethylenediamine with adipic acid, the amine has a functionality of 2. In this reaction DGEBA is bifunctional since the hydroxyl groups generated in the reaction do not participate in the reaction. But when the curing of epoxies involves the use of a Lewis acid such as BF_3, the functionality of each epoxy group is 2; that is, the functionality of DGEBA is 4. Thus the curing reactions of epoxies further illustrate the point made earlier that the functionality of a given molecule is defined for a specific reaction. By employing different reactants or varying the stoichiometry of reactants, different structures can be produced and, consequently, the properties of the final polymer can also be varied.

Polystyrene (2), polyethylene (21), polyacrylonitrile (22), poly(methyl methacrylate) (23), and poly(vinyl chloride) (24) are typical examples of linear polymers.

$$\underset{(21)}{\left[-CH_2-CH_2-\right]_n} \quad \underset{(22)}{\left[-CH_2-\underset{CN}{\underset{|}{CH}}-\right]_n} \quad \underset{(23)}{\left[-CH_2-\overset{\overset{CH_3}{|}}{\underset{\underset{\underset{\underset{CH_3}{|}}{O}}{\underset{|}{C=O}}}{\underset{|}{C}}}-\right]_n} \quad \underset{(24)}{\left[-CH_2-\underset{Cl}{\underset{|}{CH}}-\right]_n} \quad \text{(Str. 6)}$$

Substituent groups such as $-CH_3$, $-O-\overset{O}{\overset{\|}{C}}-CH_3$, –Cl, and –CN that are attached to the main chain of skeletal atoms are known as *pendant groups.* Their structure and chemical nature can confer unique properties on a polymer. For example, linear and branched polymers are usually soluble in some solvent at normal temperatures. But the presence of polar pendant groups can considerably reduce room temperature solubility. Since cross-linked polymers are chemically tied together and solubility essentially involves the separation of solute molecules by solvent molecules, cross-linked polymers do not dissolve, but can only be swelled by liquids. The presence of cross-linking confers stability on polymers. Highly cross-linked polymers are generally rigid and high-melting. Cross-links occur randomly in a cross-linked polymer. Consequently, it can be broken down into smaller molecules by random chain scission. *Ladder polymers* constitute a group of polymers with a regular sequence of cross-links. A ladder polymer, as the name implies, consists of two parallel linear strands of molecules with a regular sequence of cross-links. Ladder polymers have only condensed cyclic units in the chain; they are also commonly referred to as double-chain or double-strand polymers. A typical example is poly(imidazopyrrolone) (27), which is obtained by the polymerization of aromatic dianhydrides such as pyromellitic dianhydride (25) or aromatic tetracarboxylic acids with *ortho*-aromatic tetramines like 1,2,4,5-tetraaminobenzene (26):

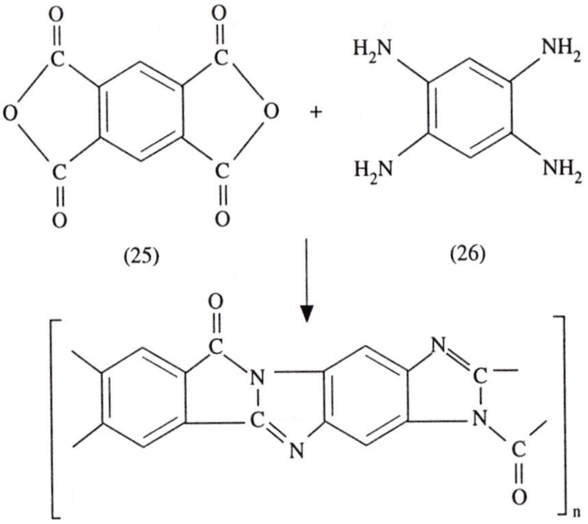

O O ‖ ‖ C C O O + H_2N NH_2 C C H_2N NH_2 ‖ ‖ O O

(25) (26)

(Str. 7)

O ‖ C N C N N C N C ‖ O n

(27)

The molecular structure of ladder polymers is more rigid than that of conventional linear polymers. Numerous members of this family of polymers display exceptional thermal, mechanical, and electrical behavior. Their thermal stability is due to the molecular structure, which in essence requires that two bonds must be broken at a cleavage site in order to disrupt the overall integrity of the molecule; when only one bond is broken, the second holds the entire molecule together.

Example 1.3: Show the polymer formed by the reaction of the following monomers. Is the resulting polymer linear or branched/cross-linked?

i. $OCN-(CH_2)_x-NCO + HO-CH_2-CH(OH)-(CH_2)_n-CH_2OH$ (Str. 8)

ii. 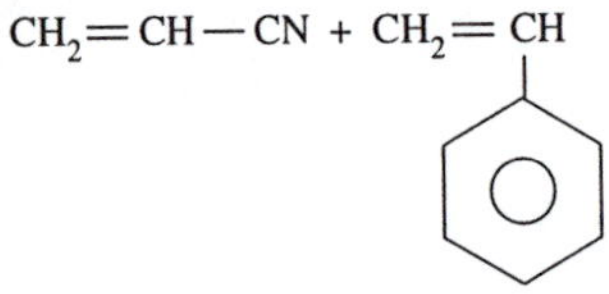 $CH_2=CH-CN + CH_2=CH-C_6H_5$ (Str. 9)

iii. 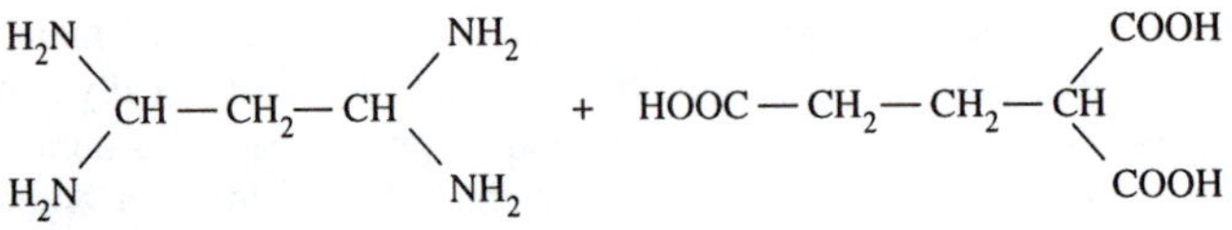 $(H_2N)_2CH-CH_2-CH(NH_2)_2 + HOOC-CH_2-CH_2-CH(COOH)_2$ (Str. 10)

iv. 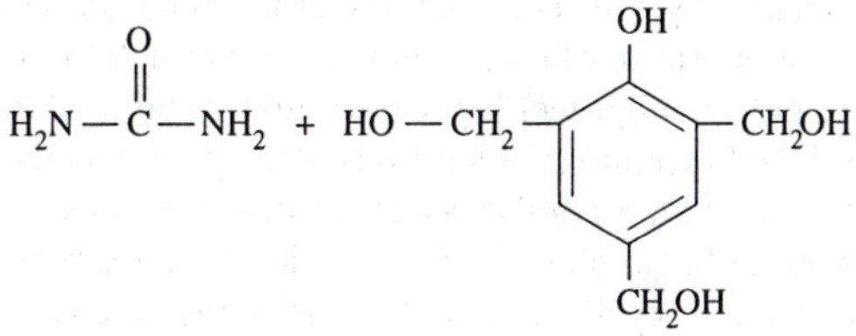 $$H_2N-\overset{\overset{\displaystyle O}{\|}}{C}-NH_2 \; + \; HO-CH_2-C_6H_2(OH)(CH_2OH)_2$$ (Str. 11)

(2,4,6-tris(hydroxymethyl)phenol: ring bearing OH, two CH_2OH groups and the $HO-CH_2$ group)

v. 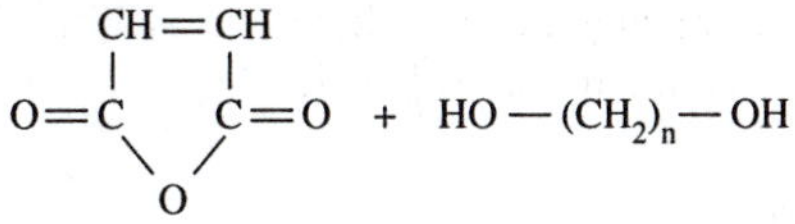 $$\text{maleic anhydride } (O{=}C-CH{=}CH-C{=}O \text{ bridged by } O) \; + \; HO-(CH_2)_n-OH$$ (Str. 12)

Solution:

i. 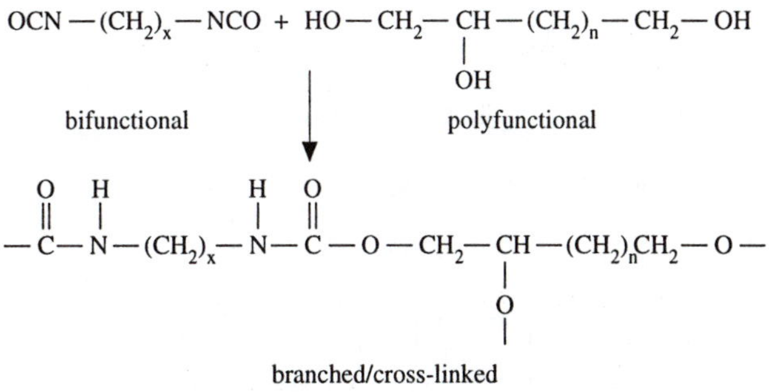 $$OCN-(CH_2)_x-NCO \; + \; HO-CH_2-\underset{\underset{\displaystyle OH}{|}}{CH}-(CH_2)_n-CH_2-OH$$ (Str. 13)

bifunctional — polyfunctional

$$\downarrow$$

$$-\overset{\overset{\displaystyle O}{\|}}{C}-\overset{\overset{\displaystyle H}{|}}{N}-(CH_2)_x-\overset{\overset{\displaystyle H}{|}}{N}-\overset{\overset{\displaystyle O}{\|}}{C}-O-CH_2-\underset{\underset{\underset{\displaystyle |}{\displaystyle O}}{|}}{CH}-(CH_2)_nCH_2-O-$$

branched/cross-linked

ii. 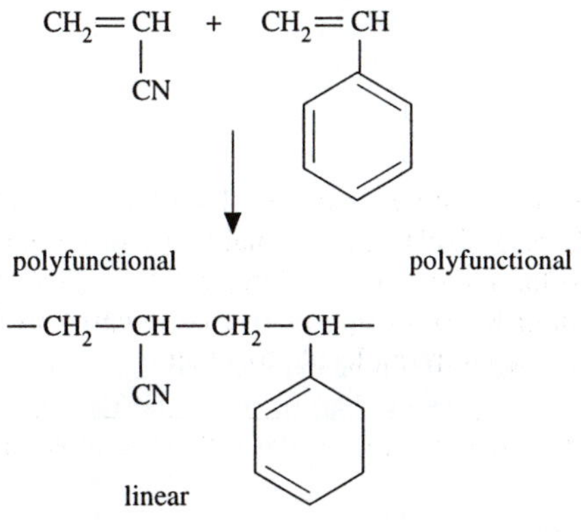 $$\underset{\underset{\displaystyle CN}{|}}{CH_2{=}CH} \; + \; \underset{\underset{\displaystyle C_6H_5}{|}}{CH_2{=}CH}$$ (Str. 14)

polyfunctional — polyfunctional

$$\downarrow$$

$$-CH_2-\underset{\underset{\displaystyle CN}{|}}{CH}-CH_2-\underset{\underset{\displaystyle C_6H_5}{|}}{CH}-$$

linear

iii. 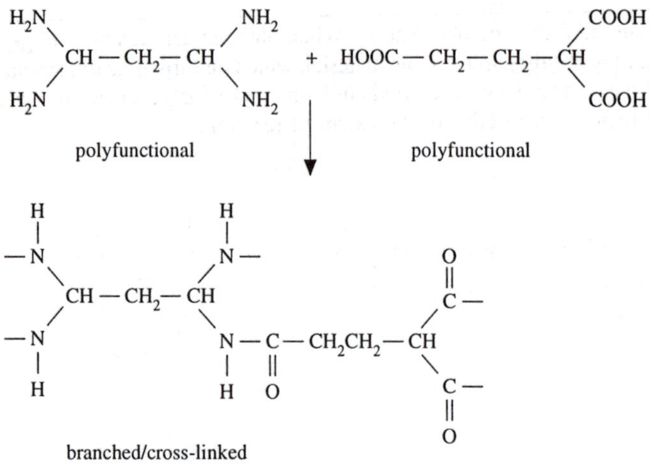 (Str. 15)

$$(H_2N)_2CH-CH_2-CH(NH_2)_2 + HOOC-CH_2-CH_2-CH(COOH)_2 \longrightarrow$$

polyfunctional polyfunctional

branched/cross-linked

iv. 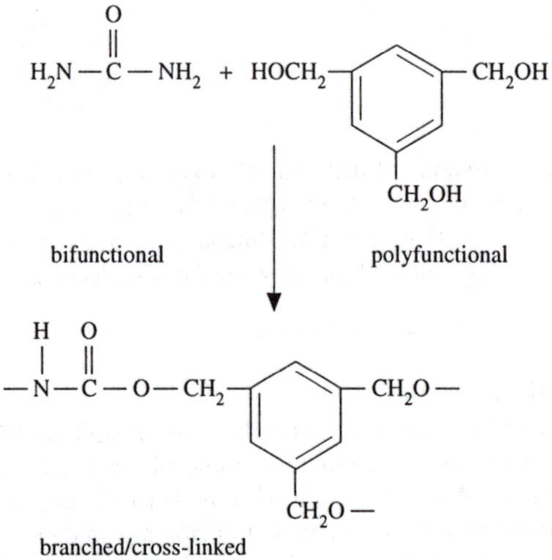 (Str. 16)

$$H_2N-\overset{O}{\overset{\|}{C}}-NH_2 + HOCH_2-C_6H_3(-CH_2OH)-CH_2OH \longrightarrow -\overset{H}{\overset{|}{N}}-\overset{O}{\overset{\|}{C}}-O-CH_2-C_6H_3(-CH_2O-)-CH_2O-$$

bifunctional polyfunctional

branched/cross-linked

The resulting secondary hydrogens in the urea linkages are capable of additional reaction depending on the stoichiometric proportions of reactants. This means that, in principle, the urea molecule may be polyfunctional (tetrafunctional).

v. 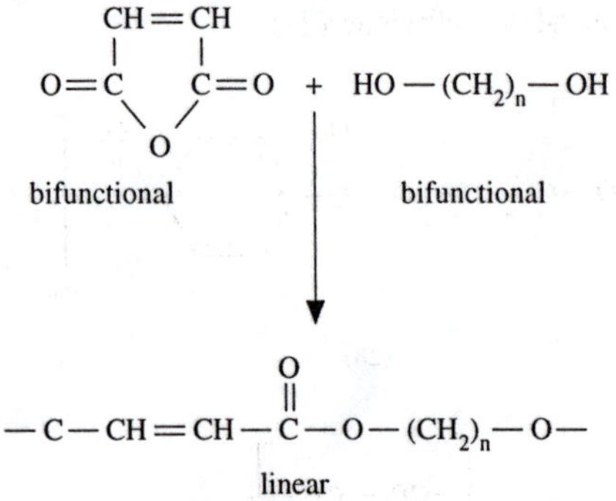 (Str. 17)

$$\text{maleic anhydride} + HO-(CH_2)_n-OH \longrightarrow -C-CH{=}CH-\overset{O}{\overset{\|}{C}}-O-(CH_2)_n-O-$$

bifunctional bifunctional

linear

Even though the resulting polymer is linear, it can be cross-linked in a subsequent reaction due to the unsaturation on the main chain – for example, by using radical initiators.

Example 1.4: Explain the following observation. When phthalic acid reacts with glycerol, the reaction leads first to the formation of fairly soft soluble material, which on further heating yields a hard, insoluble, infusible material. If the same reaction is carried out with ethylene glycol instead of glycerol, the product remains soluble and fusible irrespective of the extent of reaction.

Solution:

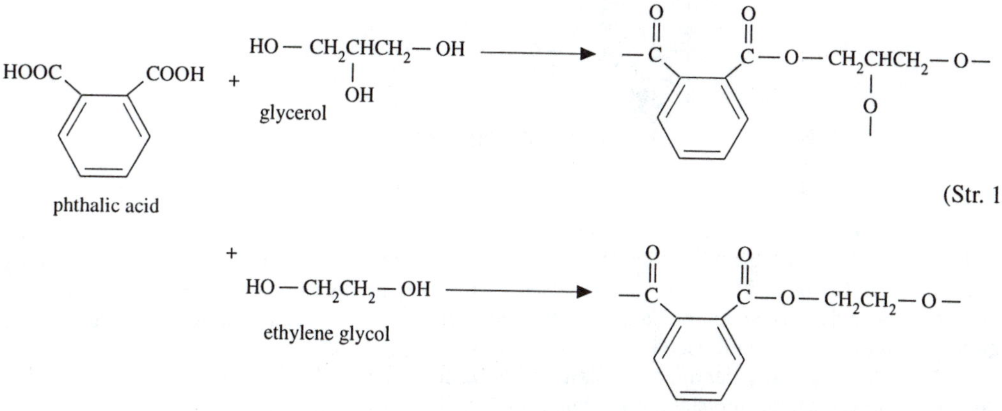

HOOC–C_6H_4–COOH (phthalic acid) + HO—CH_2CHCH_2—OH (glycerol, with OH on the central carbon) ⟶ —C(=O)–C_6H_4–C(=O)—O—CH_2CHCH_2—O— (with —O— branch on the central carbon)

(Str. 18)

\+ HO—CH_2CH_2—OH (ethylene glycol) ⟶ —C(=O)–C_6H_4–C(=O)—O—CH_2CH_2—O—

Phthalic acid and ethylene glycol are both bifunctional. Consequently, only linear polymers are produced from the reaction between these monomers. On the other hand, the reaction between phthalic acid and glycerol leads initially to molecules that are either linear, branched, or both. But since glycerol is trifunctional, cross-linking ultimately takes place between these molecules leading to an insoluble and infusible material.

2. Amorphous or Crystalline

Structurally, polymers in the solid state may be *amorphous* or *crystalline.* When polymers are cooled from the molten state or concentrated from the solution, molecules are often attracted to each other and tend to aggregate as closely as possible into a solid with the least possible potential energy. For some polymers, in the process of forming a solid, individual chains are folded and packed regularly in an orderly fashion. The resulting solid is a crystalline polymer with a long-range, three-dimensional, ordered arrangement. However, since the polymer chains are very long, it is impossible for the chains to fit into a perfect arrangement equivalent to that observed in low-molecular-weight materials. A measure of imperfection always exists. The degree of crystallinity, i.e., the fraction of the total polymer in the crystalline regions, may vary from a few percentage points to about 90% depending on the crystallization conditions. Examples of crystalline polymers include polyethylene (21), polyacrylonitrile (22), poly(ethylene terephthalate) (28), and polytetrafluoroethylene (29).

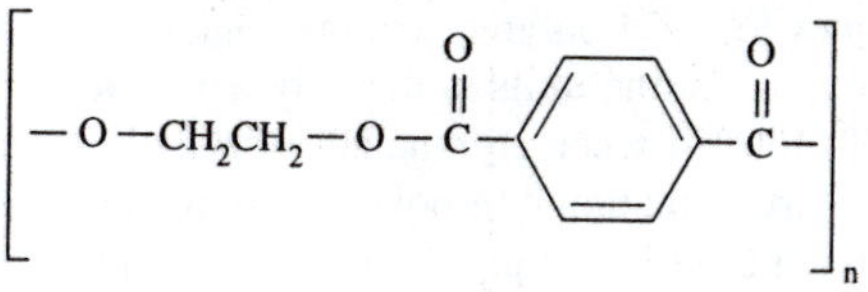

$$\left[-O-CH_2CH_2-O-\overset{O}{\overset{\|}{C}}-C_6H_4-\overset{O}{\overset{\|}{C}}- \right]_n \qquad \text{(Str. 19)}$$

(28)

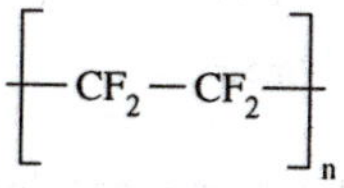

$$\left[-CF_2-CF_2- \right]_n \qquad \text{(Str. 20)}$$

(29)

In contrast to crystallizable polymers, amorphous polymers possess chains that are incapable of ordered arrangement. They are characterized in the solid state by a short-range order of repeating units. These polymers vitrify, forming an amorphous glassy solid in which the molecular chains are arranged at random and even entangled. Poly(methyl methacrylate) (23) and polycarbonate (30) are typical examples.

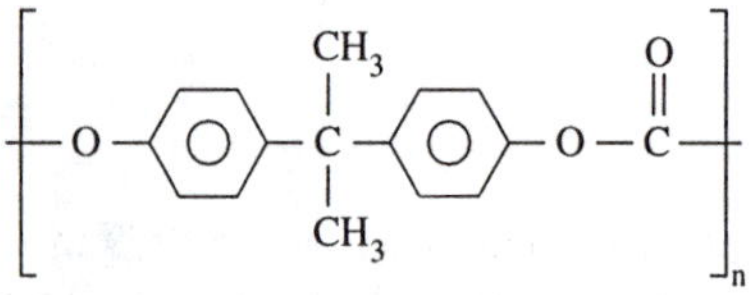

$$\left[-O-C_6H_4-C(CH_3)_2-C_6H_4-O-\overset{O}{\overset{\|}{C}}- \right]_n \qquad \text{(Str. 21)}$$

(30)

From the above discussion, it is obvious that the solid states of crystalline and amorphous polymers are characterized by a long-range order of molecular chains and a short-range order of repeating units, respectively. On the other hand, the melting of either polymer marks the onset of disorder. There are, however, some polymers which deviate from this general scheme in that the structure of the ordered regions is more or less disturbed. These are known as *liquid crystalline polymers.* They have phases characterized by structures intermediate between the ordered crystalline structure and the disordered fluid state. Solids of liquid crystalline polymers melt to form fluids in which much of the molecular order is retained within a certain range of temperature. The ordering is sufficient to impart some solid-like properties on the fluid, but the forces of attraction between molecules are not strong enough to prevent flow. An example of a liquid crystalline polymer is polybenzamide (31).

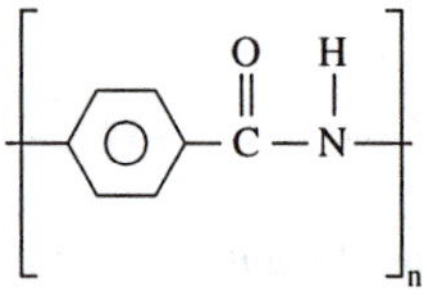

$$\left[-C_6H_4-\overset{O}{\overset{\|}{C}}-\underset{}{\overset{H}{\overset{|}{N}}}- \right]_n \qquad \text{(Str. 22)}$$

(31)

Liquid crystalline polymers are important in the fabrication of lightweight, ultra-high-strength, and temperature-resistant fibers and films such as Dupont's Kevlar and Monsanto's X-500. The structural factors responsible for promoting the above classes of polymers will be discussed when we treat the structure of polymers.

3. Homopolymer or Copolymer

Polymers may be either homopolymers or copolymers depending on the composition. Polymers composed of only one repeating unit in the polymer molecules are known as *homopolymers.* However, chemists have developed techniques to build polymer chains containing more than one repeating unit. Polymers composed of two different repeating units in the polymer molecule are defined as *copolymers.* An example is the copolymer (32) formed when styrene and acrylonitrile are polymerized in the same reactor. The repeating unit and the structural unit of a polymer are not necessarily the same. As indicated earlier, some polymers such as nylon 6,6 (5) and poly(ethylene terephthalate) (28) have repeating units composed of more than one structural unit. Such polymers are still considered homopolymers.

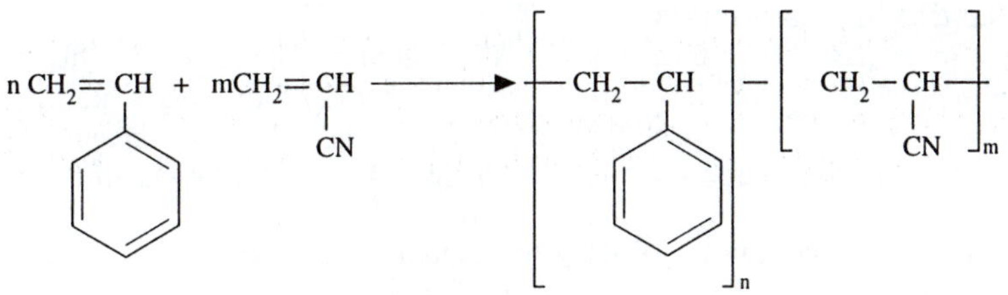

$$n\,CH_2{=}CH(C_6H_5) + m\,CH_2{=}CH(CN) \longrightarrow -\left[CH_2-CH(C_6H_5)\right]_n-\left[CH_2-CH(CN)\right]_m- \quad \text{(Str. 23)}$$

(32)

The repeating units on the copolymer chain may be arranged in various degrees of order along the backbone; it is even possible for one type of backbone to have branches of another type. There are several types of copolymer systems:

- **Random copolymer** — The repeating units are arranged randomly on the chain molecule. It we represent the repeating units by A and B, then the random copolymer might have the structure shown below:

  —AABBABABBAAABAABBA— (Str. 24)

- **Alternating copolymer** — There is an ordered (alternating) arrangement of the two repeating units along the polymer chain:

  —ABABABABABAB— (Str. 25)

- **Block copolymer** — The chain consists of relatively long sequences (blocks) of each repeating unit chemically bound together:

  —AAAAA—BBBBBBBB—AAAAAAAAA—BBBB— (Str. 26)

- **Graft copolymer** — Sequences of one monomer (repeating unit) are "grafted" onto a backbone of the another monomer type:

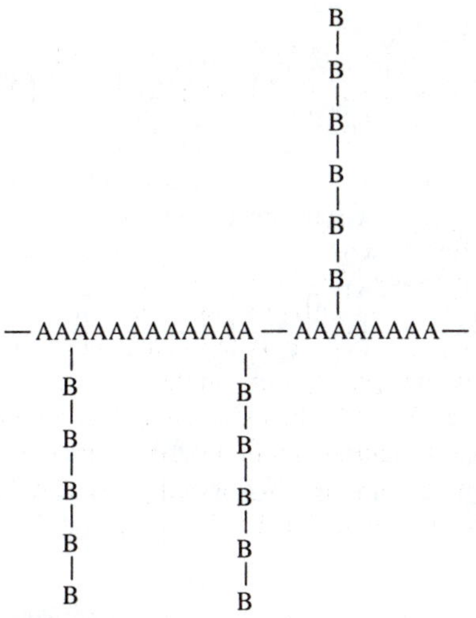

```
                                  B
                                  |
                                  B
                                  |
                                  B
                                  |
                                  B
                                  |
                                  B
                                  |
                                  B
                                  |
—AAAAAAAAAAAA—AAAAAAAA—
   |               |
   B               B
   |               |
   B               B
   |               |
   B               B
   |               |
   B               B
   |               |
   B               B
```

(Str. 27)

4. Fibers, Plastics, or Elastomers

Polymers may also be classified as fibers, plastics, or elastomers. The reason for this is related to how the atoms in a molecule (large or small) are hooked together. To form bonds, atoms employ valence electrons. Consequently, the type of bond formed depends on the electronic configuration of the atoms. Depending on the extent of electron involvement, chemical bonds may be classified as either primary or secondary.

In primary valence bonding, atoms are tied together to form molecules using their valence electrons. This generally leads to strong bonds. Essentially there are three types of primary bonds: ionic, metallic, and covalent. The atoms in a polymer are mostly, although not exclusively, bonded together by covalent bonds.

Secondary bonds on the other hand, do not involve valence electrons. Whereas in the formation of a molecule atoms use up all their valence bonds, in the formation of a mass, individual molecules attract each other. The forces of attraction responsible for the cohesive aggregation between individual molecules are referred to as secondary valence forces. Examples are van der Waals, hydrogen, and dipole bonds. Since secondary bonds do not involve valence electrons, they are weak. (Even between secondary bonds, there are differences in the magnitude of the bond strengths: generally hydrogen and dipole bonds are much stronger than van der Waals bonds.) Since secondary bonds are weaker than primary bonds, molecules must come together as closely as possible for secondary bonds to have maximum effect.

The ability for close alignment of molecules depends on the structure of the molecules. Those molecules with regular structure can align themselves very closely for effective utilization of the secondary intermolecular bonding forces. The result is the formation of a *fiber*. Fibers are linear polymers with high symmetry and high intermolecular forces that result usually from the presence of polar groups. They are characterized by high modulus, high tensile strength, and moderate extensibilities (usually less than 20%). At the other end of the spectrum, there are some molecules with irregular structure, weak intermolecular attractive forces, and very flexible polymer chains. These are generally referred to as *elastomers*. Chain segments of elastomers can undergo high local mobility, but the gross mobility of chains is restricted, usually by the introduction of a few cross-links into the structure. In the absence of applied (tensile) stress, molecules of elastomers usually assume coiled shapes. Consequently, elastomers exhibit high extensibility (up to 1000%) from which they recover rapidly on the removal of the imposed stress. Elastomers generally have low initial modulus in tension, but when stretched they stiffen. *Plastics* fall between the structural extremes represented by fibers and elastomers. However, in spite of the possible differences in chemical structure, the demarcation between fibers and plastics may sometimes be blurred. Polymers such as polypropylene and polyamides can be used as fibers and as plastics by a proper choice of processing conditions.

C. POLYMERIZATION MECHANISM

Polymers may be classified broadly as *condensation, addition,* or *ring-opening* polymers, depending on the type of polymerization reaction involved in their formation. *Condensation* polymers are formed from a series of reactions, often of condensation type, in which any two species (monomers, dimers, trimers, etc.) can react at any time leading to a larger molecule. In condensation polymerization, the stepwise reaction occurs between the chemically reactive groups or functional groups on the reacting molecules. In the process, a small molecule, usually water or ammonia, is eliminated. A typical condensation polymerization reaction is the formation of a polyester through the reaction of a glycol and a dicarboxylic acid (Equation 1.8). Examples of condensation polymers include polyamides (e.g., nylon 6,6) (5); polyesters (e.g., poly(ethylene terephthalate) (28); and urea-formaldehyde and phenol–formaldehyde resins.

$$\mathrm{nHO{-}R{-}OH + nHOOC{-}R'{-}COOH \rightleftharpoons nH{+}\left[O{-}R{-}O{-}\overset{\overset{\displaystyle O}{\|}}{C}{-}R'{-}\overset{\overset{\displaystyle O}{\|}}{C}\right]_n{-}OH + nH_2O} \tag{1.8}$$

Addition polymers are produced by reactions in which monomers are added one after another to a rapidly growing chain. The growing polymer in addition polymerization proceeds via a chain mechanism. Like all chain reactions, three fundamental steps are involved: initiation, propagation, and termination. Monomers generally employed in addition polymerization are unsaturated (usually with carbon-carbon double bonds). Examples of addition polymers are polystyrene (2), polyethylene (21), polyacrylonitrile (22), poly(methyl methacrylate) (23), and poly(vinyl chloride) (24).

As the name suggests, ring-opening polymerization polymers are derived from the cleavage and then polymerization of cyclic compounds. A broad generalization of ring-opening polymerization is shown in Equation 1.9.

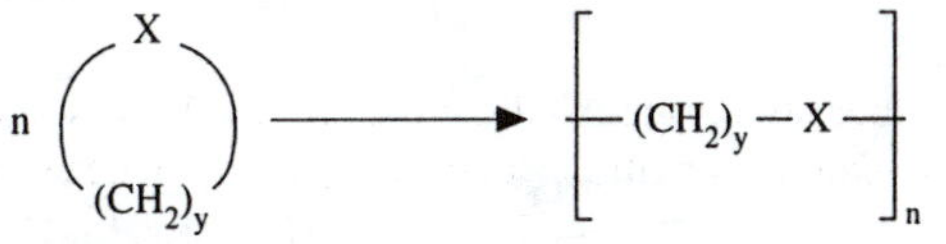

$$n \, \underbrace{\left(\overset{X}{\frown} \right)}_{(CH_2)_y} \longrightarrow \left[-(CH_2)_y - X - \right]_n \tag{1.9}$$

where X = O, S, NH, $-O-\overset{O}{\overset{\|}{C}}-$, $-\overset{H}{\overset{|}{N}}-\overset{O}{\overset{\|}{C}}-$, $-CH{=}CH-$, etc.

The nature of the cyclic structure is such that in the presence of a catalyst it undergoes equilibrium ring-opening to produce a linear chain of degree of polymerization, n. X is usually a heteroatom such as oxygen or sulfur; it may also be a group such as lactam or lactone. A number of commercially important polymers are obtained via ring-opening polymerization. Thus, trioxane (33) can be polymerized to yield polyoxymethylene (34), the most important member of the family of acetal resins, and caprolactam (35) undergoes ring-opening to yield nylon 6 (36), an important textile fiber used especially for carpets.

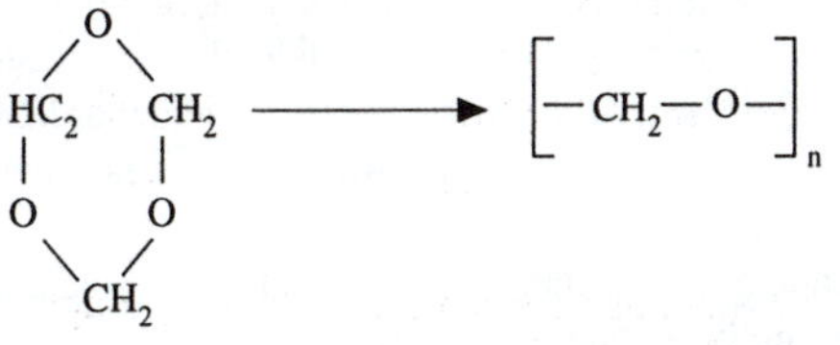

$$\text{trioxane (33): cyclic } (HC_2{-}O{-}CH_2{-}O{-}CH_2{-}O) \longrightarrow \left[-CH_2-O- \right]_n \; (34) \tag{Str. 28}$$

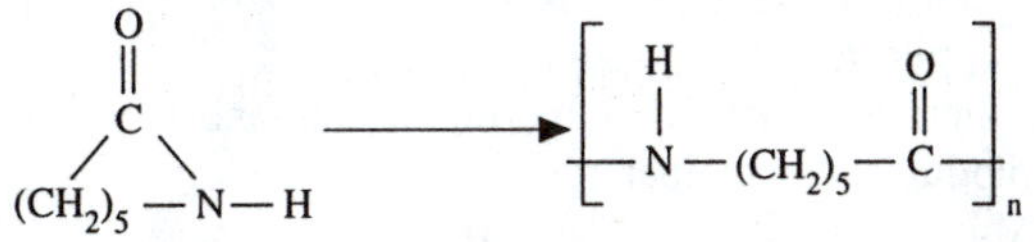

$$\text{(35): cyclic } (CH_2)_5{-}N{-}H \text{ with } \overset{O}{\overset{\|}{C}} \longrightarrow \left[-\overset{H}{\overset{|}{N}}-(CH_2)_5-\overset{O}{\overset{\|}{C}}- \right]_n \; (36) \tag{Str. 29}$$

We will discuss the various polymerization mechanisms in greater detail in Chapter 2. The original classification of polymers as either condensation or addition polymers as proposed by Carothers does not permit a complete differentiation between the two classes or polymers, particularly in view of the new polymerization processes that have been developed in recent years. Consequently, this classification has been replaced by the terms *step-reaction* (condensation) and *chain-reaction* (addition) *polymerization.* These terms focus more on the manner in which the monomers are linked together during polymerization.

D. THERMAL BEHAVIOR

For engineering purposes, the most useful classification of polymers is based on their thermal (thermo-mechanical) response. Under this scheme, polymers are classified as *thermoplastics* or *thermosets.* As the name suggests, thermoplastic polymers soften and flow under the action of heat and pressure. Upon cooling, the polymer hardens and assumes the shape of the mold (container). Thermoplastics, when compounded with appropriate ingredients, can usually withstand several of these heating and cooling cycles without suffering any structural breakdown. This behavior is similar to that of candle wax. Examples of thermoplastic polymers are polyethylene, polystyrene, and nylon.

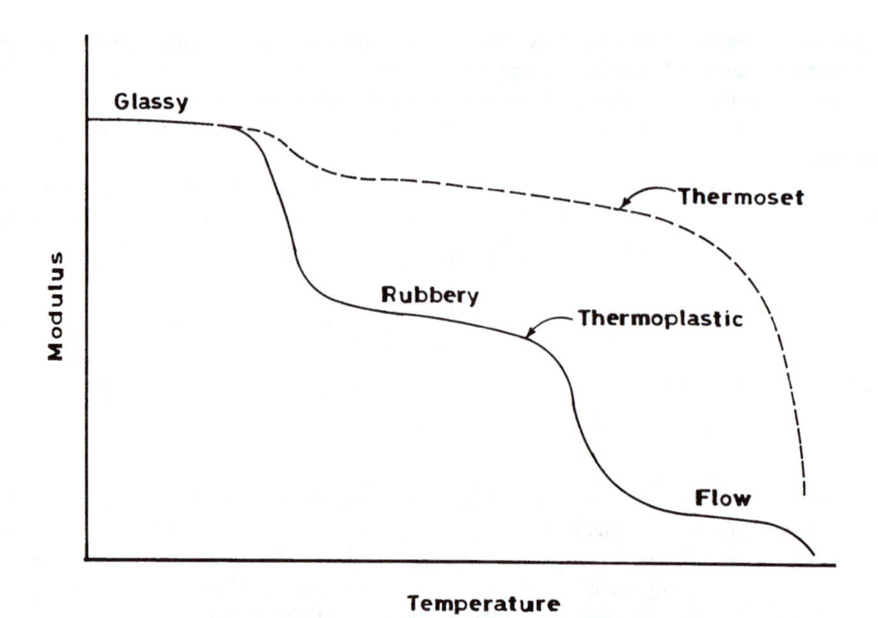

Figure 1.3 Idealized modulus–temperature curves for thermoplastics and thermosets.

A thermoset is a polymer that, when heated, undergoes a chemical change to produce a cross-linked, solid polymer. Thermosets usually exist initially as liquids called prepolymers; they can be shaped into desired forms by the application of heat and pressure, but are incapable of undergoing repeated cycles of softening and hardening. Examples of thermosetting polymers include urea–formaldehyde, phenol–formaldehyde, and epoxies.

The basic structural difference between thermoplastics and thermosets is that thermoplastic polymers are composed mainly of linear and branched molecules, whereas thermosets are made up of cross-linked systems. Recall from our previous discussion that linear and branched polymers consist of molecules that are not chemically tied together. It is therefore possible for individual chains to slide past one another. For cross-linked systems, however, chains are linked chemically; consequently, chains will not flow freely even under the application of heat and pressure.

The differences in the thermal behavior of thermoplastics and thermosets are best illustrated by considering the change in modulus with temperature for both polymers (Figure 1.3). At low temperatures, a thermoplastic polymer (both crystalline and amorphous) exists as a hard and rigid glass. As the temperature is increased, it changes from a glass to a rubbery elastomer to a viscous melt that is capable of flowing — hence this phase is also known as the flow region. (The transitions between the different phases or regions of thermal behavior are characterized by drops in the magnitude of the modulus — usually two to three orders. As we shall see later, differences exist between amorphous and crystalline thermoplastics in the details and nature of these transitions). For the thermosetting polymer, on the other hand, the modulus remains high in the rubbery region, while the flow region disappears.

E. PREPARATIVE TECHNIQUE

Polymers can be classified according to the techniques used during the polymerization of the monomer. In *bulk polymerization,* only the monomer (and possibly catalyst and initiator, but no solvent) is fed into the reactor. The monomer undergoes polymerization, at the end of which a (nearly) solid mass is removed as the polymer product. As we shall see later, bulk polymerization is employed widely in the manufacture of condensation polymers, where reactions are only mildly exothermic and viscosity is mostly low thus enhancing ready mixing, heat transfer, and bubble elimination. *Solution polymerization* involves polymerization of a monomer in a solvent in which both the monomer (reactant) and polymer (product) are soluble. *Suspension polymerization* refers to polymerization in an aqueous medium with the monomer as the dispersed phase. Consequently, the polymer resulting from such a system forms a solid dispersed phase. *Emulsion polymerization* is similar to suspension polymerization but the initiator is located in the aqueous phase (continuous phase) in contrast to the monomer (dispersed phase) in suspension polymerization. Besides, in emulsion polymerization the resulting polymer particles are considerably smaller (about ten times smaller) than those in suspension polymerization.

F. END USE

Finally, polymers may be classified according to the end use of the polymer. In this case, the polymer is associated with a specific industry (end use): diene polymers (rubber industry); olefin polymer (sheet, film, and fiber industries); and acrylics (coating and decorative materials).

IV. PROBLEMS

1.1. Show the structural formulae of the repeating units that would be obtained in the polymerization of the following monomers. Give the names of the polymers.

$$CH_2{=}CH{-}COOH \qquad \text{(Str. 30)}$$

$$CH_2{=}C(CH_3){-}C({=}O){-}O{-}CH_3 \qquad \text{(Str. 31)}$$

$$CH_2{=}CH{-}O{-}C({=}O){-}CH_3 \qquad \text{(Str. 32)}$$

$$CH_2{=}CH{-}CH_3 \qquad \text{(Str. 33)}$$

$$CH_2{=}CH{-}CN \qquad \text{(Str. 34)}$$

1.2. Show the repeating units that would be obtained from the reaction of the following monomer(s).

a. $H_2N{-}(CH_2)_5\,NH_2$ and $Cl{-}C({=}O){-}(CH_2)_5{-}C({=}O){-}Cl$ (Str. 35)

b. $HOOC{-}C_6H_4{-}COOH$ and $HO{-}(CH_2)_{10}{-}OH$ (Str. 36)

c. $HOCH_2{-}CH_2{-}CH_2OH$ and phthalic anhydride (O=C–O–C=O ring fused to benzene) (Str. 37)

d. $CH_3C_6H_3(NCO)_2$ (ring with CH_3, NCO, and NCO) and $HO{-}CH_2{-}CH_2{-}OH$ (Str. 38)

1.3. Complete the following table.

Monomer	Repeat Unit	Polymer
a.		Poly(ethyl acrylate)
b. $CH_2{=}C(CH_3)(C_6H_5)$		
c.	$-CH_2-C(CH_3)_2-$	
d.		Poly(vinylidene chloride)
e.	$-C(=O)-(CH_2)_5-N(H)-$	
f. CH_2 H_2C $C{=}O$ H_2C O CH_2-CH_2		
g.		Poly(dimethylsiloxane)
h. O CH_2 CH_2 CH_2-CH_2		

1.4. Complete the table by indicating whether the monomer(s) will form a polymer and, if so, whether the polymer formed will be linear or branched/cross-linked.

		Polymer		
			Yes	
Monomer A	**Monomer B**	**No**	**Linear**	**Branched/ Cross-linked**
a. $R-C(=O)-OH$	$HO-R'(-OH)-OH$			
b. $HOOC-R-COOH$	$HO-R'-OH$			
c. $HO-R-OH$	$R'-N=C=O$			
d. $CH_2=CH-C_6H_4-CH=CH_2$ (p-divinylbenzene)				
e. $HO-(CH_2)_5COOH$				
f. $H_2N-R(-NH_2)-NH_2$	$HOOC-R'-COOH$			
g. $O=C-CH=CH-C=O$ with ring O (maleic anhydride)	$CH_2=CH-C_6H_4-CH=CH_2$ (p-divinylbenzene)			
h. $H_2N-R-NH_2$	$OCN-R'-NCO$			
i. $CH_2=CHCOOH$				
j. CH_2O	$H_2NCH_2-(C_4H_2O)-CH_2-(C_4H_2O)-CH_2NH_2$ (furan rings)			
k. CH_2-CH_2 with bridging O (ethylene oxide)				

1.5. What is the molecular weight of the following polymers if the degree of polymerization is 1000?

a. $-N(H)-(CH_2)_5-C(=O)-$ (Str. 39)

b. 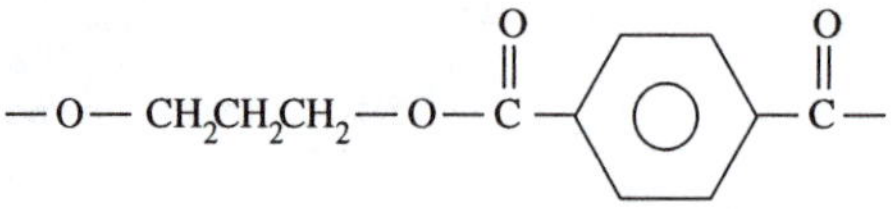 $-O-CH_2CH_2CH_2-O-C(=O)-C_6H_4-C(=O)-$ (Str. 40)

c. 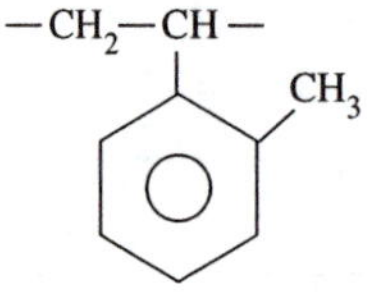 $-CH_2-CH-$ with the CH carrying a benzene ring that has a CH_3 group (Str. 41)

d. 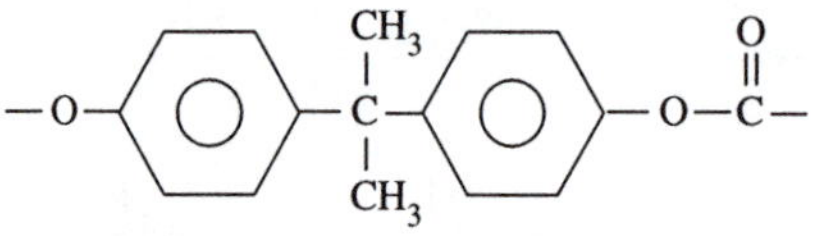 $-O-C_6H_4-C(CH_3)_2-C_6H_4-O-C(=O)-$ (Str. 42)

1.6. Draw the structural formulae of the repeating units of the following polymers.

a. Poly(ethylene succinate)
b. Poly(ethylene sebacate)
c. Poly(hexamethylene phthalate)
d. Poly(tetramethylene oxalate)

1.7. A polyester is formed by a condensation reaction between maleic anhydride and diethylene glycol. Styrene is then added and polymerized. Describe the chemical composition and molecular architecture of the resulting polymer. What would be the effect if maleic anhydride were replaced with adipic acid?

1.8. Natural rubber is a polymer of isoprene $CH_2{=}C(CH_3)-C(H){=}CH_2$ (Str. 43)

a. Show what structures can form as it polymerizes.
b. What feature of the polymer chain permits vulcanization?

1.9. An industrialist wants to set up a phenol–formaldehyde adhesive plant. He has approached you with the following phenolic compounds.

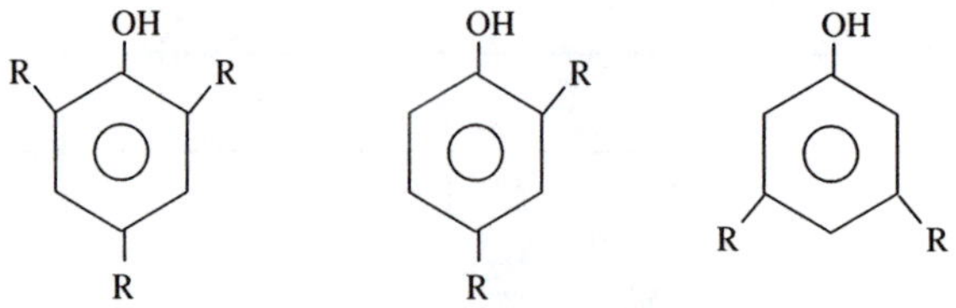

(a) phenol (OH) with R groups at both positions next to the OH and at the position opposite the OH

(b) phenol (OH) with R groups at one position next to the OH and at the position opposite the OH

(c) phenol (OH) with R groups at the two positions separated from the OH by one ring carbon (1,3,5 pattern)

(Str. 44)

Which of the compounds (a, b, or c) would you choose for reaction with formaldehyde? Explain your choice.

1.10. The following structure represents the general formula for some aliphatic amines.

$$H_2N{+}(CH_2)_2-NH{+}_n(CH_2)_2-NH_2 \quad \text{(Str. 45)}$$

What is the functionality of the corresponding amine in its reaction with diglycidyl ether of bisphenol A (DGEBA) if n = 1, 2, 3, 4?

REFERENCES

1. Frados, J., *The Story of the Plastics Industry,* Society of the Plastics Industry, New York, 1977.
2. Billmeyer, F.W., Jr., *Textbook of Polymer Science,* 3rd ed., Interscience, New York, 1984.
3. Fried, J.R., *Plast. Eng.,* 38(6), 49, 1982.
4. Fried, J.R., *Plast. Eng.,* 38(7), 27, 1982.
5. Fried, J.R., *Plast. Eng.,* 38(11), 27, 1982.
6. Fried, J.R., *Plast. Eng.,* 38(12), 21, 1982.
7. Fried, J.R., *Plast. Eng.,* 39(3), 67, 1983.
8. Kaufman, H.S., *1969/70 Modern Plastics Encyclopedia,* McGraw-Hill, New York, 1969, 29.
9. Williams, D.J., *Polymer Science and Engineering,* Prentice-Hall, Englewood Cliffs, NJ, 1971.
10. Kaufman H.S. and Falcetta, J.J., eds., *Introduction to Polymer Science and Technology,* John Wiley & Sons, New York, 1977.
11. Rudin, A., *The Elements of Polymer Science and Engineering,* Academic Press, New York, 1982.
12. Flory, P.J., *Principles of Polymer Chemistry,* Cornell University Press, Ithaca, NY, 1952.
13. Carothers, W.H., *Chem. Rev.,* 8(3), 353, 1931.
14. Wendorff, J.H., Finkelmann, H., and Ringsdorf, H., *J. Polym. Sci. Polym. Symp.,* 63, 245, 1978.
15. Braunsteiner, E.E., *J. Polym. Sci. Macromol. Rev.,* 9, 83, 1974.
16. McGrath, J.E., *Makromol. Chem. Macromol. Symp.,* 42/43, 69, 1991.

Chapter 2

Polymerization Mechanisms

I. INTRODUCTION

As discussed in Chapter 1, under a scheme proposed by Carothers, polymers are classified as addition or condensation polymers depending on the type of polymerization reaction involved in their synthesis. This classification scheme, however, does not permit a complete differentiation between the two classes of polymers. A more complete but still oversimplified scheme that is still based on the different polymerization processes places polymers into three classes: condensation, addition, and ring-opening polymers. This scheme reflects the structures of the starting monomers. Probably the most general classification scheme is based on the polymerization mechanism involved in polymer synthesis. Under this scheme, polymerization processes are classified as step-reaction (condensation) or chain-reaction (addition) polymerization. In this chapter, we will discuss the different types of polymers based on the different polymerization mechanisms.

II. CHAIN-REACTION POLYMERIZATION

Chain-reaction polymerization, an important industrial method of polymer preparation, involves the addition of unsaturated molecules to a rapidly growing chain. The most common unsaturated compounds that undergo chain-reaction polymerization are olefins, as exemplified by the following reaction of a generalized vinyl monomer.

$$n\,CH_2{=}\underset{\displaystyle R}{\underset{|}{CH}} \longrightarrow \left[-CH_2-\underset{\displaystyle R}{\underset{|}{CH}}-\right]_n \qquad (2.1)$$

The growing polymer in chain-reaction polymerization is a free radical, and polymerization proceeds via chain mechanism. Chain-reaction polymerization is induced by the addition of free-radical-forming reagents or by ionic initiators. Like all chain reactions, it involves three fundamental steps: initiation, propagation, and termination. In addition, a fourth step called chain transfer may be involved.

A. INITIATION

Initiation involves the acquisition of an active site by the monomer. This may occur spontaneously by the absorption of heat, light (ultraviolet), or high-energy irradiation. But most frequently, initiation of free-radical polymerization is brought about by the addition of small quantities of compounds called initiators. Typical initiators include peroxides, azo compounds, Lewis acids, and organometallic reagents. However, while initiators trigger initiation of the chain and exert an accelerating influence on polymerization rate, they are not exactly catalysts since they are changed chemically in the course of polymerization. An initiator is usually a weak organic compound that can be decomposed thermally or by irradiation to produce free radicals, which are molecules containing atoms with unpaired electrons. A variety of compounds decompose when heated to form free radicals. Dialkyl peroxides (ROOR), diacylperoxides (RCO–O–O–CO–R), hydroperoxides (ROOH), and azo compounds (RN=NR) are typical organic compounds that can be decomposed thermally to produce free radicals. Benzoyl peroxide, azobisisobutyronitrile, and di-*t*-butylperoxide are commonly used free-radical initiators, as illustrated in Equations 2.2–2.4.

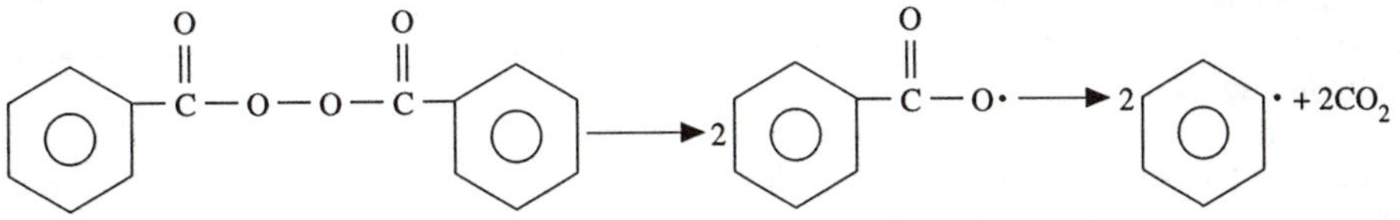

$$C_6H_5-\overset{O}{\overset{\|}{C}}-O-O-\overset{O}{\overset{\|}{C}}-C_6H_5 \longrightarrow 2\,C_6H_5-\overset{O}{\overset{\|}{C}}-O\cdot \longrightarrow 2\,C_6H_5\cdot + 2CO_2 \qquad (2.2)$$

benzoyl peroxide

$$CH_3-\underset{CN}{\overset{CH_3}{C}}-N=N-\underset{CN}{\overset{CH_3}{C}}-CH_3 \longrightarrow 2CH_3-\underset{CN}{\overset{CH_3}{C}}\cdot + N_2 \quad (2.3)$$

azobisisobutyronitrile (AIBN)

$$CH_3-\underset{CH_3}{\overset{CH_3}{C}}-O-O-\underset{CH_3}{\overset{CH_3}{C}}-CH_3 \longrightarrow 2CH_3-\underset{CH_3}{\overset{CH_3}{C}}-O\cdot \quad (2.4)$$

di-*t*-butylperoxide

The thermal decomposition of benzoyl peroxide, which takes place between 60 and 90°C, involves the homolytic cleavage of the O–O bond to yield benzoyl free radicals that may react to yield phenyl radicals and carbon dioxide. An example of photochemically induced free-radical formation is the decomposition of azo-bisisobutyronitrile by short-wavelength visible light or near-ultraviolet radiation at temperatures as low as 0°C, where no thermal initiation occurs.

In free-radical polymerization carried out in aqueous medium, the decomposition of peroxide or persulfate is greatly accelerated by the presence of a reducing system. This method of free-radical initiation is referred to as redox initiation. The initiation resulting from the thermal decomposition of organic compounds discussed above is appropriate only for polymerizations carried out at room temperature or higher. The enhanced rate of free-radical formation in redox reactions permits polymerization at relatively lower temperatures. Typical redox reactions for emulsion polymerization are shown in Equations 2.5–2.7.

$$S_2O_8^{2-} + HSO_3^- \rightarrow SO_4^{2-} + SO_4^-\cdot + HSO_3\cdot \quad (2.5)$$

$$S_2O_8^{2-} + S_2O_3^- \rightarrow SO_4^{2-} + SO_4^-\cdot + S_2O_3^-\cdot \quad (2.6)$$

$$HSO_3^- + Fe^{3+} \rightarrow HSO_3\cdot + Fe^{2+} \quad (2.7)$$

Persulfate ion initiator (e.g., from $K_2S_2O_8$) reacts with a reducing agent such as a bisulfite ion (e.g., from $NaHSO_3$) to produce radicals for redox initiation (Equations 2.5 and 2.6). Ferric ion may also be used as a source of radicals (Equation 2.7). Other redox reactions involve the use of alkyl hydroxides and a reducing agent such as ferrous ion (Equation 2.8).

$$C_6H_5-\underset{CH_3}{\overset{CH_3}{C}}-OOH + Fe^{2+} \longrightarrow C_6H_5-\underset{CH_3}{\overset{CH_3}{C}}-O\cdot + Fe^{3+} \quad (2.8)$$

As indicated earlier, free-radical polymerization of some monomers can be initiated by heating or exposing the monomers to light or high-energy irradiation such as X-rays, γ-rays, and α-rays. High-energy irradiation of monomers can be carried out either in bulk or in solution. It is certainly not as selective as photolytic initiation.

When choosing an initiator for free-radical polymerization, the important parameters that must be considered are the temperature range to be used for the polymerization and the reactivity of the radicals formed. The presence of certain promoters and accelerators and the nature of the monomer often affect the rate of decomposition of initiators. For example, the decomposition of benzoyl peroxide may be accelerated at room temperature by employing ternary or quaternary amines. Free-radical initiation processes do not require stringent exclusion of atmospheric moisture, but can be inhibited by substances such as oxygen. Free radicals are inactivated by reaction with oxygen to form peroxides or hydroperoxides. For monomers such as styrene and methylmethacrylate that are susceptible to such inhibition, initiation reactions are carried out in an oxygen-free atmosphere such as nitrogen. It must be emphasized also that organic peroxides, when subjected to shock or high temperature, can detonate. Therefore these compounds must be handled with caution.

The initiation of polymerization occurs in two successive steps. The first step involves the formation of radicals according to the processes discussed above. This may be represented broadly as:

$$\text{I-I} \rightarrow 2\text{I}\cdot \tag{2.9}$$

The second step is the addition of the initiator radical to a vinyl monomer molecule:

$$\text{I}\cdot + CH_2{=}\underset{\displaystyle R}{\overset{\displaystyle H}{\underset{|}{\overset{|}{C}}}} \longrightarrow \text{I}{-}CH_2{-}\underset{\displaystyle R}{\overset{\displaystyle H}{\underset{|}{\overset{|}{C}}}}\cdot \tag{2.10}$$

Initiator fragments have been shown by end-group analysis to become part of the growing chain. In commercial practice, 60 to 100% of all the free radicals generated do initiate polymerization.

B. PROPAGATION

During propagation, the initiated monomer described above adds other monomers — usually thousands of monomer molecules — in rapid succession. This involves the addition of a free radical to the double bond of a monomer, with regeneration of another radical. The active center is thus continuously relocated at the end of the growing polymer chain (Equation 2.11).

$$\text{I}{-}CH_2{-}\underset{\displaystyle R}{\overset{\displaystyle H}{\underset{|}{\overset{|}{C}}}}\cdot + CH_2{=}CHR \longrightarrow \text{I}{-}CH_2{-}\underset{\displaystyle R}{\underset{|}{CH}}{-}CH_2{-}\underset{\displaystyle R}{\overset{\displaystyle H}{\underset{|}{\overset{|}{C}}}}\cdot \tag{2.11}$$

Propagation continues until the growing chain radical is deactivated by chain termination or transfer as discussed below.

The substituted carbon atom is regarded as the head and the unsubstituted carbon atom the tail of the vinyl monomer. There are, therefore, three possible ways for the propagation step to occur: head-to-tail (Equation 2.11), head-to-head (Equation 2.12), and tail-to-tail (Equation 2.13). A random distribution of these species along the molecular chain might be expected. It is found, however, that head-to-tail linkages in which the substituents occur on alternate carbon atoms predominate; only occasional interruptions of this arrangement by head-to-head and tail-to-tail linkages occur. In addition, exclusive head-to-head or tail-to-tail arrangements of monomers in the chain are now known.

$$\text{I}{-}CH_2{-}\underset{\displaystyle R}{\overset{\displaystyle H}{\underset{|}{\overset{|}{C}}}}\cdot + CH_2{=}CHR \longrightarrow \text{I}{-}CH_2{-}\underset{\displaystyle R}{\overset{\displaystyle H}{\underset{|}{\overset{|}{C}}}}{-}\underset{\displaystyle R}{\overset{\displaystyle H}{\underset{|}{\overset{|}{C}}}}{-}CH_2\cdot \tag{2.12}$$

$$\text{I}{-}CH_2{-}\underset{\displaystyle R}{\overset{\displaystyle H}{\underset{|}{\overset{|}{C}}}}\cdot + CH_2{=}CHR \longrightarrow \text{I}{-}\underset{\displaystyle R}{\overset{\displaystyle H}{\underset{|}{\overset{|}{C}}}}{-}CH_2{-}CH_2{-}\underset{\displaystyle R}{\overset{\displaystyle H}{\underset{|}{\overset{|}{C}}}}\cdot \tag{2.13}$$

C. TERMINATION

In termination, the growth activity of a polymer chain radical is destroyed by reaction with another free radical in the system to produce polymer molecule(s). Termination can occur by the reaction of the polymer radical with initiator radicals (Equation 2.14). This type of termination process is unproductive and can be controlled by maintaining a low rate for initiation.

$$\mathrm{I{-}\sim{-}CH_2{-}\underset{R}{\overset{H}{C}}\cdot + I' \longrightarrow I{-}\sim{-}CH_2{-}\underset{R}{\overset{H}{C}}{-}\underset{R}{\overset{H}{C}}{-}I'} \tag{2.14}$$

The termination reactions that are more important in polymer production are combination (or coupling) and disproportionation. In termination by combination, two growing polymer chains react with the mutual destruction of growth activity (Equation 2.15), while in disproportionation a labile atom (usually hydrogen) is transferred from one polymer radical to another (Equation 2.16).

$$\mathrm{I{-}\sim{-}CH_2{-}\underset{R}{\overset{H}{C}}\cdot + \cdot\underset{R}{\overset{H}{C}}{-}CH_2{-}\sim{-}I \longrightarrow I{-}\sim{-}CH_2{-}\underset{R}{\overset{H}{C}}{-}\underset{R}{\overset{H}{C}}{-}CH_2{-}\sim{-}I} \tag{2.15}$$

$$\mathrm{I{-}\sim{-}CH_2{-}\underset{R}{\overset{H}{C}}\cdot + \cdot\underset{R}{\overset{H}{C}}{-}CH_2{-}\sim{-}I \longrightarrow I{-}\sim{-}CH_2{-}\underset{R}{\overset{H}{C}}{-}H + \underset{R}{\overset{H}{C}}{=}CH{-}\sim{-}I} \tag{2.16}$$

Coupling reactions produce a single polymer, while disproportionation results in two polymers from the two reacting polymer chain radicals. The predominant termination reaction depends on the nature of the reacting monomer and the temperature. Since disproportionation requires energy for breaking of chemical bonds, it should become more pronounced at high reaction temperatures; combination of growing polymer radicals predominates at low temperatures.

D. CHAIN TRANSFER

Ideally, free-radical polymerization involves three basic steps: initiation, propagation, and termination, as discussed above. However, a fourth step, called chain transfer, is usually involved. In chain-transfer reactions, a growing polymer chain is deactivated or terminated by transferring its growth activity to a previously inactive species, as illustrated in Equation 2.17.

$$\mathrm{I{-}\sim{-}CH_2{-}\underset{R}{\overset{H}{C}}\cdot + TA \longrightarrow I{-}\sim{-}CH_2{-}\underset{R}{\overset{H}{C}}{-}T + A\cdot} \tag{2.17}$$

The species, TA, could be a monomer, polymer, solvent molecule, or other molecules deliberately or inadvertently introduced into the reaction mixture. Depending on its reactivity, the new radical, A·, may or may not initiate the growth of another polymer chain. If the reactivity of A· is comparable to that of the propagating chain radical, then a new chain may be initiated. If its reactivity toward a monomer is less than that of the propagating radical, then the overall reaction rate is retarded. If A· is unreactive toward the monomer, the entire reaction could be inhibited. Transfer reactions do not result in the creation or destruction of radicals; at any instant, the overall number of growing radicals remains unchanged. However, the occurrence of transfer reactions results in the reduction of the average polymer chain length, and in the case of transfer to a polymer it may result in branching.

E. DIENE POLYMERIZATION

Conjugated dienes such as butadiene (1), chloroprene (2), and isoprene (3) constitute a second group of unsaturated compounds that can undergo polymerization through their double bonds.

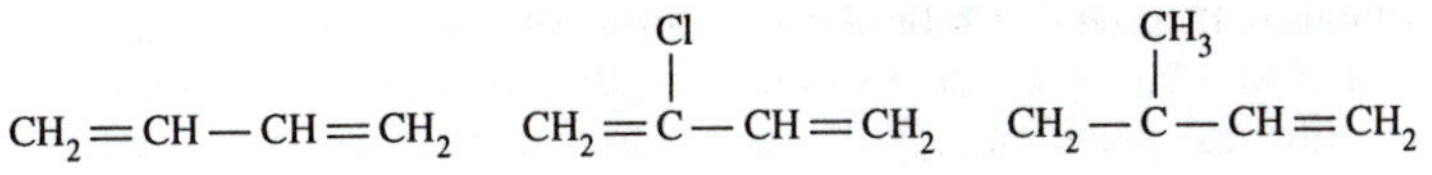

$$\underset{(1)}{CH_2{=}CH{-}CH{=}CH_2} \qquad \underset{(2)}{CH_2{=}\overset{\overset{\displaystyle Cl}{|}}{C}{-}CH{=}CH_2} \qquad \underset{(3)}{CH_2{-}\overset{\overset{\displaystyle CH_3}{|}}{C}{-}CH{=}CH_2} \qquad \text{(Str. 1)}$$

These structures contain double bonds in the 1,2 and 3,4 positions, each of which may participate independently in polymerization giving rise to 1,2 and 3,4 units. A further possibility is that both bonds are involved in polymerization through conjugate reactions, resulting in 1,4 units. These structures are shown in Equation 2.18.

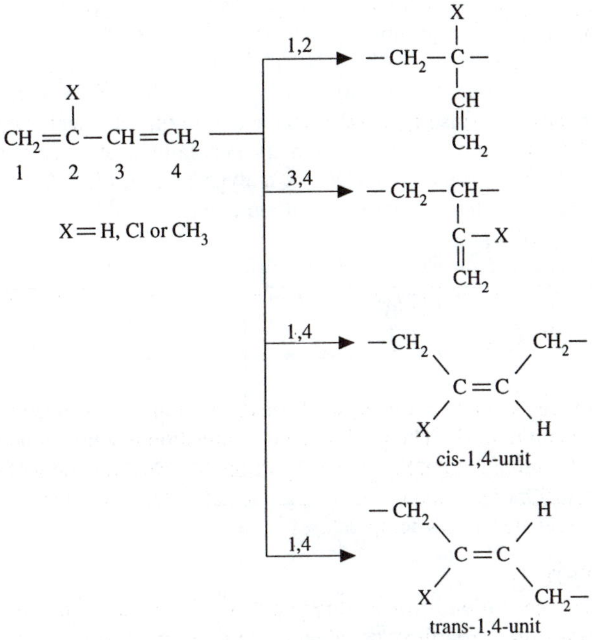

$$\underset{1\quad 2\quad 3\quad 4}{CH_2{=}\overset{\overset{\displaystyle X}{|}}{C}{-}CH{=}CH_2} \qquad X{=}H,\ Cl\ or\ CH_3$$

$$\xrightarrow{1,2} -CH_2-\overset{\overset{\displaystyle X}{|}}{\underset{\underset{\underset{\displaystyle CH_2}{\|}}{\displaystyle CH}}{\underset{|}{C}}}- \qquad \xrightarrow{3,4} -CH_2-\underset{\underset{\underset{\displaystyle CH_2}{\|}}{\displaystyle C-X}}{\underset{|}{CH}}- \qquad (2.18)$$

$$\xrightarrow{1,4} \begin{matrix} -CH_2 & & CH_2- \\ & C{=}C & \\ X & & H \end{matrix} \quad \text{cis-1,4-unit}$$

$$\xrightarrow{1,4} \begin{matrix} -CH_2 & & H \\ & C{=}C & \\ X & & CH_2- \end{matrix} \quad \text{trans-1,4-unit}$$

Diene polymerization thus gives rise to polymers that contain various isomeric units. With symmetrical dienes such as butadiene, the 1,2 and 3,4 units are identical. The 1,4 unit may occur in the *cis* or *trans* configuration. A diene polymer contains more than one of these structural units. The relative abundance of each unit in the polymer molecule depends on the nature of the initiator, experimental conditions, and the structure of the diene. The proportion of each type of structure incorporated into the polymer chain influences both thermal and physical properties. For example, butadiene can be polymerized by free-radical addition at low temperature to produce a polymer that consists almost entirely of *trans*-1,4 units and only about 20% 1,2 units. As the temperature is increased, the relative proportion of *cis*-1,4 units increases while the proportion of 1,2 structure remains fairly constant. Anionic diene polymerization with lithium or organolithium initiators like *n*-butyllithium in nonpolar solvents such as pentane or hexane yields polymers with high *cis*-1,4 content. When higher alkali metal initiators or more polar solvents are used, the relative amount of *cis*-1,4 units decreases. Stereoregularity can also be controlled by the use of coordination catalysts like Ziegler–Natta catalysts. Heterogeneous Alfin catalysts — which are combinations of alkenyl sodium compounds, alkali metal halides, and an alkoxide — give high-molecular-weight polymers with high content of *trans*-1,4 units.

As noted above, all chain-reaction polymerizations involve essentially the same number of steps. The main distinguishing feature between chain-reaction polymerizations, however, is by the initiation mechanism, which may be a free-radical, ionic (cationic or anionic), or coordination. The time between initiation and termination of a given chain is typically from a few tenths of a second to a few seconds. During this time thousands or tens of thousands of monomers add to the growing chain.

The structural unit in addition polymers is chemically identical to the monomer employed in the polymerization reaction, as exemplified by the following reaction of a generalized vinyl monomer:

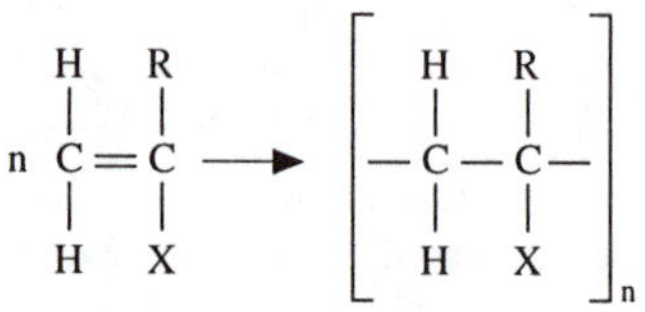

$$n\ \underset{\text{monomer}}{\overset{\begin{array}{cc}H & R\\ | & |\end{array}}{\underset{\begin{array}{cc}| & |\\ H & X\end{array}}{C{=}C}}} \longrightarrow \underset{\text{polymer}}{\left[\overset{\begin{array}{cc}H & R\\ | & |\end{array}}{\underset{\begin{array}{cc}| & |\\ H & X\end{array}}{-C-C-}}\right]_n} \qquad (2.18A)$$

Here R and X are monofunctional groups, R may be a hydrogen atom (H), or an alkyl group (e.g., $-CH_3$), while X may be any group (e.g., –Cl, –CN). The structural unit is evidently structurally identical to the starting monomer. Monomers generally employed in addition polymerizations are unsaturated (usually with double bonds). Because of the identical nature of the chemical formulas of monomers and the polymers derived from them, addition polymers generally take their names from the starting monomer — ethylene → polyethlene, propylene → polypropylene, etc. (Table 2.1). The backbone of addition polymer chains is usually composed of carbon atoms.

Table 2.1 Some Representation Addition Polymers

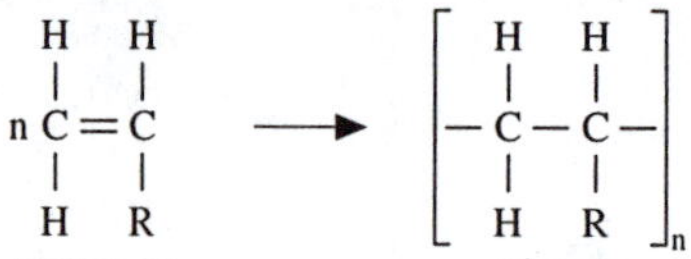

$$n\ \underset{\text{monomer}}{\overset{\begin{array}{cc}H & H\\ | & |\end{array}}{\underset{\begin{array}{cc}| & |\\ H & R\end{array}}{C{=}C}}} \longrightarrow \underset{\text{polymer}}{\left[\overset{\begin{array}{cc}H & H\\ | & |\end{array}}{\underset{\begin{array}{cc}| & |\\ H & R\end{array}}{-C-C-}}\right]_n}$$

R	Monomer	Polymer
H	Ethylene	Polyethylene
CH_3	Propylene	Polypropylene
Cl	Vinyl chloride	Poly(vinyl chloride)
CN	Acrylonitrile	Polyacrylonitrile
 (benzene ring)	Styrene	Polystyrene
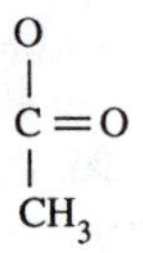 $O-C(=O)-CH_3$	Vinyl acetate	Poly(vinyl acetate)
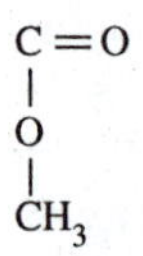 $C(=O)-O-CH_3$	Methyl acrylate	Poly(methyl acrylate)

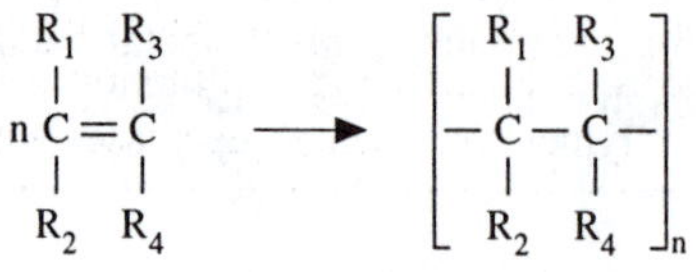

$$n\ \overset{\begin{array}{cc}R_1 & R_3\\ | & |\end{array}}{\underset{\begin{array}{cc}| & |\\ R_2 & R_4\end{array}}{C{=}C}} \longrightarrow \left[\overset{\begin{array}{cc}R_1 & R_3\\ | & |\end{array}}{\underset{\begin{array}{cc}| & |\\ R_2 & R_4\end{array}}{-C-C-}}\right]_n$$

Table 2.1 (continued) Some Representation Addition Polymers

R_1	R_2	R_3	R_4	Monomer	Polymer
H	H	CH_3	$C=O$ | O | CH_3	Methyl methacrylate	Poly(methyl methacrylate)
H	H	Cl	Cl	Vinylidene chloride	Poly(vinylidene chloride)
H	H	F	F	Vinylidene fluoride	Poly(vinylidene fluoride)
F	F	F	F	Tetrafluoro-ethylene	Polytetrafluoroethylene

Example 2.1: Explain the following observations.

a. α-Methylstyrene polymerizes much less readily than styrene.
b. Chain-transfer reactions reduce the average chain length of the polymer.

Solutions:

a. The reactivity of a vinyl monomer depends on the nature of the substituents on the monomer double bond. Substituents may either enhance the monomer reactivity by activating the double bond, depress the reactivity of the resulting radical by resonance stabilization, or provide steric hindrance at the reaction site.

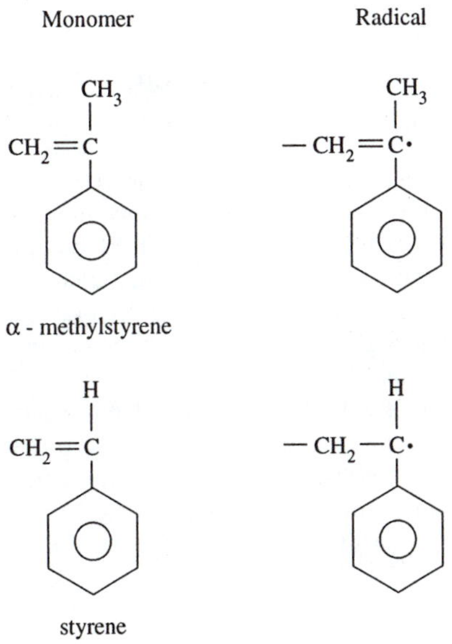

Monomer Radical

$CH_2=C(CH_3)(C_6H_5)$ α - methylstyrene — $-CH_2=C\cdot(CH_3)(C_6H_5)$

(Str. 2)

$CH_2=C(H)(C_6H_5)$ styrene — $-CH_2-C\cdot(H)(C_6H_5)$

The reactivities of α-methylstyrene and styrene radicals are essentially the same due to their similar resonance stabilization. However, the activation of the double bond by the phenyl group is compensated somewhat by the presence of the electron-donating methyl group in α-methylstyrene. The methyl group also provides steric hindrance at the reactive site. Consequently, α-methylstyrene is less reactive than styrene.

b. Chain transfer may be to a solvent, initiator or monomer. During chain-transfer reactions, for each radical chain initiated, the number of polymer molecules increases except in the case of transfer to a polymer. In other words, the average number of monomer molecules consumed by each chain radical (DP) or the average polymer chain length decreases with chain-transfer reactions.

III. IONIC AND COORDINATION POLYMERIZATIONS

As noted earlier, chain-reaction polymerization may be classified as free-radical, cationic, anionic, or coordination polymerization depending on the nature of the reactive center. The growing polymer molecule is associated with counterions in ionic (cationic and anionic) polymerization or with a coordination complex in coordination polymerization. Ionic polymerizations involve chain carriers or reactive centers that are organic ions or charged organic groups. In anionic polymerization, the growing chain end carries a negative charge or carbanions, while cationic polymerization involves a growing chain end with a positive charge or carbonium (carbenium) ion. Coordination polymerization is thought to involve the formation of a coordination compound between the catalyst, monomer, and growing chain.

The mechanisms of ionic and coordination polymerizations are more complex and are not as clearly understood as those of free radical polymerization. Here, we will briefly highlight the essential features of these mechanisms, and more details will be given in Chapter 7. Initiation of ionic polymerization usually involves the transfer of an ion or an electron to or from the monomer. Many monomers can polymerize by more than one mechanism, but the most appropriate polymerization mechanism for each monomer is related to the polarity of the monomers and the Lewis acid–base strength of the ion formed.

A. CATIONIC POLYMERIZATION

Monomers with electron-donating groups like isobutylene form stable positive charges and are readily converted to polymers by cationic catalysts. Any strong Lewis acid like boron trifluoride (BF_3) or Friedel–Crafts catalysts such as $AlCl_3$ can readily initiate cationic polymerization in the presence of a cocatalyst like water, which serves as a Lewis base or source of protons. During initiation, a proton adds to the monomer to form a carbonium ion, which forms an association with the counterion. This is illustrated for isobutylene and boron trifluoride in Equation 2.19:

$$BF_3 \cdot H_2O + CH_2{=}\underset{\displaystyle CH_3}{\overset{\displaystyle CH_3}{\underset{|}{\overset{|}{C}}}} \longrightarrow CH_3-\underset{\displaystyle CH_3}{\overset{\displaystyle CH_3}{\underset{|}{\overset{|}{C}}}}{}^{\oplus}\left[BF_3OH\right]^{\ominus} \tag{2.19}$$

Propagation involves the consecutive additions of monomer molecules to the carbonium ion at the growing chain end. Termination in cationic polymerization usually involves rearrangement to produce a polymer with an unsaturated terminal unit and the original complex or chain transfer to a monomer and possibly to the polymer or solvent molecule. Unlike free-radical polymerization, termination by combination of two cationic polymer growing chains does not occur.

Cationic polymerizations are usually conducted in solutions and frequently at temperatures as low as –80 to –100°C. Polymerization rates at these low temperature conditions are usually fast. The cation and the counterion in cationic polymerization remain in close proximity. If the intimate association between the ion pair is too strong, however, monomer insertion during propagation will be prevented. Therefore the choice of solvent in cationic polymerization has to be made carefully; a linear increase in polymer chain length and an exponential increase in the reaction rate usually occur as the dielectric strength of the solvent increases.

B. ANIONIC POLYMERIZATION

Monomers that are suitable for anionic polymerization generally contain electron-withdrawing substituent groups. Typical monomers include styrene, acrylonitrile, butadiene, methacrylates, acrylates, ethylene oxide, and lactones. The initiator in anionic polymerization may be any compound providing a strong nucleophile, including Grignard reagents and other organometallic compounds. Initiation involves the addition of the initiator to the double bond of the monomer, as illustrated for styrene and butyllithium in Equation 2.20.

$$n-C_4H_9^{-}\ Li^{+} + CH_2{=}CH_2 \longrightarrow n-C_4H_9-CH_2-CH^{-}Li^{+} \tag{2.20}$$

The reaction produces a carbanion at the head end to which is associated the positively charged lithium counterion. Propagation occurs by the successive insertion of monomer molecules by anionic attack of the carbanion. No chain transfer or branching occurs in anionic polymerization, particularly if reactions are carried out at low temperatures. Termination of the growth activity of the polymer chain takes place either by the deliberate or accidental introduction into the system of oxygen, carbon dioxide, methanol, water, or other molecules that are capable of reacting with the active chain ends. We note that in anionic polymerization as well as free-radical polymerization, the initiator or part of it becomes part of the resulting polymer molecule, attached to the nongrowing chain end. This contrasts with cationic polymerization where the catalyst is necessary for initiation and propagation, but is regenerated at the termination step.

In some systems, termination can be avoided if the starting reagents are pure and the polymerization reactor is purged of all oxygen and traces of water. This produces polymer molecules that can remain active even after all the monomer molecules are consumed. When fresh monomer is added, polymerization resumes. Such polymeric molecules are referred to as "living polymers" because of the absence of termination. Since the chain ends grow at the same rate, the molecular weight of living polymers is determined simply by the ratio of monomer concentration to that of the initiator (Equation 2.21).

$$\text{Degree of Polymerization (DP)} = \frac{[\text{monomer}]}{[\text{initiator}]} \tag{2.21}$$

Polymers produced by living polymerization are characterized by very narrow molecular weight distribution (Poisson distribution). The polydispersity D is given by Equation 2.22

$$D = \overline{M}_w/\overline{M}_n = 1 + 1/DP \tag{2.22}$$

where $\overline{M}_w$ and $\overline{M}_n$ are the weight-average molecular weight and number-average molecular weight, respectively.

As discussed in Chapter 7, the absence of termination in living polymerization permits the synthesis of unusual and unique block polymers — star- and comb-shaped polymers. Living polymerization can also be employed to introduce a variety of desired functional groups at one or both ends of polymeric chains both in homo- and block polymers. In particular, living polymerization techniques provide the synthetic polymer chemist with a vital and versatile tool to control the architecture of a polymer; complicated macromolecules can be synthesized to meet the rigid specification imposed by a scientific or technological demand.

C. COORDINATION POLYMERIZATION

As we shall see in Chapter 4, monomers with side groups asymmetrically disposed with respect to the double bond are capable of producing polymers in which the side groups have a specific stereochemical or spatial arrangement (isotactic or syndiotactic). In both cationic and anionic polymerizations, the association of initiating ion and counterion permits a preferential placement of asymmetric substituted monomers, the extent of which depends on the polymerization conditions. Unbranched and stereospecific polymers are also produced by the use of Ziegler–Natta catalysts. These are complex catalyst systems derived from a transition metal compound from groups IVB to VIIIB of the periodic table and an organometallic compound usually from a group IA or IIIA metal. A typical catalyst complex is that formed by trialkyl aluminum and titanium trichloride as shown below:

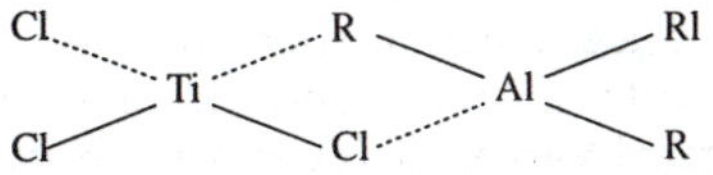

Cl R Rl
Ti Al
Cl Cl R

(Str. 3)

Monoolefins such as propylene and dienes such as butadiene and isoprene can be polymerized using Ziegler–Natta coordination catalysts. The catalysts function by forming transient π-complexes between the monomers and the transition metal species. The initiating species is a metal–alkyl complex and propagation involves the consecutive insertion of monomer molecules into a polarized titanium–carbon bond. Coordination polymerizations may be terminated by introducing poisons such as water, hydrogen, aromatic alcohols, or metals like zinc into the reacting system.

IV. STEP-GROWTH POLYMERIZATION

Step-growth polymerization involves a series of reactions in which any two species (monomers, dimers, trimers, etc.) can react at any time, leading to a larger molecule. Most step-growth polymerizations, as we shall see presently, involve a classical condensation reaction such as esterification, ester interchange, or amidization. In step-growth polymerization, the stepwise reaction occurs between pairs of chemically reactive or functional groups on the reacting molecules. In most cases, step-growth polymerization is accompanied by the elimination of a small molecule such as water as a by-product. A typical step-growth polymerization of the condensation type is the formation of a polyester through the reaction of a glycol and a dicarboxylic acid, as shown in Equation 2.23

$$\text{nHO—R—OH} + \text{nHOOC—R'—COOH} \longrightarrow \text{nH}\left[\text{R—O—}\overset{\overset{\displaystyle O}{\|}}{\text{C}}\text{—R'—}\overset{\overset{\displaystyle O}{\|}}{\text{C}}\right]_n \text{OH} + \text{nH}_2\text{O} \tag{2.23}$$

where R and R′ are the unreactive part of the molecules.

Step-growth polymerizations generally involve either one or more types of monomers. In either case, each monomer has at least two reactive (functional) groups. In cases where only one type of monomer is involved, which is known as A-B step-growth polymerization, the functional groups on the monomer are different and capable of intramolecular reactions. An example is the formation of an aliphatic polyester by the self-condensation of ω-hydroxycaproic acid (Equation 2.24).

$$\text{nHO—(CH}_2)_5\text{—}\overset{\overset{\displaystyle O}{\|}}{\text{C}}\text{—OH} \longrightarrow \left[\text{—(CH}_2)_5\boxed{\text{—}\overset{\overset{\displaystyle O}{\|}}{\text{C}}\text{—O}}\text{—}\right]_n + 2\text{nH}_2\text{O} \tag{2.24}$$

ω-hydroxycproic acid polycaprolactone

Here, each molecule contains two different functional groups: a hydroxyl group (–OH) and a carboxylic acid group (–COOH). These react to form a series of ester linkages ($-\overset{\overset{O}{\|}}{C}-O-$) shown in the shaded box. In those cases where more than one type of molecule is involved, the functional groups on each type of monomer are the same, but capable of intermolecular reaction with the other type of monomer. This is known as the A–A/B–B step-growth polymerization and is exemplified by the preparation of poly(ethylene terephthalate) and nylon 6,6 (Equations 2.25 and 2.26).

$$\text{nHO—}\overset{\overset{\displaystyle O}{\|}}{\text{C}}\text{—C}_6\text{H}_4\text{—}\overset{\overset{\displaystyle O}{\|}}{\text{C}}\text{—OH} + \text{nHO—CH}_2\text{CH}_2\text{—OH} \longrightarrow \left[\overset{\overset{\displaystyle O}{\|}}{\text{C}}\text{—C}_6\text{H}_4\text{—CH}_2\ \text{CH}_2\text{—O}\right]_n + 2\text{nHO} \tag{2.25}$$

terephthalic acid ethylene glycol

poly(ethylene terephthalate)

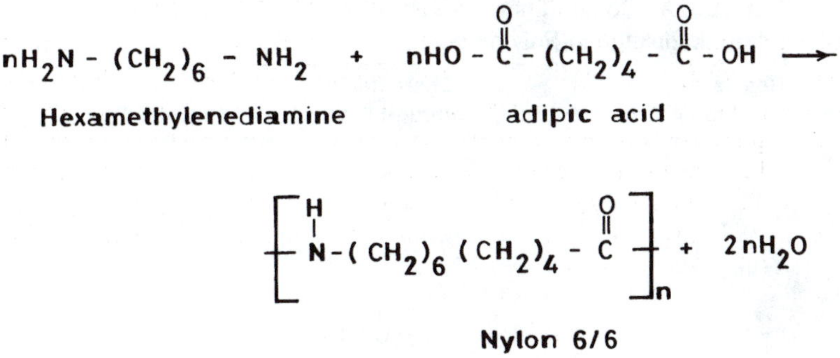

$$nH_2N-(CH_2)_6-NH_2 + nHO-\overset{O}{\overset{\|}{C}}-(CH_2)_4-\overset{O}{\overset{\|}{C}}-OH \longrightarrow$$

Hexamethylenediamine adipic acid

$$\left[\overset{H}{\overset{|}{N}}-(CH_2)_6\,(CH_2)_4-\overset{O}{\overset{\|}{C}} \right]_n + 2nH_2O \tag{2.26}$$

Nylon 6/6

In Equation 2.25, for example, poly(ethylene terephthalate) is formed from the condensation of a dicarboxylic acid and a diol.

Step-growth polymerizations can be divided into two main categories: polycondensation, in which a small molecule is eliminated at each step, as discussed above; and polyaddition, in which, as the name suggests, monomers react without the elimination of a small molecule. These are shown in Equations 2.27 and 2.28, respectively, where R and R′ are the nonreactive portions of the molecules.

$$A-R-A+B-R'-B \rightarrow A-R-R'-B+AB \tag{2.27}$$

polycondensation

$$A-R-A+B-R'-B \rightarrow A-R-AB-R'-B \tag{2.28}$$

polyaddition

An example of polyaddition-type step-growth polymerization is the preparation of polyurethane by the ionic addition of diol (1,4 butanediol) to a diisocyanate (1,6 hexane diisocyanate) (Equation 2.29).

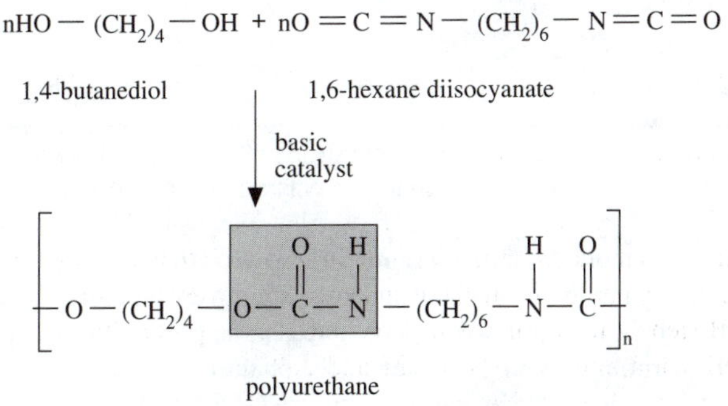

$$nHO-(CH_2)_4-OH + nO=C=N-(CH_2)_6-N=C=O$$

1,4-butanediol 1,6-hexane diisocyanate

↓ basic catalyst

$$\left[O-(CH_2)_4-O-\overset{O}{\overset{\|}{C}}-\overset{H}{\overset{|}{N}}-(CH_2)_6-\overset{H}{\overset{|}{N}}-\overset{O}{\overset{\|}{C}} \right]_n \tag{2.29}$$

polyurethane

Another example of polyaddition-type step-growth polymerization is the preparation of polyurea from the reaction of diisocyanate and diamine, as shown in Equation 2.30.

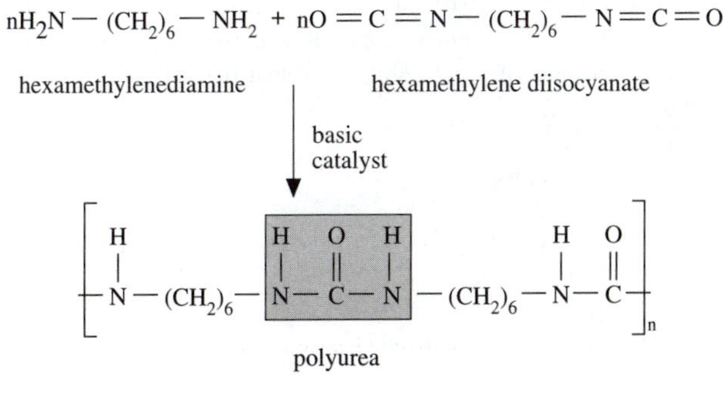

$$nH_2N-(CH_2)_6-NH_2 + nO=C=N-(CH_2)_6-N=C=O$$

hexamethylenediamine hexamethylene diisocyanate

↓ basic catalyst

$$\left[\overset{H}{\overset{|}{N}}-(CH_2)_6-\overset{H}{\overset{|}{N}}-\overset{O}{\overset{\|}{C}}-\overset{H}{\overset{|}{N}}-(CH_2)_6-\overset{H}{\overset{|}{N}}-\overset{O}{\overset{\|}{C}} \right]_n \tag{2.30}$$

polyurea

TABLE 2.2 Some Functional Groups and Their Characteristic Interunit Linkage in Polymers

Reactants Functional Group	Characteristic Interunit Linkage	Polymer Type
$-OH + -COOH$	$-\overset{O}{\overset{\parallel}{C}}-O-$	Polyester
$-NH_2 + -COOH$	$-\overset{O}{\overset{\parallel}{C}}-\overset{H}{\overset{\mid}{N}}$	Polyamide
$-OH + -NCO$	$-O-\overset{O}{\overset{\parallel}{C}}-\overset{H}{\overset{\mid}{N}}$	Polyurethane
$-NH_2 + -NCO$	$-\overset{H}{\overset{\mid}{N}}-\overset{O}{\overset{\parallel}{C}}-\overset{H}{\overset{\mid}{N}}-$	Polyurea
$-COOH + -COOH$	$-\overset{O}{\overset{\parallel}{C}}-O-\overset{O}{\overset{\parallel}{C}}-$	Polyanhydride
$-OH + -OH$	$-O-$	Polyether
$-CH-CH-$ (epoxide, bridged by O)	$-O-$	Polyether
$HO-\overset{O}{\overset{\parallel}{C}}-OH$	$-O-\overset{O}{\overset{\parallel}{C}}-O-$	Polycarbonate

The characteristic linkage (group) in each of the above classes of polymers shown in the boxes has been summarized in Table 2.2.

In contrast to addition polymers, the structural unit in step-growth polymers is not identical chemically to the structure of the starting monomer(s). Consequently, step-growth polymers derive their names from the reactive type (characteristic interunit linkage) involved in the polymerization process. In the reaction between the glycol and dicarboxylic acid, for instance, the resulting polymer is a polyester, in consonance with the general name of reactions between hydroxyl groups (–OH) and carboxylic acid groups (–COOH) (Table 2.2).

By extension of this argument, the chemical structures of condensation polymers are not readily derived from the names of the polymers. Furthermore, the backbone of condensation polymers is heterogeneous, being generally composed of carbon plus other atoms, usually nitrogen and oxygen (Table 2.3) and sometimes sulfur and silicon. As we shall see later, this has serious implications for the resultant polymer properties. The main distinguishing features between addition and condensation polymerizations are summarized in Table 2.3.

TABLE 2.3 Distinguishing Features of Chain and Step Polymerization Mechanisms

Chain Polymerization	Step Polymerization
Only growth reaction adds repeating unit one at a time of the chain.	Any two molecular species present can react.
Monomer concentration decreases steadily throughout reaction.	Monomer disappears early in reaction: at DP 10, less than 1% monomer remains.
High polymer is formed at once; polymer molecular weight changes little throughout reaction.	Polymer molecular weight rises steadily throughout reaction.
Reaction mixture contains only monomer, high polymer, and about 10^{-5} part of growing chains.	At any stage all molecular species are present in a calculable distribution.

From Billmeyer, F. W., Jr., *Textbook of Polymer Science,* 3rd ed., Interscience, New York, 1984. With permission.

Example 2.2: Unsaturated polyester resins, which are used as the matrix component of glass–fiber composites, may be obtained by the copolymerization of maleic anhydride and diethylene glycol. The low-molecular-weight product is soluble in styrene. Describe, with the aid of equations, the possible structures of the prepolymer and that of the polymer resulting from benzoyl peroxide-initiated polymerization of a solution of the prepolymer in styrene.

Solution:

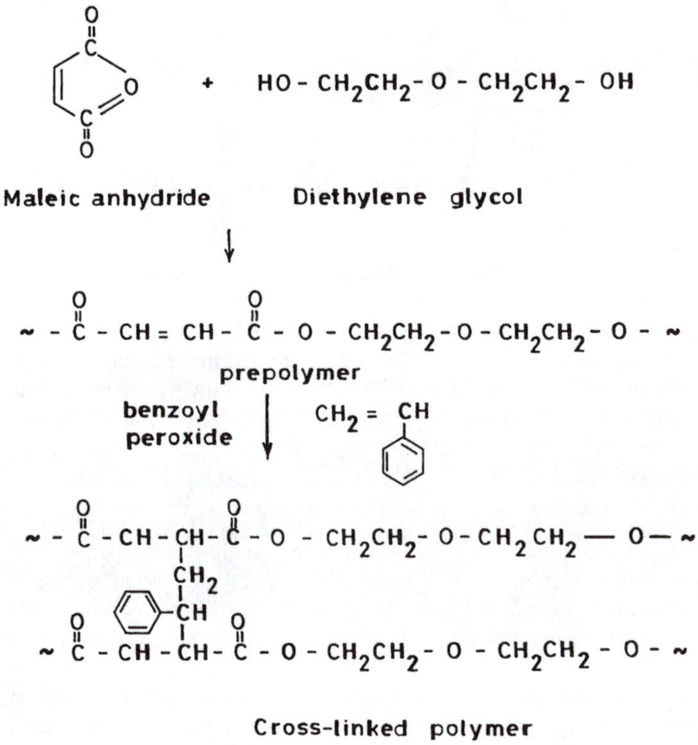

Maleic anhydride + $HO-CH_2CH_2-O-CH_2CH_2-OH$ (Diethylene glycol)

↓

$\sim -\overset{O}{\overset{\|}{C}}-CH=CH-\overset{O}{\overset{\|}{C}}-O-CH_2CH_2-O-CH_2CH_2-O-\sim$

prepolymer

↓ benzoyl peroxide, $CH_2=CH(C_6H_5)$

$\sim -\overset{O}{\overset{\|}{C}}-CH-CH-\overset{O}{\overset{\|}{C}}-O-CH_2CH_2-O-CH_2CH_2-O-\sim$

$|$
CH_2
$|$
C_6H_5-CH
$|$

$\sim \overset{O}{\overset{\|}{C}}-CH-CH-\overset{O}{\overset{\|}{C}}-O-CH_2CH_2-O-CH_2CH_2-O-\sim$

Cross-linked polymer

(Str. 4)

Maleic anhydride is unsaturated. Its reaction with diethylene glycol leads to a prepolymer with residual double bonds on the main chain. These participate in cross-linking reactions with styrene during initiation with benzoyl peroxide. The result is a network polymer.

A. TYPICAL STEP-GROWTH POLYMERIZATIONS

1. Polyesters

Polyesters form a large class of commercially important polymers. A typical polyester is poly(ethylene terephthalate) (PETP), the largest volume synthetic fiber. It is also used as film (mylar) and in bottle applications. We have already discussed in the preceding section one of the routes for the preparation of PETP. The traditional route for the production of commercial PETP is through two successive ester interchange reactions, as shown below:

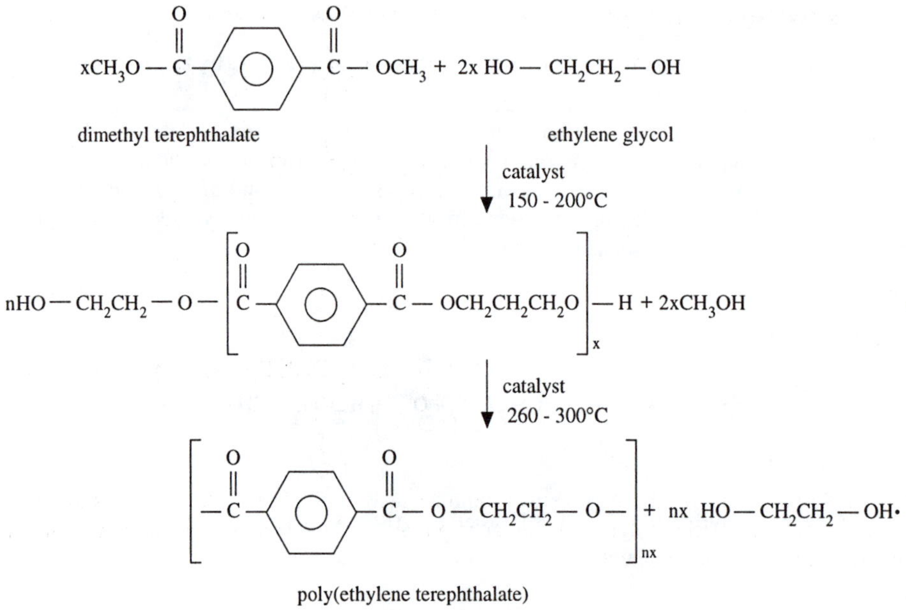

$$x\,CH_3O-\overset{O}{\overset{\|}{C}}-C_6H_4-\overset{O}{\overset{\|}{C}}-OCH_3 + 2x\,HO-CH_2CH_2-OH$$

dimethyl terephthalate ethylene glycol

$$\Big\downarrow \begin{array}{l}\text{catalyst}\\ 150-200^{\circ}C\end{array}$$

$$n\,HO-CH_2CH_2-O-\left[\overset{O}{\overset{\|}{C}}-C_6H_4-\overset{O}{\overset{\|}{C}}-OCH_2CH_2CH_2O\right]_x-H + 2xCH_3OH \qquad \text{(Str. 5)}$$

$$\Big\downarrow \begin{array}{l}\text{catalyst}\\ 260-300^{\circ}C\end{array}$$

$$\left[-\overset{O}{\overset{\|}{C}}-C_6H_4-\overset{O}{\overset{\|}{C}}-O-CH_2CH_2-O-\right]_{nx} + nx\,HO-CH_2CH_2-OH$$

poly(ethylene terephthalate)

In the first step, a 1:2 molar ratio of dimethyl terephthalate to ethylene glycol is heated at temperatures near 200°C in the presence of a catalyst such as calcium acetate. During this stage, methanol is evolved and an oligomeric product (x = 1 to 4) is obtained. The second step involves a temperature increase to about 300°C. This results in the formation of high polymer with the evolution of ethylene glycol.

Poly(ethylene terephthalate) is a linear polyester obtained from the reaction of difunctional monomers. Branched or network polyesters are obtained if at least one of the reagents is tri- or multifunctional. This can be achieved either by the use of polyols such as glycerol in the case of saturated polyesters (glyptal) or by the use of unsaturated dicarboxylic acids such as maleic anhydride in the case of unsaturated polyester. In the preparation of glyptal, glycerol and phthalic anhydride react to form a viscous liquid initially, which on further reaction hardens as a result of network formation (Equation 2.31).

$$\text{Phthalic anhydride} + HO\text{-}CH_2\underset{\large OH}{\underset{|}{C}}HCH_2\text{-}OH \;(\text{Glycerol}) \longrightarrow \sim\text{-}O\text{-}CH_2\underset{|}{C}HCH_2\text{-}O\text{-}\sim \text{ (crosslinked through O-C(=O)-C}_6\text{H}_4\text{-C(=O)-O to } \sim\text{-}O\text{-}CH_2\underset{|}{C}HCH_2\text{-}O\text{-}\sim) \tag{2.31}$$

Glyptal is used mainly as an adhesive. Glyptal modified with natural or synthetic oils is known as an alkyd resin, which is a special polyester of great importance in the coatings industry. A typical alkyd resin comprises the following reagents:

$$\text{Phthalic anhydride} + HO\text{-}CH_2\underset{\large OH}{\underset{|}{C}}HCH_2\text{-}OH \;(\text{Glycerol}) + RCOOH \;(\text{Fatty acid}) \longrightarrow \sim\text{-}O\text{-}\overset{O}{\overset{\|}{C}}\text{-}C_6H_4\text{-}\overset{O}{\overset{\|}{C}}\text{-}O\text{-}CH_2\underset{\underset{\large R}{\underset{|}{C=O}}}{\underset{\underset{|}{O}}{\underset{|}{C}}}HCH_2\text{-}O\text{-}\overset{O}{\overset{\|}{C}}\text{-}C_6H_4\text{-}\overset{O}{\overset{\|}{C}}\text{-}O\text{-}\sim \tag{2.32}$$

The fatty acid RCOOH may be derived from vegetable drying oils (e.g., soybean, linseed oils) or from nondrying oils (e.g., coconut oil).

2. Polycarbonates

Polycarbonates are a special class of polyesters derived from carbonic acid (HO–C(=O)–OH) and have the following general structure:

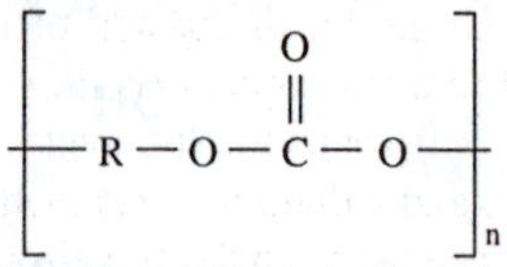

$$\left[-R-O-\overset{\overset{O}{\|}}{C}-O- \right]_n \qquad \text{(Str. 6)}$$

Polycarbonates are the second largest by volume engineering thermoplastics next to polyamides. Their preparation involves the linking together of aromatic dihydroxy compounds, usually 2,2-bis(4-hydroxyphenyl) propane or bisphenol A, by reacting them with a derivative of carbonic acid such as phosgene (Equation 2.33) or diphenyl carbonate (Equation 2.34).

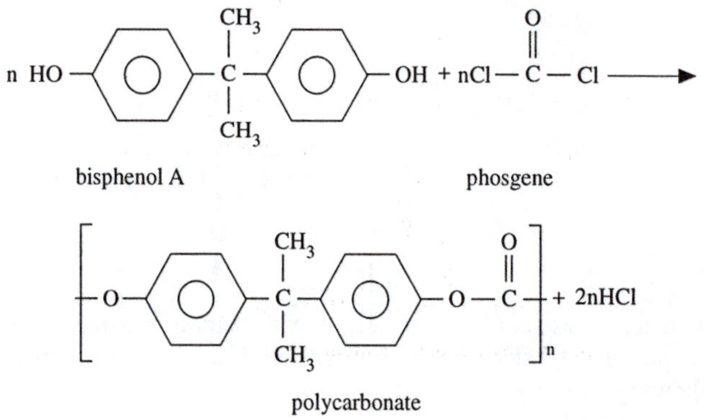

$$n\,HO-C_6H_4-\underset{CH_3}{\overset{CH_3}{C}}-C_6H_4-OH + nCl-\overset{\overset{O}{\|}}{C}-Cl \longrightarrow \qquad (2.33)$$

bisphenol A — phosgene

$$\left[-O-C_6H_4-\underset{CH_3}{\overset{CH_3}{C}}-C_6H_4-O-\overset{\overset{O}{\|}}{C}- \right]_n + 2nHCl$$

polycarbonate

$$n\,HO-C_6H_4-\underset{CH_3}{\overset{CH_3}{C}}-C_6H_4-OH + PhO-\overset{\overset{O}{\|}}{C}-OPH \longrightarrow \qquad (2.34)$$

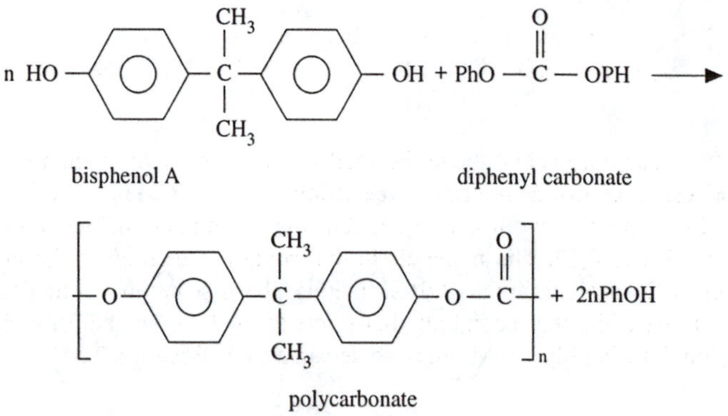

bisphenol A — diphenyl carbonate

$$\left[-O-C_6H_4-\underset{CH_3}{\overset{CH_3}{C}}-C_6H_4-O-\overset{\overset{O}{\|}}{C}- \right]_n + 2nPhOH$$

polycarbonate

The reaction of bisphenol A with phosgene involves bubbling the phosgene into a solution of bisphenol A in pyridine at 20 to 35°C and isolation of the resulting polymer by precipitation in water or methanol. In the reaction of Bisphenol A with diphenyl carbonate, a prepolymer is formed initially by heating the mixture at 180 to 220°C in vacuum. Then the temperature is raised slowly to 280 to 300°C at reduced pressure to ensure the removal of the final traces of phenol.

3. Polyamides

Polyamides, or nylons, as they are commonly called, are characterized by the presence of amide linkages (–CONH–) on the polymer main chain. Theoretically, a large number of polyamides can be synthesized based on four main synthetic routes: (1) condensation reaction between a dicarboxylic acid and a diamine,

(2) reaction between a diacid chloride and a diamine, (3) dehydration–condensation reactions of amino acids, and (4) ring-opening polymerization of lactams. Chemically, nylons may be divided into two categories: those based on synthetic routes (1) and (2); and those based on routes (3) and (4). The commercial use of nylons is centered around two products: nylon 6,6 from the first category, and nylon 6 from the second. We now expatiate our earlier discussion of the preparation of nylon 6,6, while the preparation of nylon 6 will be deferred to a subsequent section.

As with other polyamides, the classical route for the synthesis of nylon 6,6 is the direct reaction between a dicarboxylic acid (adipic acid) and a diamine (hexamethylenediamine). In practice, however, to achieve an exact stoichiometric equivalence between the functional groups, a 1:1 salt of the two reactants is prepared initially and subsequently heated at a high temperature to form the polyamide. For nylon 6,6, an intermediate hexamethylene diammonium adipate salt is formed. A slurry of 60 to 80% of the recrystallized salt is heated rapidly. The steam that is released is purged by air. Temperature is then raised to 220°C and finally to 270 to 280°C when the monomer conversion is about 80 to 90% while maintaining the steam pressure generated during polymerization at 200 to 250 psi. The pressure is subsequently reduced to atmospheric pressure, and heating is continued until completion of polymerization:

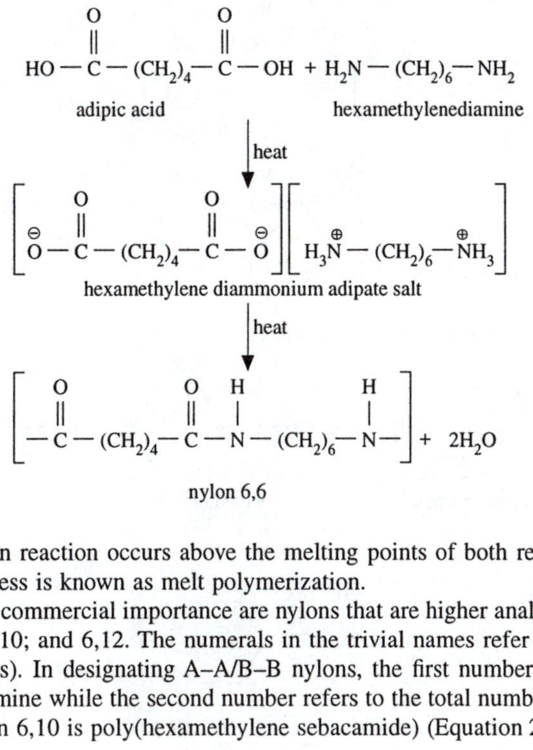

$$\underset{\text{adipic acid}}{HO-\overset{\overset{O}{\|}}{C}-(CH_2)_4-\overset{\overset{O}{\|}}{C}-OH} + \underset{\text{hexamethylenediamine}}{H_2N-(CH_2)_6-NH_2}$$

$$\downarrow \text{heat}$$

$$\underset{\text{hexamethylene diammonium adipate salt}}{\left[\overset{\ominus}{O}-\overset{\overset{O}{\|}}{C}-(CH_2)_4-\overset{\overset{O}{\|}}{C}-\overset{\ominus}{O}\right]\left[H_3\overset{\oplus}{N}-(CH_2)_6-\overset{\oplus}{N}H_3\right]} \qquad \text{(Str. 7)}$$

$$\downarrow \text{heat}$$

$$\underset{\text{nylon 6,6}}{\left[-\overset{\overset{O}{\|}}{C}-(CH_2)_4-\overset{\overset{O}{\|}}{C}-\overset{\overset{H}{|}}{N}-(CH_2)_6-\overset{\overset{H}{|}}{N}-\right]} + 2H_2O$$

Since the polymerization reaction occurs above the melting points of both reactants and the polymer, the polymerization process is known as melt polymerization.

Other polyamides of commercial importance are nylons that are higher analogs of the more common types: nylons 11; 12; 6,10; and 6,12. The numerals in the trivial names refer to the number of carbon atoms in the monomer(s). In designating A–A/B–B nylons, the first number refers to the number of carbon atoms in the diamine while the second number refers to the total number of carbon atoms in the acid. For example, nylon 6,10 is poly(hexamethylene sebacamide) (Equation 2.35).

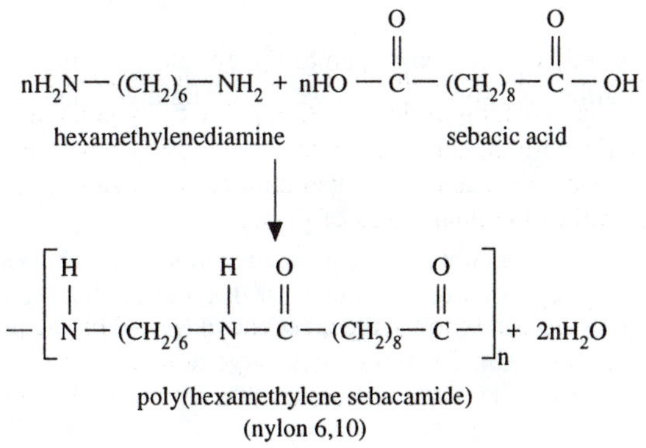

$$\underset{\text{hexamethylenediamine}}{nH_2N-(CH_2)_6-NH_2} + \underset{\text{sebacic acid}}{nHO-\overset{\overset{O}{\|}}{C}-(CH_2)_8-\overset{\overset{O}{\|}}{C}-OH}$$

$$\downarrow$$

$$\underset{\substack{\text{poly(hexamethylene sebacamide)}\\ \text{(nylon 6,10)}}}{-\left[\overset{\overset{H}{|}}{N}-(CH_2)_6-\overset{\overset{H}{|}}{N}-\overset{\overset{O}{\|}}{C}-(CH_2)_8-\overset{\overset{O}{\|}}{C}-\right]_n} + 2nH_2O \qquad (2.35)$$

In the 1960s aromatic polyamides were developed to improve the flammability and heat resistance of nylons. Poly(*m*-phenyleneisophthalamide), or Nomex, is a highly heat resistant nylon obtained from the solution or interfacial polymerization of a metasubstituted diacid chloride and a diamine (Equation 2.36).

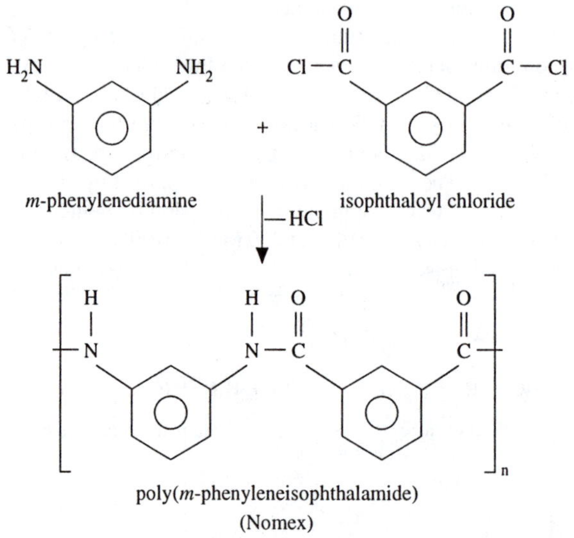

m-phenylenediamine + isophthaloyl chloride $\xrightarrow{-HCl}$ poly(*m*-phenyleneisophthalamide) (Nomex) (2.36)

The corresponding linear aromatic polyamide is Kevlar aramid which decomposes only above 500°C (Equation 2.37).

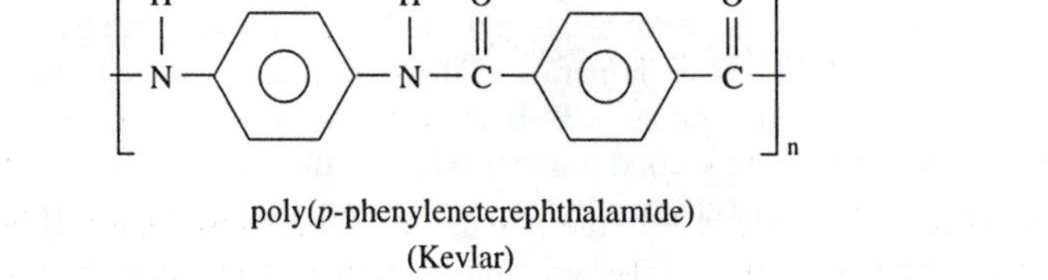

p-phenylenediamine + terephthaloyl chloride $\xrightarrow{-HCl}$ poly(*p*-phenyleneterephthalamide) (Kevlar) (2.37)

The high thermooxidative stability of Kevlar is due to the absence of aliphatic units in its main chain. The material is highly crystalline and forms a fiber whose strength and modulus are higher than that of steel on an equal weight basis.

4. Polyimides

Polyimides are condensation polymers obtained from the reaction of dianhydrides with diamines. Polyimides are synthesized generally from aromatic dianhydrides and aliphatic diamines or, in the case of aromatic polyimides, from the reaction of aromatic dianhydrides with aromatic diamines. Aromatic polyimides are formed by a general two-stage process. The first step involves the condensation of aromatic dianhydrides and aromatic diamines in a suitable solvent, such as dimethylacetamide, to form a soluble precursor or poly(amic acid). This is followed by the dehydration of the intermediate poly(amic acid) at elevated temperature:

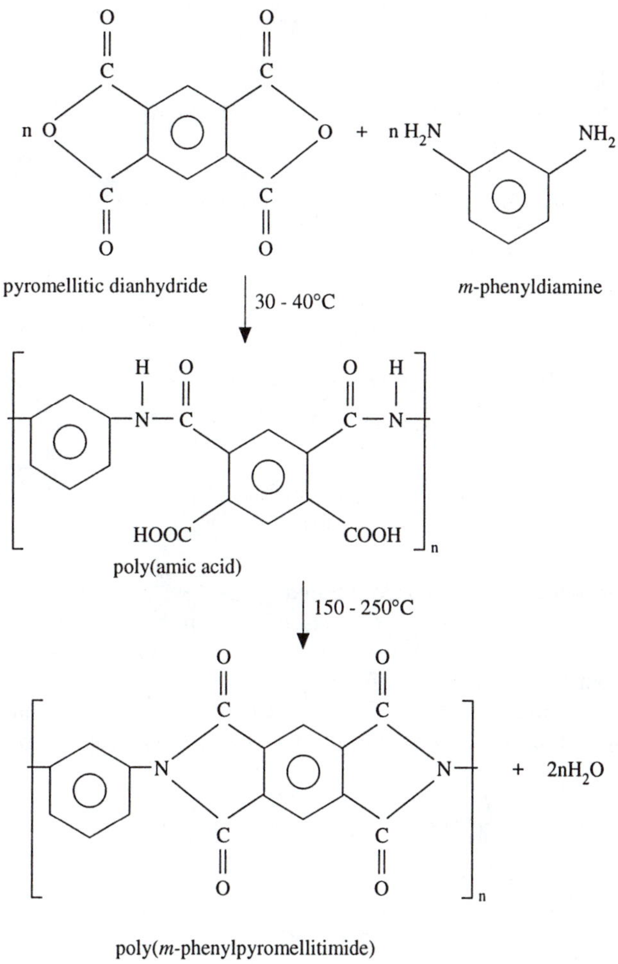

n (pyromellitic dianhydride) + n H_2N–C_6H_4–NH_2 (*m*-phenyldiamine)

↓ 30 - 40°C

poly(amic acid): $[\text{–}C_6H_4\text{–NH–CO–}C_6H_2(\text{COOH})_2\text{–CO–NH–}]_n$

↓ 150 - 250°C

poly(*m*-phenylpyromellitimide) + $2nH_2O$ (2.38)

(4)

The cured or fully imidized polyimide, unlike the poly(amic acid), is insoluble and infusible with high thermooxidative stability and good electrical-insulation properties. Thermoplastic polyimides that can be melt processed at high temperatures or cast in solution are now also available. Through an appropriate choice of the aromatic diamine, phenyl or alkyl pendant groups or main-chain aromatic polyether linkages can be introduced into the polymer. The resulting polyimides are soluble in relatively nonpolar solvents.

For polyimides to be useful polymers, they must be processable, which means that they have to be meltable. Melt processability of polyimides can be improved by combining the basic imide structure with more flexible aromatic groups. This can be achieved by the use of diamines that can introduce flexible linkages like aromatic ethers and amides into the backbone. Polyamide-imides (5) are obtained by condensing trimellitic anhydrides and aromatic diamines, while polyetherimides (6) are produced by nitro displacement reaction involving bisphenol A, 4,4′-methylenedianiline, and 3-nitrophthalic anhydride.

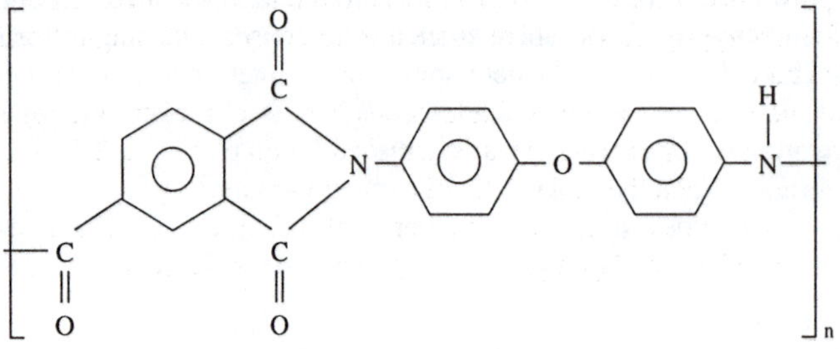

(Str. 8)

(5)

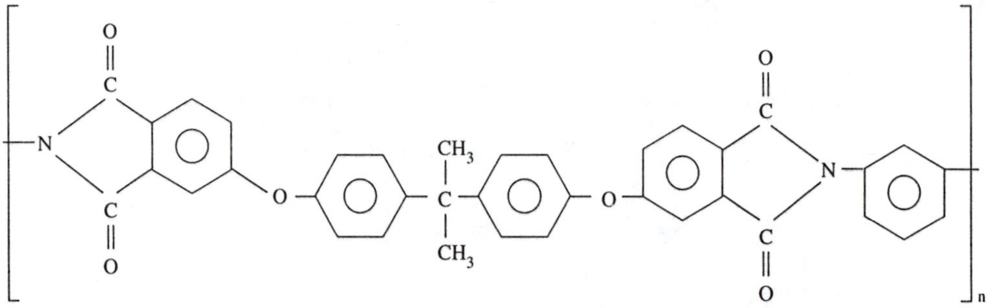

(6)

5. Polybenzimidazoles and Polybenzoxazoles

Aromatic substituents at the chains of vinyl polymers influence the behavior of these materials. Aromatic units as part of the main chain exert a profound influence on virtually all important properties of the resulting polymer. Aromatic polyamides are formed by the repetitive reaction of aromatic amino group and carboxyl group in the molar ratio of 1:1. In aromatic polyamides as well as aromatic polyesters, the chain-stiffening aromatic rings are separated from each other by three consecutive single bonds:

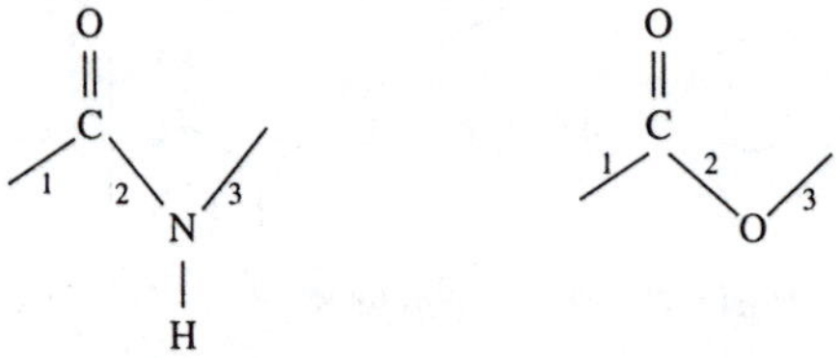

polyamide polyester

(Str. 9)

The two tetrahedral angles associated with these bonds permit some degree of chain flexibility, which limits the mechanical and thermal properties of the resulting polymers. One way of reducing flexibility and enhancing these properties is to reduce the number of consecutive single bonds between two aromatic units to two, one, or even zero. In polyethers, polysulfides, and polysulfones, as we shall see shortly, the number of consecutive single bonds has been reduced to two, and these are separated by only one tetrahedral angle:

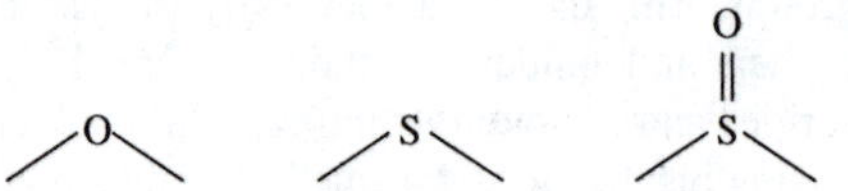

(Str. 10)

polyether polysulfide polysulfone

Polyimides, polybenzimidazole, and polybenzoxazoles are polymers where the number of these bonds has been reduced to one. Ladder polymers typify cases where there are no consecutive single bonds between aromatic moities in the main chain.

In aromatic polyimides (4), two of the three consecutive single bonds between aromatic groups in polyamides are eliminated by the formation of a new ring. This is achieved by employing a 2:1 molar ratio of aromatic carboxyl and amino groups. When the molar ratio of carboxyl groups (e.g., terephthalic acid) to amino groups (e.g., 3,3′ diaminobenzidine) is 1:2, polybenzimidazoles (7) are formed; whereas where the molar ratio of carboxyl, amino, and hydroxyl groups is 1:1:1, polybenzoxazoles (8) are formed.

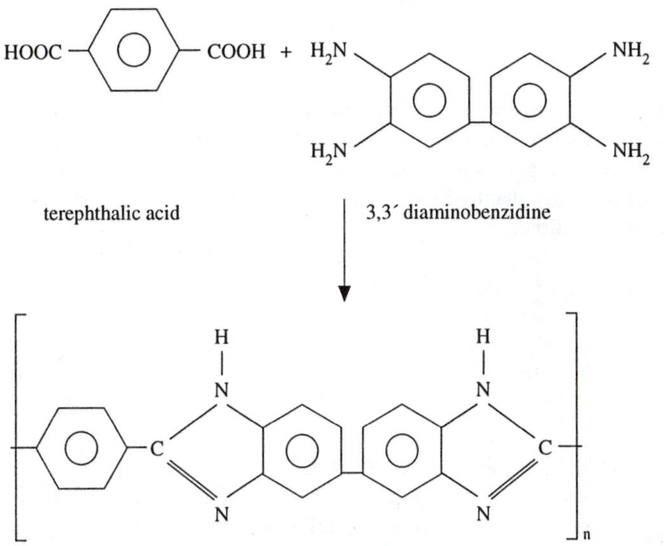

HOOC—C$_6$H$_4$—COOH + H_2N, H_2N / NH_2, NH_2 (biphenyl)

terephthalic acid 3,3′ diaminobenzidine

(2.39)

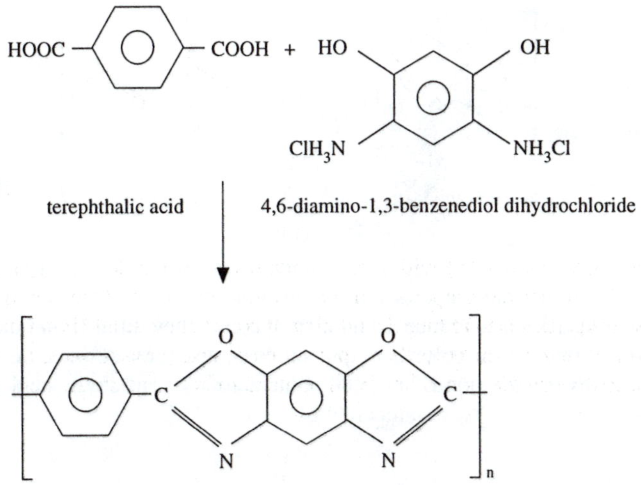

H—N, C, N ... H—N, C, N; n

polybenzimidazole

(7)

HOOC—C$_6$H$_4$—COOH + HO, OH, ClH_3N, NH_3Cl

terephthalic acid 4,6-diamino-1,3-benzenediol dihydrochloride

(2.40)

O, C, N ... O, C, N; n

polybenzoxazole

(8)

6. Aromatic Ladder Polymers

As evident from the preceding discussion, the next logical step to increase the rigidity of linear macromolecules is the elimination of single bonds in the main chain so that it is composed of only condensed cyclic units. The resulting polymer has the following generalized structure (9)

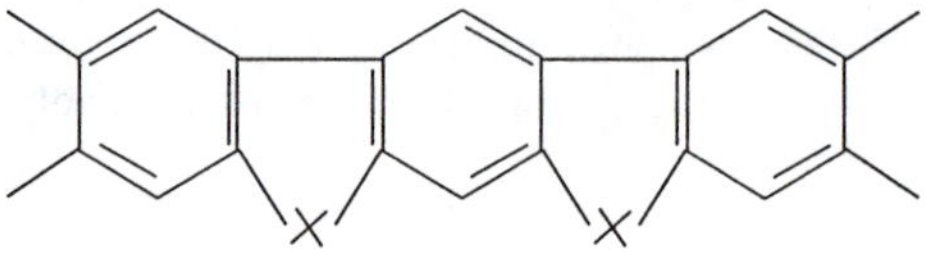

(9)

(Str. 11)

where the individual cyclic units may be either aromatic or cycloaliphatic, homocyclic or heterocyclic. Polymers of this type are known as ladder polymers. Polybenzimidazoles discussed previously are typical aromatic ladder polymers. Longer segments of ladder are present in polyimidazopyrrolones (10) prepared by polymerization of aromatic dianhydrides or aromatic tetracarboxylic acids with *ortho*-aromatic tetramines according to the following scheme:

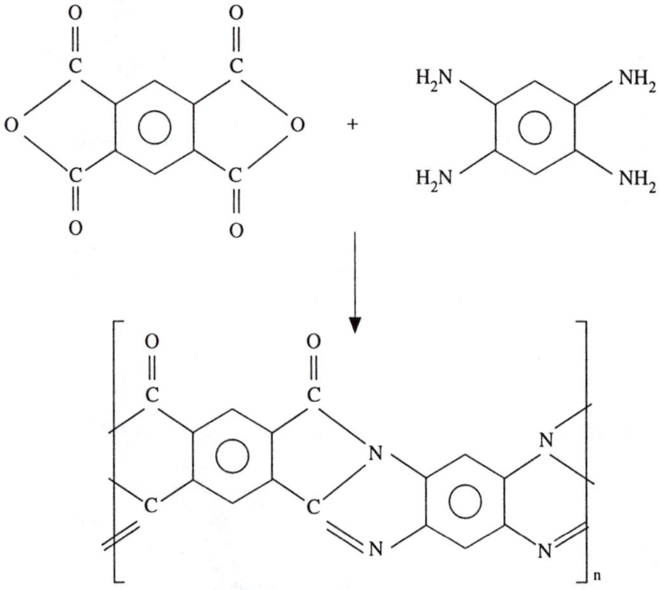

(10)

(2.41)

Polyquinoxalines (11) are another group of ladder polymers. They differ from polyimidazopyrrolones by the presence of a fused six-membered cyclic diimide structure. A possible synthetic route for polyquinoxalines involves the reaction of 1,4,5,8-naphthalene tetracarboxylic acid with aromatic tetraamines in polyphosphoric acid (PPA) at temperatures up to 220°C.

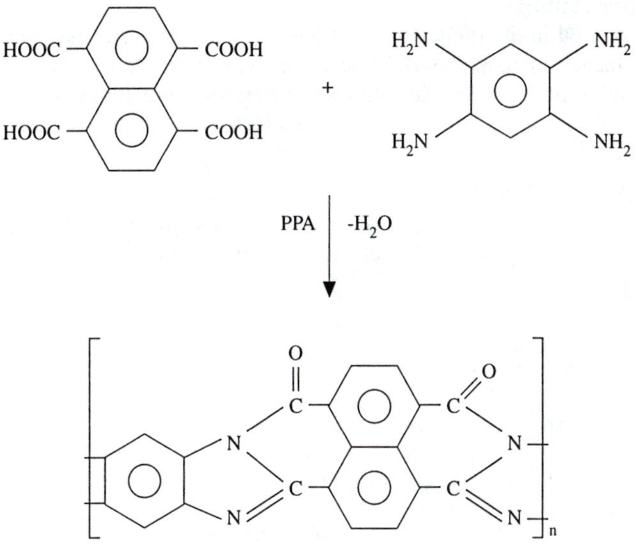

(2.42)

(11)

Ladder polymers are also referred to as "double-chain" or "double-strand" polymers because, unlike other polymers, the backbone consists of two chains. Cleavage reactions in single-chain polymers cause a reduction in molecular weight that ultimately results in a deterioration in the properties of the polymer. For this to happen in a ladder polymer, two bonds will have to be broken in the same chain residue, which is a very unlikely occurrence. Therefore, ladder polymers usually have exceptional thermal, mechanical, and electrical properties.

Example 2.3: Polyimides have been prepared from aromatic anhydrides and aliphatic diamines by melt fusion of salt from the diamine and tetracid (dianhydride). Aliphatic polypyromellitimide derived from straight-chain aliphatic diamines containing more than nine carbon atoms gave thin, flexible films, whereas those from shorter chain aliphatic diamines allowed the preparation of only thick, brittle moldings. Explain this observation.

Solution: The general structure of aliphatic polypyromellitimide is represented as follows:

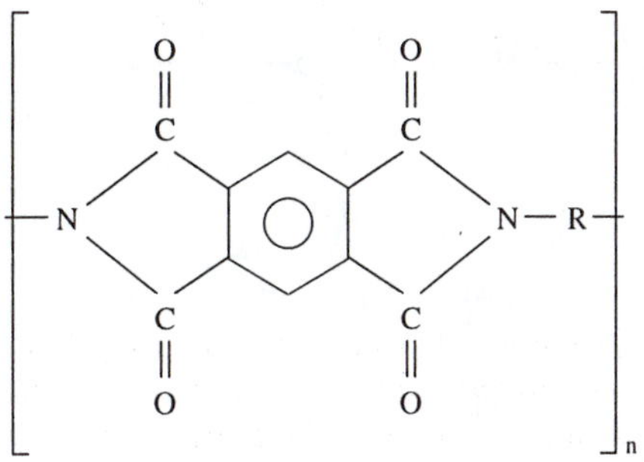

(Str. 12)

where R is an alkylene group. Polyimides are usually rigid polymers. Some flexibility can be introduced into the polymer structure by incorporating flexible groups in the backbone of the polymer such as flexible diamines. The longer the alkylene group the more flexible the diamine. It is apparent from the observation that straight-chain alkylenes with nine or more carbon atoms introduced sufficient flexibility into the resulting polymer that films made from this polymer were flexible. On the other hand, when R was less than nine carbon atoms, the resulting polymers were still too rigid and, consequently, brittle.

---

7. Formaldehyde Resins

Formaldehyde is employed in the production of aminoplasts and phenoplasts, which are two different but related classes of thermoset polymers. Aminoplasts are products of the condensation reaction between either urea (urea–formaldehyde or UF resins) or melamine (melamine–formaldehyde or MF resins) with formaldehyde. Phenoplasts or phenolic (phenol–formaldehyde or PF) resins are prepared from the condensation products of phenol or resorcinol and formaldehyde.

a. Urea–Formaldehyde Resins

Urea–formaldehyde resin synthesis consists basically of two steps. In the first step, urea reacts with aqueous formaldehyde under slightly alkaline conditions to produce methylol derivatives of urea (12, 13).

$$\underset{\text{urea}}{H_2N-\overset{\overset{\displaystyle O}{\|}}{C}-NH_2} + \underset{\text{formaldehyde (aqueous)}}{HO-CH_2-OH} \longrightarrow \underset{(12)}{H_2N-\overset{\overset{\displaystyle O}{\|}}{C}-NH-CH_2-OH} \text{ (monomethylol urea)} \quad (2.43)$$

$$\xrightarrow{HO-CH_2-OH} \underset{(13)}{HO-CH_2-NH-\overset{\overset{\displaystyle O}{\|}}{C}-NH-CH_2-OH} \text{ (dimethylol urea)}$$

In the second step, condensation reactions between the methylol groups occur under acidic conditions, leading ultimately to the formation of a network structure (14):

$$-N-\overset{\overset{\displaystyle O}{\|}}{C}-NH-CH_2OH + HOCH_2-NH-\overset{\overset{\displaystyle O}{\|}}{C}-N-$$

$$\longrightarrow -N-\overset{\overset{\displaystyle O}{\|}}{C}-NH-CH_2-O-CH_2-NH-\overset{\overset{\displaystyle O}{\|}}{C}-N- \quad (2.44)$$

$$\longrightarrow \underset{(14)}{-\overset{|}{N}-\overset{\overset{\displaystyle O}{\|}}{C}-NH-CH_2-NH-\overset{\overset{\displaystyle O}{\|}}{C}-\overset{|}{N}-}$$

b. Melamine–Formaldehyde Resins

Production of melamine–formaldehyde polymers involves reactions essentially similar to those of UF resins, that is, initial production of methylol derivatives of melamine, which on subsequent condensation, ultimately form methylene bridges between melamine groups in a rigid network structure (15).

$$\underset{\text{melamine}}{C_3N_3(NH_2)_3} + \underset{\text{formaldehyde}}{CH_2O} \longrightarrow C_3N_3[N(CH_2OH)_2]_3 \longrightarrow \underset{\mathbf{15}}{\cdots\text{–}N\text{–}C_3N_3\text{–}N\text{–}CH_2\text{–}N\text{–}C_3N_3\text{–}N\text{–}\cdots} \tag{2.45}$$

c. Phenol–Formaldehyde Resins

Phenolic resins are prepared by either base-catalyzed (resoles) or acid-catalyzed (novolacs) addition of formaldehyde to phenol. In the preparation of resoles, phenol and excess formaldehyde react to produce a mixture of methylol phenols. These condense on heating to yield soluble, low-molecular-weight prepolymers or resoles (16). On heating of resoles at elevated temperature under basic, neutral, or slightly acidic conditions, a high-molecular-weight network structure or phenolic rings linked by methylene bridges (17) is produced.

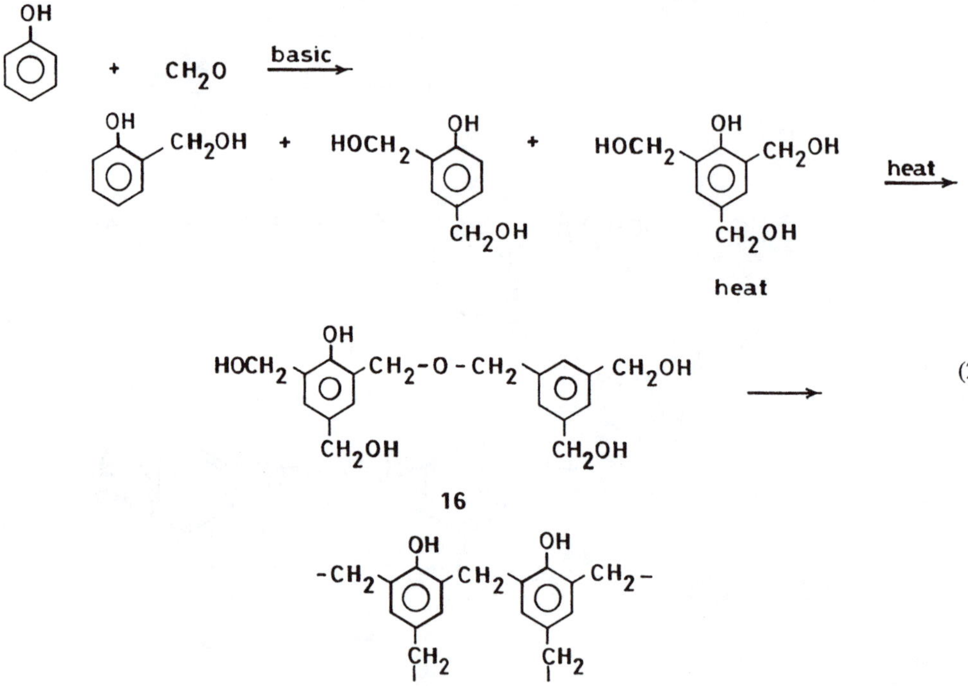

OH + CH_2O basic →

OH CH_2OH + $HOCH_2$ OH CH_2OH + $HOCH_2$ OH CH_2OH CH_2OH heat →

heat

$HOCH_2$ OH CH_2-O-CH_2 CH_2OH CH_2OH CH_2OH → (2.46)

16

-CH_2 OH CH_2 OH CH_2- CH_2 CH_2

17

Novolacs are low-molecular-weight, fusible but insoluble prepolymers (18) prepared by reaction of formaldehyde with molar excess of phenol. Novolacs, unlike resoles, do not contain residual methylol groups. A high-molecular-weight network polymer similar to that of resoles is formed by heating novolac with additional formaldehyde, paraformaldehyde, or hexamethylenetetramine.

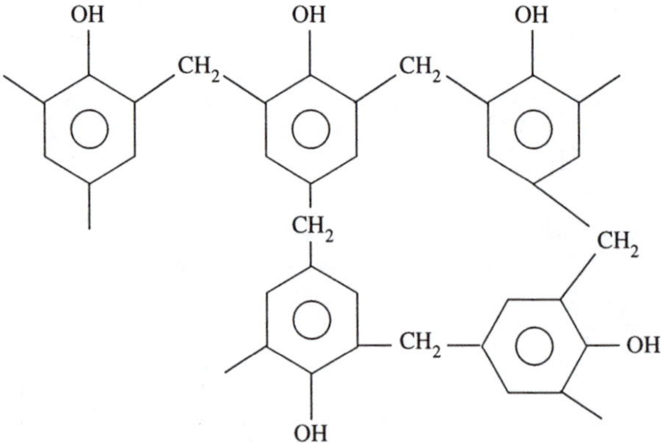

OH OH OH CH_2 CH_2 CH_2 CH_2 CH_2 OH OH (Str. 13)

(18)

We now discuss a number of polymerization reactions that are of the step-reaction type but which may not necessarily involve condensation reactions.

8. Polyethers

As we said earlier, the introduction of aromatic units into the main chain results in polymers with better thermal stability than their aliphatic analogs. One such polymer is poly(phenylene oxide), PPO, which has many attractive properties, including high-impact strength, resistance to attack by mineral and organic acids, and low water absorption. It is used, usually blended with high-impact polystyrene (HIPS), to ease processability in the manufacture of machined parts and business machine enclosures.

Poly(phenylene oxide) (19) is obtained from free-radical, step-growth, oxidative coupling of 2,6-dimethylphenol (2,6-xylenol). This involves passing oxygen into a reaction mixture containing 2,6-xylenol, cuprous chloride, and pyridine.

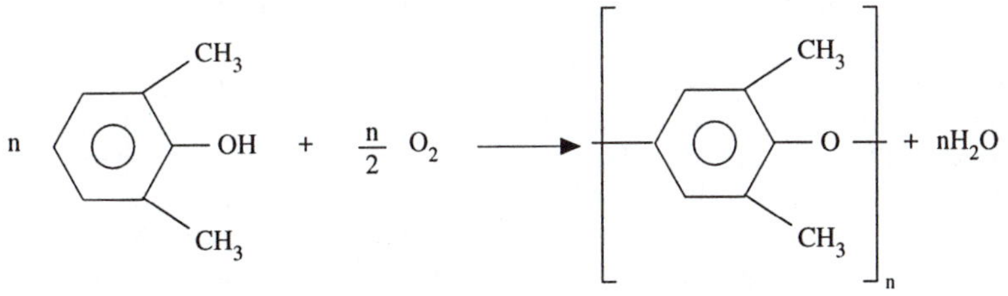

$$n\ HO{-}C_6H_3(CH_3)_2 + \frac{n}{2}\ O_2 \longrightarrow \left[C_6H_2(CH_3)_2{-}O \right]_n + nH_2O \qquad (2.47)$$

(19)

9. Polysulfides

Aromatic polythioethers or polysulfides are closely related to polyethers in structure and properties. A typical aromatic polysulfide is poly(phenylene sulfide) (PPS) (20), which is used as electrical insulators and structural parts in the building of engines and vehicles. Poly(phenylene sulfide) is prepared by the condensation reaction between *p*-dichlorobenzene and sodium sulfide:

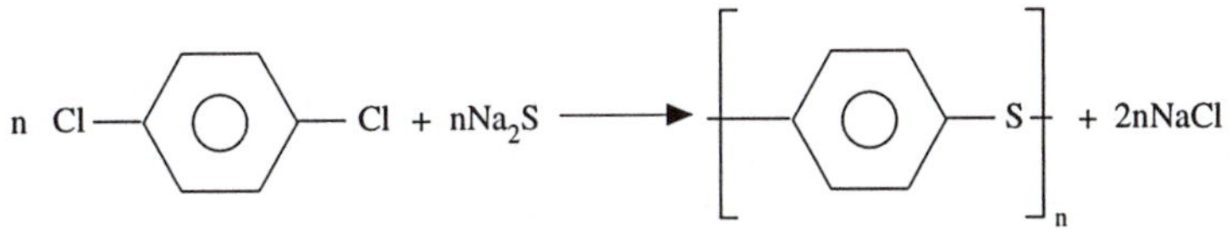

$$n\ Cl{-}C_6H_4{-}Cl + nNa_2S \longrightarrow \left[C_6H_4{-}S \right]_n + 2nNaCl \qquad (2.48)$$

(20)

10. Polysulfones

Another family of linear aromatic polymers is the polysulfones. They are tough, high-temperature-resistant engineering thermoplastics. Polysulfones may be synthesized by the nucleophilic substitution of alkali salts of biphenates with activated aromatic dihalides. A typical example is the preparation of bisphenol A polysulfone (21) from the reaction of disodium salt of bisphenol A with dichlorodiphenyl sulfone:

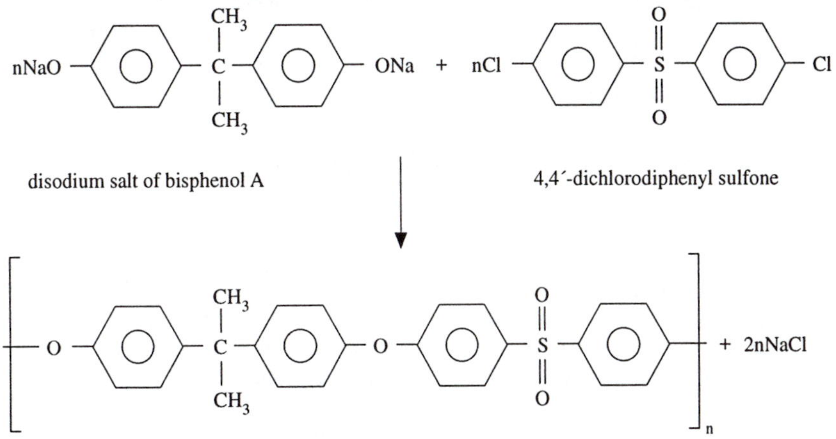

$$nNaO{-}C_6H_4{-}C(CH_3)_2{-}C_6H_4{-}ONa + nCl{-}C_6H_4{-}SO_2{-}C_6H_4{-}Cl$$

disodium salt of bisphenol A — 4,4′-dichlorodiphenyl sulfone

$$\downarrow \qquad (2.49)$$

$$\left[O{-}C_6H_4{-}C(CH_3)_2{-}C_6H_4{-}O{-}C_6H_4{-}SO_2{-}C_6H_4 \right]_n + 2nNaCl$$

(21)

The polymerization reaction involves the initial preparation of the disodium salt of bisphenol A by the addition of aqueous NaOH to bisphenol A in dimethyl sulfoxide (DMSO). A solution of dichlorodiphenyl sulfone is added, and polymerization is carried out at 160°C.

V. RING-OPENING POLYMERIZATION

While ring-opening polymerization shares certain features with condensation and addition polymerization mechanisms, it differs from each of them in at least one important respect. In the first place, in contrast to condensation polymerization, no small molecule is split off in ring-opening polymerization. Second, unlike olefin polymerization, the driving force for ring-opening polymerization is not derived from the loss of unsaturation.

A significant number of polymers has been produced from the ring-opening of cyclic organic compounds, including epoxides such as ethylene and propylene oxides and epichlorohydrin and other cyclic ethers like trioxane and tetrahydrofuran. Other important systems include cyclic esters (lactones), cyclic amides (lactams), cycloolefins, and siloxane. Ring-opening polymerization involves essentially an initial ring-opening of the cyclic monomer followed by polyaddition. The resulting polymers are normally linear. Their structural units usually have the same composition as the monomer. Major applications of polymers obtained from ring-opening polymerization are in coatings, fibers, elastomers, adhesives, and thermoplastics- and thermoset-based composite systems.

Ring-opening polymerizations may be represented broadly by Equation 2.50

$$\text{cyclo}[X-(CH_2)_y] \longrightarrow \left[-(CH_2)_y-X-\right]_n \qquad (2.50)$$

where X may be a heteroatom such as O, S, or a group like NH, –O–CO–, –NH–CO, or –C=C–. Not all cyclic compounds can undergo ring-opening polymerization. It is therefore understood from Equation 2.50 that the cyclic structure is capable of ring-opening polymerization to produce a linear chain of degree of polymerization, n. The nature of X is such that it provides a mechanism for a catalyst or initiator to form the initiating coordination intermediate with the cyclic ring. Table 2.4 shows a number of commercially important polymers produced by ring-opening polymerization together with the associated polymerization initiation mechanism.

A. POLY(PROPYLENE OXIDE)

The polymerization of propylene oxide represents an important example of industrial ring-opening polymerization. It involves an attack of the least sterically hindered carbon by the hydroxyl anion to produce the alkoxide (Equation 2.51). This produces essentially linear polymer molecules.

$$KOH + CH_2\overset{O}{\frown}CH-CH_3 \xrightarrow{\Delta} \qquad (2.51)$$

$$HO-CH_2-\underset{CH_3}{\underset{|}{CH}}-O^{\ominus}K^{\oplus} \xrightarrow[\text{then } H^+]{\text{Propagation}} \left[CH_2-CH-O\right]_n$$

Poly(propylene oxide) glycols are utilized extensively as soft segments in urethane foams, which, among other applications, are used as automobile seats. It is frequently necessary to modify the growing species in propylene oxide polymerization with ethylene oxide in order to produce a polymer with acceptable reactivity with isocyanates and urethane product with desirable properties.

Table 2.4 Examples of Commercially Important Ring-Opening Polymerizations

Monomer	Polymer	Mechanism
O, CH_2, CH_2, O, O, CH_2 (six-membered ring) trioxane	$[-CH_2-O-]_n$ polyoxymethylene	Cationic
O, CH_2-CH_2 (three-membered ring) ethylene oxide	$-[CH_2-CH_2-O-]_n$ poly(ethylene oxide)	Anionic, cationic, coordination
O, CH_2, CH_2, CH_2-CH_2 (five-membered ring) tetrahydrofuran	$-[(CH_2)_4-O-]_n$ poly(tetramethylene oxide)	Cationic
O, CH_2, NH, CH_2, CH_2, CH_2-CH_2 (ring) Caprolactam	$[-N(H)-(CH_2)_5-C(=O)-]_n$ Polycaprolactam (nylon 6)	Hydrolytic, anionic
$C(=O)$, CH_2, O, CH_2, CH_2, CH_2-CH_2 (ring) caprolactone	$[-O-(CH_2)_5-C(=O)-]_n$ polycaprolactone	Anionic, cationic
Si, O, O, Si, Si, O, O, Si (eight-membered ring) Dimethysiloxane (cyclic tetramer)	$[-Si(CH_3)_2-O-]_n$ poly(dimethylsiloxane)	Anionic, cationic

From McGrath, J. E., *Makromol. Chem. Macromol. Symp.*, 42/43, 69, 1991. With permission.

B. EPOXY RESINS

Epoxy resins are normally prepared by the base-catalyzed reaction between an epoxide such as epichlorohydrin and a polyhydroxy compound such as bisphenol A:

$$HO-C_6H_4-C(CH_3)_2-C_6H_4-OH + \underset{\text{epichlorohydrin}}{\overset{O}{CH_2-CH}-CH_2Cl} \longrightarrow$$

bisphenol A — epichlorohydrin

$$HO-C_6H_4-C(CH_3)_2-C_6H_4-O-CH_2-CH_2-\overset{OH}{\overset{|}{CH}}-CH_2Cl \cdot \xrightarrow[-NaCl]{NaOH}$$

$$HO-C_6H_4-C(CH_3)_2-C_6H_4-O-CH_2-\overset{O}{CH-CH_2} \qquad (2.52)$$

The molar ratio of epichlorohydrin to bisphenol A can range from as high as 10:1 to as low as 1.2:1. This produces resins ranging from liquid to semisolid to solid and varying molecular weights and softening points. The products are oligomers or prepolymers, which are hardly used as such; they have pendant hydroxyl groups and terminal epoxy groups. The epoxy prepolymer can be cross-linked or cured by reaction with a number of reagents, including primary and secondary amines.

C. POLYCAPROLACTAM (NYLON 6)

The industrial manufacture of nylons involves either water-initiated (hydrolytic) or a strong base-initiated (anionic) polymerization of caprolactam. Polymerization by cationic initiation is also known, but monomer conversion and attainable molecular weights are inadequate for practical purposes and as such there is no commercial practice of this process.

Hydrolytic polymerization of caprolactam is the most important commercial process for the production of nylon 6. The following synthetic scheme outlines hydrolytic polymerization of caprolactam:

$$\underbrace{HN-(CH_2)_5-C=O}_{\text{ring}} + H_2O \xrightleftharpoons{250°C} H_2N-(CH_2)_5-COOH \xrightleftharpoons{250°C} H_3N^+(CH_2)_5-COO^- \qquad (2.53)$$

$$H_2N-(CH_2)_5-COOH \rightleftharpoons H\left[\overset{H}{\overset{|}{N}}-(CH_2)_5-\overset{O}{\overset{||}{C}}\right]OH + n\text{-}1H_2O \qquad (2.54)$$

Water opens the caprolactam ring producing aminocaproic acid, which is believed to exist as the zwitterion. The zwitterion interacts with and initiates the step polymerization of the monomer, with the ultimate generation of linear polymer molecules. In other words, the polymerization process involves an initial ring-opening of the monomer that is followed by the step polymerization.

The hydrolytic polymerization process may be carried out as a continuous operation or it may be operated batchwise. It involves heating caprolactam in an essentially oxygen-free atmosphere in the presence of water to temperatures in the range 250 to 270°C for periods ranging from 12 h to more than

24 h. Most of the water used to initiate the reaction is removed during the process after about 80 to 90% conversion. The overall polymerization involves various equilibria and does not result in complete conversion of the caprolactam. The quantity of the residual monomer depends on the reaction temperature, which under industrial conditions amounts to 8 to 10%. In addition, there are about 3% of predominantly cyclic low-molecular-weight oligomers. These impurities adversely affect the subsequent processing and end-use performance of the polymer and must, therefore, be removed from it. This is achieved either by hot-water extraction or by vacuum evaporation.

The second approach for the commercial synthesis of nylon 6, which accounts for up to 10% of the volume of the polymer, is the base-initiated anionic polymerization of caprolactam. A small but important number of applications utilize this process, which is characterized by high conversion rates. It involves two techniques: "high-temperature" and "low-temperature" polymerizations. The high-temperature polymerization is carried out at temperatures above the melting point of nylon 6 (i.e., 220°C); whereas the low temperature polymerization involves temperatures in the range 140 to 180°C, which are above the melting point of caprolactam but below the melting point of the resulting polycaprolactam. The polymerization catalysts are strong bases such as sodium hydride or a Grignard reagent. To obtain satisfactory low-temperature polymerization, a coinitiator such as *N*-acyl caprolactam or acyl-urea is employed in addition to the strong base.

Sodium or magnesium caprolactam salt is produced by reaction of sodium hydride or a Grignard reagent with caprolactam. A rapid polymerization occurs when the acylated lactam, the catalyst, and monomer are mixed at temperature at or greater than 140°C. This polymerization is usually carried out in a two-stream reactor in which one stream contains the catalyst dissolved in the monomer and the second stream contains the initiator dissolved separately in the monomer:

Stream 1:

$$\underset{\text{monomer}}{\mathrm{HN{-}C(=O){-}(CH_2)_5}\ (\text{ring})} + \underset{(<1\ \text{mol\%})}{\mathrm{NaH}} \xrightarrow[\text{dry}]{100°\mathrm{C}} \underset{\text{catalyst}}{\mathrm{Na^+N^-{-}C(=O){-}(CH_2)_5}\ (\text{ring})} + \tfrac{1}{2}\,\mathrm{H_2} \tag{2.55}$$

Stream 2:

$$\underset{\text{initiator}}{\mathrm{CH_3{-}C(=O){-}N{-}C(=O){-}(CH_2)_5}\ (\text{ring})} + \underset{\text{monomer}}{\mathrm{NH{-}C(=O){-}(CH_2)_5}\ (\text{ring})} \xrightarrow{100°\mathrm{C}} \text{Homogeneous Solution} \tag{2.56}$$

$$\text{Monomer + Catalyst + Initiator} \xrightarrow[>140°\mathrm{C}]{\Delta} \mathrm{CH_3{-}C(=O)}\left[\mathrm{N(H){-}(CH_2)_5{-}C(=O)}\right]_n\mathrm{N{-}C(=O){-}(CH_2)_5}\ (\text{ring}) \tag{2.57}$$

Unlike the high-temperature process where about 8 to 10% cyclics are generated, the equilibrium monomer content of nylon 6 resulting from polymerization at temperature lower than 200°C (i.e., low-temperature polymerization) can be less than 2%. The polymer therefore does not usually require any additional purification. Also, the maximum rate of crystallization of nylon 6 falls within the range of temperature employed in low-temperature polymerization. Consequently, the resulting polymers are characterized by a high degree of crystallinity. Nylon 6 objects of any desired shape are obtainable in essentially a single-stage polymerization. Therefore, this process is frequently referred to as "cast nylon 6" or, in current usage, Reaction Injection-Molded (RIM) Nylon 6.

Example 2.4: Ethylene oxide polymerizes readily to high conversions under either anionic or cationic conditions. Tetrahydrofuran can be induced to polymerize in the presence of phosphorous or antimony pentafluorides as catalysts. Tetrahydropyran is unreactive under polymerization conditions. Explain these observations.

Solution:

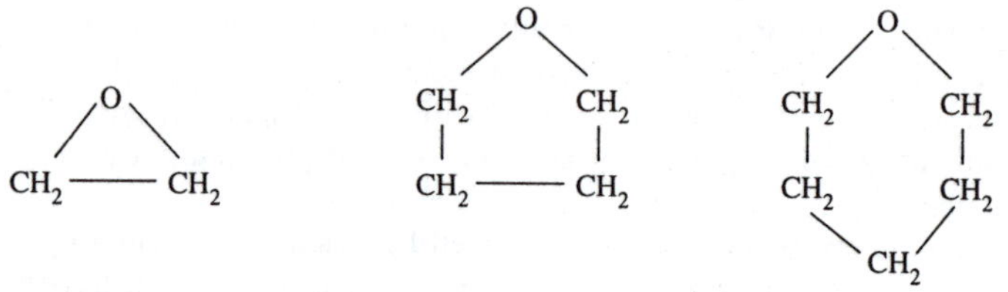

O

CH_2—CH_2

ethylene oxide

O

CH_2 CH_2

CH_2—CH_2

tetrahydrofuran

O

CH_2 CH_2

CH_2 CH_2

CH_2

tetrahydropyran

(Str. 14)

From the structures of these compounds, it is apparent that the ease of polymerization is related to the degree of ring strain. Ethylene oxide is a highly strained three-membered ring. The need to release this strain provides the driving force for polymerization and is reflected in the ease with which ethylene oxide undergoes polymerization. The ring strain decreases in the five-membered tetrahydrofuran and is absent in the six-membered tetrahydropyran.

VI. PROBLEMS

2.1. Explain the following observations:

a. Butadiene polymerizes much less readily than vinyl acetate.
b. Increasing initiator concentration usually leads to lower average molecular weight.

2.2. The use of phthalic anhydride along with maleic anhydride reduces the brittleness of resin described in Example 2.2 in the text. Why?

2.3. Melt polymerization of useful aliphatic polypyromellitimides is limited to those polyimides with melting points sufficiently low that they remain molten under the polymerization conditions. Which of the two diamines (ethylenediamine or decamethylenediamine) is more likely to form a useful polymer?

2.4. Polyheteroaromatics can be produced from aromatic amino carboxylic acids with the following general structure:

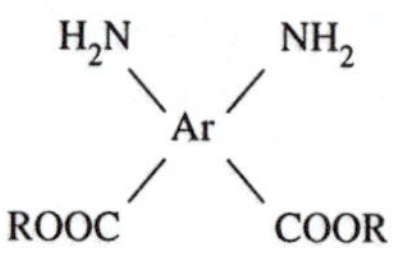

H_2N NH_2

Ar

ROOC COOR

(Str. 15)

Comment on the relative thermal stabilities of the polymers made from compounds where Ar is either A, B, or C.

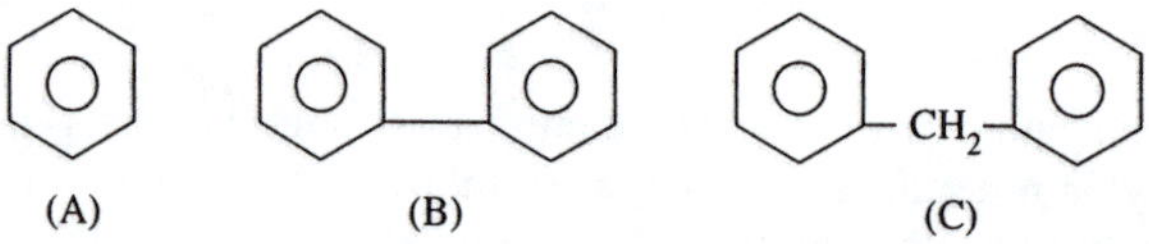

(A) (B) (C) CH_2

(Str. 16)

2.5. Several higher homologs of phenol such as *p*-tertiary butyl phenol, *p*-tertiary amyl phenol and *p*-tertiary octyl-phenol are used to make oil-soluble phenolic resins for varnishes and other coatings systems. What structural features of these compounds make oil solubility of the resins possible?

2.6. Sketch, on the same graph, the molecular weight conversion curves for A, free-radical polymerization; B, condensation polymerization; and C, living polymerization. Explain the basis of your sketch.

REFERENCES

1. Billmeyer, F.W., Jr., *Textbook of Polymer Science,* 3rd ed., Interscience, New York, 1984.
2. Prane, J.A., *Introduction to Polymers and Resins,* Federation of Societies for Coatings Technology, Philadelphia, 1986.
3. Allcock, H.R. and Lampe, F.W., *Contemporary Polymer Chemistry,* 2nd ed., Prentice-Hall, Englewood Cliffs, NJ, 1990.
4. Reimschuessel, H.K., *J. Polym. Sci.: Macromol. Rev.,* 12, 65, 1977.
5. Handerson, J.F. and Szwarc, M., The Use of Living Polymers in the Preparation of Polymer Structures of Controlled Architecture.
6. Sroog, C.E., *J. Polym. Sci. Macromol. Rev.,* 11, 161, 1976.
7. Noren, G.K. and Stille, J.K., *J. Polym. Sci. (D) Macromol. Rev.,* 5, 385, 1971.
8. Braunsteiner, E.E., *J. Polym. Sci. Macromol. Rev.,* 9, 83, 1974.
9. Webster, O.W., *Science,* 251, 887, 1991.
10. McGrath, J.E., *Makromol. Chem. Macromol. Symp.,* 42/43, 69, 1991.
11. Fried, J.R., *Plast. Eng.,* 38(6), 49, 1982.
12. Fried, J.R., *Plast. Eng.,* 38(10), 27, 1982.
13. Fried, J.R., *Plast. Eng.,* 38(12), 21, 1982.
14. Fried, J.R., *Plast. Eng.,* 39(3), 67, 1983.
15. Fried, J.R., *Plast. Eng.,* 39(5), 35, 1983.
16. Chruma, J.L. and Chapman, R.D., *Chem. Eng. Prog.,* p. 49, January 1985.

Chapter 3

Chemical Bonding and Polymer Structure

I. INTRODUCTION

You will recall from your elementary organic chemistry that the physical state of members of a homologous series changes as the molecular size increases. Table 3.1 briefly outlines this for members of the alkane series with the general formula [C_nH_{2n+2}]. From Table 3.1, it is obvious that moving from the low- to the high-molecular-weight end of the molecular spectrum, members of the series change progressively from the gaseous state through liquids of increasing viscosity (decreasing volatility) to low melting solids and ultimately terminate in high-strength solids. Polymers belong to the high-molecular-weight end of the spectrum. In the following discussion, we will attempt to illustrate how the unusual properties of high polymers are developed. To do this, it will be convenient to consider the chemical and structural aspects of polymers at three different levels:

1. The chemical structure (atomic composition) of the monomer (primary structure)
2. The single polymer chain (secondary level)
3. Aggregation of polymer chains (tertiary structure)

But before we proceed into extensive discussion of these aspects, we must first consider the molecular forces operative in polymers. After all, this is fundamental to understanding polymer structures.

II. CHEMICAL BONDING

The electronic structure of atoms determines the type of bond between the atoms concerned. As we said earlier, chemical bonds may be classified as *primary* or *secondary* depending on the extent of electron involvement. Valence electrons are involved in the formation of primary bonds. This results in a substantial lowering of the potential energies. Consequently, primary bonds are quite strong. On the other hand, valence electrons are not involved in the formation of secondary bonds — leading to weak bonds. Primary and secondary bonds can be further subdivided:

1. Primary bonds
   a. Ionic
   b. Covalent
   c. Metallic
2. Secondary bonds
   a. Dipole
   b. Hydrogen
   c. Induction
   d. van der Waals (dispersion)

We now discuss briefly some of these bonds that occur in polymers.

A. THE IONIC BOND

The so-called inert gases — Ne, Ar, and Kr — have completely filled s and p outermost orbitals, resulting in a spherical distribution of electrons. The inertness of these elements (gases) suggests that their electronic configuration confers stability. It is indeed observed that all elements seek to achieve this stable inert gas electronic configuration. They do this by either losing, gaining, or sharing electrons.

The mutual satisfaction of the need to attain the inert gas electronic configuration by those elements that lose electrons (electropositive elements) and those that gain electrons (electronegative elements) leads to the ionic bond. To illustrate this, consider sodium chloride [NaCl]. Sodium (with low ionization energy) can easily lose the outermost 3s electron to achieve the stable inert gas configuration. Chlorine,

Table 3.1 Change of State with Molecular Size for the Alkane $[C_nH_{2n+2}]$ Series

No. of Carbon Atoms	Molecular State
1	Methane — boiling point –162°C
2–4	Natural gas — liquefiable
5–10	Gasoline, diesel fuel — highly volatile, low viscosity liquid
$10–10^2$	Oil, grease — nonvolatile, high viscosity liquid
$10_2–10^3$	Wax — low melting solid
$10^3–10^6$	Solid — high strength

on the other hand (with large electron affinity), can achieve a stable electronic configuration by gaining an extra electron. The loss of an electron by sodium results in a positively charged sodium ion, while the gain of an extra electron by the chlorine atom results in a negatively charged chloride ion:

$$Na + Cl \rightarrow Na^+ + Cl^- \quad \text{(Str. 1)}$$

The bonding force in sodium chloride is a result of the electrostatic attraction between the two ions.

Ionic bonds are not common features in polymeric materials. However, divalent ions are known to act as cross-links between carboxyl groups in natural resins. The relatively new class of polymers known as ionomers contain ionic bonds, as will be discussed later.

B. THE COVALENT BOND

In the previous section, we saw that the stable inert gas electronic configuration can be achieved by electropositive elements through ionization. For elements in the central portion of the periodic table, ionic bonding is impossible because a large amount of energy would be required to ionize the valence electrons. However, stable electronic configuration can be attained by the sharing of valence electrons. Bonds formed by electron sharing are called *covalent* bonds. Consider the formation of methane from hydrogen and carbon. The carbon atom has four unpaired electrons in its outer electron shell, while hydrogen has one electron. By sharing electrons, one from each atom per bond, a stable octet is obtained for the carbon atom and a stable pair for each hydrogen atom. The result is the methane molecule:

```
                 H                 H
                 ..                |
C + 4H -------> H:C:H     or    H—C—H        (Str. 2)
                 ..                |
                 H                 H
```

This is the predominant bond in polymers. Covalent bonds can be single, double, or triple depending on the number of electron pairs. Typical values of bond strengths, expressed as dissociation energy, for covalent bonds that commonly occur in polymers are summarized in Table 3.2.

Recall that dissociation energy is that required to break the bond. It has a direct relationship with the thermal stability of polymers. Note also that while atoms are free to rotate about single bonds (flexible), they remain spatially fixed (rigid) for double and triple bonds.

C. DIPOLE FORCES

Molecules are electrically neutral, but will have a permanent dipole if the centers of the positive and negative charges do not coincide; this arises if the electrons shared by two atoms spend more time on one of the atoms due to differences in electronegativity. This can be illustrated by considering a diatomic molecule such as hydrogen chloride, HCl. Because chlorine is more electronegative than hydrogen, the shared pair of electrons between the chlorine atom and the hydrogen atom is drawn closer to the chlorine atom. Consequently, the chlorine atom has net negative charge while the hydrogen atom has a net positive charge:

Table 3.2 Properties of Some Primary Covalent Bonds in Polymers

Type of bond	Bond length in Å	Average dissociation energy (kcal/mol)	Type of bond	Bond length in Å	Average dissociation energy (kcal/mol)
C–C	1.54	83	C=S	1.71	124
C=C	1.34	147	C–Cl	1.77	79
C=C	1.20	194	N–H	1.01	93
C–H	1.09	99	N–O	1.15	57
C–O	1.43	84	N–Si	1.74	—
C=O	1.23	171	O–H	0.96	111
C–N	1.47	70	O–O	1.48	33
C=N	1.27	147	O–Si	1.64	88
C=N	1.16	213	S–S	2.04	51
C–S	1.81	62	S–H	1.35	81

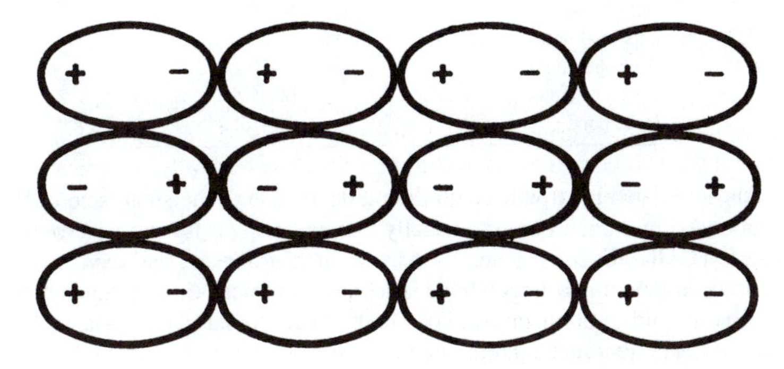

Figure 3.1 Dipole–dipole interaction between polar molecules.

$$\overset{\delta+}{\mathrm{H}} \dashrightarrow \overset{\delta-}{\mathrm{Cl}} \qquad \text{(Str. 3)}$$

Any diatomic molecule in which there is a separation of positive and negative charge is said to be polar. As we shall see in Section III.A, in molecules containing more than two atoms the polarity of the molecule is determined by the bond angles. Polar molecules therefore have a small separation of charge, and this sets up a permanent dipole. Dipoles interact through coulombic forces, which can become quite significant at molecular distances. Polar molecules are held together in the solid state by the interaction between oppositely charged ends of the molecules. The interaction forces between these molecules are called dipole-interaction forces or dipole–dipole interaction (Figure 3.1). This type of molecular orientation is generally opposed by thermal agitation. Consequently, the dipole–dipole interaction is temperature dependent. As we shall see later, dipole forces play a significant role in determining the tertiary structure and, hence, properties of some polymers.

D. HYDROGEN BOND

A particularly important kind of dipole interaction is the hydrogen bond. This is the bond between a positively charged hydrogen atom and a small electronegative atom like F, O, or N. The anomalous properties of water, for example, are associated with the hydrogen bonding between water molecules (Figure 3.2). The difference in electronegativities between hydrogen (2.1) and oxygen (3.5) causes the bonding electrons in H_2O to shift markedly to the oxygen atom so that the hydrogens behave essentially as bare protons. Hydrogen bonding is limited primarily to compounds containing F, N, and O because the small size of hydrogen permits these atoms to approach the hydrogen atom in another molecule very closely. For example, in spite of the similarity in electronegativities between Cl and N (3.0 for both), HCl with the larger chlorine atoms shows hardly any tendency to form hydrogen bonds.

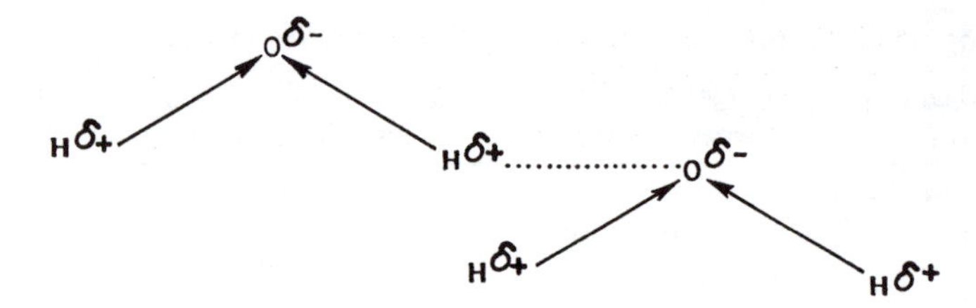

Figure 3.2 Hydrogen bonding between two water molecules.

Table 3.3 Relative Interaction Energies for Different Types of Bonds Found in Polymers

Nature of Interaction	Interaction Energy (kJ/mol)
Dipole–induced dipole	≤2
van der Waals	0.08–4.0
Dipole–dipole	≤20
Hydrogen bond	≤50
Covalent bond	60–600
Ionic bond	560–1000

Hydrogen bonds are relatively stronger than dipole bonds due to the small size of the hydrogen ion (Table 3.3). In polymers, hydrogen bonding usually occurs between functional groups in the same or different molecules. The hydrogen is generally part of such groups as carboxyl, hydroxyl, amine, or amide, while the other atom in the hydrogen bond is frequently oxygen (in carbonyl, ethers, or hydroxyls) or nitrogen (in amines, amides, urethanes, urea). The hydrogen bond plays a vital role in the structure and properties of polymers, particularly proteins.

E. INDUCTION FORCES

Every dipole has an electric field associated with it. This electric field is capable of inducing relative displacements of the electrons and nuclei in neighboring molecules. The result is that the surrounding molecules become polarized, i.e., possess induced dipoles. Intermolecular forces, called induction forces, exist between the permanent and induced dipole. Induction forces are weak and temperature independent. The ease with which molecules can be polarized — referred to as polarizability — varies.

F. VAN DER WAALS (DISPERSION) FORCES

From the above, it would be expected that the gases He, Ne, Ar, and Kr are incapable of forming any type of bonds (ionic, covalent, or metallic). In fact, these so-called inert gases derive the name from that usual stability (considerable reluctance to undergo reactions). However, at sufficiently low temperature these gases are known to condense to form solids. Similarly, molecules such as methane [CH_4], carbon dioxide [CO_2], and hydrogen [H_2] have all the valency requirements fulfilled and should, in principle, be incapable of forming bonds. Yet, these also solidify at sufficiently low temperatures. It is therefore apparent that some form of intermolecular force exists in these materials.

Electrons are usually in constant motion about their nuclei. At any particular instant, the centers of negative charge of the electrons cloud may not coincide with those of the nuclei. Consequently, instantaneous (fluctuating) dipoles exist even in nonpolar materials. If the orientations of fluctuating dipoles in neighboring molecules are in proper alignment, intermolecular attractions occur. These attractive forces are referred to as van der Waals (dispersion) forces. Van der Waals forces are present in all molecules and, as we shall see later, they contribute significantly to the bonding in polymers. Table 3.3 shows the relative magnitudes of the different interaction energies, while typical melting points for various compounds are shown in Table 3.4.

Table 3.4 Typical Melting Temperatures for Some Substances with Different Types of Chemical Bonding

Type of Bond	Substance	Melting Temperature (°C)
Ionic	Na F	988
	NaCl	801
	Na Br	740
	Na I	660
	NaO	2640
	CaO	2570
	Sr O	2430
	Ba O	1923
	Al_2O_3	3500
Covalent	Ge	958
	GaAs	1238
	Si	1420
	SiC	2600
	Diamond	3550
Metallic	Na	98
	Al	660
	Cu	1083
	Fe	1535
	W	3370
Van der Waals	Ne	–249
	Ar	–189
	CH_4	–184
	Kr	–157
	CL_2	–103
Hydrogen	HF	–92
	H_2O	0

Example 3.1: Explain the trend in the melting points of the following:

Compound/Element	Melting Point
KF	46
Na	97.5
F_2	–219.6
Polyethylene	135

Solution:

Compound/Element	Chemical Bonding	Bond Type
KF	Ionic	Primary
Na	Metallic	Primary
F_2	van der Waals	Secondary
Polyethylene (PE)	van der Waals	Secondary

Primary bonds are stronger than secondary bonds. Within the primary bonds ionic bonds are generally stronger than metallic bonds, particularly for univalent metals. Both fluorine and PE molecules are held by van der Waals forces. In the case of fluorine molecules, these forces are readily overcome by thermal agitation and, consequently, fluorine is a gas at room temperature. However, because of the macromolecular sizes of PE molecules, in the aggregate the van der Waals forces become very large. This, coupled with extensive physical entanglements, results in a high melting point.

III. PRIMARY STRUCTURE

Primary structure refers to the atomic composition and chemical structure of the monomer — the building block of the polymer chain. An appreciation of the nature of the monomer is fundamental to understanding the structure–property relationship of polymers. The chemical and electrical properties of a polymer are directly related to the chemistry of the constituent monomers. The physical and mechanical properties of polymers, on the other hand, are largely a consequence of the macromolecular size of the polymer, which in itself is related to the nature of the monomer. By definition, a polymer is a chain of atoms hooked together by primary valence bonds. Therefore, basic to understanding the structure of the monomer vis-á-vis the structure and properties of the resulting polymer is a fundamental understanding of:

- The nature of bonds in monomers (chemical bonding)
- The type of monomers that are capable of forming polymers (functionality of monomers)
- The mode of linking of monomers (polymerization mechanisms)
- The chemical composition of monomers and the properties conferred on monomers as a result of their chemical composition

We have discussed chemical bonding, monomer functionality, and polymerization mechanisms in previous sections. Our attention now focuses on the chemical composition of monomers.

A. POLARITY OF MONOMERS

The chemical composition and atomic arrangement of an organic molecule confer certain properties on the molecule. One such property is the polarity of the molecule. We now discuss this briefly.

The ionic compound sodium chloride is formed by an electron transfer from sodium (leaving behind a positively charged ion) to chlorine (leaving a negatively charged chloride ion). A diatomic molecule with such a pair of equal but opposite charges possesses a permanent dipole moment and is said to be *polar*. Sodium chloride, like all ionic substances where complete charge transfer has occurred, is highly polar. This polarity is responsible for the electrostatic attraction between adjacent ions in solid sodium chloride.

Covalent molecules, on the other hand, are formed by the sharing of electrons between the constituent atoms. In a diatomic molecule formed from two *like* atoms (e.g., H_2), the electron pair linking the two atoms is equally shared and the molecule is said to be *nonpolar.* But when molecules are formed from two unlike atoms (e.g., hydrogen fluoride, HF), the distribution of the electron cloud is concentrated on the more electronegative atoms (fluorine, in this case). Here again, as in ionic compounds, there is a separation of positive and negative charge and the molecule is said to be polar. However, since no complete charge transfer has taken place in this case, the polarity (of covalent molecules) is less than that of ionic compounds Even among covalent molecules, the degree of polarity varies depending on the electronegativities (electron-attracting ability) of the constituent atoms. The electronegativities of atoms commonly occurring in organic molecules are shown in Table 3.5. It is evident from the table that groups like C–Cl, C–F, –CO–, –CN, and –OH are polar.

In a polyatomic molecule, the polarity is a vector sum of all the dipole moments of the groups within the molecule. This depends on the spatial distribution (symmetry) of the groups within the molecule. To illustrate this, let us consider two triatomic molecules: water [H_2O] and carbon dioxide [CO_2]. Both the OH and CO groups are polar. But while the H_2O molecule is polar, CO_2 is a nonpolar molecule. The structure of CO_2 is linear, resulting in a cancellation of the dipole moments. However, H_2O has a triangular structure and, consequently, possesses an overall dipole moment (Figure 3.3).

Carbon tetrachloride, CCl_4, is another molecule that is nonpolar even though it has four polar C–Cl bonds. The nonpolar nature of CCl_4 is due to the symmetrical distribution of the four chlorines around the carbon atom. Replacement of one of the chlorine atoms by hydrogen destroys symmetry. The resulting

Table 3.5 Electronegativities of Some Elements

Atom	H	C	N	O	F	SI	S	Cl
Electronegativity	2.1	2.5	3.0	3.5	4.0	1.8	2.5	3.0

From Pauling, L., *The Nature of the Chemical Bond,* Cornell University Press, Ithaca, NY, 1960. With permission.

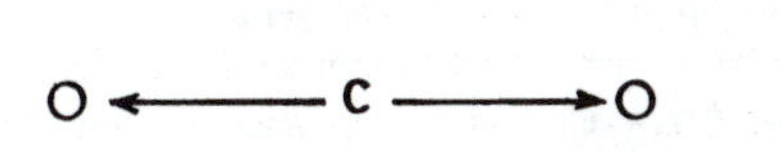

Non-polar; dipoles from polar bonds cancel due to symmetry

O
105°
H H

Polar; dipoles from polar bonds do not cancel

Figure 3.3 Effect of symmetry on polarity of molecules.

molecule, chloroform [$CHCCl_3$], is polar. Monomers such as ethylene and propylene are nonpolar, and so are the polymers derived from them. On the other hand, the polar monomers vinyl chloride [$CH_2{=}CHCl$] and acrylonitrile [$CH_2{=}CHCN$] result in polar polymers. However, the symmetrical monomers vinylidene chloride and vinylidene fluoride lead to nonpolar polymers (Table 3.6). The characteristic interunit linkages in condensation polymers, for example, –CO.O– (ester), –CO.NH– (amide), –HN–CO–NH– (urea), and –O.CO–NH– (urethane) are polar. Polarity, as we shall see later, affects the intermolecular attraction between chain molecules, and thus the regularity and symmetry of polymer structure. Naturally, properties such as the solubility and electrical nature of polymers, which depend on polymer structure, are intimately related to polarity.

IV. SECONDARY STRUCTURE

To be able to understand polymer properties, we must be able to develop a physical picture of what these long molecules are really like. This is what we refer to as the secondary structure, i.e., the size and shape of an isolated single molecule. The size of the polymer is best discussed in terms of molecular weight. The shape of the polymer molecule (molecular architecture) will be influenced naturally by the nature of the repeating unit and the manner in which these units are linked together. It is therefore convenient to consider polymer shape in two contexts:

- Configuration — Arrangement fixed by primary valence bonds; can be altered only through the breaking or reforming of chemical bonds
- Conformation — Arrangement established by rotation about primary valence bonds

A. CONFIGURATION

As we saw earlier, a polymer molecule may be linear, branched, or cross-linked depending on the functionality of the monomers used. But let us look more closely at the polymer chain. If repeating units along the chain are chemically and sterically regular, then the polymer is said to possess structural regularity. To consider structural regularity, we need to define two terms: *recurrence regularity* and *stereoregularity.*

Recurrence regularity refers to the regularity with which the repeating unit occurs along the polymer chain. This may be illustrated by examining the polymers resulting from monosubstituted vinyl monomers. Here there are three possible arrangements:

- Head-to-tail configuration

$$\underset{\displaystyle\;\;\,|\\ \;\;\,X}{CH_2{=}CH} \dashrightarrow \underset{|\\ X}{CH_2{-}CH}{-}\underset{|\\ X}{CH_2{-}CH}{-}\underset{|\\ X}{CH_2{-}CH}{-} \tag{3.1}$$

- Head-to-head configuration

$$\underset{|\\ X}{CH_2{=}CH} \dashrightarrow {-}CH_2{-}\underset{|\\ X}{CH}{-}\underset{|\\ X}{CH}{-}CH_2{-} \tag{3.2}$$

Table 3.6 Polarity of Monomers and Their Associated Polymers

Monomer	Polarity	Polymer	Polarity
$CH_2{=}CH_2$ Ethylene	Nonpolar	$\text{+}CH_2-CH_2\text{+}$ Polypropylene	Nonpolar
$CH_2{=}CH_2$ with pendant CH_3 Propylene	Nonpolar	$\text{+}CH_2-CH_2\text{+}$ with pendant CH_3 Polypropylene	Nonpolar
$CH_2{=}CH$ with pendant Cl Vinyl chloride	Polar	$[-CH_2-CH-]$ with pendant Cl Poly(vinyl chloride)	Polar
$CH_2{=}CCl_2$ Vinylidene chloride	Nonpolar	$\text{+}CH_2-C\text{+}$ with two pendant Cl Poly(vinylidene chloride)	Nonpolar
$CH_2{=}CH$ with pendant CN Acroylonitrile	Polar	$\text{+}CH_2-CN\text{+}$ with pendant CN Polyacrylonitrile	Polar
$CF_2{=}CF_2$ Tetrafluoroethylene (symmetrical)	Nonpolar	$\text{+}CF_2-CF_2\text{+}$ Polytetrafluorethylene (Teflon)	Nonpolar

- Tail-to-tail configuration

$$\underset{\displaystyle X}{CH_2{=}\overset{}{CH}} \dashrightarrow -\underset{\displaystyle X}{CH}-CH_2-CH_2-\underset{\displaystyle X}{CH}- \qquad (3.3)$$

The last two configurations do not appear in any measurable extent in known polymers.

Stereoregularity refers to the spatial properties of a polymer molecule. To discuss this, let us consider two examples.

1. Diene Polymerization

You will recall that the propagation step in the polymerization of diene monomers (monomers with two double bonds) can proceed by either of two mechanisms: 1,2, and 1,4 additions. In 1,2 addition the resulting polymer unsaturation is part of the pendant group, while in 1,4 addition the unsaturation is part of the backbone. In the latter case, the backbone has a rigid structure and rotation is not free around it. Therefore, two different configurations, known as *cis* and *trans,* are possible. For example, 1,4-polyisoprene:

$$\begin{matrix} -CH_2 & & CH_2- \\ & C=C & \\ CH_3 & & H \end{matrix} \qquad \text{(Str. 4)}$$

cis-1,4-polyisoprene (natural rubber)

$$\begin{matrix} -CH_2 & & H \\ & C=C & \\ CH_3 & & CH_2- \end{matrix} \qquad \text{(Str. 5)}$$

trans-1,4-polyisoprene (gutta-percha)

2. Tacticity

Polymers of monosubstituted olefins [CH_2=CHXl contain a series of asymmetric carbon atoms along the chain. For this type of polymers, in a planar zigzag form, three arrangements are possible, namely:

- Isotactic — All the substituent groups, R, on the vinyl polymer lie above (or below) the plane of the main chain.

R R R R R

(Str. 6)

- Syndiotactic — Substituent groups lie alternately above and below the plane.

R R R

R R

(Str. 7)

- Atactic — Random sequence of position of substituent occurs along the chain.

R R R

R R

(Str. 8)

B. CONFORMATION

In addition to the molecular shape fixed by chemical bonding, variations in the overall shape and size of the polymer chain may occur due to rotation about primary valence bonds (conformation). A polymer molecule may assume a large or limited number of conformations depending on:

- Steric factors
- Whether the polymer is amorphous or crystalline
- Whether the polymer is in a solution state, molten state, or solid state

To amplify the discussion, let us consider the possible arrangements of a single isolated polymer chain in dilute solution. We start with a short segment of the chain consisting of four carbon atoms (Figure 3.4).

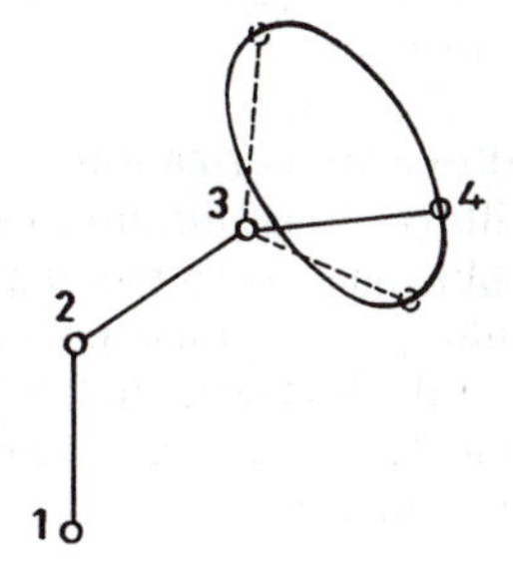

Figure 3.4 A segment of polymer chain showing four successive chain atoms; the first three of these define a plane, and the fourth can lie anywhere on the indicated circle which is perpendicular to and dissected by the plane.

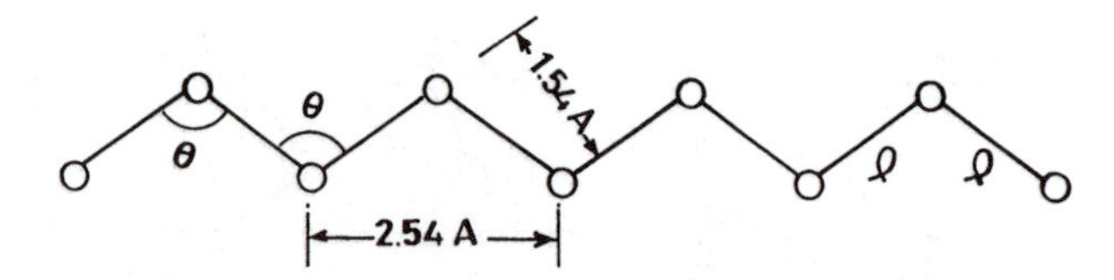

Figure 3.5 The fully extended all-*trans* conformation of a carbon–carbon chain.

We define a plane by three of the carbon atoms in this segment and allow free rotation about the carbon–carbon bond. In this case, the fourth carbon atom can be anywhere on the circle indicated in the figure. Because of steric hindrance, some positions will certainly be more probable than others. Each successive carbon atom on the chain can similarly take any of the several positions in a circle based, randomly, on the position of the preceding atom. For a chain consisting of thousands of carbon atoms, it can thus be seen that the number of conformations is literally infinite. One of these conformations of particular interest is that in which each successive carbon atoms lies in the same plane in the *trans* location with respect to earlier carbon atoms in the chain — thus forming a fully extended plane of zigzag arrangement of carbon atoms (Figure 3.5).

This represents one of the two extreme shapes of a polymer chain, the other being the completely random coil. The planar zigzag conformation exists in some crystalline polymers or in highly oriented amorphous polymers. Typical examples are simple molecules like PE, PVC, and polyamides, where the small size of the pendant group does not complicate alignment and packing. In those polymers with large and bulky side groups like PP and PS (in general, isotactic and syndiotactic polyolefins), it is impossible sterically to accommodate the pendant groups in the planar zigzag. Consequently, the entire main chain is rotated in the same direction to form either a right- or left-handed helix. This occurs exclusively in the crystalline form of stereoregular polymers with bulky side groups (Figure 3.6).

The other extreme of the conformation spectrum that may be assumed by the polymer chain is the completely random coil. Polymers that are in solution, in melt, or amorphous in the solid state assume this conformation. Between these two extremes (planar zigzag and random coil conformation) the number of conformation shapes that a polymer chain can assume is virtually limitless. This, of course, assumes that there is free rotation about single bonds. In practice, however, there is no such thing as completely free rotation. All bonds have to overcome certain rotational energy barriers whose magnitude depends on such factors as steric hindrance, dipole forces, etc. (Figure 3.7).

The thermal energy of the molecular environment provides the energy required to overcome the rotational energy barrier. Consequently, the shape (flexibility) of a polymer molecule is temperature dependent. At sufficiently high temperatures, the polymer chain constantly wiggles, assuming a myriad of random coil conformations. As we shall see later, the flexibility of polymer molecules, which is a function of substituents on the backbone, has a strong influence on polymer properties.

C. MOLECULAR WEIGHT

The terms *giant molecule, macromolecule,* and *high polymer* are used to describe a polymer molecule to emphasize its large size. We noted earlier that the same bonding forces (intra- and intermolecular) operate in both low- and high-molecular-weight materials. However, the unique properties exhibited by polymers and the difference in behavior between polymers and their low-molecular-weight analogs are attributable to their large size and flexible nature.

Important mechanical properties (tensile and compressive strengths, elongation at break, modulus, impact strength) and other properties (softening point, solution and melt viscosities, solubility) depend on molecular weight in a definite way. At very low molecular weights, hardly any strength, for example, is developed. Beyond this MW or DP, there is a steep rise in the performance until a certain level, beyond which the properties change very little with increase in molecular weight. Finally, an asymptotic value is reached (Figure 3.8). The curve in Figure 3.8 is general for all polymers. Differences exist only in numerical details. Optical and electrical properties, color, and density show a less marked dependence on molecular weight.

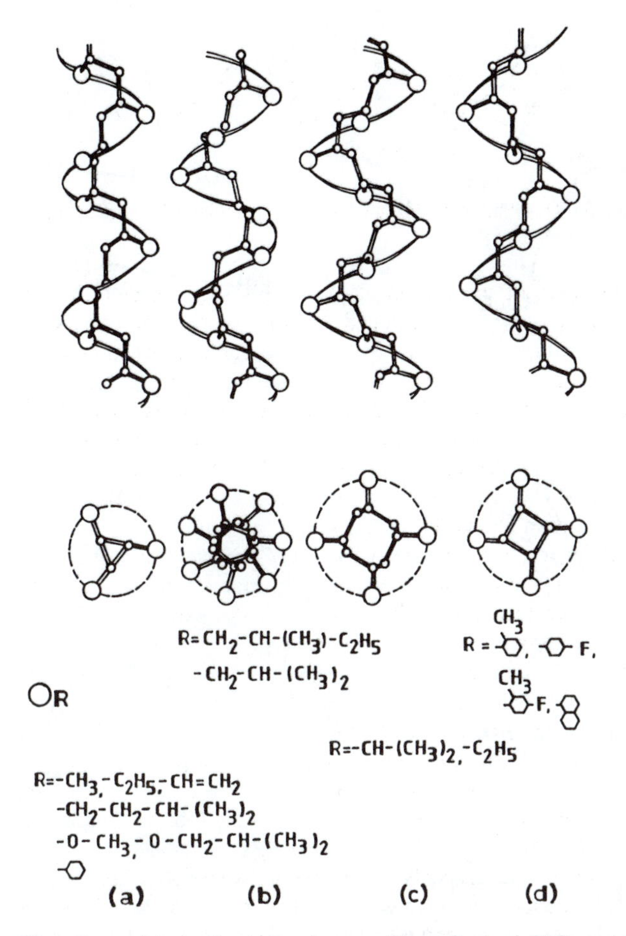

Figure 3.6 Helical conformations of isotactic vinyl polymers. (From Gaylord, N.G. and Mark, H., *Linear and Stereoregular Addition Polymers,* Interscience, New York, 1959. With permission.)

Example 3.2: Which of the following materials will be most suitable for the manufacture of thermoplastic sewage pipe? Explain your answer very briefly.

$$(-CH_2-CH_2-)_{50} \qquad (-CH_2-CH_2-)_{5000} \qquad (CH_2-CH_2-)_{500{,}000}$$

A B C (Str. 9)

Solution:

Material	Molecular Weight
A	1.4×10^3
B	1.4×10^5
C	1.4×10^7

Material B will be most suitable. The molecular weight of A is too low, and the material will not have developed the physical properties necessary to sustain the mechanical properties that a plastic pipe must withstand. On the other hand, the molecular weight of material C is relatively too high to permit easy processing.

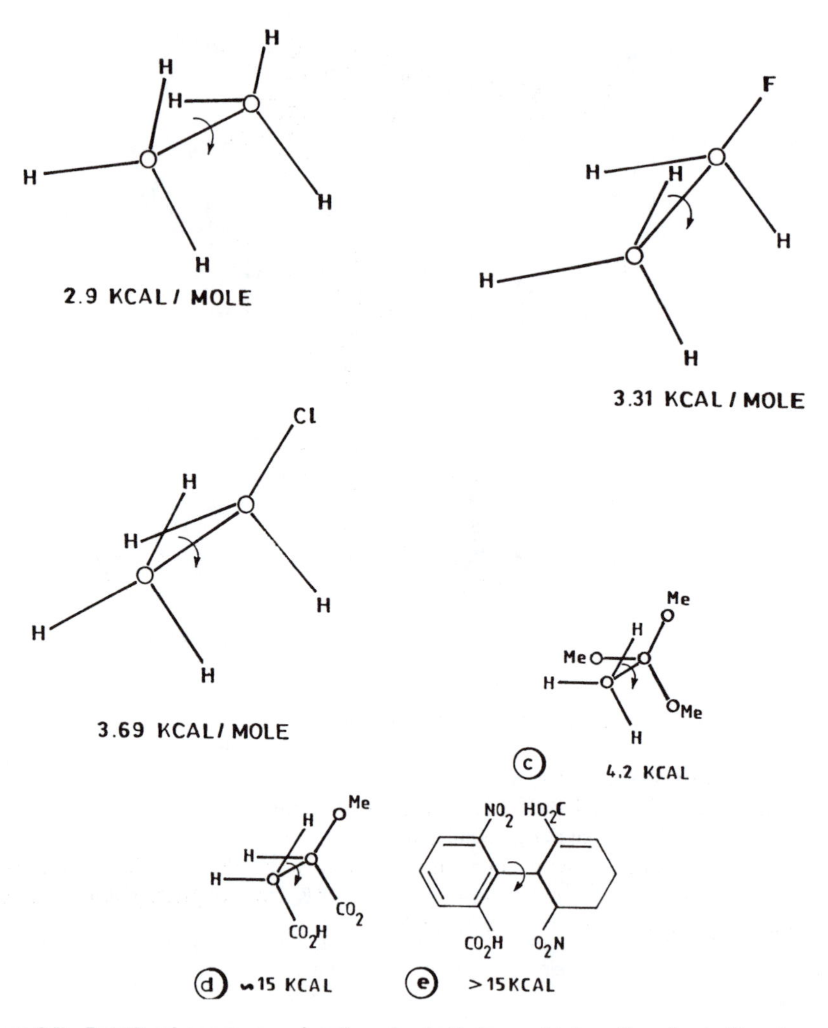

Figure 3.7 Rotational energy as a function of substitution and interaction of substituent groups.

We also noted earlier that irrespective of the polymerization mechanism, the formation of polymer is a purely random occurrence. Consequently, unlike biological systems, synthetic polymers do not consist of identical molecules, but represent a mixture of many systems each of which has a different molecular weight. In order to characterize polymers, therefore, we use the molecular weight distribution (MWD) curve, which represents a plot of the percentage (frequency) of a particular species against its molecular weight (Figure 3.9).

As a result of the existence of different sizes of molecular species in a polymeric material, we cannot strictly speak of the molecular weight of a polymer. Instead we use molecular weight averages to express the size of synthetic polymers. Different average molecular weights exist. The most common ones in use are number-average molecular weight, $\overline{M}_n$, and weight-average molecular weight, $\overline{M}_w$. Others are the z-average molecular weight, $\overline{M}_z$, and viscosity-average molecular weight, $\overline{M}_v$. Below are the relevant formulas for computing these average molecular weights (Equations 3.4–3.7).

$$\overline{M}_n = \frac{W}{\sum_{i=1}^{\infty} N_i} = \frac{\sum_{i=1}^{\infty} N_i M_i}{\sum_{i=1}^{\infty} N_i} = \frac{\sum_{i=1}^{\infty} W_i}{\sum_{i=1}^{\infty} W_i/M_i} = \frac{1}{\sum w_i/M_i} \tag{3.4}$$

$$\overline{M}_w = \sum_{i=1}^{\infty} w_i M_i = \frac{\sum_{i=1}^{\infty} w_i M_i}{W} = \frac{\sum_{i=1}^{\infty} N_i M_i^2}{\sum_{i=1}^{\infty} N_i M_i} \tag{3.5}$$

$$\overline{M}_v = \left[\sum_{i=1}^{\infty} w_i M_i^a\right]^{1/a} = \left[\frac{\sum_{i=1}^{\infty} N_i M_i^{a+1}}{\sum_{i=1}^{\infty} N_i M_i}\right]^{1/a} \tag{3.6}$$

$$\overline{M}_z = \frac{\sum_{i=1}^{\infty} N_i M_i^3}{\sum_{i=1}^{\infty} N_i M_i^2} = \frac{\sum_{i=1}^{\infty} w_i M_i^2}{\sum_{i=1}^{\infty} w_i M_i} \tag{3.7}$$

where N_i = number of molecules having molecular weight M_i
W = total weight
N = total number of molecules
w_i = weight fraction of molecules having molecular weight M_i
W_i = weight of molecule having molecular weight M_i
a = constant in Mark–Houwink equation (η) Km^a in which the intrinsic viscosity (η) and the molecular weight M are related through constants K and a for given polymer/solvent system; a is at least 0.5 and mostly less than 0.8.

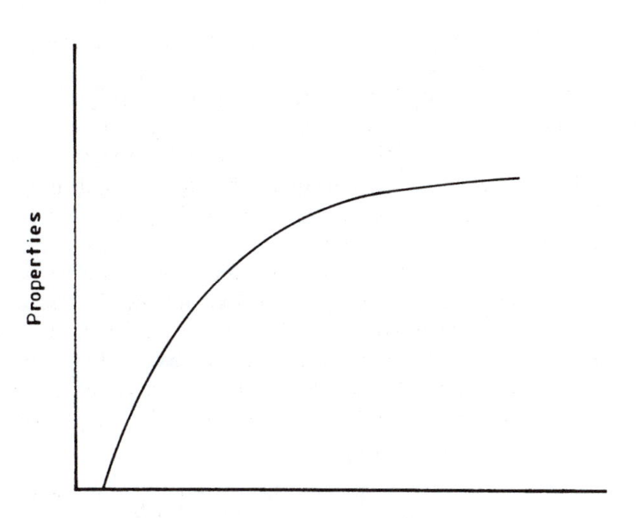

Figure 3.8 Change of physical properties with molecular weight.

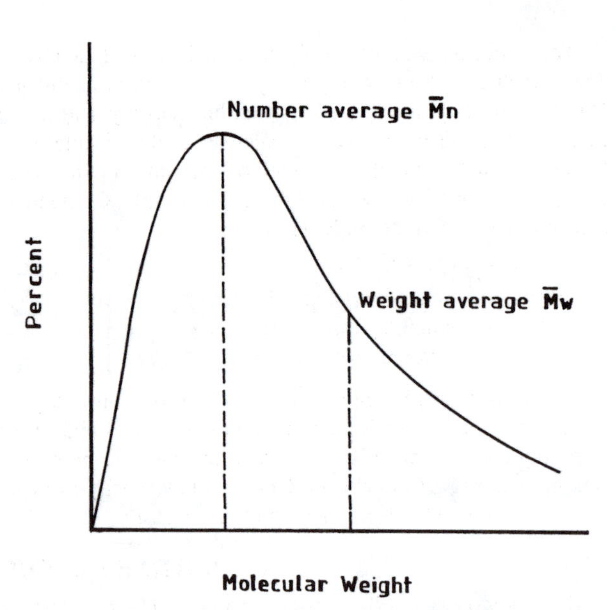

Figure 3.9 Molecular weight distribution curve.

Example 3.3: The following data were obtained in a determination of the average molecular weight of a polymer.

Molecular Weight	Weight (g)
60,000	1.0
40,000	2.0
20,000	5.0
10,000	2.5

a. Compute the number average and the weight average-molecular weights, $\overline{M}_n$ and $\overline{M}_w$.
b. What is the polydispersity of the polymer and how many molecules are in 1 gram of polymer?

Solution:

a. $\frac{\sum W_i}{\sum W_i/M_i}$ where W_i = weight species i.

$$\overline{M}_n = \frac{1+2+5+2.5}{1/60\times10^3 + 2/40\times10^3 + 5/20\times10^3 + 2.5/10^4}$$

$$= 18,600 \text{ g/mol}$$

$$\overline{M}_W = \sum w_i M_i; \quad w_i = \frac{W_i}{W} = \text{weight fraction of species i}$$

$$= \frac{10^4}{10.5}[6+2\times4+5\times2+2.51]$$

$$= 25,200 \text{ g/mol}$$

b. Polydispersity $= \overline{M}_w/\overline{M}_n = \frac{25,200}{18,600} = 1.35$

$$\text{Molecules/g} = N_I/\overline{M}_n = \frac{6.02\times10^{23}}{18,600} = 3.25\times10^{19}$$

The molecular weight of polymers can be determined by a number of physical and chemical methods. These include (1) end group analysis, (2) measurement of colligative properties, (3) light scattering, (4) ultracentrifugation, (5) dilute solution viscosity, and (6) gel permeation chromatography (GPC). The first four methods permit a direct calculation of molecular weight without the need to resort to calibration by another method; that is, the methods are, in principle, absolute. The last two methods require proper calibration to obtain the value of molecular weight. Colligative properties are determined by the following measurements on dilute polymer solutions:

- Vapor pressure lowering
- Boiling point elevation (ebulliometry)
- Freezing point depression (cryoscopy)
- Osmotic pressure (osmometry)

The number-average weight, $\overline{M}_n$, is observed from end-group analysis, colligative property measurements, and gel permeation chromotography. The weight-average molecular weight, $\overline{M}_w$, is determined from light scattering, ultracentrifugation and gel permeation chromatography. z-average molecular weight, $\overline{M}_z$, is determined from GPC, while viscosity-average molecular weight, $\overline{M}_v$, can be determined from measurements of polymer solution viscosity.

V. TERTIARY STRUCTURE

A given polymeric solid material is an aggregate of a large number of polymer molecules. Depending on the molecular structure, the process of molecular aggregation occurs essentially by either of two possible arrangements of molecules, leading to either a crystalline or amorphous material. However, irrespective of the type of molecular arrangement, the forces responsible for molecular aggregation are the intermolecular secondary bonding forces. The overall bonding energies due to secondary bonding forces range from 0.5 to 10 kcal/mol compared with those of primary bonding forces, which are of the order 50 to 100 kcal/mol. But when molecules are large enough, the attractive forces resulting from the secondary intermolecular bonding forces may build up to such a level that, in some cases, they become greater than the primary valence forces responsible for intramolecular bonds. The magnitude of these secondary bonding forces, coupled with the high physical entanglement between chains, dictates many polymer properties. Tertiary structure is concerned with the nature of the intermolecular secondary bonding forces and with structural order of the resulting polymer.

A. SECONDARY BONDING FORCES (COHESIVE ENERGY DENSITY)

As we said earlier, secondary bonds consist of dipole, induction, van der Waals, and hydrogen bonds. Dipole forces result from the attraction between permanent dipoles associated with polar groups. Induction forces arise from the attraction between permanent and induced dipoles, while van der Waals (dispersion) forces originate from the time-varying perturbations of the electronic clouds of neighboring atoms. Hydrogen bonds are very important in determining the properties of such polymers as polyamides, polyurethanes, and polyureas. In general, the magnitude of the bond energies decreases from hydrogen bond to dipole bond to van der Waals (dispersion) forces.

A quantitative measure of the magnitude of secondary bonding forces is the cohesive energy density (CED), which is the total energy per unit volume needed to separate all intermolecular contacts and is given by:

$$\text{CED} = \frac{\Delta E_v}{V_L} \tag{3.8}$$

where ΔE_v = molar energy of vaporization
V_L = molar volume of the liquid

It can be shown from the Classius–Clapeyron equation that

$$\Delta E_v = \Delta H_v - RT \tag{3.9}$$

where ΔH_v = molar heat of vaporization
T = absolute temperature (K)

Consequently,

$$CED = \frac{\Delta E_v}{V_L} = \frac{\Delta H_v - RT}{V_L} \quad (3.10)$$

For liquids of low molecular weight the energy necessary to separate molecules from one another is evaluated from the heat of evaporation or from the dependence of vapor pressure on temperature. Since polymers cannot be evaporated, the cohesive energy density is estimated indirectly by dissolution in liquids of known cohesive energy density. To do this, we employ the relation between the cohesive energy density and solubility parameter (Equation 3.11).

$$CED = \delta^2 \quad (3.11)$$

where δ = solubility parameter. As a first approximation and in the absence of strong interactions such as hydrogen bonding, a polymer δ_2 will dissolve in a solvent δ_1 if

$$\delta_1 - \delta_2 \leq 1.7 - 2.0.$$

Values of solubility parameters and cohesive energy of some polymers are given in Table 3.7. The value of E_{coh} is also dependent on the molar volume. For polymers the appropriate volume is that occupied by each repeat unit in the solid state. Thus E_{coh} represents the cohesive energy per repeat unit volume, V_R. These simple relations as stated before, however, are not exact; stronger interactions change the validity of Equation 3.11. However, significant practical predictions can be made from the values in Table 3.7, such as what solvents will dissolve a given polymer.

Table 3.7 Cohesive Energy of Polymers

Polymers	d ($cal^{1/2}/cm^{3/2}$)		V_R (cm^3/mol)	E_{coh} (from δ) (cal/mol)	
	From	To		From	To
Polyethylene	7.7	8.35	32.9	1,950	2,290
Polypropene	8.3	9.2	49.1	3,300	4,160
Polyisobutene	7.8	8.1	66.8	4,060	4,300
Polyvinylchloride	9.4	10.8	45.2	3,990	5,270
Polyvinylidene chloride	9.9				
Polyvinyl bromide	9.5		63.8	4,850	
Polyvinylfluoroethylene	6.2		50.0	1,920	
Polychlorotrifluoroethylene	7.2	7.9	61.8	3,200	3,860
Polyvinyl alcohol	12.6	14.2	35.0	5,560	7,060
Polyvinyl acetate	9.35	11.05	72.2	6,310	8,820
Polyvinyl propionate	8.8		90.2	6,900	
Polystyrene	8.5	9.3	98.0	7,080	8,400
Polymethyl acrylate	9.7	10.4	70.1	6,600	7,500
Polyethyl acrylate	9.2	9.4	86.6	7,330	7,650
Polypropyl acrylate	9.05		103.0	8,440	
Polybutyl acrylate	8.8	9.1	119.5	9,360	9,900
Polyisobutyl acrylate	8.7	11.0	114.2	9,020	14,420
Poly-2,2,3,3,4,4,4, heptafluorobutyl acrylate	6.7		148.0	6,640	
Polymethyl methacrylate	9.1	19.8	86.5	7,160	14,170
Polyethyl methacrylate	8.9	9.15	102.4	8,110	8,570
Polybutyl methacrylate	8.7	9.0	137.2	10,380	11,110
Polyisobutyl methacrylate	8.2	10.5	135.7	9,120	14,960
Poly-*tert*,butyl methacrylate	8.3		138.9		9,570
Polyethoxyethyl methacrylate	9.0	9.9	145.6	11,790	15,270
Polybenzyl methacrylate	9.8	10.0	152.0	14,600	15,200
Polyacrylonitrile	12.5	15.4	44.8	7,000	10,620

Table 3.7 (continued) Cohesive Energy of Polymers

Polymers	d ($cal^{1/2}/cm^{3/2}$) From	To	V_R (cm^3/mol)	E_{coh} (from δ) (cal/mol) From	To
Polymethacrylonitrile	10.7		63.9	7,320	
Poly-*a*-cyanomethyl	14.0	14.5	82.1	16,090	17,260
Polybutadiene	0.1	8.6	60.7	3,900	4,490
Polyisoprene	7.9	10.0	75.7	4,730	7,570
Polychloroprene	8.2	9.25	71.3	4,790	6,100
Polyepichlorohydrin	9.4		69.7	6,160	
Polyethylene terephthalate	9.7	10.7	143.2	13,470	16,390
Polyhexamethylene adipamide	13.6		208.3		138,500
Poly(δ-aminocaprylic acid)	12.7		135.9		21,920
Polyformaldehyde	10.2	11.0	25.0	2,600	3,030
Polytetramethylene oxide	8.3	8.55	74.3	5,120	5,430
Polyethylene sulfide	9.0	9.4	47.9	3,880	1,230
Polypropylene oxide	7.5	9.9	57.6	3,240	5,650
Polystyrene sulfide	9.3		115.8		10,020
Polydimethyl siloxane	7.3	7.6	75.6	4,030	4,300

From Van Krevelen, D.W., *Properties of Polymers*, Elsevier, Amsterdam, 1972. With permission.

Example 3.4: The table below shows the density and enthalpy of vaporization [ΔH_{vap}] of two solvents: methylethyl ketone and acetone.

Solvent	Density (g/cm^3)	ΔHv· (cal/g)
Methylethyl ketone	0.8	106
Acetone	0.8	125

Which is a better solvent for polystyrene at room temperature? The CED for polystyrene is 75 cal/cm³. Assume room temperature is 27°C.

Solution: Basis = 1 g of solvent*

$$CED = \frac{\Delta H_{v'} - RT}{V_L} = \frac{\Delta H_{v\cdot} - RT}{1/\rho_{sol}}$$

$$= \rho_{sol}\left(\Delta H_v - RT\right)$$

where ρ_{sol} = density of solvent. In the SI system

$$R = 8.303 \times 10^3 \, \frac{Nm}{(K)kg\ mol}$$

$$= 8.303 \times 10^3 = J/(K)\,(kg\ mol)$$

$$R = 1.984\ cal/(K)\,(g\ mol)$$

$$= 1.984\ cal/(g); \quad T = 27 + 273 = 300\ K$$

* Units of ΔHv are in cal/g; therefore, the units of the term RT must be consistent with this.

Recall 1 cal = 4.184 J, 1 kg mol = 10^3 g mol.

$$RT = (1.984/M) \times 300 \text{ cal/g} = 595.2/M \text{ cal/g}$$

where M = molecular weight of solvent.

$$\text{Acetone } CH_3\text{–}\overset{\overset{\displaystyle O}{\|}}{C}\text{–}CH_3, \quad M = 58 \text{ g/g mol}$$

$$\text{Methylethyl ketone } CH_3\text{–}\overset{\overset{\displaystyle O}{\|}}{C}\text{–}CH_2\text{–}CH_2, \quad M = 72 \text{ g/mol}$$

Methylethyl ketone:

$$\begin{aligned} CED &= 0.8\left(g/cm^3\right)\left(106 - \frac{595.2}{72}\text{ cal}\right) \\ &= 0.8\,[106 - 8.27]\text{ cal/cm}^3 \\ &= 78.18\text{ cal/cm}^3 \end{aligned}$$

Acetone:

$$\begin{aligned} CED &= 0.8\left(g/cm^3\right)\left(125 - \frac{595.2}{58}\text{ cal}\right) \\ &= 0.8\,[125 - 10.26] \\ &= 91.79\text{ cal/cm}^3 \end{aligned}$$

Since the CED of methylethyl ketone is closer to that of polystyrene than that of acetone, methylethyl ketone should be a better solvent for polystyrene than acetone.

B. CRYSTALLINE AND AMORPHOUS STRUCTURE OF POLYMERS

As discussed earlier, when a polymer is cooled from the melt or concentrated from a dilute solution, molecules are attracted to each other forming a solid mass. In doing so, two arrangements are essentially possible:

- In the first case, the molecules vitrify, with the polymer chains randomly coiled and entangled. The resulting solid is amorphous and is hard and glassy.
- In the second case, the individual chains are folded and packed in a regular manner characterized by three-dimensional long-range order. The polymer thus formed is said to be crystalline.

We must recall, however, that polymers are made up of long molecules; therefore, the concept of crystallinity in polymers must be viewed slightly differently from that in low-molecular-weight substances. Complete parallel alignment is never achieved in polymeric systems. Only certain clusters of chain segments are aligned to form crystalline domains. These domains, as we shall see shortly, do not have the regular shapes of normal crystals. They are much smaller in size, contain many more imperfections, and are connected with the disordered amorphous regions by polymer chains that run through both the ordered and the disordered segments. Consequently, no polymer is 100% crystalline.

1. Crystallization Tendency

Secondary bonding forces, as we saw earlier, are responsible for intermolecular bonding in polymers. You will recall also that these forces are effective only at very short molecular distances. Therefore, to maximize the effect of these forces in the process of aggregation of molecules to form a crystalline solid mass, the molecules must come as close together as possible. The tendency for a polymer to crystallize, therefore, depends on the magnitude of the inherent intermolecular bonding forces as well as its structural features. Let us now discuss these in further detail.

Table 3.8 Properties of Polyisoprene Isomers

Isomer	Structure	Properties
1,4-*cis*=polyisoprene (heavea rubber)	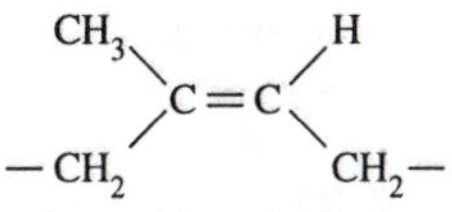 CH_3(—C=C—)H; $-CH_2$, CH_2- (cis)	Soft, pliable, easily soluble rubber; has a high retractive force; used for making vehicle tires
1,4-*trans*-polyisoprene (gutta-percha)	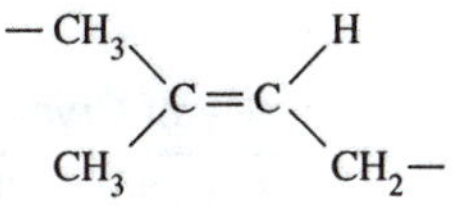 $-CH_3$(—C=C—)H; CH_3, CH_2- (trans)	Tough, hard; used as golf ball covers

2. Structural Regularity

We have just said that in the process of association of polymer molecules to form a solid mass, molecules must come as close together as possible. It follows that any structural features of polymer molecules that can impede this process will necessarily detract from crystallinity. Polyethylene is perhaps the simplest molecule to consider in this case. Polyethylene is nonpolar, and the intermolecular attraction is due to the relatively weak van der Waals forces. The chains can readily assume a planar zigzag conformation characterized by a sequence of *trans* bonds and can therefore produce short identity periods along the polymer chain length. The rotation around the C–C bond is inhibited by an energy barrier of about 2.7 kcal/mol of bonds. Thus, even though polyethylene molecules are held together by weak van der Waals forces, the high structural regularity that permits close packing of the chains coupled with the limited chain flexibility leads to an unexpectedly high melting point (T_m = 135°C), relatively high rigidity, and low room-temperature solubility. However, as irregularities are introduced into the structure, as with low-density polyethylene (LDPE), the value of these properties shows a significant reduction. The crystalline melting point of polyethylene, for example, is reduced 20 to 25°C on going from the linear to the branched polymer.

Regularity per se is not sufficient to ensure crystallizability in polymers. The spatial regularity and packing are important. To illustrate this, let us consider two examples of stereoregular polymers. Table 3.8 shows the properties of two isomers: *cis*- and *trans*-polyisoprene. It is obvious from the table that the stereoregular *trans* form is more readily packed and crystallizable and has properties of crystalline polymers.

The second example is the stereoregularity displayed by monosubstituted vinyl polymers of olefins. As we saw earlier, these types of polymers can occur in three forms of tacticity: isotactic, syndiotactic, and atactic. Isotactic and syndiotactic polymers possess stereoregular structures. Generally these polymers are rigid, crystallizable, high melting, and relatively insoluble. On the other hand, atactic polymers are soft, low melting, easily soluble, and amorphous.

3. Chain Flexibility

In the preceding discussion, we consistently emphasized that close alignment of polymer molecules is a vital prerequisite for the effective utilization of the intermolecular bonding forces. During crystallization, this alignment and uniform packing of chains are opposed by thermal agitation, which tends to induce segmental, rotational and vibrational motions. The potential energy barriers hindering this rotation range from 1 to 5 kcal/mol, the same order of magnitude as molecular cohesion forces. It is to be expected, therefore, that those polymers whose chains are flexible will be more susceptible to this thermal agitation than those with rigid or stiff chain structure.

The flexibility of chain molecules arises from rotation around saturated chain bonds. With a chain of $-CH_2-$ units as a basis, it is interesting to consider how variations on this unit will affect rotation of adjacent units and, hence, chain flexibility. Studies of this type have led to the following general conclusions:

- Rapid conformational change due to ease of rotation around single bonds occurs if such groups as (-–CO–O–), (–O–CO–O–), and (–C–N–) are introduced into the main chain. If they are regular and/or if there exist considerable intermolecular forces, the materials are crystallizable, relatively high melting, rigid, and soluble with difficulty. However, if they occur irregularly along the polymer chain, they are amorphous, soft, and rubbery materials.

The flexibility of chain molecules arises from rotation around saturated chain bonds. With a chain of $-CH_2-$ units as a basis, it is interesting to consider how variations on this unit will affect rotation of adjacent units and, hence, chain flexibility. Studies of this type have led to the following general conclusions:

- Rapid conformational change due to ease of rotation around single bonds occurs if such groups as (--CO-O-), (-O-CO-O-), and (-C-N-) are introduced into the main chain. If they are regular and/or if there exist considerable intermolecular forces, the materials are crystallizable, relatively high melting, rigid, and soluble with difficulty. However, if they occur irregularly along the polymer chain, they are amorphous, soft, and rubbery materials.

Table 3.9 Effect of Chain Flexibility of Crystalline Melting Point

Polymer	Repeating Unit	T_m (°C)
Polyethylene	$-CH_2-CH_2-$	135
Polyoxyethylene	$-CH_2-CH_2-O-$	65
Poly(ethylene suberate)	$-O(CH_2)_2-OCO-(CH_2)_6CO-$	45
Nylon 6,8	$-NH(CH_2)_6NHCO(CH_2)_6CO-$	235
Poly (*p*-xylene)	$-CH_2-C_6H_4-CH_2-$	400

- Ether and imine bonds and double bonds in the *cis* form reduce the energy barrier for rotation of the adjacent bonds and "soften" the chain by making polymers less rigid, more rubbery, and more readily soluble than the corresponding chain of consecutive carbon–carbon atoms. If such "plasticizing" bonds are irregularly distributed along the polymer chain length, crystallization is inhibited.
- Cyclic structures in the backbone and polar group such as $-SO_2-$, and –CONH– drastically reduce flexibility and enhance crystallizability.

Table 3.9 illustrates the effect of these factors on crystalline melting point.

4. Polarity

When molecules come together and aggregate into a crystalline solid, a significant cohesion between neighboring chains is possible. Consequently, polymer molecules with specific groups that are capable of forming strong intermolecular bonding, particularly if these groups occur regularly without imposing valence strains on the chains, are crystallizable. You will recall from our earlier discussion that such groups as

$$-\underset{|}{\overset{\text{H}}{\text{N}}}-\overset{\text{O}}{\overset{\|}{\text{C}}}- \text{ (amide); } -\underset{|}{\overset{\text{H}}{\text{N}}}-\overset{\text{O}}{\overset{\|}{\text{C}}}-\text{O}- \text{ (urethane); and } -\overset{\text{H}}{\text{N}}-\overset{\text{O}}{\overset{\|}{\text{C}}}-\overset{\text{H}}{\text{N}}- \text{ (urea)} \qquad \text{(Str. 10)}$$

provide sites for hydrogen bonding whose energy ranges from 5 to 10 kal/mol. In nylon 6 or 6,6, for example, the regular occurrence of amide linkages leads to a highly crystalline, high melting polymer.

Molecules whose backbone contains –O– units or with polar side groups (–CN, –Cl, –F, or $-NO_2$) exhibit polar bonding. The bonding energies of such dipoles or polarizable units are in the range between hydrogen bonding and van der Waals bonding. If these groups occur regularly along the chain (isotactic and syndiotactic), the resulting polymers are usually crystalline and have higher melting points than polyethylene (Table 3.10).

Table 3.10 Effect of Polarity on Crystallizability

Polymer	Repeat Unit	T_{em} (°C)
Polyethylene	$-CH_2-CH_2-$	135
Nylon 6	$-\overset{H}{\overset{\vert}{N}}-\overset{O}{\overset{\Vert}{C}}-(CH_2)_5-$	223
Nylon 6,6	$-\overset{H}{\overset{\vert}{N}}-(CH_2)_6-\overset{H}{\overset{\vert}{N}}-\overset{O}{\overset{\Vert}{C}}-(CH_2)_4-\overset{O}{\overset{\Vert}{C}}-$	265
Polyoxymethylene	$-CH_2-O-$	180
Poly(vinyl chloride)	$-CH_2-\underset{Cl}{\underset{\vert}{CH}}-$	273
Polyacrylonitrile	$-CH_2-\underset{CN}{\underset{\vert}{CH}}-$	317

Given our earlier argument that the presence of –O– units in a chain backbone enhances flexibility, the fact that the melting point of polyoxymethylene (180°C) is higher than that of polyethylene (135°C) (Table 3.10) seems contradictory. However, the dipole character of the C–O–C group produces polar forces between adjacent chains that act over a longer range and are stronger than van der Waals forces. Thus, for polyoxymethylene the induced flexibility is more than offset by the increased bonding forces resulting from polarity.

5. Bulky Substituents

The vibrational and rotational mobility of intrinsically flexible chains can be inhibited by bulky substituents; the degree of stiffening depends on the size, shape, and mutual interaction of the substituents. For example, vinyl polymers with small substituents such as polypropylene [$-CH_3$] and polystyrene [$-C_6H_5$] can crystallize if these pendant groups are spaced regularly on the polymer chain as in their isotactic and syndiotactic forms. In their atactic forms, the randomly disposed pendant groups prevent the close packing of the chains into crystalline lattice. The atactic forms of these polymers are therefore amorphous. Large or bulky substituents, on the other hand, increase the average distance between chains and, as such, prevent the effective and favorable utilization of the intermolecular bonding forces. Thus polymers like poly(methyl acrylate) and poly(vinyl acetate) with large pendant groups $-\overset{O}{\overset{\Vert}{C}}-O-CH_3$ and $O-\overset{O}{\overset{\Vert}{C}}-CH_3$, respectively, cannot crystallize even if the pendant groups are spaced regularly (isotactic and syndiotactic forms). Table 3.11 shows the change in the crystalline melting point with increased length of the polymer side chain. Note that for the polyolefins, polyethylene, polypropylene, poly(1-butene), and poly(1-pentene), the melting point shows a maximum for polypropylene. A large pendant group in close proximity of the main chain stiffens the chain. However, when the size of the pendant group is such that the packing distance between the chains in the solid state is increased, the forces of interaction between chains decrease and so does the melting point. The presence of an aromatic side group in polystyrene considerably stiffens the chain, which has a stable helix form in the solid state. The helices pack efficiently to allow greater interchain interaction.

The above discussion clearly indicates that stereo regularity, chain flexibility, polarity, and other steric factors have profound influence on crystallizability and melting points and, hence, as we shall see later, play an important role in the thermal and mechanical behavior of polymers.

Table 3.11 Stiffening of Polymer Chains by Substituents

Polymer	Repeat Unit	T_m (°C)
Polyethylene	$-CH_2-CH_2-$	135
Polypropylene	$-CH_2-CH(CH_3)-$	176
Poly(1-butene)	$-CH_2-CH(CH_2CH_3)-$	125
Poly(1-pentene)	$-CH_2-CH(CH_2CH_2CH_3)-$	75
Polystyrene	$-CH_2-CH(C_6H_5)-$	240

Example 3.5: Explain the following observations.

a. Atactic polystyrene can be oriented (have its chain aligned by stretching at a temperature above its T_g, but does not crystallize; rubber on the other hand, both crystallizes and becomes oriented when it is stretched.
b. Poly(vinyl alcohol) is made by the hydrolysis of poly(vinyl acetate) because vinyl alcohol monomer is unstable. The extent of reaction may be controlled to yield polymers with anywhere from 0 to 100% of the original acetate groups hydrolyzed. At room temperature, pure poly(vinyl) acetate), i.e., 0% hydrolysis, is insoluble in water. However, as the extent of hydrolysis is increased, the polymers become more water soluble up to 87% hydrolysis, after which further hydrolysis decreases water solubility.
c. Toluene and xylene have approximately the same cohesive energy density (CED), but xylene is a more convenient solvent for polyethylene.

Solution:

a. Atactic polystyrene has an irregular structure and therefore will not crystallize when oriented. Natural rubber (*cis*-1,4-polyisoprene) has a relatively small pendant group and as such is crystallizable.

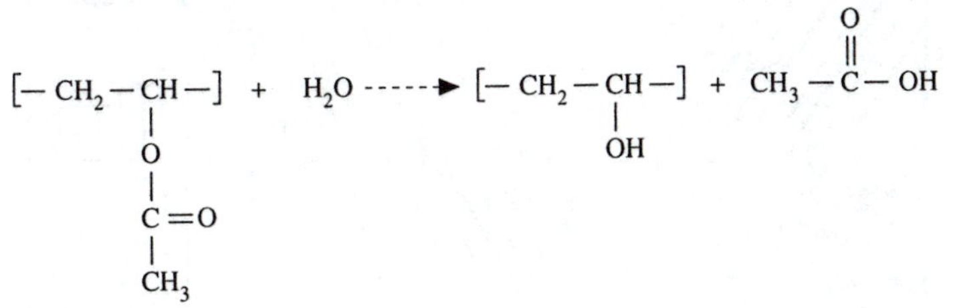

$$[-CH_2-\underset{\underset{\underset{\underset{\underset{\underset{CH_3}{|}}{C=O}}{|}}{O}}{|}}{CH}-] + H_2O \dashrightarrow [-CH_2-\underset{\underset{OH}{|}}{CH}-] + CH_3-\overset{\overset{O}{\|}}{C}-OH$$

poly(vinyl acetate) poly(vinyl alcohol) acetic acid (Str. 11)

As the hydroxyl group becomes available following the hydrolysis, there is a corresponding increase in the water solubility of the product, poly(vinyl alcohol), as a result of intermolecular hydrogen bond formation. At an advanced stage of hydrolysis the CED of poly(vinyl alcohol) increases to such an extent that water solubility decreases, from the arguments in Section V.A.

c. Toluene has a boiling point of about 111°C, while xylene has a boiling point of 138 to 144°C. As a result of the highly crystalline nature of polyethylene it will dissolve in solvents only at temperatures close to its melting point (about 135°C). When the boiling points of toluene and xylene are compared, xylene is obviously a preferable solvent for polyethylene.

C. MORPHOLOGY OF CRYSTALLINE POLYMERS

Most polymers are partially crystalline. Evidence for this emerged in the1920s from X-ray diffraction studies. X-ray diffraction patterns of some polymers, in contrast to those of simple crystalline solids, showed sharp features, associated with regions of three-dimensional order, superimposed on a diffuse background characteristic of amorphous, liquidlike substances. The interpretation of these patterns was that polymers are semicrystalline, consisting of small, relatively ordered regions — the crystallites embedded in an otherwise amorphous matrix. This interpretation led to the "fringed micelle" model of crystalline polymers. The fringed micelle concept, which enjoyed popularity for many years, held that, since polymer chains are very long, they passed successively through the crystallites and amorphous regions (Figure 3.10). The chains were thought to run parallel to the longer direction of the crystallites. Although the fringed micelle model of polymer morphology seemed to explain many of the properties of semicrystalline polymers, it has now been abandoned in favor of more ordered and complex models. This change is partly as a result of developments in the field of electron microscopy.

The morphology of crystalline polymers — that is, the size, shape, and relative magnitude of crystallites — is rather complex and depends on growth conditions such as solvent media, temperature, and growth rate. In discussing polymer crystalline morphology, our initial focus is on molecular packing. This concerns how the polymer chains (with an extended conformation of either planar zigzag or helix) are packed into the unit cell, which is the fundamental element of a crystal structure. This is followed by discussion of the morphologic features of the polymer single crystal and those of polymers crystallized from the melt.

1. Crystal Structure of Polymers

The fully extended planar zigzag (*trans* conformation) is the minimum energy conformation for an isolated section of polyethylene or paraffin hydrocarbon. The energy of the *trans* conformation is about 800 cal/mol less than that of the gauche form. Consequently, the *trans* form is favored in polymer crystal structures. Typical polymers that exhibit this *trans* form include polyethylene, poly(vinyl alcohol), syndiotactic forms of poly(vinyl chloride) and poly(1,2-butadiene), most polyamides, and cellulose. Note that *trans* conformation is different from the *trans* configuration discussed in Section IV.A.

In some cases, however, steric hindrance causes the main chain to assume a minimum energy conformation other than the *trans* form. Some of these variations may be mere distortions of the fully extended planar zigzag conformation, as in most polyesters, polyisoprenes, and polychloroprene. In other

Figure 3.10 Fringed micelle model. (From Bryant, W.W.D., *J. Polym. Sci.*, 2, 547, 1947. With permission.)

cases, in order to relieve the strain due to the presence of bulky substituents, the main chain rotates and assumes the helical conformation in the crystalline phase. This is the case for most isotactic polymers and 1,1 disubstituted ethylenes.

a. Polyethylene

The unit cell in polyethylene is a parallelepiped with a rectangular cross-section and lattice parameters: a = 7.41 Å; b = 4.94 Å; and c = 2.55 Å (orthorhombic crystal system) (Figure 3.11). By convention, the polymer chains, in passing through the unit cell, lie parallel to the lattice translation vector c. The lattice parameter c (magnitude of c) depends on the crystallographic repeat unit (Bravais lattice or crystal system). In polyethylene, the crystallographic repeat unit contains one chemical repeat unit. The packing of the repeat units in the unit cell is shown in Figure 3.12.

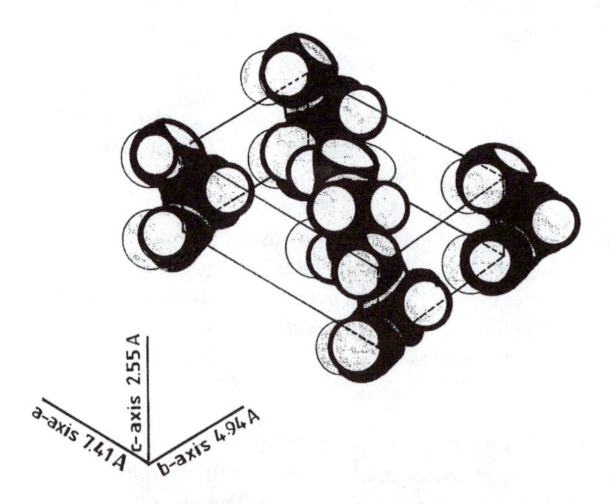

Figure 3.11 Arrangement of chains in the unit cell of polyethylene. (From Geil, P.H., *Polymer Single Crystals*, John Wiley & Sons, New York, 1933. With permission.)

b. Poly(ethylene terephthalate)

The conformation is a slight distortion of the planar zigzag. The benzene ring lies nearly in the plane of zigzag, but the main chains are no longer exactly planar. They make a slight angle with the planar zigzag (Figure 3.13). The unit cell has lattice constants:

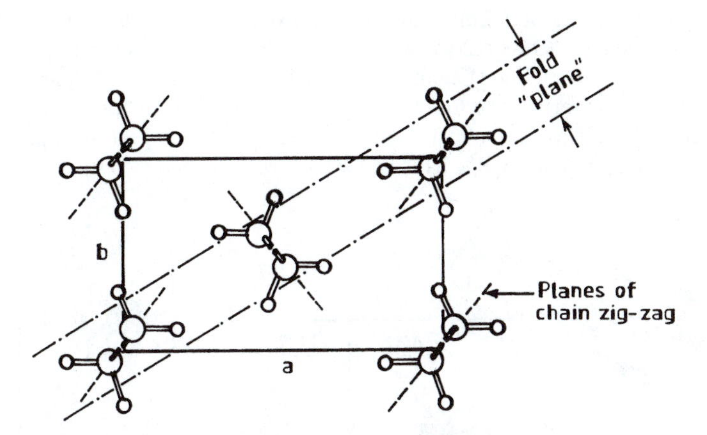

Figure 3.12 Packing in the crystal structure of polyethylene as viewed along the c-axis. (From Natta, G. and Corradini, P., *Rubber Chem. Technol.,* 33, 703, 1960. With permission.)

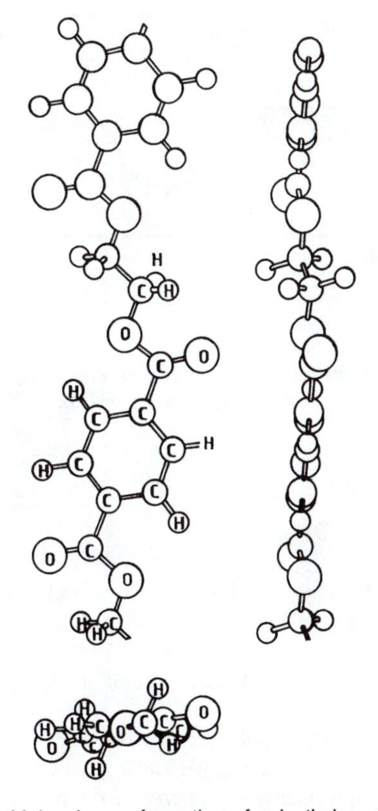

Figure 3.13 Molecular conformation of polyethylene terephthalate.

a = 4.56 Å α = 98.5°
b = 5.94 Å β = 118°
c = 10.75 Å δ = 112°

c. Polypropylene

We have first seen that polyethylene exists in the planar zigzag conformation. Polypropylene can be considered as having a linear polyethylene backbone, but with the H atom on every other carbon atom replaced by a methyl [$-CH_3$] group. Polypropylene can exist in either atactic (noncrystallizable) form or in the crystallizable syndiotactic or isotactic forms. For the isotactic form, because of the size of the pendant [$-CH_3$] group (relative to the H atom, in polyethylene), the backbone can no longer exist in the planar zigzag form; it must rotate. The lowest energy state is attained by a regular rotation of 120° by each chemical repeat unit. This means that there are three chemical repeat units per turn. These pack into a monoclinic crystal system (Figure 3.14) whose unit cell has parameters:

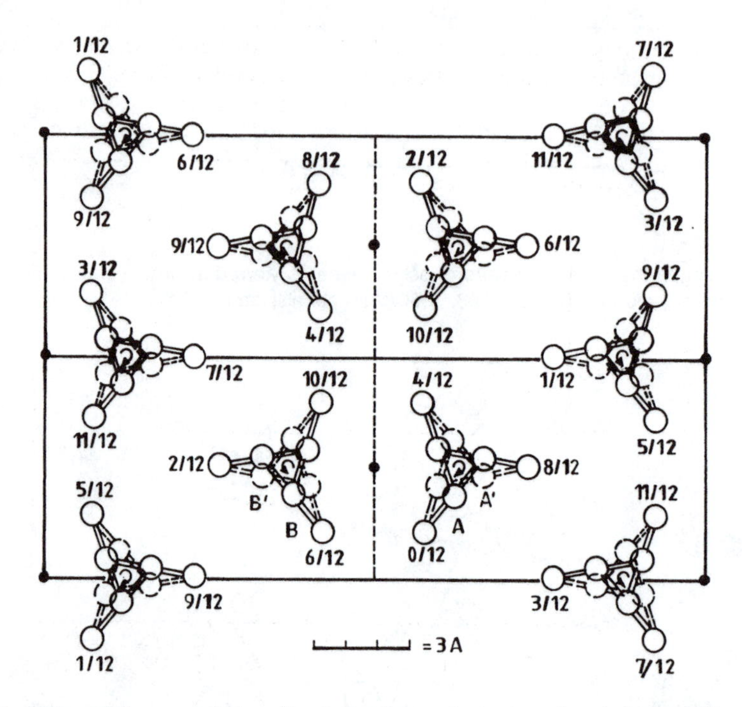

Figure 3.14 Projection of the monoclinic unit cell of polypropylene along the chain-axis. (From Natta, G. and Corradini, P., *Nuovo Cimento Suppl.*, 15(1), 40, 1960. With permission.)

a = 6.65 Å + 0.05 Å
b = 20.96 + 0.15 Å   β = 99.° 20′ + 1°
c = 6.50 + 0.04 Å

d. Degree of Crystallinity

We noted earlier that polymers, by virtue of their large size and in contrast to low-molecular-weight materials, are incapable of 100% crystallinity. To visualize this mentally, the term *semicrystalline* is frequently used to describe crystalline polymers. One of the most useful and practical concepts in the characterization of semicrystalline polymers is the degree of crystallinity. Let us now consider how this can be estimated.

We start by treating the semicrystalline polymer as a two-phase system with a distinct demarcation between the crystalline and amorphous material. We know, of course, that this is not strictly true. Now suppose P_m is an actual or measured intensive property of the polymer, while P_c and P_a are the same property due, respectively, to the crystalline and amorphous materials (components) in the same state as exists in the polymer. Then the degree of crystallinity can be deduced from the individual contributions of the crystalline and amorphous components to the measured property:

$$p_m = \phi\, p_c + (1-\phi)\, p_a \tag{3.12}$$

By rearrangement, Equation 3.12 becomes

$$\phi = \frac{\rho_a - \rho_m}{\rho_a - \rho_c} \tag{3.13}$$

The degree of crystallinity may be derived this way by measurement of a material property such as specific volume, specific heat, enthalpy, and electrical resistivity:

Example 3.6: For the polyethylene of Figure 3.15 (lower curve) calculate the fraction of crystalline material at 20°C assuming the coefficient of expansion for amorphous material is the same above and below T_m.

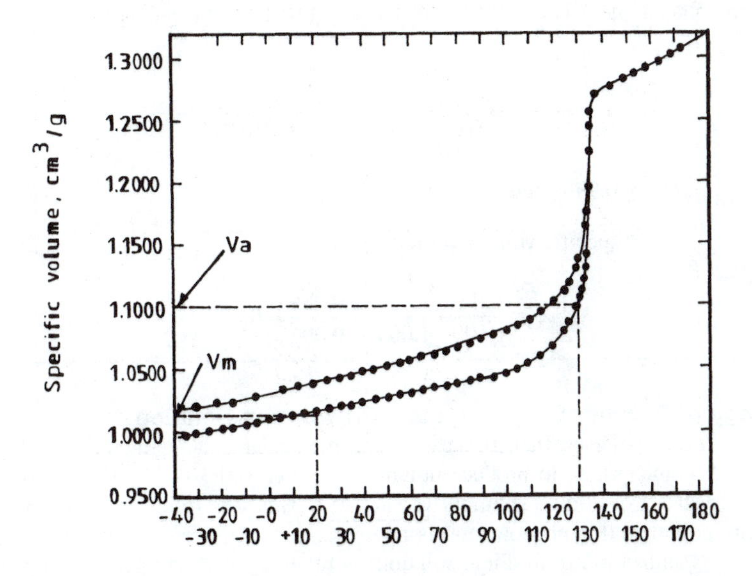

Figure 3.15 Specific volume–temperature relations for linear polyethylene (Marlex 50). Specimen slowly cooled from melt to room temperature prior to experiments (0) and specimen crystallized at 130°C for 40 days and then cooled to room temperature prior to experiment (0). (From Mardelkern, L., *Rubber Chem. Technol.*, 32, 1392, 1959. With permission.)

Solution:

$$\phi = \frac{V_a - V_m}{V_a - V_c}$$

where V_a = specific volume due to the amorphous component
V_m = measured specific volume
V_c = specific volume due to the crystalline component

At 130°C, the crystallization temperature, the specific volume would be due to the amorphous phase. From Figure 3.15, V_a = 1.100 and V_m = 1.06 (at 20°C). To calculate the specific volume of the ideal polymer remember that

Density of a perfect crystal = ρ where

$$\rho = \frac{\text{Mass}}{\text{Volume}} = \frac{nM}{N_A V}$$

where n = number of polymer repeat units per unit cell

M = molecular weight of repeat unit unit
N_A = Avogadro's number
V = volume of a unit cell

For polyethylene, the unit cell is orthorhombic with lattice parameters:

a = 7.41 Å
b = 4.94 Å
c = 2.55 Å

Besides, each unit cell contains two repeat units, one at the center, four at the corners, but with each shared with four other unit cells. That is a total contribution of one repeat unit by the corners (see Figure 3.12).

$$\rho = \frac{2 \times 28}{\left(6.023 \times 10^{23}\right)\left(7.41 \times 10^{-8}\right)\left(4.94 \times 10^{-8}\right)\left(2.55 \times 10^{-8}\right)}$$

$$= 0.996 \text{ g/cm}^2$$

$$V_C = \text{specific volume} = 1/\rho = 1.004$$

$$\phi = \frac{1.100 - 1.016}{1.100 - 1.004} = \frac{0.084}{0.096} = 0.875$$

---

2. Morphology of Polymer Single Crystals Grown from Solution

For a long time, it was believed that, because of the molecular entanglements of polymer chains in solution, it would be impossible to produce polymer single crystals. The first report of the growth of polymer single crystals from dilute solution was in 1953. This was followed by several other reports and for so many polymers this phenomenon is now regarded as universal. Growth of polymer single crystals requires crystallization from dilute solutions at relatively high temperatures by cooling from a temperature above the crystalline melting point. Different morphologies result depending on polymer type and growth conditions.

a. Lamellae

All polymer single crystals have the same general appearance. Under an electron microscope, they appear as thin, flat platelets that are 100 to 120 Å thick and several microns in lateral dimensions. This lamellar nature of polymer single crystals has been found to be fundamental. Growth of the crystal normal to lamellar surface occurs by the formation of additional lamellae of the same thickness as the basal lamellae; thick crystals are usually multilamellar.

b. Chain Folding

Figure 3.16 is an electron micrograph of crystals of polyethylene obtained by cooling a dilute solution (0.1%, in tetrachloroethylene). Such electron microscopy and diffraction studies have confirmed not only the lamellar nature of single crystals but have also revealed that the polymer molecules are oriented normal (or, in some cases, very nearly normal) to the lamellar surface.

Since polymer molecules are generally 1000 to 10,000 Å long and lamellae are only 100 Å thick, it follows that chains must fold repeatedly on themselves. For polyethylene, for example, it has been demonstrated that only about five chain carbon atoms are required for the chain to fold on itself.

The plane on which the regular folding of chains occurs defines the *fold plane,* while the thickness of the lamella is regarded as the *fold period* (Figure 3.16). Chain ends may either terminate within the crystal, forming a defect, or be excluded from the crystal, forming cilia. In some cases, irregular folding and branch points can also occur.

Let us now consider how this fold conformation fits into the overall morphological features of the polymer single crystal. Figure 3.17 is a schematic representation of the top surface of an idealized model of a diamond-shaped polyethylene single crystal as seen along the (001) (c axis). The curved lines,

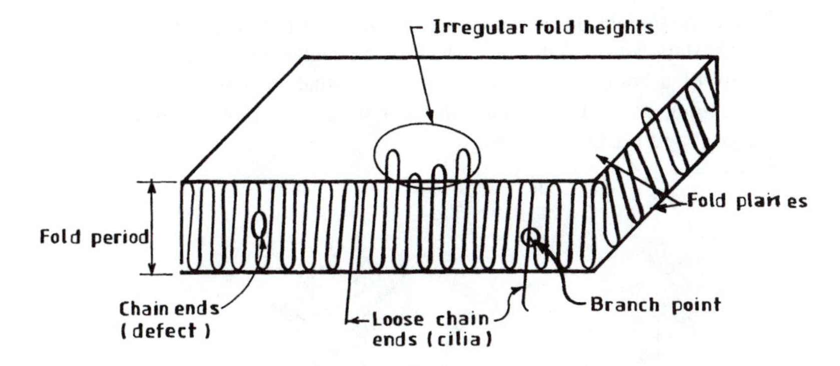

Figure 3.16 Schematic diagram of chain folding showing conformational imperfections.

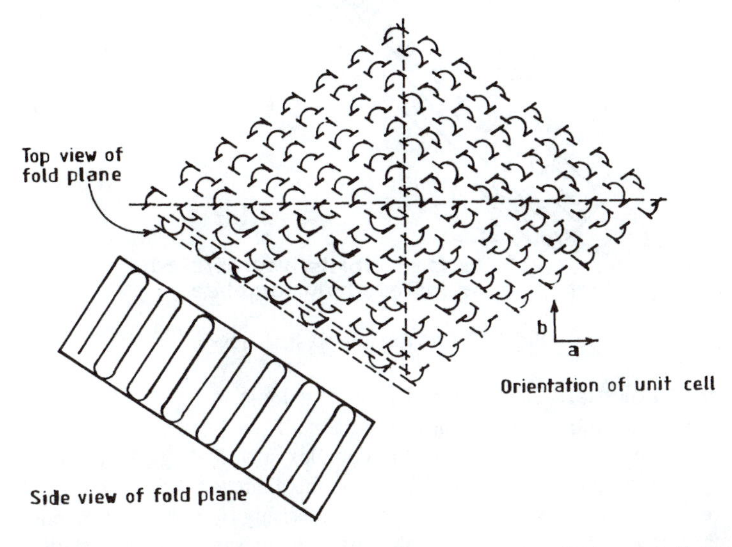

Figure 3.17 Fold packing in a polyethylene single crystal. (From Reneker, D.H. and Geil, P.H., *J. Appl. Phys.*, 31, 1916, 1960. With permission.)

which represent the chain folds, terminate on solid lines representing the plane of zigzag. Note the following features:

- The alternating parallel orientation of the planes of zigzag indicates that the chains twist in addition to folding within their fold plane.
- The crystal is divided into four quadrants (broken lines) and, as we shall see shortly, each sector slopes away from the apex of the pyramidal structure. The fold planes in each quadrant are parallel to the outside edge of that quadrant. It follows that the entire crystal is composed of four triangular quadrants that contain rows of fold planes.

c. Hollow Pyramidal Structure

Ridges are formed by pleats of extra material deposited along one of the diagonals of each diamond-shaped crystal. Obviously, therefore, crystals of polyethylene are not simply flat lamellae. Experimental evidence has shown that they may exist in solution as hollow pyramids. These pyramids may or may not be corrugated depending on the crystallization conditions.

The hollow pyramidal structure is due to the packing of the folded chains in which successive planes of folded molecules are displaced from their neighbors by an integral of repeat distances. In some cases, the fold and fold period are regular, and the displacement of adjacent fold planes is uniform. This results in the formation of a planar pyramid. In other cases, however, the direction of displacement is reversed periodically. In this case, corrugated pyramids are formed.

Besides the structures already discussed, more complex morphologies may be obtained from the growth of polymer crystals from solutions. The structure that emerges from the crystallization of a polymer is a function of a complex interaction of factors that include the type of solvent, solution temperature, concentration, and polymer molecular weight. Some examples of these structures include spiral growth, dendrites, and hedrites.

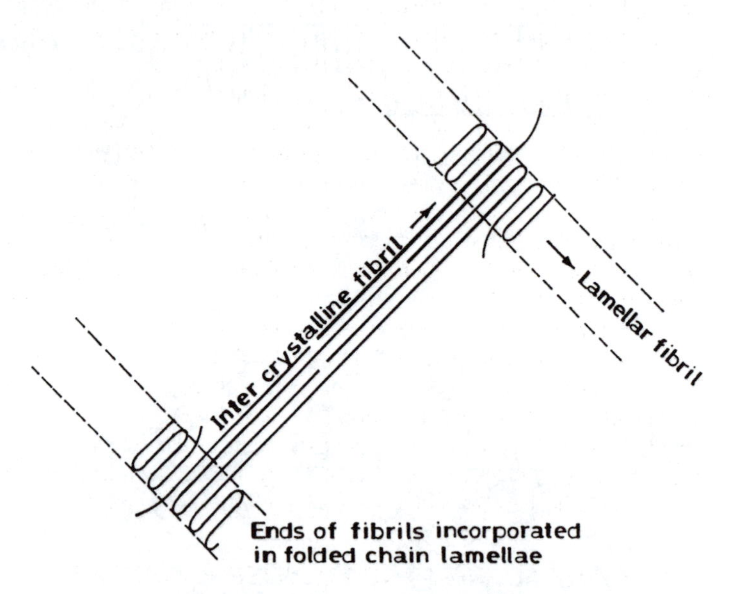

Figure 3.18 The intercrystalline fibril.

3. Morphology of Polymers Crystallized from the Melt

The most prominent structural feature of polymers crystallized from the melt is the spherulite. The spherulite is not a single crystal, but an extremely complex spherical aggregate of lamellae ranging in size from about 0.1 μ to possibly a few millimeters in diameter. Under a polarizing microscope, spherulites show characteristic dark Maltese cross patterns arising from the birefringent effects associated with the molecular orientation of lamellae morphology. When a sample of a crystallizable polymer like polyethylene, nylon 6,6, or poly(ethylene terephthalate) is heated above its melting temperature and then supercooled by about 10 to 15°C, spherulite growth is indicated at several centers. In the case of homogenous nucleation these centers arise spontaneously in the melt, while for heterogeneous nucleation the nucleation center is a foreign body. During growth, spherulites expand radially at a constant linear rate until the growth fronts from neighboring spherulites impinge.

Electron microscopy and electron diffraction studies have revealed that for almost all polymers, spherulites are composed of lamellar structure. Each lamella is a flat ribbon, and, like in simple crystals, chains that are folded are oriented perpendicular to the surface of the lamella. The growth nucleus for crystallization or spherulite development is thought to be a simple crystal that develops by the formation of a multilayer stack. Thereafter, one axis of each lamella extends forming a lamellar fibril. These lamellar fibrils now grow radially from a central nucleus, but have a tendency to twist, diverge, and branch during growth. Since individual lamella do not increase in lateral dimensions, their characteristic branching via screw dislocations is a space-filling process. Lamellae usually have dimensions of 1 μ in length and 100 Å in thickness.

As crystallization proceeds, the growth fronts of two different spherulites meet, and the lamellae extend across spherulite boundaries into uncrystallized material available, thus holding the material together. In addition, interlamellar fibrils tie two or more lamellae together and also bridge spherulite themselves.

One important feature of the interlamellar fibril is that the chain molecules lie parallel to the length of the fibril in contrast to the situation existing in the lamella where chains are oriented at right angles (Figure 3.18). This implies that each interlamellar fibril is an extended chain crystal. The formation of interlamellar links is thought to originate from the inclusion of one chain molecule in two different and possibly widely separated lamellae. This provides the nucleus for subsequent deposition of other molecules, thus producing the intercrystalline ties.

The picture that emerges from the above discussion of the development of spherulitic structure is that spherulites represent the crystalline portion of a sample growing at the expense of the noncrystallizable material. The amorphous regions therefore constitute the residual elements of disorder resulting from the fact that the noncrystallizable material in the original melt — which includes catalyst residues, nonstereoregular chains (e.g., atactic chain segments), short-chain components, plasticizer molecules, and chain ends (low-molecular-weight chains) — is unable to disentangle and rearrange itself into the ordered arrays required in the crystalline state. It appears that as spherulite growth proceeds the noncrystallizable material diffuses ahead of the growth front. However, spherulites themselves, while predominantly crystalline, do contain defects such as chain ends, dislocations, and chemical impurities. The defect materials segregate and separate the radiating lamellae and contribute to the overall amorphous content. Consequently, polymers have a wide variety of crystallinity, which varies from 0% in noncrystallizable polymers like atactic PMMA to almost 100% for highly crystallizable polymers like polytetrafluoroethylene and linear polyethylene. For a particular crystallizable polymer, the degree of crystallinity depends on spherulite growth conditions, which determine the size and extent of perfection of the crystals. For example, it is known that higher degrees of lamellar perfection can be obtained generally at high crystallization temperatures (in the neighborhood of the T_m polymers and after prolonged periods at these temperatures.

Given the long-chain nature of polymer molecules, it is obvious that the process of crystallization involves extensive molecular translation from the high degree of disorder characteristic of the melt to the highly ordered state. Also, this must occur in a time that is short relative to the time required for crystallization. Consequently, the degree of crystallinity in most polymers is a function of the rate of crystallization. After rapid crystallization, the amorphous content of the polymer sample is increased. On the other hand, if a molten polymer is crystallized slowly, the crystals develop in a more perfect manner and tend to exclude impurities that could interfere with the ordering process.

VI. CRYSTALLINITY AND POLYMER PROPERTIES

To conclude this discussion, we need to examine how crystallinity is related to polymer properties. As we have seen above, polymers are semicrystalline, which means that they are composed of amorphous and crystalline phases. Since the amorphous phase can exist in the rubbery or glassy state, the overall effect of the semicrystalline nature of crystalline polymers depends, in the first place, on the state of the amorphous phase or the temperature of use. For example, the modulus of crystalline polymers is only about an order of magnitude higher than the modulus of an amorphous polymer in the glassy state, whereas it is about four orders of magnitude higher than the modulus of the amorphous in the rubbery state. This suggests therefore that modification of polymer properties due to crystallization will be more pronounced for a polymer with amorphous component in the rubbery state than for one whose amorphous phase is in the glassy state. For example, the modulus of rubber can be increased dramatically by induced crystallization. However, for polystyrene — whose amorphous component is glassy at room temperature — crystallization, if induced, has a negligible effect on its modulus, which is already high. By similar arguments, it can be seen that any polymer property that is different for both the amorphous and crystalline components of the polymer will be determined by the relative amounts of these two components as well as their form and distribution (i.e., by the polymer morphology). It is therefore obvious that for engineering design of polymers, control of the properties of a semicrystalline polymer resolves into control of its morphology or spherulite development process. The size and degree of perfection of spherulites are controlled by the crystallization conditions that exist during the unit operations involved in the production of a polymer product.

Example 3.7: Poly(ethylene terephthalate) is cooled rapidly from 300°C (state 1) to room temperature (state 2). The resulting material is rigid and perfectly transparent. The sample is then heated to 100°C and maintained at that temperature during which time it gradually becomes translucent (state 3). It is then cooled down to room temperature and is again found to be rigid, but is now translucent rather than transparent (state 4). For this polymer, $T_m = 267°C$ and $T_g = 69°C$. Describe the molecular mechanisms responsible for the behavior of the polymer in each state. Sketch a general specific-volume vs. temperature curve for this polymer indicating T_g, T_m, and the locations of states 1–4 described above.

Solution:

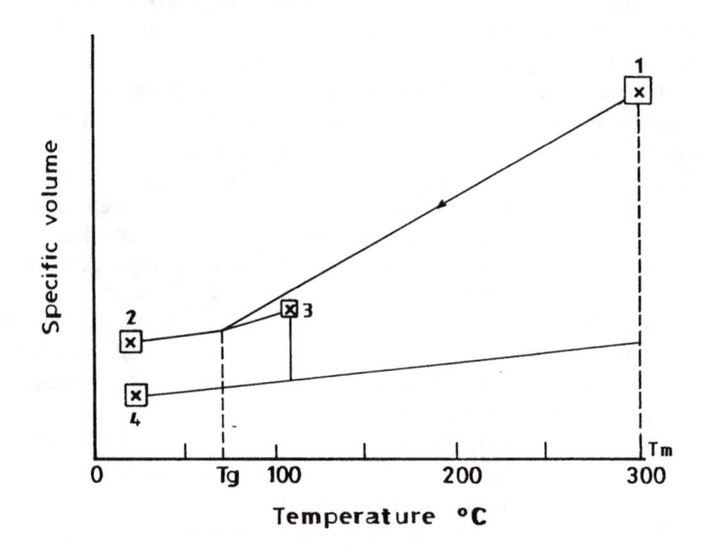

Crystallinity is important in determining optical properties because the refractive index of the crystalline region is always higher than that of the amorphous component irrespective of whether the amorphous component is in the glassy or rubbery state. This difference in refractive indices of the component phases leads to high scattering and consequently, the translucency or haziness of semicrystalline polymers. For a purely amorphous polymer, this does not occur, and hence amorphous polymers are usually transparent. Therefore the state of polyethylene terephthalate can be explained as follows:

State	Polymer Properties
1	Polymer is in the melt form and therefore completely amorphous.
2	As a result of rapid cooling, polymer molecules are unable to align themselves for crystallization. Polymer is therefore amorphous, glassy, and transparent.
3	At 100°C which is higher than the T_g of the polymer, some molecular (segmental) mobility of polymer is now possible. Given sufficient time at this temperature, molecular alignment occurs for crystallization to take place. Polymer is now semicrystalline since crystallization cannot be 100%. The differing refractive indices of the amorphous and crystalline components result in translucency.
4	When cooled to this state, polymer retains its semicrystalline nature and hence its translucency.

VII. PROBLEMS

3.1 Explain the following observations:

a. Polyisobutylene has a much higher oxidation resistance than natural rubber.
b. Polyethylene is a better material for transformer insulation than poly(*p*-chlorostyrene).
c. At room temperature, poly(methylmethacrylate) is transparent, while polyethylene is translucent. At temperatures above 135°C, both polymers are transparent.
d. Polyethylene assumes a planar zigzag conformation, while polyisobutylene has a helical conformation.
e. The percentage of moisture absorption of the following polyamides:
   Nylon 6: 1.3–1.9
   Nylon 12: 0.25–0.3
f. Small amounts of chlorination (10 to 50 wt%) of polymethylene ($–CH_2–$), a saturated paraffin, lower its softening temperature. However, with higher amounts of chlorination (>70 wt%) the softening temperature of the polymer increases again.
g. The addition of glass fibers increases the heat distortion temperature of crystalline polymers like polyethylene and polypropylene, but not that of amorphous polymers like polycarbonate and polysulfone.

h. Both polyethylene and gasoline are hydrocarbons, but high-density polyethylene (HDPE) is used for making automobile gasoline tanks.
i. Teflon is used in nonstick pans.
j. Nylon 6 has a higher melting point than polyethylene.

3.2. A sample of styrene monomer is heated until it is partially polymerized. In order to determine the quantity of polymer formed, the solution of polystyrene in unreacted styrene monomer is poured into a large volume of methanol. The polymer precipitates (methanol has a considerably higher cohesive energy density than polystyrene) and is readily dried and weighed. The CED of water is even higher than that of methanol. Why is water not used in place of methanol for precipitating the polymer? Could the quantity of polymer be determined by distilling off the unreacted monomer and weighing the residue? Explain.

3.3. Two hundred grams of polymer consist of the fractions shown in the following table:

Fraction	Mass (g)	Molecular Weight (g/gmol)
I	100	2×10^3
II	50	2×10^4
III	50	1×10^5

What are the values of $\overline{M}_n$, $\overline{M}_w$, and the polydispersity of the sample.

3.4. Three samples of a polymer were mixed thoroughly without reaction as shown below. Calculate M_n and M_w.

Sample	$\overline{M}_n$	$\overline{M}_w$	Mass in Mixture (g)
A	1.2×10^5	4.5×10^5	200
B	5.6×10^5	8.9×10^5	200
C	10.0×10^5	10.0×10^5	100

3.5. The number-average degrees of polymerization required for good mechanical properties by polymers V and P are 2000 and 1500, respectively. It is known, however, that one of these polymers is poly(vinyl chloride), while the other is poly(vinylidene chloride). On the basis of this information assign polymers V and P. Explain your decision.

3.6. In most cases, free-radical polymerization of a monomer containing a C=C double bond results in a noncrystalline polymer. Explain. Give three examples of monomers that yield crystalline polymers by free-radical polymerization.

3.7. Bristow and Watson[1] reported the following data at 25°C for a 1:39 copolymer of acrylonitrile and butadiene.

Solvent	ΔH_{vap} (cal/mol)	V_1 (cm³/mol)	V_2 (vol fraction) (polymer in gel)
2,2,4-trimethylpentane	8,396	166.0	0.9925
n-Hexane	7,540	131.6	0.9737
CCl_4	7,770	97.1	0.6862
$CHC;_3$	7,510	80.7	0.1510
Dioxane	8,715	85.7	0.2710
CH_2Cl_2	7,004	64.5	0.1563
$CHBr_3$	10,385	87.9	0.1781
Acetonitrile	7,976	52.9	0.4219

From Bristow and Watson, *Trans. Faraday Soc.*, 54, 1731, 1958. With permission.

Calculate the cohesive energy density (CED) and δ_2 for each solvent. Plot $(V_2)^{-1}$ vs. δ_1 to determine δ_2 and CED for the polymer. If the heat of mixing is given by

$$\Delta H_m - V_2 \left(1 - V_2\right)\left(\delta_1 - \delta_2\right)^2,$$

calculate the heat of mixing equal volumes of uncross-linked polymer and acetonitrile.

3.8. Explain the following:

a. Polyethylene and polypropylene produced with stereo-specific catalysts are each fairly rigid, translucent plastics, while a 65:35 copolymer of the two produced in exactly the same manner is a soft, transparent rubber.
b. A plastic is commercially available that is similar in appearance and mechanical properties to polyethylene and polypropylene in (a) but consists of 65% ethylene and 35% propylene units. The two components of this plastic cannot be separated by any physical or chemical means without degrading the polymer.

3.9. The polymers of amino acids are termed *nylon n* where n is the number of consecutive carbon atoms in the chain. Their general formula is

$$\left[-\overset{\displaystyle H}{\overset{|}{N}} - \overset{\displaystyle O}{\overset{\|}{C}} - (CH_2)_{n-1} \right]_x \qquad \text{(Str. 12)}$$

The polymers are crystalline and will not dissolve in either water or hexane at room temperature. They will, however, reach an equilibrium level of absorption when immersed in each liquid. Describe *how* and *why* water and hexane absorption will vary with n.

3.10. Explain why a styrene–butadiene copolymer with solubility parameter $\delta = 8.1$ is insoluble in both pentane ($\delta = 7.1$) and ethylene acetate ($\delta = 9.1$), but will dissolve in a 1:1 mixture of the two solvents.

3.11. The urea derivatives of the following amines were made by refluxing the amines with stoichiometric amounts of urea. Explain the observed crystallization tendencies and melting points of the resulting urea derivatives of amines.

Amine	Crystallization Tendency
$H_2N-(CH_2)_6-NH_2$ Hexamethylenediamine	Crystals formed on cooling the reflux product; mp 200°C
$H_2N-CH_2-N(CH_2CH_2NH_2)-CH_2-CH_2-NH_2$ Tris(2-aminoethylamine)	Crystals formed on cooling and leaving the reflux product overnight; mp = 165°C
$H_2N-CH_2CH_2O-CH_2CH_2O-CH_2CH_2-NH_2$ Triethyleneoxide-diamine	Crystals formed on cooling and leaving the reflux product for 3 weeks; mp = 135°C
$CH_3CH_2-C(CH_2-[OCH_2\,CH(CH_3)-]_x-NH_2)(CH_2-[OCH_2\,CH(CH_3)-]_z-NH_2)-CH_2[O-CH_2CH(CH_3)]_y-NH_2$ Poly(oxypropylene) triamine x + y + z = 5.3	Viscous liquid

3.12. For a crystallizable polymer the degree of crystal imperfection is much higher in bulk-crystallized material than in material crystallized from dilute solution. However, the level of crystal perfection can be significantly enhanced if the polymer melt is left for sufficient time at relatively high temperature.

REFERENCES

1. Gaylord, N.G. and Mark, H., *Linear and Stereoregular Addition Polymers,* Interscience, New York, 1959.
2. Van Krevelen, D.W., *Properties of Polymers,* Elsevier, Amsterdam, 1972.
3. Bryant, W.W.D., *J. Polym. Sci.,* 2, 547, 1947.
4. Geil, P.H., *Polymer Single Crystals,* John Wiley & Sons, New York, 1933.
5. Natta, G. and Corradini, P., *Rubber Chem. Technol.,* 33, 703, 1960.
6. Daubency, R. dep, Brian, C.W., and Brown, J.C., *Proc. R. Soc. London,* 226A, 531, 1954.
7. Natta, G. and Corradini, P., *Nuovo Cimento Suppl.,* 15(1), 40, 1960.
8. Reneker, D.H. and Geil, P.H., *J. Appl. Phys.,* 31, 1916, 1960.
9. Ranby, B.G., Morehead, F.F., and Walter, N.M., *J. Polym. Sci.,* 44, 349, 1968.
10. Geil, P.H. and Reneker, D.H., *J. Polym. Sci.,* 51, 569, 1975.
11. Maxwell, B., Modifying polymer properties mechanically, in *Polymer Processing,* Fear, J.V.D., Ed., Chemical Engineering Process Symp., Ser. 60(49), 10, 1964.
12. Keith, H.D., Padden, F.J., and Vadinsky, R.G., *J. Polym. Sci.,* 4(A-2), 267, 1966.
13. Bristow and Watson, *Trans. Faraday Soc.,* 54, 1731, 1958.
14. Mardelkern, L., *Rubber Chem. Technol.,* 32, 1392, 1959.
15. Billmeyer, F.W., Jr., *Textbook of Polymer Science,* 2nd ed., John Wiley & Sons, New York, 1971.
16. Kaufman, H.S. and Falcetta, J.J., Eds., *Introduction to Polymer Science and Technology,* John Wiley & Sons, New York, 1977.
17. Williams, D.J., *Polymer Science and Engineering,* Prentice-Hall, Englewood Cliffs, NJ, 1971.

18. Sharples, A., *Introduction to Polymer Crystallization,* St. Martin's Press, New York, 1966.
19. Pauling, L., *The Nature of the Chemical Bond,* Cornell University Press, Ithaca, NY, 1960.

Chapter 4

Thermal Transitions in Polymers

I. INTRODUCTION

When a block of ice is heated, its temperature increases until at a certain temperature (depending on the pressure) it starts to melt. No further increase in temperature will be observed until all the ice has melted (solid becomes liquid). If heating is continued, the same phenomenon is observed as before and as the liquid starts to boil (liquid turns to vapor). It is pertinent to make two observations here:

- Water exists in three distinct physical states — solid, liquid, and gas (vapor).
- Transitions between these states occur sharply at constant, well-defined temperatures.

The thermal behavior of all simple compounds, such as ethanol or toluene, is analogous to that of water. However, the transitions in polymers are somewhat different and certainly more complex. In the first place, those molecules large enough to be appropriately termed polymers do not exist in the gaseous state. At high temperatures, they decompose rather than boil since what we would consider conventionally as their "boiling points" are generally higher than their decomposition temperatures. Second, a given polymeric sample is composed of a mixture of molecules having different chain lengths (molecular weights). In contrast to simple molecules, therefore, the transition between the solid and liquid forms of a polymer is rather diffuse and occurs over a temperature range whose magnitude (of the order of 2 to 10°C) depends on the polydispersity of the polymer (Figure 4.1). On melting, polymers become very viscous (viscoelastic) fluids, not freely flowing as in the case of low-molecular-weight materials.

In addition, there is a still more fundamental difference between the thermal behavior of polymers and simple molecules. To understand, first recall that molecular motion in a polymer sample is promoted by its thermal energy. It is opposed by the cohesive forces between structural segments (groups of atoms) along the chain and between neighboring chains. These cohesive forces and, consequently, thermal transitions in polymers depend on the structure of the polymer. In this regard, two important temperatures at which certain physical properties of polymers undergo drastic changes have been identified:

- The glass transition temperature, T_g
- The crystalline melting point, T_m

If a polymer is amorphous, the solid-to-liquid transition occurs very gradually, going through an intermediate "rubbery" state without a phase transformation. The transition from the hard and brittle glass into a softer, rubbery state occurs over a narrow temperature range referred to as the glass transition temperature. In the case of a partially crystalline polymer, the above transformation occurs only in the amorphous regions. The crystalline zones remain unchanged and act as reinforcing elements thus making the sample hard and tough. If heating is continued, a temperature is reached at which the crystalline zones begin to melt. The equilibrium crystalline melting point, T_m, for polymers corresponds to the temperature at which the last crystallite starts melting. Again, in contrast to simple materials, the value of T_m depends on the degree of crystallinity and size distribution of crystallites. The general changes in physical state due to changes in temperature and molecular weight are shown in Figure 4.2 for amorphous and crystalline polymers.

The thermal behavior of polymers is of considerable technological importance. Knowledge of thermal transitions is important in the selection of proper processing and fabrication conditions, the characterization of the physical and mechanical properties of a material, and hence the determination of appropriate end uses. For example, the glass transition temperature of rubber determines the lower limit of the use of rubber and the upper limit of the use of an amorphous thermoplastic. We take up discussion of these transition temperatures in succeeding sections.

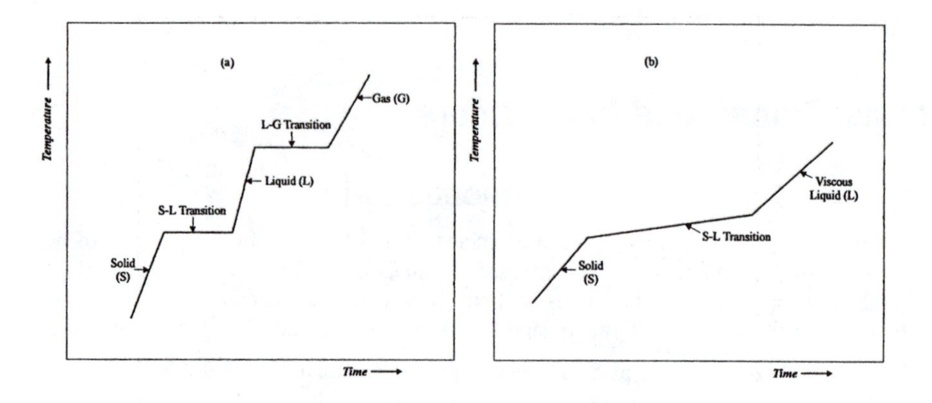

Figure 4.1 Relative thermal responses of simple molecules (a) and polymers (b).

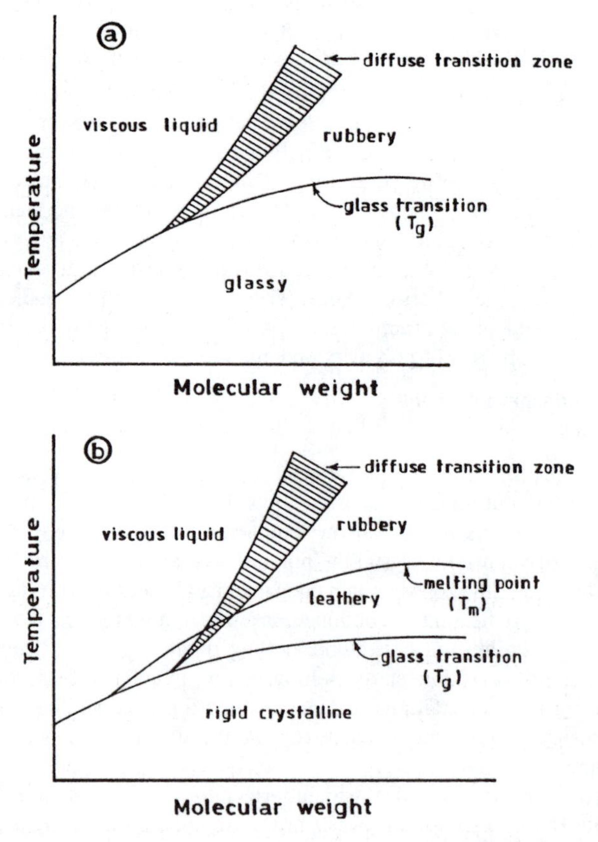

Figure 4.2 Temperature–molecular weight diagram. (a) For amorphous polymer; (b) for crystalline polymer.

II. THE GLASS TRANSITION

To illustrate the concept of glass transition, let us consider the specific volume–temperature behavior for both amorphous (ABCD) and crystalline (ABEF) polymers, as shown in Figure 4.3. As the amorphous polymer (line ABCD) is heated from the low-temperature region (region D), the volume expands at a

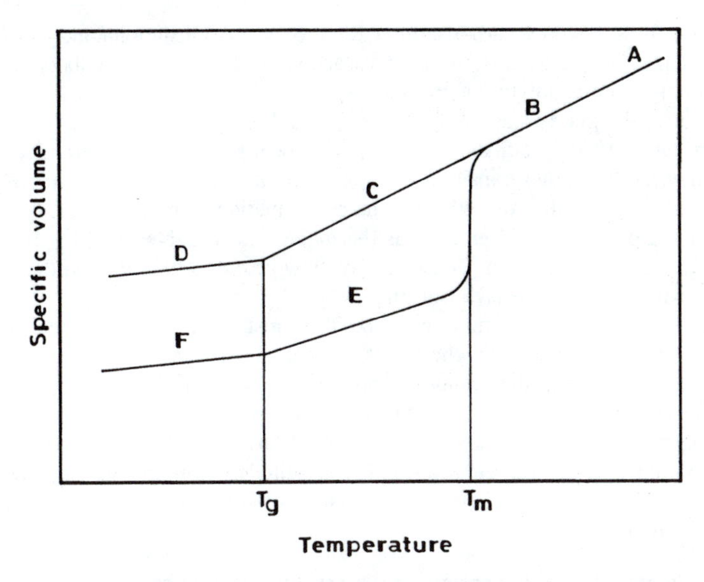

Figure 4.3 Specific volume–temperature curves for a semicrystalline polymer. (A) Liquid region; (B) viscous liquid with some elastic response; (C) rubbery region; (D) glassy region; (E) crystallites in a rubbery matrix; (F) crystallites in a glassy matrix.

constant rate. At a characteristic temperature, T_g, the rate of volume expansion increases suddenly to a higher constant level, i.e., there is a change in the slope of the volume–temperature curve from a lower to a higher volume coefficient of expansion. At the same time, there is an abrupt change in physical behavior from a hard, brittle, glassy solid below T_g (region D) to a soft, rubbery material above T_g (region C). On further heating, the polymer changes gradually from the rubbery state to a viscous liquid (region B) whose viscosity decreases with increasing temperature until decomposition sets in.

For a crystalline polymer, the changes at T_g are less drastic. This is because these changes are restricted mainly to the amorphous domains while the crystalline zones remain relatively unaffected. Between the glass transition (T_g) and the melting temperature (T_m) (region E) the semicrystalline polymer is composed of rigid crystallites immersed (dispersed) in a rubbery amorphous matrix. In terms of mechanical behavior, the polymer remains rigid, pliable, and tough. At the melting temperature, the crystallites melt, leading to a viscous state (region B). Above T_m the crystalline polymer, like the amorphous polymer, exists as a viscous liquid.

A. MOLECULAR MOTION AND GLASS TRANSITION

In polymers, *intra*molecular bonds are due to primary valence bonds (covalent) while the *inter*molecular attractions usually are due to secondary bonding forces. The intermolecular forces are opposed by thermal agitation, which induces vibration, rotation, and translation of a molecular system. Atomic vibrations exist at all temperature levels. The stability of the molecular system depends on the vibration energy of the chemical bonds. In polymers, thermal degradation occurs when the energy of vibration exceeds the primary bonding between atoms, while the transitional phenomena associated with crystalline melting point, the glass transition temperature, and the polymer deformations are related to rotation and vibration of molecular chains.

Bearing this in mind, let us consider what happens on a molecular scale when an amorphous polymer is heated from below its glass transition temperature. At very low temperatures — i.e., in the glassy state — chain segments are frozen in fixed positions; atoms undergo only low-amplitude vibratory motion about these positions. As the temperature is increased, the amplitude of these vibrations becomes greater, thereby reducing the effectiveness of the secondary intermolecular bonding forces. Consequently, the cooperative nature of the vibrations between neighboring atoms is enhanced. At the glass transition temperature, chain ends and a substantial number of chain segments have acquired sufficient energy to overcome intermolecular restraints and undergo rotational and translational motion. Therefore, the glass transition temperature is referred to as the onset of large-scale cooperative motion of chain segments (of the order of 20 to 50 consecutive carbon atoms). Rotational and translational modes of motion provide important mechanisms for energy absorption. This accounts for glassy-to-rubbery transition and the tough nature of an amorphous polymer above its glass transition temperature.

Below the T_g, or in the glassy state, only atoms or small groups of atoms such as short sections of the main chain or pendant/side groups move against the local restraints of intermolecular interactions. This movement may result in other transitions, which are designated α, β, γ, etc., in order of decreasing temperature. The fully extended chain, which is the conformation of minimum energy, is the preferred conformation at low temperatures. Therefore, as the molecules straighten out, the free volume, as we shall see in the next section, decreases. Consequently, flow becomes difficult and the polymer assumes the characteristic hard and brittle behavior of glasses.

As we said above, the molecular motion of the T_g is restricted only to segmental motion; entire molecular motion is as yet precluded by chain entanglements. However, above the T_g, or in the rubbery state, there is a sharp increase in the number of possible conformations. The molecular motion in the rubbery state requires more free volume, and this rise in the relative free volume leads to the observed higher volume expansion coefficient above the T_g. As heating is continued into the liquid region, molecules acquire increased thermal energy, and the amplitudes of associated molecular motions also increase. Translation, or slip of entire molecules, becomes possible; large changes in conformation occur and elasticity virtually disappears.

B. THEORIES OF GLASS TRANSITION AND MEASUREMENT OF THE GLASS TRANSITION TEMPERATURE

The fundamental nature of the glass transition is still unclear. It is a complex process that involves equilibrium, thermodynamic and kinetic factors. The various theories of the glass transition, however, have used either the thermodynamic or the kinetic approach. The thermodynamic approach is based on entropy considerations of the glassy state, while the kinetic theory of the glass transition considers the relaxation phenomena associated with the glass transition. Each approach gives only a partial explanation to the observed behavior of polymers. We now briefly discuss these theories along with the free volume theory.

1. Kinetic Theory

The kinetic concept of glass transition considers the glass transition as a dynamic phenomenon since the position of the T_g depends on the rate of heating or cooling. It predicts that the value of T_g measured depends on the time scale of the experiment in relation to that of the molecular motions arising from the perturbation of the polymer system by temperature changes. A number of models have been proposed to correlate these molecular motions with changes in macroscopic properties observed in the experiment. One approach considers the process of vitrification (glassification) as a reaction involving the movement of chain segments (kinetic units) between energy states. For the movement of a chain segment from one energy state to another to occur, a critical "hole" or empty space must be available. To create this hole sufficient energy must be available to overcome both the cohesive forces of the surrounding molecules and the potential energy barrier associated with the rearrangement. The temperature at which the number of holes of sufficient size is great enough to permit flow is regarded as the T_g. This theory permits a description of the approach to thermodynamic equilibrium. When a polymeric material above T_g is cooled, there is sufficient molecular motion for equilibrium to be achieved. However, the rate of approach to equilibrium, and hence the T_g, depends on the cooling rate employed in the experiment.

2. Equilibrium Theory

The equilibrium concept treats the ideal glass transition as a true second-order thermodynamic transition, which has equilibrium properties. The ideal state, of course, cannot be obtained experimentally since its realization would require an infinite time. According to the theory of Gibbs and DiMarzio,[1] the glass transition process is a consequence of the changes in conformational entropy with changes in temperature. The reduced level in molecular reorganization observed near the transition temperature is attributed to the reduction in the number of available conformations as the temperature is lowered. The equilibrium conformational entropy becomes zero when a thermodynamic second-order transition is reached ultimately. Thereupon, the conformations are essentially "frozen in" since the time required for conformational changes becomes virtually infinite. The glass transition temperature, T_g, therefore approaches the true transition temperature as the time scale of experiment becomes longer. Based on this reasoning and using a statistical thermodynamics treatment that utilizes a quasi-lattice theory, Gibbs and DiMarzio[1] developed quantitative predictions of the second-order phase transition that are in agreement with experiment.

3. Free Volume Theory

A most useful and popular theory of glass transition is the "free volume" model of Fox and Ferry and, later, of Williams, Landell, and Ferry.[2] This theory considers the free volume, V_f, of a substance as the difference between its specific volume, $\bar{V}$, and the space actually occupied by the molecules, V_0, where V_0 is expressed as:

$$V_o = V' + \alpha_g T \tag{4.1}$$

where V' = the extrapolated volume of glass at absolute zero
α_g = thermal expansion coefficient of the glass

This model further defines the free volume fraction, f, at temperature T as

$$\begin{aligned} f &= V_f/\bar{V} \\ &= f_g + \alpha_f (T - T_g) \end{aligned} \tag{4.2}$$

and

$$\alpha_f = \alpha_l - \alpha_g = \frac{df}{dT} = \frac{1}{V_f}\frac{dV_f}{dT} \tag{4.3}$$

where f_g = free volume fraction at T_g
α_l = thermal expansion coefficient above T_g
α_g = thermal expansion coefficient below T_g

For most amorphous polymers, the free volume fraction at the glass transition temperature is found to be a constant, with a value of 0.025. Amorphous polymers, when cooled, are therefore supposed to become glassy when the free volume fraction attains this value. Thereupon no significant further change in the free volume will be observed.

Many important physical properties of polymers (particularly amorphous polymers) change drastically at the glass transition temperature. The variations of these properties with temperature form a convenient method for determining T_g. Some of the test methods include the temperature variation of specific volume (dilatometry) as discussed in Section II, refractive index (refractometry), and specific heat (calorimetry, DSC or DTA). Others include temperature-induced changes in vibrational energy level (infrared spectroscopy), proton environment (nuclear magnetic resonance or NMR), dipole moment (dielectric constant and loss), elastic modulus (creep or stress relaxation), and mechanical energy absorption (dynamic mechanical analysis or DMA). Discussion of details of these test methods is beyond the scope of this volume.

C. FACTORS AFFECTING GLASS TRANSITION TEMPERATURE

We have seen from the previous discussion that at the glass transition temperature there is a large-scale cooperative movement of chain segments. It is therefore to be expected that any structural features or externally imposed conditions that influence chain mobility will also affect the value of T_g. Some of these structural factors include chain flexibility; stiffness, including steric hindrance, polarity, or interchain attractive forces; geometric factors; copolymerization; molecular weight, branching; cross-linking; and crystallinity. External variables are plasticization, pressure, and rate of testing.

Table 4.1 Effect of Chain Flexibility on T_g

Polymer	Repeat Unit	T_h (°C)
Polyethylene	$-CH_2-CH_2-$	–120
Polydimethylsiloxane	$-Si(CH_3)_2-O-$	–123
Polycarbonate	$-O-C_6H_4-C(CH_3)_2-C_6H_4-O-C(=O)-$	150
Polysulfone	$-O-C_6H_4-C(CH_3)_2-C_6H_4-O-C_6H_4-SO_2-C_6H_4-$	190
Poly(2,6-dimethyl-1,4-phenylene oxide)	$[-C_6H_2(CH_3)_2-O-]$	220

1. Chain Flexibility

Chain flexibility is determined by the ease with which rotation occurs about primary valence bonds. Polymers with low hindrance to internal rotation have low T_g values. Long-chain aliphatic groups — ether and ester linkages — enhance chain flexibility, while rigid groups like cyclic structures stiffen the backbone. These effects are illustrated in Table 4.1. Bulky side groups that are stiff and close to the backbone cause steric hindrance, decrease chain mobility, and hence raise T_g (Table 4.2).

The influence of the side group in enhancing chain stiffness depends on the flexibility of the group and not its size. In fact, side groups that are fairly flexible have little effect within each series; instead polymer chains are forced further apart. This increases the free volume, and consequently T_g drops. This is illustrated by the polymethacrylate series (Table 4.3).

2. Geometric Factors

Geometric factors, such as the symmetry of the backbone and the presence of double bonds on the main chain, affect T_g. Polymers that have symmetrical structure have lower T_g than those with asymmetric structures. This is illustrated by two pairs of polymers: polypropylene vs. polyisobutylene and poly(vinyl chloride) vs. poly(vinylidene chloride) in Table 4.4. Given our discussion above on chain stiffness, one would have expected that additional groups near the backbone for the symmetrical polymer would enhance steric hindrance and consequently raise T_g. This, however, is not the case. This "discrepancy" is due to conformational requirements. The additional groups can only be accommodated in a conformation with a "loose" structure. The increased free volume results in a lower T_g.

Another geometric factor affecting T_g is *cis–trans* configuration. Double bonds in the *cis* form reduce the energy barrier for rotation of adjacent bonds, "soften" the chain, and hence reduce T_g (Table 4.5).

3. Interchain Attractive Forces

Recall from our earlier discussion that intermolecular bonding in polymers is due to secondary attractive forces. Consequently, it is to be expected that the presence of strong intermolecular bonds in a polymer chain, i.e., a high value of cohesive energy density, will significantly increase T_g. The effect of polarity, for example, can be seen from Table 4.6. The steric effects of the pendant groups in series (CH_3, –Cl,

Table 4.2 Enhancement of T_g by Steric Hindrance

Polymer	Repeat Unit	T_g(°C)
Polyethylene	$-CH_2-CH_2-$	–120
Polypropylene	$-CH_2-CH(CH_3)-$	–10
Polystyrene	$-CH_2-CH(C_6H_5)-$	100
Poly(α-methylstyrene)	$-CH_2-C(CH_3)(C_6H_5)-$	192
Poly(*o*-methylstyrene)	$-CH_2-CH(C_6H_4\text{-}o\text{-}CH_3)-$	119
Poly(*m*-methylstyrene)	$-CH_2-CH(C_6H_4\text{-}m\text{-}CH_3)-$	72
Poly(α-vinyl naphthalene)	$-CH_2-CH(C_{10}H_7)-$	135
Poly(vinyl carbazole)	$-CH_2-CH(N\text{-carbazolyl})-$	208

and –CN) are similar, but the polarity increases. Consequently, T_g is increased in the order shown in the table. The same effect of increased T_g with increasing CED can be observed when one considers going from the intermolecular forces in poly(methyl acrylate), an ester, through the strong hydrogen bonds in poly(acrylic acid) to primary ionic bonds in poly(zinc acrylate) (Table 4.7).

Recall again that secondary bonding forces are effective only over short molecular distances. Therefore, any structural feature that tends to increase the distance between polymer chains decreases the cohesive energy density and hence reduces T_g. This effect has already been clearly demonstrated in the polyacrylate series where the increased distance between chains due to the size of the alkyl group, R, reduced T_g.

Table 4.3 Decrease of T_g with Increasing Flexibility of Side Chains for Polymethacrylate Series

Generalized Formula	R	T_g (°C)
$-CH_2-C(CH_3)(C(=O)-O-R)-$	methyl	105
	ethyl	65
	n-propyl	35
	n-butyl	21
	n-hexyl	–5
	n-octyl	–20
	n-dodecyl	–65

Table 4.4 Effect of Symmetry of T_g

Polymer	Repeat Unit	T_g(°C)
Polypropylene	$-CH_2-CH(CH_3)-$	–10
Polyisobutylene	$-CH_2-CH(CH_3)_2-$	–70
Poly(vinyl chloride)	$-CH_2-CH(Cl)-$	87
Poly(vinylidene chloride)	$-CH-C(Cl)_2-$	–17

Table 4.5 Relative Effects of *cis–trans* Configuration on T_g

Polymer	Repeat Unit	T_g(°C)
Poly(1,4-*cis*-butadiene)	$-CH_2-CH=CH-CH_2-$ (cis)	–108
Poly(1,4-*trans*-butadiene)	$-CH_2-CH=CH-CH_2-$ (trans)	–83

Table 4.6 Effect of Polarity on T_g

Polymer	Repeat Unit	Dielectric Constant at 1kHz	T_g(°C)
Polypropylene	$-CH_2-CH(CH_3)-$	2.2–2.3	–10
Poly(vinyl chloride)	$-CH_2-CH(Cl)-$	3.39	87
Polyacrylonitrile	$-CH_2-CH(CN)-$	5.5	103

Table 4.7 Effect of Polarity on the T_g of Some Acrylic Polymers

Polymer	Repeat Value	T_g(°C)
Polymethylacrylate	$-CH_2-CH(C(=O)-O-CH_3)-$	3
Poly(acrylic acid)	$-CH_2-CH(C(=O)-O-H)-$	106
Poly(zinc acrylate)	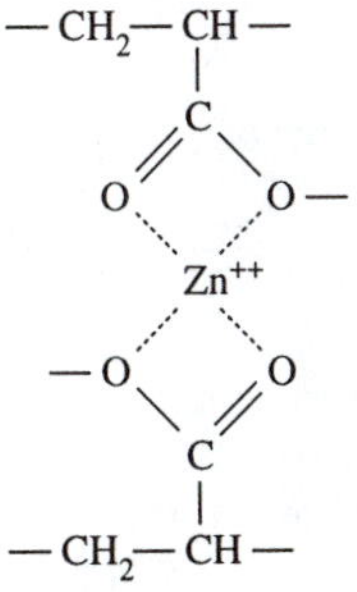 $-CH_2-CH(C(=O)O^-)-$ ··· Zn^{++} ··· $-CH_2-CH(C(=O)O^-)-$	>400

4. Copolymerization

The transition temperatures T_g and T_m are important technological characteristics of polymers. It is desirable — in fact, valuable — to be able to control either T_g or T_m independent of each other. This, however, is often impossible. Polymer chemists have circumvented this problem to some extent by polymer modification via copolymerization and polyblending. These procedures have become powerful tools for tailoring polymer systems for specific end uses.

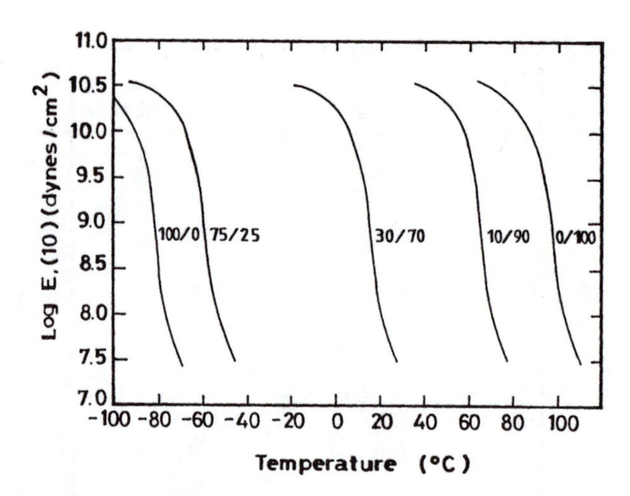

Figure 4.4 E_r vs. fraction ratios of butadiene–styrene copolymers. (From Tobolsky, A.V., *Properties and Structure of Polymers,* John Wiley & Sons, New York, 1960. With permission of Dorothy Tobolsky.)

A copolymer system may be characterized either by the geometry of the resulting polymer — that is, the arrangement of the different monomers (random, alternating, graft, or block) — or by the compatibility (miscibility) of the two monomers.

a. *Isomorphous Systems (Homogeneous Copolymers or Compatible Polyblends)*

In isomorphous systems, the component monomers occupy similar volumes and are capable of replacing each other in the crystal system. The resulting copolymer, irrespective of its geometry, is necessarily homogeneous, and polyblends of the individual homopolymers or copolymers have similar transition properties. Copolymerization merely shifts the T_g to the position intermediate between those of the two homopolymers; it does not alter the temperature range or the modulus within the transition region (Figure 4.4). This shift is illustrated in Figure 4.4, which shows the modulus temperature curves for polybutadiene (100/0) and polystyrene (0/100) and for various compositions of butadiene–styrene copolymer.

For this system, if the glass transitions (T_{g1} and T_{g2}) of the individual homopolymers (1 and 2) are known, it is possible to estimate the T_g of the copolymer (or polyblend) using the relation

$$T_g = V_1 T_{g2} + V_2 T_{g2} \tag{4.4}$$

where V_1 and V_2 are the volume fractions of components 1 and 2, respectively. This is shown schematically in Figure 4.5 (line 1).

b. *Nonisomorphous Systems*

In nonisomorphous systems, the specific volumes of the monomers are different. In this case, the geometry of the resulting polymer becomes important.

Random or alternating — For these copolymers, the composition is necessarily homogeneous (no phase separation) and, as discussed above, the glass transitions are intermediate between those of the two homopolymers. The increased disorder resulting from the random or alternating distribution of monomers enhances the free volume and consequently reduces T_g below that predicted by Equation 4.4 (line 2, Figure 4.5). The T_g of the copolymer whose components have weight fractions W_1 and W_2 and glass transitions T_{g1} and T_{g2}, respectively, can be calculated from the relation

$$\frac{1}{T_g} = \frac{W_1}{T_{g_1}} + \frac{W_2}{T_{g_2}} \tag{4.5}$$

Examples of this type are methyl methacrylate–acrylonitrile, styrene–methyl methacrylate, and acrylonitrile–acrylamide copolymers. It is also possible that monomers involved in the copolymerization process

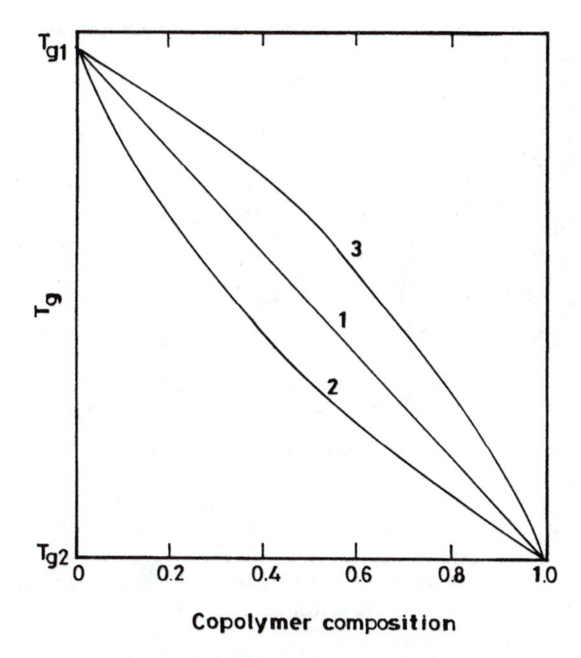

Figure 4.5 Variation in glass transition temperature with copolymer composition (schematic).

(as in the copolymers methylacylate–methylmethacrylate and vinylidene chloride–methylacrylate) introduce significant interaction between chains. In this case the T_g will be enhanced relative to the predicted value (Figure 4.5, line 3).

Block and graft copolymers (incompatible copolymers) — For block or graft copolymers in which the component monomers are incompatible, phase separation will occur. Depending on a number of factors — for example, the method of preparation — one phase will be dispersed in a continuous matrix of the other. In this case, two separate glass transition values will be observed, each corresponding to the T_g of the homopolymer. Figure 4.6 shows this behavior for polyblends of polystyrene (100) and 30/70 butadiene–styrene copolymer (0).

Example 4.1: What is the T_g of butadiene–styrene copolymer containing 10 vol% styrene?

Solution: Butadiene and styrene form a completely compatible random copolymer. Therefore the following relation is applicable: $T_g = V_1T_{g1} + V_2T_{g2}$.

$$\text{Assume } 1 = \text{polybutadiene}$$

$$2 = \text{polystyrene}$$

$$T_{g1} = -80°C, T_{g2} = 100°C$$

$$T_g = 0.90\,(-80) + 0.10\,(100)$$

$$= -62°C$$

---

5. Molecular Weight

Since chain end segments are restricted only at one end, they have relatively higher mobility than the internal segments, which are constrained at both ends. At a given temperature, therefore, chain ends provide a higher free volume for molecular motion. As the number of chain ends increases (which means a decrease in M_n), the available free volume increases, and consequently there is a depression of T_g. The effect is more pronounced at low molecular weight, but as M_n increases, T_g approaches an asymptotic value. An empirical expression relating the inverse relations between T_g and $\overline{M}_n$ is given by Equation 4.6.

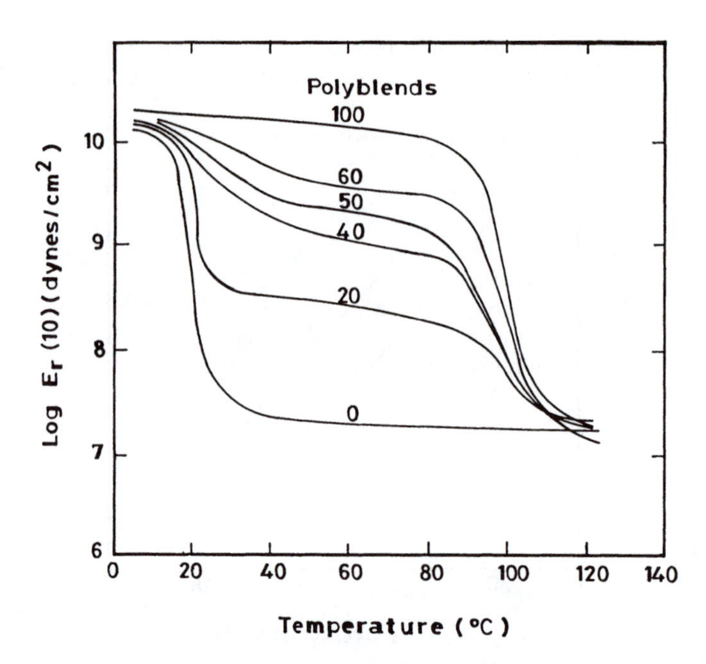

Figure 4.6 $E_r(10)$ vs. temperature for polyblends of polystyrene and a 30/70 butadiene–styrene copolymer. Numbers on the curves are the weight percent of polystyrene in the blend. (From Tobolsky, A.V., *Properties and Structure of Polymers,* John Wiley & Sons, New York, 1960. With permission of Dorothy Tobolsky.)

$$T_g = T_g^{\infty} = K/\overline{M}_n \tag{4.6}$$

where T_g^{∞} = T_g of an infinite molecular weight
K = a constant

For polystyrene it has been found that T_g^{∞} = 100°C while K is about 2×10^5.

Example 4.2: What is the T_g of polystyrene of $\overline{M}_n = 3000$?

Solution: From above, $T_g^{\infty} = 100$, $K = 2 \times 10^5$.

$$T_g = 100 - \frac{2 \times 10^5}{3000}$$

$$= 33°C$$

---

6. Cross-Linking and Branching

By definition, cross-linking involves the formation intermolecular connections through chemical bonds. This process necessarily results in reduction in chain mobility. Consequently, T_g increases. For lightly cross-linked systems like vulcanized rubber, T_g shows a moderate increase over the uncross-linked polymer. In this case, T_g and the degree of cross-linking have a linear dependence, as shown by the following approximate empirical equation.

$$T_g - T^0 = \frac{3.9 \times 10^4}{M_c} \tag{4.7}$$

where T_g = the glass transition temperature of the uncross-linked polymer having the same chemical composition as the cross-linked polymer
M_c = the number-average molecular weight between cross-linked points

For highly cross-linked systems like phenolics and epoxy resins, the glass transition is virtually infinite. This is because the molecular chain length between cross-links becomes smaller than that required for cooperative segmental motion.

Like long and flexible side chains, branching increases the separation between chains, enhances the free volume, and therefore decreases T_g.

7. Crystallinity

In semicrystalline polymers, the crystallites may be regarded as physical cross-links that tend to reinforce or stiffen the structure. Viewed this way, it is easy to visualize that T_g will increase with increasing degree of crystallinity. This is certainly not surprising since the cohesive energy factors operative in the amorphous and crystalline regions are the same and exercise similar influence on transitions. It has been found that the following empirical relationship exists between T_g and T_m.

$$\frac{T_g}{T_m} = \begin{Bmatrix} 1/2 \text{ for symmetrical polymers} \\ 2/3 \text{ for unsymmetrical polymers} \end{Bmatrix} \tag{4.8}$$

where T_g and T_m are in degrees Kelvin.

8. Plasticization

Plasticity is the ability of a material to undergo plastic or permanent deformation. Consequently, plasticization is the process of inducing plastic flow in a material. In polymers, this can be achieved in part by the addition of low-molecular-weight organic compounds referred to as plasticizers (see Chapter 9). Plasticizers are usually nonpolymeric, organic liquids of high boiling points. Plasticizers are miscible with polymers and, in principle, should remain within the polymer. Addition of plasticizers to a polymer, even in very small quantities, drastically reduces the T_g of the polymer. This is exemplified by the versatility of poly(vinyl chloride) which, if unmodified, is rigid, but can be altered into a flexible material by the addition of plasticizers such as dioctylphthalate (DOP).

The effect of plasticizer in reducing T_g can be interpreted in several ways. Plasticizers function through a solvating action by increasing intermolecular distance, thereby decreasing intermolecular bonding forces. Alternatively, the addition of plasticizers results in a rapid increase in chain ends and hence an increase in free volume. A plasticized system may also be considered as a polyblend, with the plasticizer acting as the second component. In this case, our earlier relations for polyblends would apply (Equations 4.1 and 4.2). Since plasticizers generally have very low T_g, between –50°C and –160°C, addition of small amounts of the plasticizer would be expected to result in a substantial decrease in the T_g of a polymer. This is illustrated in Figure 4.7 for a poly(vinyl chloride)–diethylhexyl succinate system.

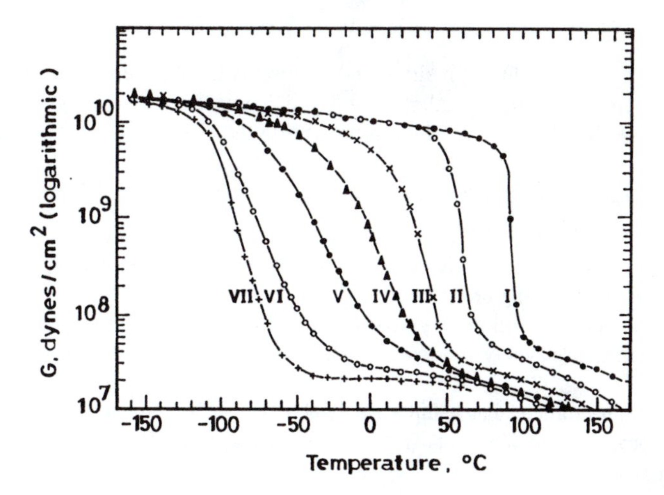

Figure 4.7 Shear modules, G vs. temperature, measured for a time scale of approximately 1 s, poly(vinyl chloride) plasticized with diethylhexyl succinate. I, 100% monomer; II, 91%; III, 79%; IV, 70.5%; V, 60.7%; VI, 51.8%; VII, 40.8%. (From Schneider, K. and Wolf, K., *Kolloid Z.*, 127, 65, 1952.)

Observe that in addition to the reduction in T_g, there is a considerable broadening of the transition region in plasticized PVC.

Example 4.3: Toluene behaves as a plasticizer for polystyrene. Estimate T_g of a polystyrene sample containing 20 vol% toluene.

Solution: Since toluene is completely compatible with polystyrene, we may use the relation

$$T_g = V_A T_{gA} + V_B T_{gB}$$

where T_{gA}, T_{gB} = glass transition temperature of polystyrene and toluene, respectively
V_A, V_B = volume fraction of polystyrene and toluene, respectively

Assuming T_{ga} = melting point of toluene = –50°C.

$$T_g = (1 - V_B)\, T_{gA} + V_B\,(T_{gB})$$

$$= 0.8 \times 100 \times 0.2\,(-50)$$

$$= 70°C$$

III. THE CRYSTALLINE MELTING POINT

Melting involves a change from the crystalline solid state into the liquid form. For low-molecular-weight (simple) materials, melting represents a true first-order thermodynamic transition characterized by discontinuities in the primary thermodynamic variables of the system such as heat capacity, specific volume (density), refractive index, and transparency. Melting occurs when the change in free energy of the process is zero; that is,

$$\Delta G_m = \Delta H_m - T_m \Delta g_m = 0 \tag{4.9}$$

or

$$T_m = \frac{\Delta H_m}{\Delta S_m} \tag{4.10}$$

where ΔH_m = enthalpy change during melting and represents the difference between cohesive energies of molecules in the crystalline and liquid states
ΔS_m = entropy change during melting representing the change in order between the two states

This concept has been extended to melting in crystalline polymeric systems. We must remember, however, that in the case of crystalline polymers:

- The macromolecular nature of polymers and the existence of molecular weight distribution (polydispersity) lead to a broadening of T_m.
- The process of crystallization in polymers involves chain folding. This creates inherent defects in the resulting crystal. Consequently, the actual melting point is lower than the ideal thermodynamic melting point.
- Because of the macromolecular nature of polymers and the conformational changes associated with melting, the process of melting in polymer is more rate sensitive than that in simple molecules.
- No polymer is 100% crystalline.

The factors that determine crystallization tendency have been dealt with earlier (Chapter 3). We simply recap them here.

- Structural regularity — For the effective utilization of the secondary intermolecular bonding forces during the formation of a crystalline polymer, close alignment of polymer molecules is a prerequisite. Any structural feature of a polymer chain that impedes this condition must necessarily detract from crystallinity.
- Chain flexibility — In the process of aggregation to form a crystalline solid, polymer molecules are opposed by thermal agitation, which induces segmental rotational and vibrational motion. Polymers with flexible chains are more susceptible to this agitation than those with stiff backbones. Consequently, chain flexibility reduces the tendency for crystallization.
- Intermolecular bonding — Since secondary bonding forces are responsible for intermolecular bonding, polymer molecules with specific groups that promote enhanced intermolecular interaction and whose structural features lead to identity periods are more crystallizable.

As we said earlier, melting is a true first-order thermodynamic transition involving a phase change and is associated with discontinuities in primary extensive thermodynamic properties. In principle, therefore, any property whose values are different for the crystalline and amorphous states provides a convenient method for measuring the crystalline melting point. Methods for measuring the crystalline melting point include dilatometry, calorimetry, and thermal analysis; dynamic techniques (mechanical dielectric, nuclear magnetic resonance); stress relaxation; and creep.

A. FACTORS AFFECTING THE CRYSTALLINE MELTING POINT, T_M

Bearing in mind the peculiar nature of polymers, melting in crystalline polymers can be considered a pseudoequilibrium process that may be described by the free energy equation:

$$T_m = \frac{\Delta H_m}{\Delta S_m} \qquad (4.10)$$

In this case, ΔH_m represents the difference in cohesive energies between chains in the crystalline and liquid states, while ΔS_m represents the difference in the degree of order between polymer molecules in the two states. ΔH_m is generally independent of the molecular weight. But, as would be expected, polar groups on the chain — particularly if disposed regularly on the chain so as to encourage regions of extensive cooperative bonding — would enhance the magnitude of ΔH_m. ΔS_m depends not only on molecular weight, but also on structural factors like chain stiffness. Chains that are flexible in the molten state would be capable of assuming a relatively larger number of conformations than stiff chains and hence result in a large ΔS_m. We now discuss these factors that affect T_m in greater detail.

1. Intermolecular Bonding

The cohesive forces in polymers involve the secondary bonding forces ranging from the weak van der Waals forces through the much stronger hydrogen bonds. In some cases, these forces even include primary ionic bonds. Figure 4.8 shows the variation of T_m for a homologous series of various types of polymers. With polyethylene as a reference and neglecting for the moment possible fine details in trends, we observe that:

- The melting points approach that of polyethylene as the spacing between polar groups increases.
- For the same number of chain atoms in the repeat unit, polyureas, polyamides, and polyurethanes have higher melting points than polyethylene, while polyesters have lower.

As would be expected, the decrease in the cohesive energy density associated with the decrease in the density of sites for intermolecular bonding (increased space between polar groups) leads to a reduction in the melting points.

Van Krevelen and Hoftyzer[6] have calculated the contributions of the characteristic groups in various polymers to Y_m, a quantity they termed *molar melt transition function* (identified with ΔH_m in Equation 4.7). These are shown in Table 4.8. For the same number of chain backbone atoms, chain flexibility (ΔS_m) will not be significantly different for the various polymers. From Table 4.8, while the absolute values calculated for the characteristic interunit groups may not be significant, the trend in their magnitudes definitely corresponds to that of the melting points of the various types of polymers. In Figure 4.8, we note specifically that the melting points for polyesters are lower than the T_m of polyethylene.

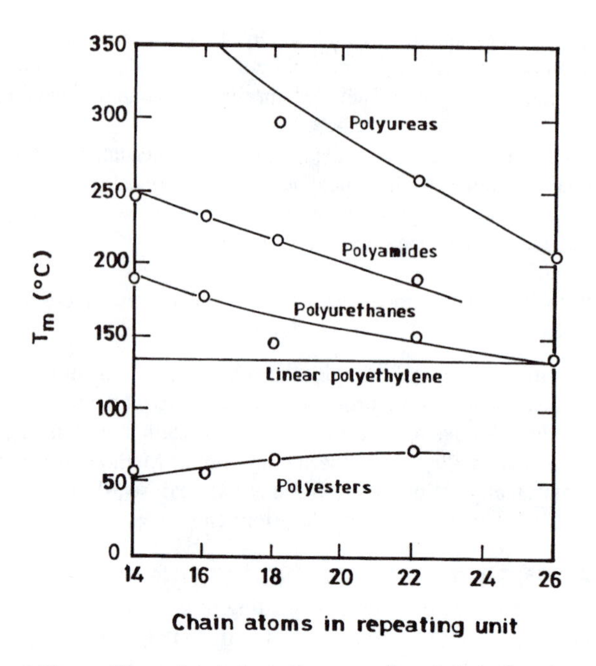

Figure 4.8 Trend of crystalline melting points in homologous series of aliphatic polymers. (From Hill, R.H. and Walker, E.E., *J. Polym. Sci.*, 3, 609, 1948. With permission.)

Table 4.8 Group Contributions to the Melting Point

Polymer	Characteristic Group	Contribution to Y_m
Polyester	—C(=O)—O—	1160
Polyamide	—C(=O)—N(H)—	2560
Polyurethane	—O—C(=O)—N(H)—	2430
Polyurea	—N(H)—C(=O)—N(H)—	3250

From Van Krevelen, D.W. and Hoftyzer, P.J., *Properties of Polymers: Correlations with Chemical Structure,* Elsevier, Amsterdam, 1976. With permission.

This is because the enhanced flexibility resulting from the presence of oxygen atoms in the polyester chains considerably offsets the weak polar bonding from ester linkages. This is demonstrated in Table 4.9. The melting points of the nylons reflect the density of the hydrogen-bond-forming amide linkages. The densities of the interunit linkages in polycaprolactone (ester units) and polycaprolactam (amide units) are the same. However, the amide units are more polar than the ester units. Consequently, polycaprolactam has a much higher T_m than polycaprolactone.

Table 4.9 Effect of Intermolecular Bonding on T_m

Polymer	Characteristic Group	Melting Temperature (T_m)(°C)
Polycaprolactone	$\left[O-(CH_2)_5-\overset{O}{\overset{\parallel}{C}} \right]_n$	61
Polycaprolactam (nylon 6)	$\left[\overset{O}{\overset{\parallel}{C}}-(CH_2)_5-\overset{H}{\overset{\vert}{N}} \right]_n$	226
Poly(hexamethylene adipamide) (nylon 6,6)	$\left[\overset{O}{\overset{\parallel}{C}}-(CH_2)_4-\overset{O}{\overset{\parallel}{C}}-\overset{H}{\overset{\vert}{N}}-(CH_2)_6-\overset{H}{\overset{\vert}{N}} \right]_n$	265
Nylon 12	$\left[\overset{O}{\overset{\parallel}{C}}-(CH_2)_{11}-\overset{H}{\overset{\vert}{N}} \right]_n$	179

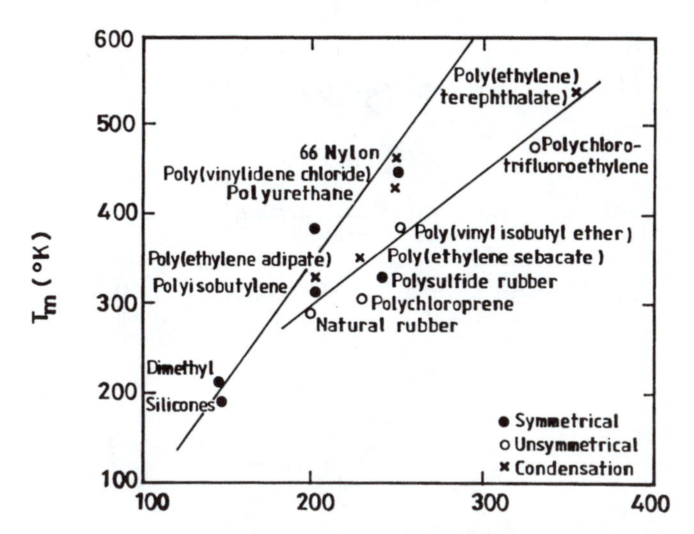

Figure 4.9 Relation between T_m and T_g for various polymers. (From Bayer, R.F., *J. Appl. Phys.*, 25, 585, 1954. With permission.)

2. Effect of Structure

The structural dependence of the crystalline melting temperature is essentially the same as that for the glass transition temperature. The only difference is the effect of structural regularity, which has a profound influence on crystallizability of a polymer. T_g is virtually unaffected by structural regularity. From a close examination of data for semicrystalline polymers it has been established that the ratio T_g/T_m (K) ranged from 0.5 to 0.75. The ratio is found to be closer to 0.5 in symmetrical polymers (e.g., polyethylene and polybutadiene) and closer to 0.75 in unsymmetrical polymers (e.g., polystyrene and polychloroprene). This behavior is shown in Figure 4.9.

3. Chain Flexibility

Polymers with rigid chains would be expected to have higher melting points than those with flexible molecules. This is because, on melting, polymers with stiff backbones have lower conformational entropy changes than those with flexible backbones. As we saw earlier, chain flexibility is enhanced by the presence of such groups as –O– and –(CO·O)– and by increasing the length of ($-CH_2-$) units in the main chain. Insertion of polar groups and rings restricts the rotation of the backbone and consequently reduces conformational changes of the backbone, as illustrated by the following polymers (Table 4.10).

Table 4.10 Effect of Chain Flexibility to T_m

Polymer	Repeat Unit	T_m(°C)
Polyethylene	$-CH_2-CH_2-$	135
Polypropylene	$-CH_2-CH(CH_3)-$	165
Polyethylene oxide	$-CH_2-CH_2-O-$	66
Poly(propylene oxide)	$-CH_2-CH(CH_3)-O-$	75
Poly(ethylene adipate)	$-O-CH_2-CH_2-O-C(=O)-(CH_2)_4-C(=O)-$	50
Poly(ethylene terephthalate)	$-O-CH_2CH_2-O-C(=O)-C_6H_4-C(=O)-$	265
Poly (diphenyl-4,-4 diethylene carboxylate)	$-O-CH_2CH_2-O-C(=O)-C_6H_4-C_6H_4-C(=O)-$	355
Polycarbonate	$-O-C_6H_4-C(CH_3)_2-C_6H_4-O-C(=O)-$	270
Poly(*p*-xylene)	$-CH_2-C_6H_4-CH_2-$	380
Polystyrene (isotactic)	$-CH_2-CH(C_6H_5)-$	240
Poly(*o*-methylstyrene)	$-CH_2-CH(C_6H_4\text{-}o\text{-}CH_3)-$	>360
Poly(*m*-methylstyrene)	$-CH_2-CH(C_6H_4\text{-}m\text{-}CH_3)-$	215

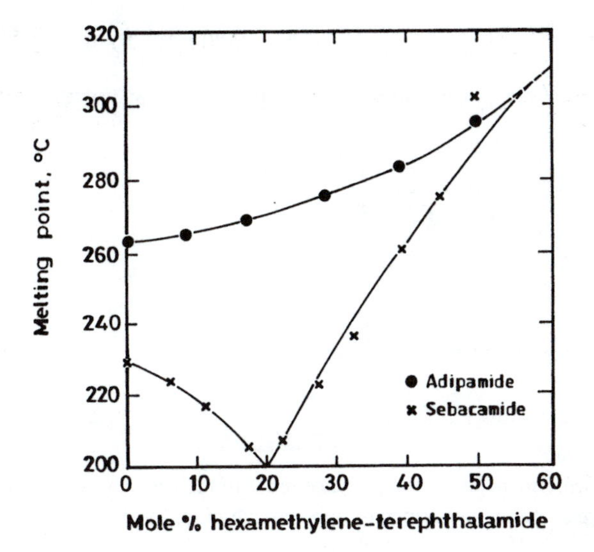

Figure 4.10 Melting points of copolymers of hexamethylene adipamide and terephthalamide, and of hexamethylene sebacamide and terephthalamide. (From Edgar, D.B. and Hill, R.J., *J. Polym. Sci.*, 8(1), 1, 1952. With permission.)

4. Copolymerization

The effect of copolymerization on T_m depends on the degree of compatibility of the comonomers. If the comonomers have similar specific volumes, they can replace each other in the crystal lattice (i.e., isomorphous systems), and the melting point will vary smoothly over the entire composition range. On the other hand, if the copolymer is made from monomers each of which forms a crystalline homopolymer, the degree of crystallinity and the crystalline melting point decrease as the second constituent is added to either of the homopolymers. In this case, the T_m of the copolymer (i.e., the reduction in the melting point, T_m° of the homopolymer due to the addition of the second constituent) is given by Equation 4.11.[8]

$$\frac{1}{T_m} = \frac{1}{T_m^\circ} - \frac{R}{\Delta H_m} \ln x \tag{4.11}$$

where ΔH_m and X are, respectively, the heat of fusion and mole fraction of the homopolymer or crystallizing (i.e., major) component (Figure 4.10). It is obvious from the foregoing that it is impossible to attempt to raise the crystalline melting point of a polymer by copolymerizing with small amounts of a monomer with a high melting point except for isomorphous systems, which are rare in vinyl polymers.

Block and graft copolymers with sufficiently long homopolymer chain sequences crystallize and exhibit properties of both homopolymers and consequently have two melting points, one for each type of chain segment.

Example 4.4: What is the melting point of a copolymer of ethylene and propylene with 90 mol% ethylene?

Solution: From the foregoing discussion:

$$\frac{1}{T_m} = \frac{1}{T_m^\circ} - \frac{R}{\Delta H_m} \ln x$$

$$T_m^\circ = \text{melting point of PE}$$

$$= 135°C = 408\ K$$

$$\Delta H_m \text{ for PE} = 66.5 \text{ cal/g} = 7.80 \times 10^3 \text{ J/mol}$$

$$\left(\text{Note molecular weight of PE}\left[\text{repeat unit}\right] = 28\right)$$

$$\frac{1}{T_m} = \frac{1}{408} - \frac{8.314}{7.80 \times 10^3} \ln 0.9$$

$$= 25.63 \times 10^{-4}$$

$$T_m = 390 \text{ K} = 117°\text{C}$$

IV. PROBLEMS

4.1. Arrange the following materials in the probable order of their increasing crystalline melting points and justify your answer. Assume the degree of polymerization, n, for the polymers is the same.

a. 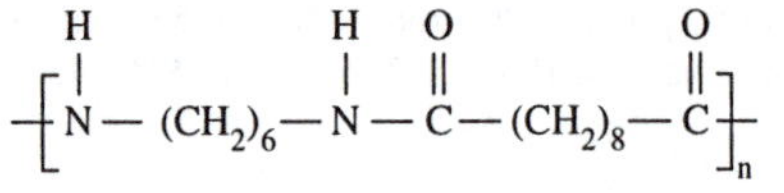 $\left[\text{N(H)}-(CH_2)_6-\text{N(H)}-\text{C(=O)}-(CH_2)_8-\text{C(=O)}\right]_n$

b. $\left[\text{N(H)}-(CH_2)_6-\text{N(H)}-\text{C(=O)}-(CH_2)_4-\text{C(=O)}\right]_n$

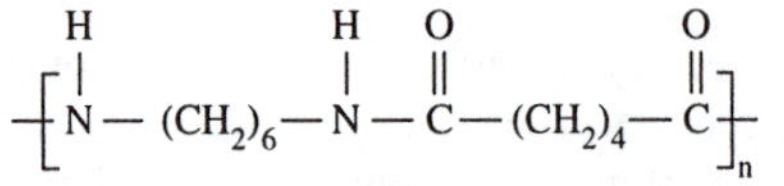

c. $\left[\text{N(H)}-(CH_2)_6-\text{N(H)}-\text{C(=O)}-\text{N(H)}-CH_2-C_6H_4-CH_2-\text{N(H)}-\text{C(=O)}\right]_n$

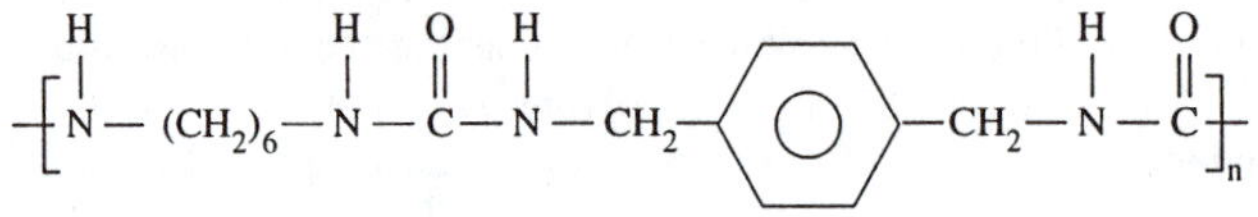

d. 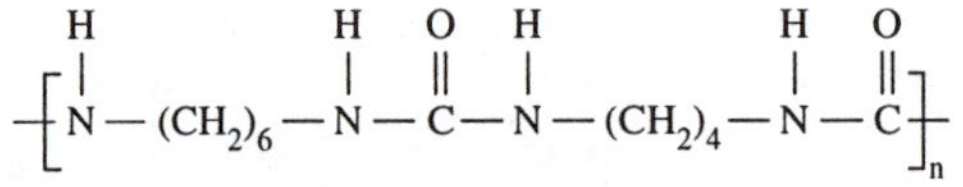 $\left[\text{N(H)}-(CH_2)_6-\text{N(H)}-\text{C(=O)}-\text{N(H)}-(CH_2)_4-\text{N(H)}-\text{C(=O)}\right]_n$

e. 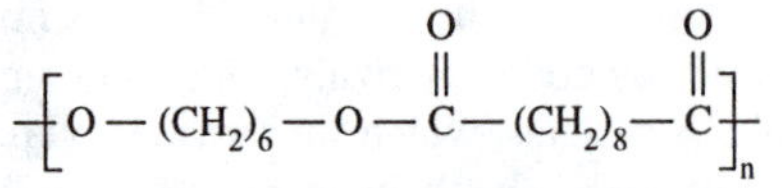 $\left[\text{O}-(CH_2)_6-\text{O}-\text{C(=O)}-(CH_2)_8-\text{C(=O)}\right]_n$

4.2. The solubility parameters of poly(vinyl chloride) (PVC) and dibutyl sebacate are 9.7 and 9.2, respectively. What amount (in volume percent) of dibutyl sebacate will be required to make PVC a flexible polymer at room temperature? Assume that the T_g of dibutyl sebacate is –100°C and that room temperature is 25°C.

4.3. Arrange the following linear polymers in orders of decreasing crystalline melting points. Explain the basis of your decision.

a. Poly(ethylene adipate)
b. Poly(ethylene terephthalate)
c. Poly(hexamethylene adipate)
d. Poly(ethylene adipamide)

4.4. Cross-linking polystyrene with divinyl benzene increased its T_g by 7.5°C. What is the number of styrene residues between cross-links?

4.5. The T_g of polystyrene is 100°C. What is its melting temperature, T_m?

4.6. The heat of fusion of the repeating unit for a homopolymer that melts at 250°C is 2500 cal/mol. Predict the melting point of a random copolymer of this polymer with 25 mol% of a comonomer.

4.7. Which of the following pairs of polymers will have a higher glass transition temperature, T_g? Explain your choice.

a. Poly(2-chloroethyl methacrylate) or poly(*n*-propyl methacrylate)
b. Poly(*n*-butyl methacrylate) or poly(2-methoxyethyl methacrylate)

4.8. What is the free volume fraction of polystyrene at 150°C if its volume coefficient of expansion is 60.0×10^{-6} cm/cm^3 °C?

4.9. Which of each of the following pairs has a higher T_μ? Why?

a. Polyethylene or a random copolymer of polyethylene and polypropylene
b. Poly(vinyl chloride) or polytetrafluoroethylene
c. Nylon 6 or nylon 11

REFERENCES

1. Gibbs, J.H. and DiMarzio, E.A., *J. Chem. Phys.* 28, 373, 1955; 28, 807, 1958.
2. Williams, M.L., Landell, R.F., and Ferry, J.A., *J. Am Chem. Soc.*, 77, 3701, 1955.
3. Tobolsky, A.V., *Properties and Structure of Polymers,* John Wiley & Sons, New York, 1960.
4. Schneider, K. and Wolf, K., *Kolloid Z.,* 127, 65, 1952.
5. Hill, R.H. and Walker, E.E., *J. Polym. Sci.,* 3, 609, 1948.
6. Van Krevelen, D.W. and Hoftyzer, P.J., *Properties of Polymers: Correlations with Chemical Structure,* Elsevier, Amsterdam, 1976.
7. Bayer, R.F., *J. Appl. Phys.,* 25, 585, 1954.
8. Flory, P.J., *Principles of a Polymer Chemistry,* Cornell University Press, Ithaca, New York, 1953, chap. 13.
9. Edgar, O.B. and Hill, R.J., *J. Polym. Sci.,* 8(1), 1, 1952.
10. Fried, J.R., *Plast. Eng.,* 38(7), 27, 1982.
11. Chruma, J.L. and Chapman, R.D., *Chem. Eng. Prog.,* 8, 49, 1985.
12. Billmeyer, F.W., Jr., *Textbook of Polymer Science,* 3rd ed., John Wiley & Sons, New York, 1984.
13. Williams, D.J., *Polymer Science and Engineering,* Prentice-Hall, Englewood Cliffs, NJ, 1971.
14. Kaufman, H.S. and Falcetta, J.J., eds., *Introduction to Polymer Science and Technology,* John Wiley & Sons, New York, 1977.

Chapter 5

Polymer Modification

I. INTRODUCTION

In Chapter 2, we discussed the various polymerization mechanisms. In principle, with the large number of monomers available, a virtually endless number of polymers can be obtained by various polymerization reactions. Quite frequently, a given monomer is converted into the corresponding homopolymer. However, in order to meet increasingly stringent and specific end-use requirements, novel polymeric materials have also been developed through modification of existing polymers. Development of tailor-made materials by polymer modification is usually less costly and can be achieved more readily than through synthesis of new polymers.

For a polymer to be useful, it must be able to function properly in a given application. The performance of a polymer is determined primarily by the composition and structure of the polymer molecule. These control the physical, chemical, and other characteristics of the polymer material. Therefore modification of the composition of the structural units represents one of the main approaches to the modification of polymer behavior. In addition to the chemical nature and composition of the structural units that constitute the polymer backbone, molecular architecture also contributes to the ultimate properties of polymeric products. Thus polymer modification can be accomplished by employing one or more of the following techniques:

- Copolymerization of more than one monomer
- Control of molecular architecture
- Postpolymerization polymer reactions involving functional/reactive groups introduced deliberately into the polymer main chain or side groups

The above property modification techniques are associated with the control of the chemical, composition, and structural nature of the polymer, which is effected largely during the polymerization process. However, few polymers are used technologically in their chemically pure form. Virtually all commercially available polymeric materials are a combination of one or more polymeric systems with various additives designed, with due consideration to cost factors, to produce an optimum property and/or process profile for specific applications. Modification of polymers through the use of chemical additives and reinforcements through alloying and blending procedures and by composite formation is discussed in Chapter 9. In this chapter, we restrict our discussion to polymer modification techniques associated with chemical phenomena brought about either during or after polymerization.

II. COPOLYMERIZATION

Macromolecular design and architecture through copolymerization of monomers has led to a number of commercially important polymers. Copolymer composition can be varied over wide limits, resulting in a wide range of property/process performance. A copolymer may be composed of comparable amounts of the constituent monomers. The properties of the resulting copolymer will be substantially different from those of the parent homopolymers. On the other hand, the copolymer may contain only a very small amount of one of the monomers. In this case, the gross physical characteristics of the copolymer probably approximate those of the homopolymer of the major constituent, while the minor constituent confers specific chemical properties on the copolymer. We now illustrate these principles by discussing some examples.

A. STYRENE–BUTADIENE COPOLYMERS

Polybutadiene is an elastomeric material with good elastic properties and outstanding toughness and resilience. However, it has relatively poor resistance to oils, solvents, oxidation, and abrasion. Polystyrene, on the other hand, is relatively chemically inert and is quite resistant to alkalis, halide acids, and oxidizing and reducing agents. It is also very easy to process. But then, polystyrene is quite brittle with a low heat-deflection temperature (82 to 88°C). Styrene and butadiene copolymers provide a good illustration of the considerable latitude in the variation of polymer properties that can be achieved by a careful manipulation of the composition of the copolymer and the distribution of these components. Styrene and butadiene can be copolymerized to produce either random, graft, or block copolymers. Styrene–butadiene random copolymer exhibits a homogeneous single phase and has properties intermediate to those of the parent homopolymer. On the other hand, block or graft copolymers of styrene and butadiene form heterogeneous multiphase systems whose properties are not simply characteristic of each homopolymer, but are dictated by the multiphase character of the copolymer.

Most of the defects of polybutadiene homopolymer can be overcome by the incorporation of 28% styrene into the copolymer. This is due essentially to the great rigidity and other beneficial properties of the styrene molecule. The enhanced properties of SBR and its ease of processability makes SBR the preferred material over natural rubber in such applications as belting, hose, and molded goods and in vulcanized sheet and flooring. Rubber shoe soles are made almost exclusively from SBR. Copolymers containing about 25% styrene are also useful adhesives, especially in the form of aqueous dispersion or in solution. If the ratio of styrene to butadiene is in the range 60:40 and higher, the copolymer is nontacky and is used in hot-melt adhesives and latex paints. For example, emulsion copolymers composed of 74% styrene and 25% butadiene (by weight) find wide applications in paints. In those applications, the hardness of polystyrene is partially retained, but its brittleness is modified by the presence of butadiene.

1. Styrene–Butadiene Rubber (SBR) (Random Copolymer)

SBR is produced by the free-radical polymerization of styrene and butadiene, and, as such, the resulting copolymer is necessarily random. Also, the structure is irregular. Consequently, SBR is not crystallizable. Butadiene can undergo either 1,2 or 1,4 polymerization. The structure of the resulting SBR copolymer is shown in Figure 5.1.

Commercial SBR is produced by either emulsion or solution copolymerization of butadiene and styrene. Emulsion copolymerization is either a cold (41°F) or hot (122°F) process. The copolymers from the hot and cold processes have principal differences in molecular weight, molecular weight distribution, and microstructure, as shown in Table 5.1. The solution copolymerization process for the production of

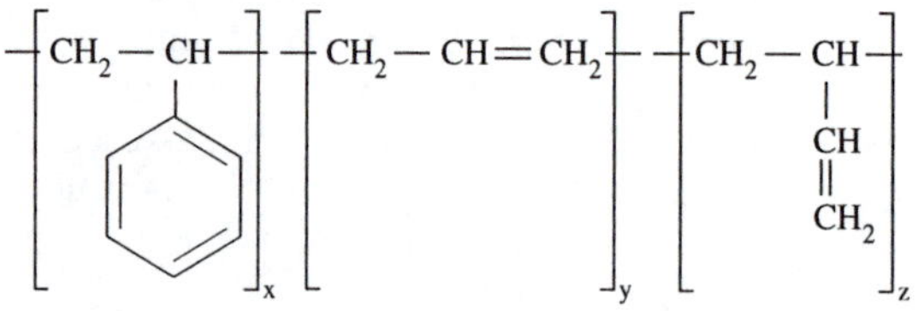

$$\left[CH_2-CH(C_6H_5) \right]_x \left[CH_2-CH=CH_2 \right]_y \left[CH_2-CH(CH=CH_2) \right]_z$$

Figure 5.1 Structure of styrene–butadiene rubbers.

Table 5.1 Property Differences between Hot and Cold SBR

Property	Hot	Cold
Molecular weight		
Viscosity average, $\overline{M}_v$	150–400,000	280,000
Weight average, $\overline{M}_w$	250–450,000	500,000
Number average, $\overline{M}_n$	30–100,000	110–260,000
Microstructure (%)		
1,4 (*cis*)	15	18
1,4 (*trans*)	58	69
1,2 (vinyl)	27	23

From Stricharczuk, P.T. and Wright, D.E., *Handbook of Adhesives,* 2nd ed., Skeist, I., Ed., Van Nostrand Reinhold, New York, 1977. With permission.

SBR involves the use of alkyllithium catalysts. Solution SBR generally has a higher molecular weight, narrower molecular weight distribution, and higher *cis*-diene content than emulsion SBR.

Example 5.1: Generally, hot SBR is better suited to adhesive formulation than cold SBR. Explain.

Solution: From Table 5.1, hot SBR has a lower molecular weight and a broader molecular weight distribution than cold SBR. The lower molecular weight fraction provides "quick stick," while the higher molecular weight fraction provides shear strength. On the other hand, narrow molecular weight distribution and higher *trans*-structure of cold SBR make it less adaptable to adhesives.

Example 5.2: Explain the following observations:

a. Tire treads made from solution SBR are superior to those from emulsion SBR.
b. The ozone resistance of SBR is superior to that of natural rubber, but when cracks start in SBR they grow much more rapidly.

Solution:

a. Solution SBR has a higher molecular weight, narrower molecular weight distribution, and higher *cis*-1,4-polybutadiene content than emulsion SBR. These qualities reduce tread wear and enhance crack resistance of the tire.
b. The presence of styrene in SBR reduces the amount of unsaturation relative to natural rubber. Consequently, the ozone resistance of SBR is higher than that of natural rubber. Natural rubber has a *cis*-1,4 configuration, which makes it a tough material resistant to crack growth. SBR is mainly *trans*-1,4, which is less impact resistant.

---

2. Styrene–Butadiene Block Copolymers

Styrene–butadiene block copolymers belong to a new class of polymers called thermoplastic elastomers (TPE). Products made from these polymers have properties similar to those of vulcanized rubbers, but they are made from equipment used for fabricating thermoplastic polymers. Vulcanization is a slow and energy-intensive thermosetting process. In contrast, the processing of thermoplastic elastomers is rapid and involves cooling the melt into a rubberlike solid. In addition, like true thermoplastics, scrap from TPE can be recycled.

Styrene–butadiene block copolymer belongs to the A-B-A type thermoplastic elastomer. The principal structure of this type of polymer involves the thermoplastic rubber molecules terminated by the hard, glassy end blocks. The A and B copolymer block segments are incompatible and, consequently, separate spontaneously into two phases. Thus in the solid state, the styrene–butadiene (S-B-S) thermoplastic elastomer has two phases: a continuous polybutadiene rubber phase and the dispersed glassy domains of polystyrene. The styrene plastic end blocks, called domains, act as cross-links locking the rubber phase in place.

In commercial thermoplastic S-B-S rubber, the end-block phase is present in a smaller proportion with a styrene-to-butadiene (end-block-to-midblock) ratio in the range 15:85 to 40:60 on weight basis. The useful temperature range of S-B-S copolymer lies between the T_g of polybutadiene and polystyrene. Below the T_g of polybutadiene, the elastomeric midblocks become hard and brittle. Above the T_g of polystyrene, the domains soften and cease to act as cross-links for the soft midblocks. Between the T_g of both homopolymers, however, the hard styrene domains prevent the flow of the soft elastomeric butadiene midsegments through a network similar to vulcanized rubber. Therefore, within normal use temperature, S-B-S block copolymer retains the thermoplasticity of styrene blocks and the toughness and resilience of the elastomer units.

B. ETHYLENE COPOLYMERS

Low-density polyethylene (LDPE) is produced under high pressures and temperatures. It finds applications in film and sheeting uses and in injection-molded products such as insulated wires and cables. Its physical properties are dictated by three structural variables: density, molecular weight, and molecular weight distribution. As density increases, barrier properties, hardness, abrasion, heat, and chemical resistance, strength, and surface gloss increase. On the other hand, decreasing density results in enhanced

Table 5.2 Some Ethylene Copolymers

R	Copolymer
$-\overset{O}{\overset{\Vert}{C}}-O-Me$	Ethylene–methyl acrylate (EMA)
$-\overset{O}{\overset{\Vert}{C}}-O-Et$	Ethylene–ethyl acrylate (EEA)
$-O-\overset{O}{\overset{\Vert}{C}}-Me$	Ethylene–vinyl acetate (EVAc)

toughness, stress-crack resistance, clarity, flexibility, and elongation and in reduced creep and mold shrinkage. Melt index (MI) is a measure of molecular weight; it decreases with increasing molecular weight. Increasing MI improves clarity, surface gloss, and mold shrinkage. Decreasing MI leads to improved creep and heat resistance, toughness, melt strength, and stress-cracking. Narrower molecular weight distribution gives better impact strength, but reduced resin processability. A broader molecular weight distribution is more shear sensitive and, consequently, leads to shear viscosity at high shear rates and is thus easier to process.

By copolymerizing ethylene with polar α-olefins, it is possible to produce a variety of materials ranging from rubbery to low melting products — suitable for hot-melt adhesives — to those that demonstrate unusual toughness and flexibility. This class of copolymers can be represented by the general formula

$$\left[CH_2 - CH_2 \right]_x \left[CH_2 - \underset{\displaystyle R}{\underset{|}{CH}} \right]_y$$

where R is a polar group shown in Table 5.2. The introduction of comonomers with a polar pendant group, R, produces a highly branched random copolymer but with increased interchain interaction. Thus, relative to the homopolymer, these copolymers have enhanced flexibility, toughness, stress-cracking resistance, oil and grease resistance, clarity, and weatherability. Some of these are illustrated for EVAc in Table 5.3. EEA, for example, also has the flexibility of plasticized vinyl without the thermal instability and plasticizer migration problems associated with PVC. In general, the range of properties of the copolymer can be varied by varying the proportion and molecular weight of the comonomer.

C. ACRYLONITRILE-BUTADIENE-STYRENE (ABS)

ABS is the generic name of a family of engineering thermoplastics produced by a combination of three monomers: acrylonitrile, butadiene, and styrene. The overall property balance of the terpolymer is a

Table 5.3 Properties of Rotational-Molding Polyethylene Resins

Property	EVAc Copolymer	LDPE
Melt index, g/10 min	1.5	1.5
Density, g/cm^3	0.933	0.925
Tensile strength, psi	1600	1450
Elongation, %	500	450
Flexural modulus, psi	50,000	35,000
Environmental stress-crack resistance, h	50	20
Tensile impact strength, ft-lb/in.^{2}	32	25
Dart impact strength at –20°F, ft-lb	32	20

From Wake, W.C., *Adhesion and the Formulation of Adhesives,* Applied Science Publishers, London, 1976.

result of the contribution of the unique characteristics of each monomer. Polymer chemical resistance and heat and aging stability depend on acrylonitrile, while its toughness, impact resistance, and property retention at low temperature are developed through butadiene. Copolymer rigidity, glossy surface appearance, and ease of processability are contributions from styrene. The terpolymer properties are controlled by manipulation of the ratio and distribution of the three components.

ABS resins consist essentially of two phases: a rubbery phase dispersed in a continuous glassy matrix of styrene–acrylonitrile copolymer (SAN) through a boundary layer of SAN graft. The dispersed rubbery phase is rubber polymerized from butadiene. Styrene and acrylonitrile are graft-polymerized to the rubber thus forming the boundary layer between the dispersed rubber phase and the continuous glassy matrix. Increased molecular weight of SAN improves product strength and ease of processability, while the concentration, size, and distribution of the rubber particles influence product toughness and impact strength. By a careful variation of the parameters controlling the phases, a family of ABS with a broad range of properties has been developed.

D. CONDENSATION POLYMERS

A large number of commercially important condensation polymers are employed as homopolymers. These include those polymers that depend on crystallinity for their major applications, such as nylons and fiber-forming polyesters, and the bulk of such important thermosetting materials like phenolics and urea–formaldehyde resins. In many applications, condensation polymers are used as copolymers. For example, fast-setting phenolic adhesives are resorcinol-modified, while melamine has sometimes been incorporated into the urea–formaldehyde resin structure to enhance its stability. Copolyesters find application in a fairly broad spectrum of end uses.

The diverse end-use pattern of copolyesters derives from the rather broad structural modifications that are in general possible with condensation polymers. Addition polymers consist of a chain of carbon atoms on their backbones. Modification of these polymers is limited essentially to the manipulation of the pendant group. On the other hand, for condensation polymers, in addition to the modification of the pendant structure, there is the benefit of the additional possible structural alteration of the backbone. In the case of polyesters and copolyesters a variety of monomers offers this possibility: aromatic, alicyclic, and unsaturated diols and dibasic acids. As discussed in Chapter 4, the presence of aromatic or alicyclic rings on the chain backbone enhances chain stiffness, while the presence of oxygen facilitates chain rotation and consequently increases chain flexibility. Also, by a suitable choice of monomers, the spacing between the polar ester groups and hence the cohesive energy density can be controlled. The numerous and varied applications of copolyesters are a consequence of the literally infinite combinations of dibasic acids and diols that are possible. Some of these applications include surface coatings, polyurethane polyester–polyether thermoplastic elastomers, laminating resins, and thermosetting molding compounds.

Most polymers that function properly at ambient temperature quite frequently have limited performance at sustained elevated temperatures. This invariably limits the utility of polymeric materials. The low thermal stability is generally due to decreased crystallinity and/or thermal decomposition. Polymer chemists have, through some ingenious ways, synthesized polymer — such as aromatic polyimides and the so-called ladder polymers — specifically designed for high-temperature applications. However, it has also been possible to modify polymers to improve their thermal stability and hence extend their range of utility. A few examples of condensation polymers illustrate this point.

1. Acetal Copolymer

The elemental composition of the backbone of a polymer determines its thermal stability. Most olefin polymers with carbon–carbon backbone degrade thermally by random chain scission. They are usually difficult to stabilize thermally through copolymerization. On the other hand, heterochain addition polymers degrade thermally by depolymerization or successive unzipping of monomers. If a comonomer is incorporated into the polymer backbone to interrupt the unzipping action, improved thermal stability of the polymer can be achieved. This principle is illustrated by the thermal stabilization of acetal polymers. Acetal homopolymer or polyoxymethylene is a formaldehyde polymer that is characterized by a good balance of physical, mechanical, chemical, and electrical properties. However, the polymer has hemiacetal hydroxy end groups that render the polymer thermally unstable. One way of stabilizing the homopolymer is by the use of appropriate additives. Another approach is to transform the unstable end groups into stable end groups by "end capping" through acetylation or methylation. The polymer reaction is carried

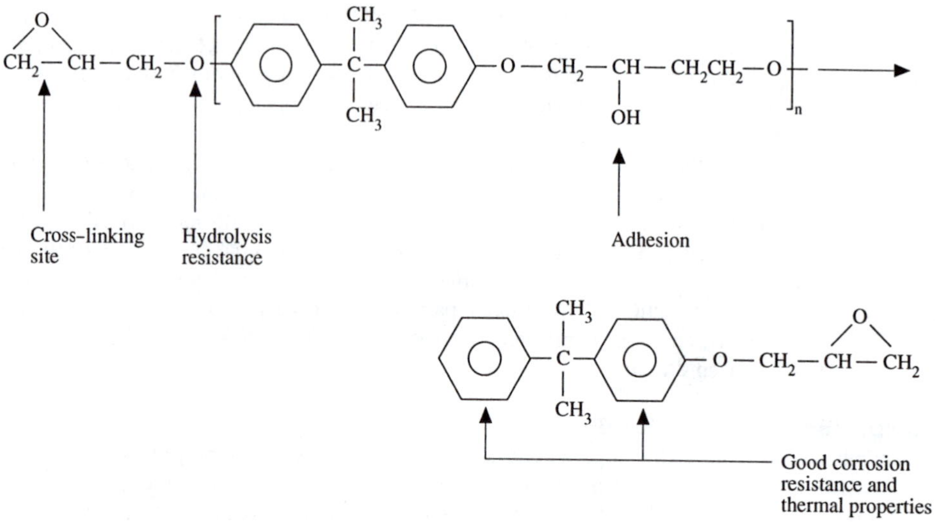

Cross-linking site
Hydrolysis resistance
Adhesion
Good corrosion resistance and thermal properties

Figure 5.2 Diglycidyl ether of bisphenol A (DGEBA).

out in the solid state to produce a polymer that is sufficiently thermally stable to be fabricated by melt processing.

In contrast to acetal homopolymer, acetal copolymers have built-in heat stabilization. They are prepared by copolymerization of trioxane with small amounts of comonomer, usually cyclic ethers like ethylene oxide or 1,3-diozolane.

$$(CH_2O)_3 + \underset{\text{O}}{CH_2{-}CH_2} \longrightarrow -CH_2O-CH_2O-CH_2O-CH_2CH_2O- \quad \text{(Str. 1)}$$

This results in the random distribution of C–C bonds in the polymer chain. The depolymerization of ethylene oxide units is much more difficult than that of oxymethylene units. Thus copolymerization confers thermal stability on the acetal copolymer. The copolymer exhibits good property retention when exposed to hot air at temperatures up to 220°F or water as hot as 180°F for long periods of time. For intermittent use, higher temperatures can be tolerated.

2. Epoxies

Epoxies are polymer materials characterized by the presence of reactive terminal epoxide groups. A frequently used epoxy resin is diglycidyl ether of bisphenol A (DGEBA) (Figure 5.2). Epoxy resins, because of their versatility, are used in a variety of applications in protective coatings, adhesives, laminates, and reinforced plastics and in electrical and electronic devices. Epoxy resins, however, exhibit lower heat resistance than phenolics due to the lower aromatic units in their structures. A combination of the desirable characteristics of phenolics and epoxies is obtained in epoxy–novolak, which is a type of multifunctional epoxy resin based on the modification of epoxy resin with novolak phenolics (Figure 5.3). In this system, the phenolic component confers thermal stability, while the epoxide group provides sites for cross-linking. The strength–temperature profiles for three unmodified epoxies are shown in Figure 5.4A. It can be observed that none of these adhesives shows any appreciable strength retention beyond 100°C. However, the phenolic-modified epoxy adhesive system exhibits good strength retention even up to 300°C (Figure 5.4B).

3. Urea–Formaldehyde (UF) Resins

Another example of the improvement of properties of condensation polymers through copolymerization is illustrated by recent work with urea–formaldehyde (UF) resins.[3,4] The preponderance of wood adhesive

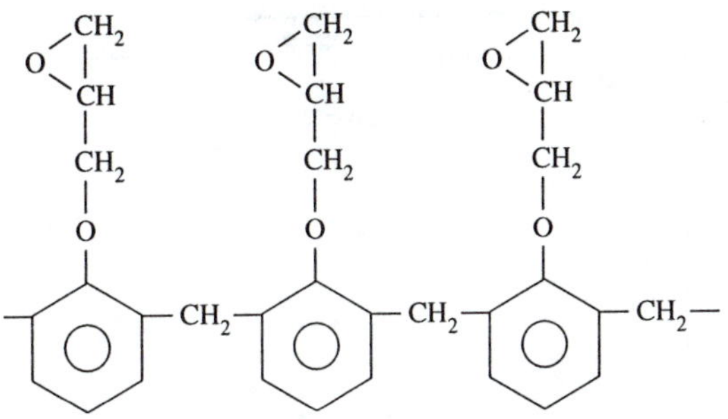

Figure 5.3 Structure of novolak-modified epoxy resin.

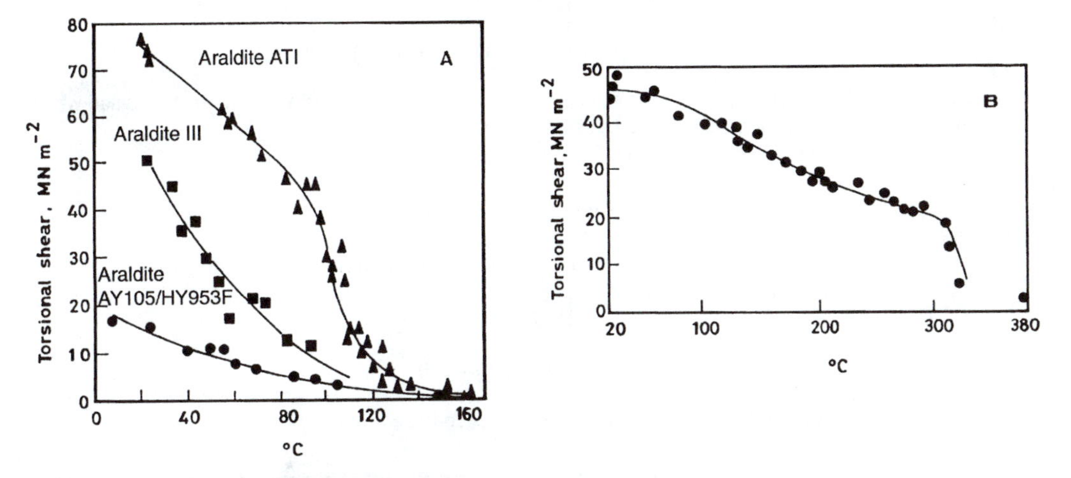

Figure 5.4 Comparison of temperature dependence of adhesive strength. (A) Three Araldite epoxy adhesives; (B) Epoxy-phenolic copolymer. (From Wake, W.C., *Adhesion and the Formulation of Adhesives,* Applied Science Publishers, London, 1976.)

bonding is done with urea–formaldehyde and phenol–formaldehyde (PF) adhesives. Compared to PF resins, bonding with UF resins is cheaper and can be carried out under a wider variety of conditions. However, PF-bonded wood products can be employed in outdoor structural applications, while the poor stability of UF-bounded wood products limits their use to interior, nonstructural applications. Evidence of these limitations includes strength losses of UF-bonded solid wood joints, irreversible swelling of UF-bonded composite panels, and formaldehyde release. Two bond degradation processes viewed currently as responsible for the poor durability of UF are hydrolytic scission and stress rupture. Some of the molecular structural factors that contribute to these processes are (1) low and nonuniform distribution of cross-links in cured UF resins, which lead to extreme sensitivity to small losses in cross-link density arising from hydrolytic scission or stress rupture, and (2) the brittleness of the cured resin due partly to the inherent rotational stiffness of urea molecules leading to inability to respond reversibly to stresses arising from cure shrinkage and moisture-related wood swelling and shrinking. To minimize these defects, urea derivatives of flexible di- and trifunctional amines were incorporated into UF resin structure through copolymerization. Figure 5.5 shows the improved response to cyclic wet–dry stress of joints bonded with a modified UF resin compared with joints bonded with unmodified resin. The amine modifier employed in this case was the urea derivative of propylene oxide-based triamine with the structure shown in Figure 5.6.

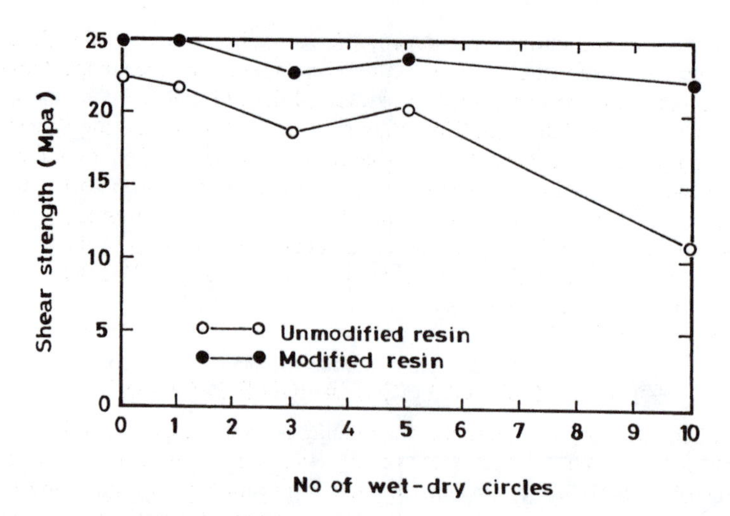

Figure 5.5 Enhanced soak–dry stress resistance of solid wood joints bonded with UF-resin modified with the urea derivative of a flexible propylene oxide-based triamine.

$$
\begin{array}{c}
\qquad\qquad\qquad CH_2-[O-CH_2CH(CH_3)]_x\ NH_2 \\
\qquad\quad | \\
CH_3-CH_2-C-CH_2-[-O-CH_2CH(CH_3)]_y\ NH_2 \\
\qquad\quad | \\
\qquad\qquad\qquad CH_2-[O-CH_2CH(CH_3)]_z\ NH_2
\end{array}
$$

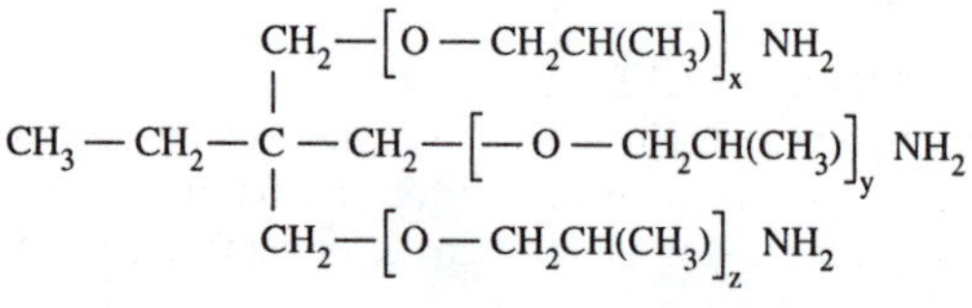

$$(x + y + z = 5.3)$$

Figure 5.6 Structure of propylene oxide-based triamine modifier.

Example 5.3: Internal stresses are developed as a result of shrinkage during adhesive cure. These stresses can be determined by measuring the deflection of an aluminum strip coated on one side with a thin film of the adhesive (Figure E5.3). The maximum in the curve for the unmodified resin was found to coincide with the fragmentation of the adhesive film. Explain the observed deflections under ambient cure of two such aluminum strips coated, respectively, with a modified and an unmodified UF resin. The modifier is the urea derivative of propylene oxide-based triamine shown in Figure 5.6.

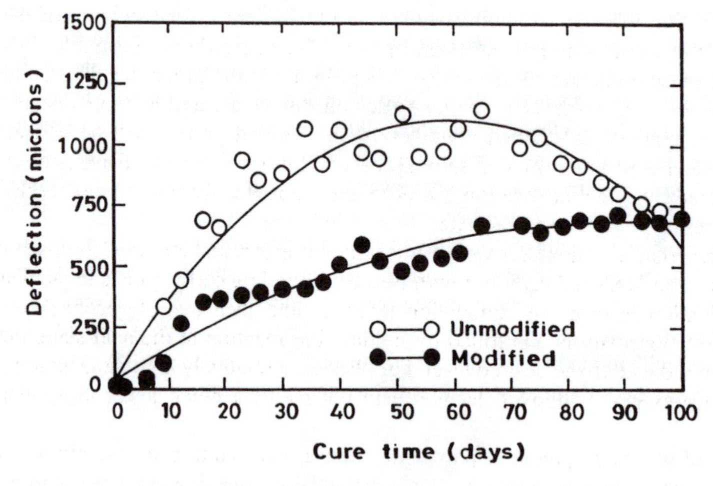

Figure E5.3 Relative internal stresses generated during cure by unmodified and modified UF adhesives.

Solution: As a result of the intrinsic rotational stiffness of the urea molecule, UF resins develop a considerable amount of internal stresses during cure. These stresses are seen to build up with time under ambient cure. However, for the unmodified resin after 60 days of ambient cure, the magnitude of the stresses resulted in a breakup of the resin. This relieved the internal stress, and consequently the deflection of the aluminum strip began to decline. On the other hand, with the modified resin, the flexibility of the modifying agent incorporated into the resin cure. Thus the aluminum strip coated with the modified resin structure provides an internal mechanism for relieving some of the stresses generated during resin showed much smaller deflection for the same duration of cure. Apparently even after 100 days of ambient cure, the internal stress generated in the modified resin had not reached a level where it became greater than the cohesive strength of the cured resin.

III. POSTPOLYMERIZATION REACTIONS

As indicated earlier, another powerful tool for upgrading polymer properties is the postpolymerization reaction of preformed polymers. These reactions may occur on reactive sites dispersed in the polymer main chain. Such reactions include chain extensions, cross-linking, and graft and block copolymer formation. The reactions may also occur on reactive sites attached directly or via other groups/chains to the polymer backbone. Reactions of this type are halogenation, sulfonation, hydrolysis, epoxidation, surface, and other miscellaneous reactions of polymers. In both cases these types of reactions transform existing polymers into those with new and/or improved properties.

A. REACTIONS OF POLYSACCHARIDES

Even though the treatment of polysaccharide chemistry falls outside the scope of this volume, reactions of the polysaccharide cellulose provide an excellent illustration of how the modification of a polymer, in this case a natural polymer, can transform an otherwise intractable material into one that can be readily fabricated.

1. Cellulose Derivatives

Cellulose is a polysaccharide — a natural polymer composed of glucosidic rings linked through oxygen bridges (Figure 5.7). The repeating unit has three hydroxyl groups and an acetal linkage. The cyclic nature of the repeating units results in stiff molecules. In addition, the β-(1→4) links between the anhydro-D-glucose units confer linearity on cellulose molecules. These factors coupled with the strong hydrogen-bond formation through the hydroxyl groups make cellulose a highly crystalline polymer. As a result of the extremely high interchain bonding between cellulose molecules, cellulose is insoluble and infusible. It burns rather than melts. To make cellulose processable it is necessary to reduce its melting point below its decomposition temperature. This is achieved by derivatization, which essentially breaks down the intrinsic hydrogen bonding while also interfering with the crystalline nature of the polymer.

In the preparation of cellulose derivatives a controlled degree of substitution of the three hydroxyls is necessary. Complete reaction of the three hydroxyls is generally not desirable. In addition, the fine structure of cellulose dictates both the course of reaction and the properties of the end products. When reacting with cellulose, reagents normally attack the most accessible (noncrystalline) regions first. Therefore, if the reaction is stopped prematurely, the reacted groups will be concentrated in certain regions rather than distributed randomly within the cellulose structure. For example, 30% acetylated cellulose prepared by direct acetylation has different properties from that prepared by 70% hydrolysis of the acetyl groups in cellulose triacetate.

The most important cellulose derivatives are cellulose esters and ethers. Cellulose esters are prepared by the reaction of an activated cellulose with the corresponding carboxylic acid, acid anhydride, or acid halide. Esterification is taken to completion (triester) and then hydrolyzed back to the desired free hydroxyl content. Viscosity is controlled by holding the reaction at the acid stage until the molecular weight has been reduced to the desired level. For plastics, a relatively high molecular weight is desirable, while for applications as adhesives, lacquer, and hot melts, a much lower molecular weight is more suitable.

The history of the development of polymers is intimately tied to cellulose nitrate, also referred to as nitrocellulose. This inorganic cellulose ester is obtained by reacting cellulose with a mixture of nitric acid and sulfuric acid for about 30 min.

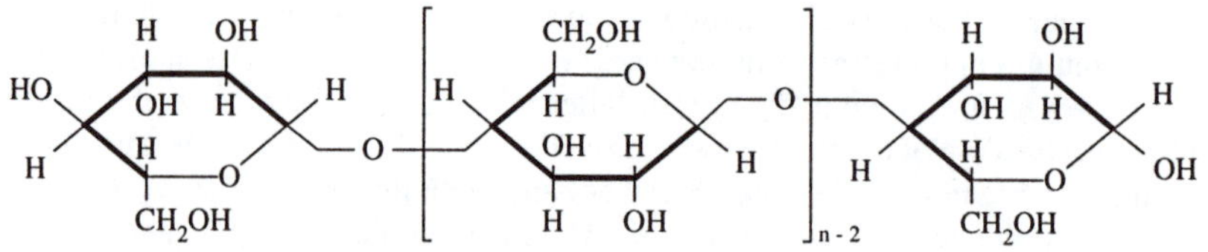

Figure 5.7 Structure of cellulose.

$$\text{Cell–OH} + \text{HNO}_3 \rightleftarrows \text{Cell–O–NO}_2 + \text{H}_2\text{O} \qquad \text{(Str. 2)}$$

Some degradation accompanies the nitration process. Products with different nitrate contents are obtained. Fully nitrated products have a nitrogen content of about 14.8%. The properties and applications of the nitrated material depend on its degree of nitration; plastics and lacquer-grade material contain 10.5 to 12% nitrogen, which corresponds to the dinitrate roughly. Products with nitrogen content greater than 12.5% are used for explosives.

Cellulose ethers like methyl cellulose are prepared by reacting sodium cellulose (obtained by the action of aqueous sodium hydroxide on cellulose) with the corresponding alkyl halide.

$$\text{CellOH} + \text{RCl} + \text{NaOH} \rightarrow \text{Cell–O–R} + \text{NaCl} + \text{H}_2\text{O} \qquad \text{(Str. 3)}$$

The ether content and the viscosity of the product are controlled by controlling the reaction conditions. In some cases, cellulose ethers are made by reacting sodium cellulose with ethylene or propylene oxide to yield the corresponding hydroxyalkyl ether. Such hydroxy-terminated cellulose ethers are of considerable interest.

Ethyl cellulose is the most important of the cellulose ethers. Commercial ethyl cellulose, which has about 2.4 to 2.5 ethoxy groups per glucose residue is a molding material that is heat stable and has low flammability and high impact strength. It is flexible and tough even at low temperatures; it has a relatively high water absorption, which is, however, lower than that of cellulose acetate.

2. Starch and Dextrins

Like cellulose, starch is also a polysaccharide that on complete hydrolysis yields glucose units. However, there are two significant differences between starch and cellulose. Unlike those in cellulose, the anhydro-D-glucose units in starch are linked through α-(1 $\rightarrow$ 4) glycosidic bonds. Also, in contrast to the completely linear structure of cellulose molecules, the structure of starch is a mixture of linear amylose molecules and branched amylopectin chains. Granular starch has no adhesive use. Therefore for adhesive and most applications, starch must be dispersed in solution by cooking in water until the granules swell and rupture. Such aqueous dispersions of starch are extremely viscous. In order to reduce the viscosity of such starch dispersions, starch must be modified through processes that involve partial depolymerization and/or rearrangement of molecules. To effect this conversion, starch must be subjected to hydrolytic, oxidative, or thermal processes, which yield three classes of products: acid-converted (thin boiling) starches; oxidized starches; and dextrins.

Conversion processes cause weakening and/or solubilization of starch granules, resulting in products that are more readily dispersible and of lower viscosities. Thin boiling starches are produced by reactions of granular starch with warm mineral acids such as HCl or H_2SO_4. Oxidized starches are made by the reaction of starch with sodium hypochlorite. These reactions introduce carbonyl and carboxyl groups and cleave the glycosidic linkages. Dextrins are degradation products of starch produced by heating starch in the presence or absence of hydrolytic agents. Depending on the conditions of conversion, three types of dextrins are produced: white dextrins, yellow (canary) dextrins, and British gums. The conversion mechanism is complex, but is thought to involve hydrolytic breakdown of starch molecules into smaller fragments followed by their rearrangement/repolymerization into a branched polymer structure (Figure 5.8).

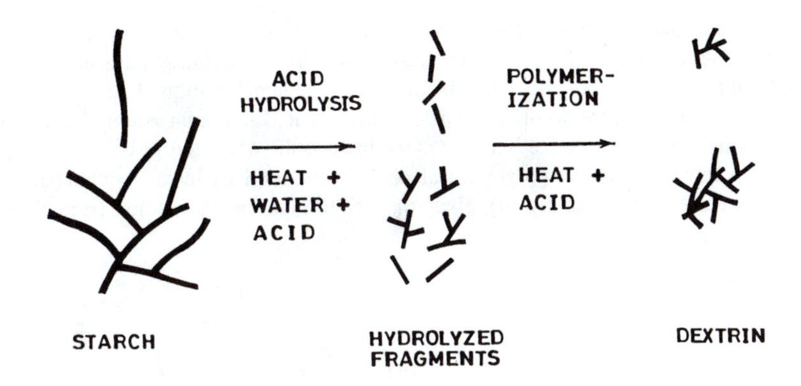

Figure 5.8 Hydrolysis and repolymerization during dextrination of starch. (From Jarowenko, W., *Handbook of Adhesives,* 2nd ed., Skeist, I., Ed., Van Nostrand Reinhold, New York, 1977. With permission.)

Example 5.4: Explain the following observations:

a. Cellulose acetate-molded and extruded articles are used extensively where extreme moisture resistance is not required. On the other hand, plastic sheets and films made from cellulose propionate, cellulose butyrate, or cellulose-acetate-butyrate are hydrophobic.
b. Products from cellulose ethers with the same degree of substitution as cellulose ester have similar properties but are more resistant to hydrolysis.

Solution:

a. Cellulose acetate is an ester formed by the reaction of acetic acid with the polyhydric alcohol cellulose. Esterification reactions are typically equilibrium reactions.

$$\text{Cell–OH} + \text{R–COOH} \rightleftarrows \text{Cell–O} - \text{CO} - \text{R} + H_2O \qquad \text{(Str. 4)}$$

In the presence of water (moisture) the reverse reaction can occur and is indeed favored at ordinary conditions for organic acids. Consequently, cellulose esters from organic acids like acetic acid and higher homolog are prepared only by the removal of water as it is formed. The resulting product is moisture sensitive, the degree of which decreases with the progressive hydrocarbon nature of R. Therefore while cellulose acetate ($R = –CH_3$) is susceptible to hydrolysis, cellulose propionate ($R = –CH_2–CH_2–CH_3$) and cellulose butyrate [$R = –(CH_2)_3–CH_3$] are hydrophobic.

b. The formation of cellulose ethers, unlike that of cellulose esters, is not reversible. Cellulose ethers are therefore less sensitive to hydrolysis than cellulose esters.

$$\text{Cell} - \text{OH} + \text{RCl} + \text{NaOH} \rightarrow \text{Cell} - \text{O} - \text{R} + \text{NaCl} + H_2O \qquad \text{(Str. 5)}$$

B. CROSS-LINKING

In Chapter 2 we indicated that the formation of a polymer requires that the functionality of the reacting monomer(s) must be at least 2. Where the functionality of one of the monomers is greater than 2, then a cross-linked polymer is formed. Thermosets like phenol–formaldehyde, urea–formaldehyde, and epoxy resins develop their characteristic properties through cross-linking. In this section our discussion is confined to those polymeric systems designed with latent cross-linkability that under appropriate conditions can be activated to produce a polymer with desirable properties.

1. Unsaturated Polyesters

Unsaturated polyesters represent an excellent illustration of a polymeric system with a latent ability to cross-link. Unsaturated polyesters are, as the name implies, unsaturated polyester prepolymers mixed with cross-linking monomers and catalysts. The resulting mixture is normally a viscous liquid that can be poured, sprayed, or shaped into the desired form and then transformed into a thermosetting solid by cross-linking.

The unsaturated polyester prepolymers are obtained from the condensation of polyhydric alcohols and dibasic acids. The dibasic acid consists of one or more saturated acid and/or unsaturated acid. The saturated acid may be phthalic anhydride, adipic acid, or isophthalic acid, while the unsaturated acid is usually maleic anhydride or fumaric acid. The polyhydric alcohols in common use include glycol (such as ethylene glycol, propylene glycol, diethylene glycol), glycerol, sorbitol, and pentaerythritol (Equation 5.1).

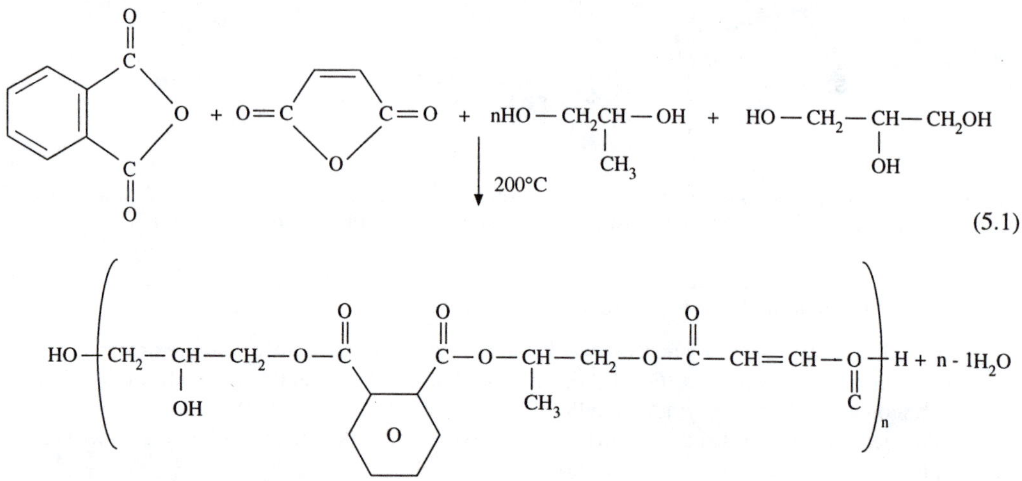

$$\text{(phthalic anhydride)} + \text{(maleic anhydride)} + n\,HO-CH_2CH(CH_3)-OH + HO-CH_2-CH(OH)-CH_2OH \xrightarrow{200°C}$$

$$HO\left(CH_2-CH(OH)-CH_2-O-\overset{O}{\overset{\|}{C}}-C_6H_4-\overset{O}{\overset{\|}{C}}-O-CH(CH_3)-CH_2-O-\overset{O}{\overset{\|}{C}}-CH=CH-\underset{C}{\underset{\|}{O}}\right)_n H + n-1H_2O \qquad (5.1)$$

The reactive monomer responsible for the cross-linking reaction is normally styrene, vinyl toluene, methyl methacrylate, or diallyl phthalate. The unsaturated polyester prepolymers are, by themselves, relatively stable at ambient temperatures. When they are added to the cross-linking monomers, however, the blend is extremely unstable. To prevent premature gelation and to control the cross-linking process, inhibitors like hydroquinone are added to the blend. The cross-linking process is usually initiated by an organic peroxide added by the end user. The peroxide may be activated by heat. For room-temperature cure, accelerators and activators such as cobalt salts are also added. The cross-linking reaction can be represented as follows (Equation 5.2):

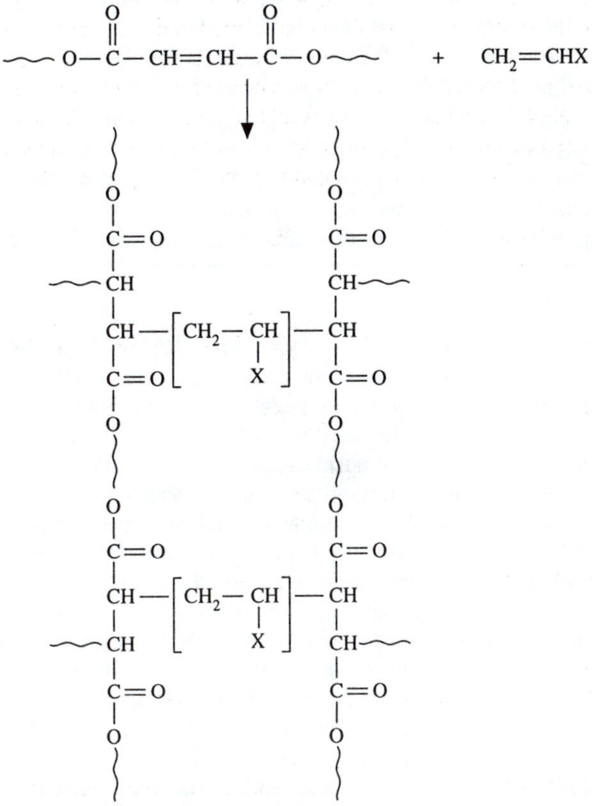

$$\sim O-\overset{O}{\overset{\|}{C}}-CH=CH-\overset{O}{\overset{\|}{C}}-O\sim \quad + \quad CH_2=CHX$$

$$\downarrow$$

$$\begin{array}{ccc}
\wr & & \wr \\
O & & O \\
| & & | \\
C=O & & C=O \\
| & & | \\
\sim CH & & CH\sim \\
| & & | \\
CH- & \left[\begin{array}{c} CH_2-\underset{X}{\underset{|}{CH}} \end{array}\right] & -CH \\
| & & | \\
C=O & & C=O \\
| & & | \\
O & & O \\
\wr & & \wr \\
O & & O \\
| & & | \\
C=O & & C=O \\
| & & | \\
CH- & \left[\begin{array}{c} CH_2-\underset{X}{\underset{|}{CH}} \end{array}\right] & -CH \\
| & & | \\
\sim CH & & CH\sim \\
| & & | \\
C=O & & C=O \\
| & & | \\
O & & O \\
\wr & & \wr
\end{array} \qquad (5.2)$$

Alkyd resins constitute a special class of unsaturated polyesters. They are used predominantly as surface coatings (organic paints) and, to a limited extent, as molding compounds. Unlike the related polyester molding compounds (used in the manufacture of automotive parts like distributor caps, ignition coil towers, and similar electrical parts, the cross-linking monomer in alkyd molding compounds is nonvolatile, e.g., allyl phthalate. The organic peroxide catalyst decomposes on the application of heat and initiates cross-linking yielding on infusible, insoluble solids.

Alkyd resins used as surface coatings are usually modified by the addition of fatty acids derived from mineral and vegetable oils. Those modified with unsaturated acids such as oleic acid (I) or linoleic acid (II) are the air-drying type cured through oxidation by atmospheric oxygen. Those types of surface coatings that dry through chemical reactions are known as varnishes. Others that dry through solvent evaporation are referred to as lacquers.

$$CH_3-(CH_2)_7-CH=CH-(CH_2)_7-COOH \qquad \text{(Str. 6)}$$

oleic acid (I)

$$CH_3-(CH_2)_7-CH-CH-CH-CH-(CH_2)_5-COOH \qquad \text{(Str. 7)}$$

linoleic acid (II)

Example 5.5: An unsaturated polyester based entirely on maleic anhydride as the dibasic acid, when cross-linked with styrene, forms a hard and rigid polymer. However, that based on maleic anhydride mixed with some phthalic anhydride forms a less rigid polymer. Explain.

Solution: As we shall see in Chapter 8, the copolymerization of maleic anhydride and styrene shows a strong tendency toward alternation ($r_1 r_2 = 0.06$). Consequently, a maleic anhydride-based unsaturated polyester cross-linked with styrene forms a large number of short cross-links, resulting in a rigid polymer. With the addition of phthalic anhydride, the tendency toward alternation is reduced due to the occasional intervention of phthalic anhydride. The reduction of cross-link density results in a less rigid polymer.

2. Vulcanization

Elastomers such as 1,4-diene polymers — polychloroprene, polybutadiene, and polyisoprene — and copolymers — butadiene–acrylonitrile and butadiene–styrene — must possess the capability of undergoing large deformations under stress and the essential property of rapid and complete recovery from such deformations on the removal of the imposed stress. This means, in molecular terms, that the long-range movements of the polymer molecules must be restrained at the same time that high local segmental mobility is permitted. To acquire this property, an elastomer needs to be lightly cross-linked. The process by which such a network of cross-links is introduced into an elastomer is called vulcanization. It transforms an elastomer from a weak thermoplastic mass without useful properties into a strong, elastic, tough rubber. Vulcanization decreases the flow of an elastomer and increases its tensile strength and modulus, but preserves its extensibility. Vulcanization, discovered by Goodyear in 1939, achieved by heating the elastomer solely with sulfur is a slow and inefficient process. It can be speeded up considerably and the sulfur waste reduced substantially by the addition of small amounts of organic and inorganic compounds known as accelerators. Many accelerators themselves require the presence of activators or promoters to function optimally. Some accelerators used include sulfur-containing compounds and a few nonsulfur-containing compounds, as shown in Table 5.4.

Activators are normally metallic oxides such as zinc oxide. They function best in the presence of soap such as metallic salt of stearic acid. The use of accelerators and activators has increased the efficiency of cross-linking in some cases to less than two sulfur atoms per cross-link.

a. *Rubbers*[33]

The early synthetic rubbers were diene polymers such as polybutadiene. Diene elastomers possess a considerable degree of unsaturation, some of which provide the sites required for the light amount of cross-linking structurally necessary for elastomeric properties. The residual double bonds make diene elastomers vulnerable to oxidative and ozone attack. To overcome this problem, saturated elastomers like butyl rubber and ethylene–propylene rubber (EPR) were developed. These rubbers were, unfortunately, not readily vulcanized by conventional means. To enhance cure, it was therefore necessary to

Table 5.4 Some Acclerators Used in Vulcanization.

Accelerator	Structure
2-mercaptobenzothiazole	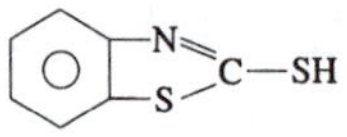 (benzene ring fused: N=C—SH, S)
Tetramethylthioureadisulfide	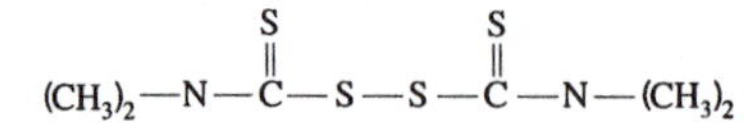 $(CH_3)_2—N—C(=S)—S—S—C(=S)—N—(CH_3)_2$
Diphenylguanidine	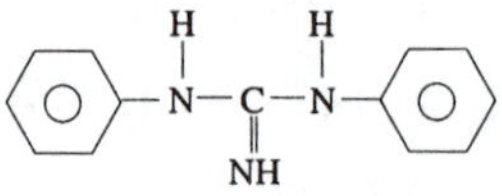 $C_6H_5—NH—C(=NH)—NH—C_6H_5$
Zinc butyl xanthate	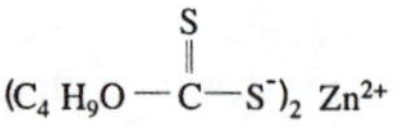 $(C_4H_9O—C(=S)—S^-)_2\ Zn^{2+}$

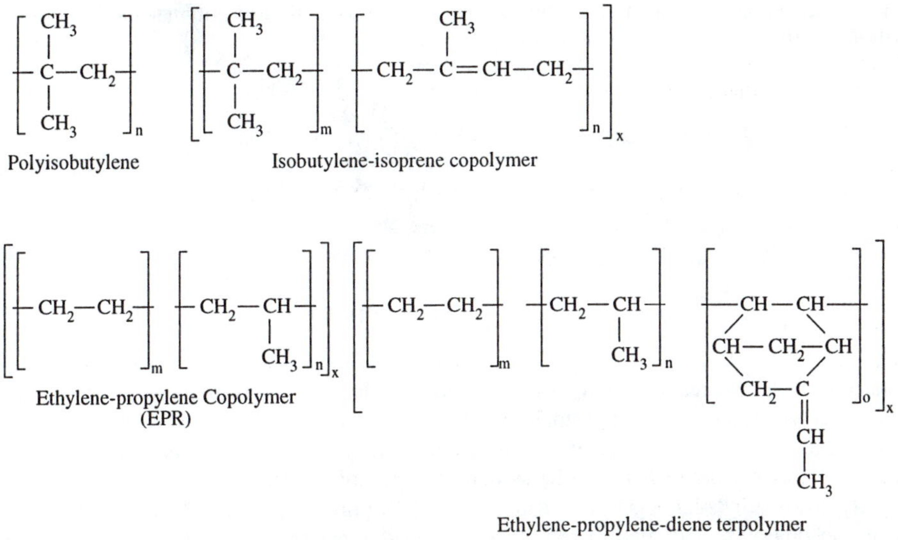

$$\left[-\underset{CH_3}{\overset{CH_3}{\mid}}{C}-CH_2- \right]_n$$

Polyisobutylene

$$\left[\left[-\underset{CH_3}{\overset{CH_3}{\mid}}{C}-CH_2- \right]_m \left[-CH_2-\overset{CH_3}{\overset{\mid}{C}}=CH-CH_2- \right]_n \right]_x$$

Isobutylene-isoprene copolymer

$$\left[\left[-CH_2-CH_2- \right]_m \left[-CH_2-\underset{CH_3}{\underset{\mid}{CH}}- \right]_n \right]_x$$

Ethylene-propylene Copolymer (EPR)

$$\left[\left[-CH_2-CH_2- \right]_m \left[-CH_2-\underset{CH_3}{\underset{\mid}{CH}}- \right]_n \left[\text{dicyclopentadiene unit with } =CH-CH_3 \text{ side group} \right]_o \right]_x$$

Ethylene-propylene-diene terpolymer

Figure 5.9 Creation of vulcanization sites for butyl and ethylene–propylene rubbers through copolymerization.

incorporate some unsaturation into the structure of these polymers through copolymerization of unsaturated monomers. The usual monomers for butyl rubber and EPR are shown in Figure 5.9.

The comonomer shown for EPR is dicyclopentadiene. Other comonomers are ethylidene norbornene (obtained from the reaction of butadiene with cyclopentadiene) and 1,4-hexadiene. Each of the diene copolymers with ethylene and propylene produces different characteristics in the final elastomer. Dicyclopentadiene yields a branched polymer with slow cure rates because of the low reactivity of the second double bond. With ethylidene norbornene, both linear and branched polymers can be produced and sulfur vulcanization is improved. The polymer resulting from the incorporation of 1,4-hexadiene into EPR possesses good processing characteristics and heat resistance. Since EPR is considered saturated, it usually requires peroxide cure, while EPDM polymers are cured normally by conventional sulfur vulcanization. The type of cure is dictated by the polymer and the expected processing technique (extrusion, calendering, molding, etc.). The commercial cross-linking of diene polymers is carried out almost exclusively by heating with sulfur (Equation 5.3).

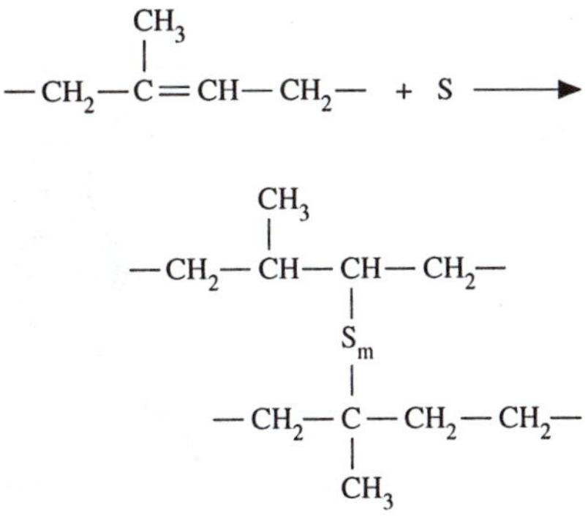

$$-CH_2-\overset{CH_3}{\overset{\mid}{C}}=CH-CH_2- \; + \; S \longrightarrow \begin{array}{c} -CH_2-\overset{CH_3}{\overset{\mid}{CH}}-CH-CH_2- \\ \mid \\ S_m \\ \mid \\ -CH_2-\underset{CH_3}{\underset{\mid}{C}}-CH_2-CH_2- \end{array} \qquad (5.3)$$

b. Polyolefins and Polysiloxanes

While the first vulcanization involved heating the elastomer with sulfur, it has since been recognized that neither heat nor sulfur is imperative for the vulcanization process. Vulcanization or cure can be effected by nonsulfur-containing compounds, including peroxides, nitro compounds, quinones, or azo compounds. Polyethylene, ethylene–propylene copolymers, and polysiloxanes are cross-linked by compounding them with a peroxide and heating them. The process involves the formation of polymer radicals followed by the coupling of these radicals (Equations 5.4–5.6).

$$\text{ROOR} \longrightarrow 2\,\text{RO}\cdot \tag{5.4}$$

$$\text{RO}\cdot + -\text{CH}_2-\text{CH}_2- \longrightarrow \text{ROH} + -\text{CH}_2\dot{\text{C}}\text{H}- \tag{5.5}$$

$$2-\text{CH}_2\dot{\text{C}}\text{H}- \longrightarrow \begin{array}{l} -\text{CH}_2-\text{CH}- \\ \qquad\quad\;| \\ -\text{CH}_2-\text{CH}- \end{array} \tag{5.6}$$

The efficiency of this process is usually less than one cross-link per peroxide molecule decomposed. To increase the cross-linking efficiency, small amounts of unsaturation are introduced into the polymer structure. We have already discussed EPDM polymers, which are essentially diene monomers copolymerized with ethylene–propylene (EPR) polymers. For polysiloxanes, copolymerization of small amounts of vinyl–methylsilanol greatly enhances cross-linkability (Equation 5.7). The unsaturation introduced into an otherwise saturated structure provides additional sites for cross-linking through chain reaction.

$$\begin{array}{c} \text{CH}_2 \\ \| \\ \text{CH} \\ | \\ \text{HO}-\text{Si}-\text{OH} \\ | \\ \text{CH}_3 \end{array} + \begin{array}{c} \text{CH}_3 \\ | \\ \text{HO}-\text{Si}-\text{OH} \\ | \\ \text{CH}_3 \end{array} \longrightarrow -\text{O}-\begin{array}{c} \text{CH}_2 \\ \| \\ \text{CH} \\ | \\ \text{Si} \\ | \\ \text{CH}_3 \end{array}-\text{O}-\begin{array}{c} \text{CH}_3 \\ | \\ \text{Si} \\ | \\ \text{CH}_3 \end{array}-\text{O}- \tag{5.7}$$

C. HYDROLYSIS[11]

A primary use of poly(vinyl acetate) is in the production of water-based emulsion paints in addition to its wide use in emulsion-type and hot-melt adhesives. It is also used in the production of poly(vinyl alcohol) and poly(vinyl acetal) polymers, which cannot be prepared directly since their monomers are unknown.

Poly(vinyl alcohol) (PVA) is prepared by the hydrolysis (or more correctly alcoholysis) of poly(vinyl acetate) with methanol or ethanol. The reaction is catalyzed by both bases and acids, but base catalysis is normally employed because it is faster and free of side reactions (Equation 5.8).

$$\left[\text{CH}_2-\underset{\underset{\underset{\underset{\text{CH}_3}{|}}{\text{C}=\text{O}}}{|}}{\underset{\text{O}}{|}}}{\text{CH}}\right]_n + n\text{CH}_3\text{OH} \xrightarrow{\text{H}^+ \text{ or OH}^-} \left[\text{CH}_2-\underset{\text{OH}}{\underset{|}{\text{CH}}}\right]_n + n\text{CH}_3\overset{\overset{\text{O}}{\|}}{\text{C}}-\text{OCH}_3 \tag{5.8}$$

Because of its water solubility, poly(vinyl alcohol) is used as a thickening agent for various emulsion and suspension systems. With its high hydroxyl content, PVA is used widely as water-soluble adhesive with excellent binding capacity for cellulosic materials like paper.

Partially hydrolyzed poly(vinyl acetate) contains both hydroxyl and acetate groups. When the OH groups in partially hydrolyzed poly(vinyl acetate) are condensed with aldehydes, acetal units are formed. The resulting polymer contains acetal, hydroxyl, and acetate groups and is known as the poly(vinylacetals) (Equation 5.9).

$$-CH_2-\underset{\displaystyle OH}{\underset{|}{CH}}-CH_2-\underset{\displaystyle OH}{\underset{|}{CH}}-CH_2-\underset{\displaystyle OH}{\underset{|}{CH}}-CH_2-\underset{\displaystyle O-C(=O)-CH_3}{\underset{|}{CH}} \xrightarrow{RCHO}$$

$$\underbrace{\left[CH_2-CH-CH_2-CH\right]}_{\text{acetal}\;(\text{O},\text{O bridged by } C(R)(H))} \quad \underbrace{\left[CH_2-\underset{\displaystyle OH}{\underset{|}{CH}}\right]}_{\text{alcohol}} \quad \underbrace{\left[CH_2-\underset{\displaystyle O-C(=O)-CH_3}{\underset{|}{CH}}\right]}_{\text{acetate}} \tag{5.9}$$

The reaction of butyraldehyde or formaldehyde results in poly(vinyl butyral) or poly(vinyl formal), respectively. The most important of the poly(vinylacetals) is by far poly(vinyl butyral). The residual OH groups in poly(vinyl acetal) can condense with methylol groups in PF, MF, and UF resins. For example, poly(vinyl formal) cured with phenolic resins was first used as a structural adhesive for metals. Poly(vinyl butyral) finds important application as the interlayer in laminated automotive and aircraft safety glass. Laminated glass, also made from poly(vinyl butyral), is used in buildings for controlled transmission of light and heat through reduction of glare, heat loss, and UV, thus providing aesthetic appeal. In these applications, the residual hydroxyl groups provide the needed strength and adhesion to glass.

D. BLOCK AND GRAFT COPOLYMER FORMATION[6,7]

Block and graft copolymer formation involves the reaction of a previously formed homopolymer or copolymer with fresh monomers. Consequently, normal methods of polymerization can be utilized. For vinyl monomers two general methods are employed: (1) the polymerization of a monomer in the presence of a polymer by the initiation of growth through chain transfer; and polymerization of a monomer in the presence of a polymer containing reactive sites. The first procedure necessarily leads to the formation of graft copolymers. In the second case, block copolymers are formed from polymers with terminal active sites while graft copolymerization occurs on active sites located on the polymer backbone or pendant groups. In both of these methods, the polymerization of a monomer in the presence of a polymer results in a mixture of products: (1) the initial homopolymer that did not participate in the reaction; (2) the homopolymer of the fresh monomer; (3) the cross-linked parent homopolymer through graft polymerization or the branched block copolymer; and (4) the desired copolymer. The composition of the product mixture will depend on the nature of the polymer, monomer, and initiator. For example, the polymerization of methyl methacrylate in the presence of polystyrene yields appreciable amounts of graft copolymer with benzoyl peroxide as the initiator. However, if AIBN is the initiator, the amount of graft copolymer formed is considerably smaller. In contrast, the polymerization of vinyl acetate in the presence of poly(methyl methacrylate) yields appreciable amounts of the graft copolymer irrespective of the nature of the initiator. In any case, while it is possible in principle to obtain a pure product by employing general separation processes, this is often problematic for industrial purposes.

1. Block Copolymerization[8,9]

The preparation of block copolymers requires the presence of terminal reactive groups. A variety of techniques have been used to achieve this. We briefly discuss a few of them.

Block copolymers of butyl acrylate–styrene and acrylonitrile–styrene have been prepared by irradiating butyl acrylate or acrylonitrile containing a photosensitive initiator (e.g., 1-azo-bis.1-cyanocyclohexane) with an intensive UV radiation. This creates a radical-rich monomer that when mixed with styrene yields the appropriate block copolymer.

In situ block polymer formation is obtained by using a mixture of water-soluble and oil-soluble monomers. An example involves dissolution of acrylic acid or methacrylic acid in water followed by

Table 5.5 Various Block Copolymers Prepared by Mechanical Rupture of Polymer Chains in Presence of Monomer

Polymer Unit	Monomer
Acrylamide	Acrylonitrile
Isobutylene	Acrylonitrile
	Styrene
	Vinylidene chloride
Methyl acrylate	Vinyl chloride
	Vinyllidene chloride
Methacrylonitrile	Acrylonitrile
Methacrylonitrile-*co*-vinyl chloride	Methyl methacrylate
Methyl methacrylate	Methacrylonitrile
	Styrene
	Vinylidene chloride
Methyl methacrylate-*co*-methacrylonitrile	Styrene
Styrene	Methyl methacrylate
	Vinylidene chloride
Vinyl acetate	Vinyl chloride
Vinyl chloride	Methyl vinyl ketone

From Fetters, E.M., Ed., *Chemical Reactions of Polymers,* Interscience, New York, 1964. With permission.

styrene emulsification in the water. With persulfate initiator, polymerization of acid occurs in the aqueous phase until the growing chains diffuse into the micelle containing styrene monomer, whereupon styrene chain growth ensues resulting in the formation of acrylic/methacrylic acid–styrene block copolymer.

When a polymer chain is ruptured mechanically, terminal-free radicals can be generated, and these can be utilized to initiate block copolymerization. Under an inert atmosphere, block copolymers can be produced by cold milling, or mastication of two different polymers or of a polymer in the presence of a second monomer. This generally results in the formation of graft copolymers in addition to the block copolymers since radicals can be located in nonterminal positions by chain transfer. However, predominant yield of block copolymers is obtained by milling monomer-swollen polymers. The success of this technique depends on the physical state of the polymer. Generation of radical is favored if the polymer exists at or near the glassy state; otherwise, polymer flow rather than bond rupture will occur. Table 5.5 shows some block copolymers prepared by this technique.

When polymer chains are terminated by labile end groups, block polymers can be produced by irradiation. An example is the irradiation of terminal-brominated styrene dissolved in methyl methacrylate to yield styrene–methyl methacrylate block copolymer (Equation 5.10).

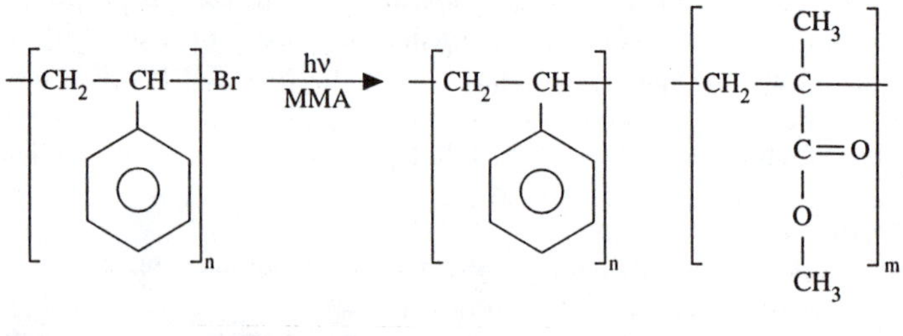

$$\left[CH_2 - \underset{C_6H_5}{CH} \right]_n Br \xrightarrow[MMA]{h\nu} \left[CH_2 - \underset{C_6H_5}{CH} \right]_n \left[CH_2 - \underset{\underset{\underset{CH_3}{O}}{C=O}}{\overset{CH_3}{C}} \right]_m \quad (5.10)$$

A common technique of preparing block copolymers is the introduction of peroxide groups into the polymer backbone or as stable end groups. The polymers are then mixed with fresh monomer, and the peroxide groups are decomposed under appropriate conditions to yield block copolymers. For example, polymeric phthaloyl peroxide was polymerized to a limited extent with styrene. The resulting polymer was mixed with methyl methacrylate. On decomposition, the internal and terminal peroxide groups formed radicals that initiated the polymerization of methyl methacrylate, as shown below in Equation 5.11.

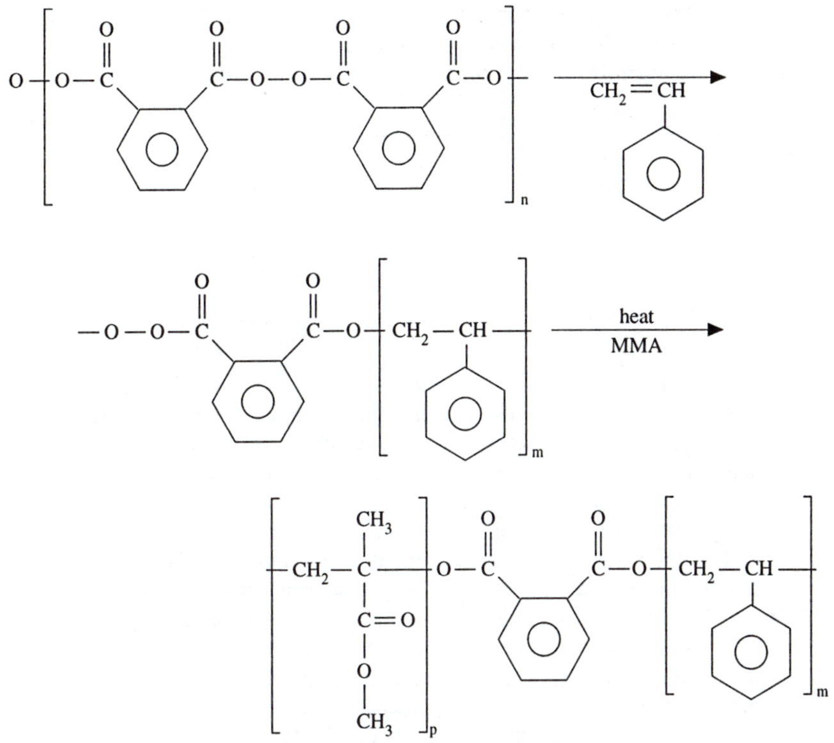

$$\xrightarrow{CH_2{=}CH(C_6H_5)} \qquad \xrightarrow[\text{MMA}]{\text{heat}} \qquad (5.11)$$

2. Graft Copolymerization[8]

There are essentially three approaches to the preparation of graft copolymers via free radical mechanism:

1. Chain transfer to a saturated or unsaturated polymer
2. Activation by photochemical or radiative methods
3. Introduction and subsequent activation of peroxide and hydroperoxide groups

Graft copolymer formation by chain transfer requires three components: a polymerizable monomer, a polymer, and an initiator. By definition, the initiator serves to create active sites either on the polymer or on the monomer. The amount of initiator added to a polymerization system is generally low. Therefore, if this is the sole source for the initiation of radical sites on the backbone of the polymer by way of chain transfer, only a relatively few branches will be produced. Consequently, to achieve a reasonable amount of grafting, the polymer itself must be capable of generating radicals through hydrogen abstraction and hence promote chain transfer. As we shall see in Chapter 8, the relative reactivity ratios dictate that for copolymerization to occur the nature of the monomer must be such that it can react with the polymer radical. This means that not all monomers can be grafted onto a polymer backbone, and, by the same argument, not all polymers can react with a given monomer. For example, whereas vinyl acetate can be grafted onto poly(vinyl chloride) or polyacrylonitrile, hardly any grafting occurs in the case of vinyl acetate or vinyl chloride and polystyrene; styrene or vinyl chloride and poly(vinyl acetate); or styrene and poly(vinyl chloride).

Sites that are susceptible to chain transfer may be introduced into the polymer structure during synthesis or as a postreaction. An example of the latter case involves the introduction of a reactive mercaptan group by the reaction of thioglycolic acid or hydrogen sulfide with a copolymer of methyl methacrylate and glycidyl methacrylate (Equation 5.12).

$$-CH_2-\overset{CH_3}{\underset{C=O-O-CH_3}{C}}-CH_2-\overset{CH_3}{\underset{C=O-O-CH_2-\underset{CH_2}{CH}>O}{C}}- \xrightarrow{HS-CH_2-COOH} -CH_2-\overset{CH_3}{\underset{C=O-O-CH_3}{C}}-CH_2-\overset{CH_3}{\underset{C=O-O-CH_2-\underset{HS-CH_2-COO-CH_2}{\overset{}{CH}}(-OOC-CH_2-SH)}{C}}- \tag{5.12}$$

Monomers such as styrene, acrylate, and methacrylate have been grafted onto this polymer, with a relatively high yield of pure graft copolymer. A commercially important graft copolymer resulting from transfer to unsaturated polymer is impact polystyrene, which is butadiene grafted to polystyrene. The nature of the transfer process in this case results in a copolymer with the rubber (polybutadiene) graft existing as a dispersed phase within a styrene matrix. Such a copolymer morphological structure reduces the brittleness of polystyrene, and this enhances its impact resistance.

Active sites can be generated on a polymer to initiate graft copolymerization by ultraviolet light and high-energy irradiation. The process may involve (1) simultaneous irradiation of polymer in contact with monomer; (2) irradiation of polymer predipped in the monomer; (3) pre-irradiation of the polymer in the absence of air (O_2) followed by exposure to the monomer; and (4) pre-irradiation of the polymer in air to form peroxide and subsequent decomposition in the presence of the monomer. The first two techniques are known as mutual irradiation, as distinct from pre-irradiation.

Photolytic activation of polymer growth is based on the ability of certain functional groups to absorb radiation. This generally involves polymers with pendant halogen atoms or carbonyl groups, which are highly susceptible to activation by UV radiation. When sufficient energy is absorbed, bond scission occurs resulting in the generation of radicals. Monomers such as acrylonitrile, methyl methacrylate, and vinyl acetate have been grafted onto poly(vinyl methyl ketone) by this technique. Other examples include grafting of styrene and methyl methacrylate on brominated styrene. Mutual irradiation with UV in the presence of a photosensitizer has also been used to graft methyl methacrylate and acrylamide to natural rubber.

Generation of active sites for radical graft copolymerization using X-rays or γ-rays is less selective. In this case, hydrogen cleavage and carbon–carbon bond scission occur resulting in homopolymerization of the vinyl monomer, cross-linking in addition to the grafting reaction. Table 5.6 shows some graft copolymers produced by natural irradiation with high-energy radiation and the properties of the resulting graft copolymer.

Unstable compounds such as peroxides and hydroperoxides are usual initiators for vinyl polymerization reactions. These groups can be introduced into a solid polymer by pre-irradiation in the presence of air or oxygen. Thermal decomposition of these groups in the presence of monomer yields graft copolymers. Some examples include grafting of acrylonitrile, styrene, and methyl methacrylate to polyethylene and polypropylene.

Table 5.6 Properties of Graft Copolymers Produced by Mutual Irradiation of Monomer

Polymer	Monomer	Property Changes
Polyethylene	Acrylonitrile	Increased solvent resistance and softening temperature; excellent adhesion to polar materials
	Vinyl chloride	Retention of superior electrical properties while increasing softening point to 215°C
	Acrylate methacrylate	Following hydrolysis, surface becomes permanently conductive; prevention of accumulation of static charges; good adhesion to materials like cellulose, glass, and metals
	Styrene	Increased melt viscosity
Polytetrafluoroethylene Polychlorotrifluoroethylene	Styrene	Enhanced adhesion, enhanced elimination of plastic flow, and increased ultimate strength

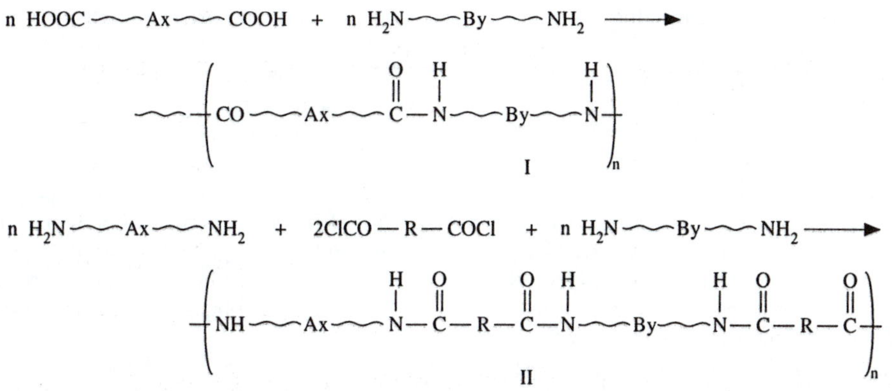

$$n\ \mathrm{HOOC}\sim\sim\mathrm{Ax}\sim\sim\mathrm{COOH} + n\ \mathrm{H_2N}\sim\sim\mathrm{By}\sim\sim\mathrm{NH_2} \longrightarrow$$

$$\sim\sim\left(\mathrm{CO}\sim\sim\mathrm{Ax}\sim\sim\overset{\overset{\displaystyle O}{\|}}{\mathrm{C}}-\overset{\overset{\displaystyle H}{|}}{\mathrm{N}}\sim\sim\mathrm{By}\sim\sim\overset{\overset{\displaystyle H}{|}}{\mathrm{N}}\right)_n$$

I

$$n\ \mathrm{H_2N}\sim\sim\mathrm{Ax}\sim\sim\mathrm{NH_2} + 2\mathrm{ClCO-R-COCl} + n\ \mathrm{H_2N}\sim\sim\mathrm{By}\sim\sim\mathrm{NH_2} \longrightarrow$$

$$\left(\mathrm{NH}\sim\sim\mathrm{Ax}\sim\sim\overset{\overset{\displaystyle H}{|}}{\mathrm{N}}-\overset{\overset{\displaystyle O}{\|}}{\mathrm{C}}-\mathrm{R}-\overset{\overset{\displaystyle O}{\|}}{\mathrm{C}}-\overset{\overset{\displaystyle H}{|}}{\mathrm{N}}\sim\sim\mathrm{By}\sim\sim\overset{\overset{\displaystyle H}{|}}{\mathrm{N}}-\overset{\overset{\displaystyle O}{\|}}{\mathrm{C}}-\mathrm{R}-\overset{\overset{\displaystyle O}{\|}}{\mathrm{C}}\right)_n$$

II

Figure 5.10 Formation of block copolymers from condensation polymers: (I) from two prepolymers with mutually reactive terminal end groups; (II) by the use of a coupling agent.

So far, our discussion has been restricted to chain block and graft copolymerization. This is largely because the practical utility of copolymerization is more elaborate in chain polymerization than step polymerization. Also, in step copolymerization, block copolymers are generally preferred to the other types of copolymers. Therefore only block step-polymerization copolymers are discussed here and only in a very limited scope to illustrate the principles involved in their preparation.

Block copolymers have been prepared from condensation polymers by coupling low-molecular-weight homopolymer with appropriate reactive end groups. These prepolymers may possess terminal end groups that are different but mutually reactive. If, however, the two polymer blocks have the same or similar terminal groups, a bifunctional coupling agent capable of reacting with such groups is employed to prepare the block copolymers. Step-polymerization block copolymer can also be prepared via interchange reactions. Figure 5.10 illustrates these general principles.

There are a variety of commercially important condensation block copolymers. For example, alkyd–amine resins are cross-linked block copolymers resulting from the reaction of hydroxyl and carboxyl groups on alkyd resins with the methylol groups of urea- or melamine-(amine)-formaldehyde resins. Epoxy resins contain secondary hydroxy and terminal epoxy groups that are capable of reacting with other functionalities to produce block copolymers. Reaction with the amine groups (primary and secondary) in polyamide yields block copolymers ranging from rubbery and resilient to hard, tough, and shock-resistant solids. Epoxy resins can also react with polysulfide liquid polymers to give products varying in property from brittle to highly flexible. Formaldehyde resins (phenol–formaldehyde, urea–formaldehyde, and melamine–formaldehyde) react with epoxy resins to yield block copolymers used in coatings and adhesives.

E. SURFACE MODIFICATION[10]

Surface reactions of polymers, in strict terms, should include polymer interactions with its environment, referred to as aging or weathering. The discussion here, however, focuses on the deliberate chemical modifications of surface properties. Surface reactions, by definition, do not include bulk homogeneous reactions of polymers, but refer to those confined to the surface and those that generally do not significantly alter the physical properties of the substrate. Surface reactions have been used commercially to improve feel, washability, dye retention, antistatic properties, and abrasion resistance of fibers and to enhance printability, solvent resistance, adhesion, and permeability to liquids and vapors of films. Surface modification of polymers can be effected by three general procedures: interpolymerization by energy irradiation or chemical means and by conventional chemical reactions of polymers. Interpolymerization, as has been discussed, results in the formation of graft or block copolymers. The chemical treatment of polymers, including such reactions as oxidation, halogenation, and other standard chemical reactions, are briefly reviewed.

Surface oxidation reactions have been carried out on a number of polymers, particularly polyethylene. Surface oxidation techniques include the use of corona discharge, ozone, hydrogen peroxide, nitrous acid, alkaline hypochloride, UV irradiation, oxidizing flame, and chromic acid. The reactions lead initially to the formation of hydroperoxides, which catalyze the formation of aldehydes and ketones and, finally, acids and esters. Surface oxidation treatment has been used to increase the printability of polyethylene and poly(ethylene terephthalate) and to improve the adhesion of polyethylene and polypropylene to polar polymers and that of polytetrafluoroethylene to pressure-sensitive tapes. Surface-oxidized polyethylene, when coated with a thin film of vinylidene chloride, acrylonitrile, and acrylic acid terpolymers becomes impermeable to oxygen and more resistant to grease, oil, abrasion, and high temperatures. The greasy feel of polyethylene has also been removed by surface oxidation.

Alkali and acid treatments have also been used to modify surface properties of polymers: sulfonated polyethylene films treated first with ethylenediamine and then with a terpolymer of vinylidene chloride, acrylonitrile, and acrylic acid exhibited better clarity and scuff resistance and reduced permeability. Permanently amber-colored polyethylene containers suitable for storing light-sensitive compounds have been produced by treating fluorosulfonated polyethylene with alkali. Poly(ethylene terephthalate) dipped into trichloroacetic/chromic acid mixture has improved adhesion to polyethylene and nylons. Antifogging lenses have been prepared by exposing polystyrene films to sulfonating conditions. Acid and alkali surface treatments have also been used to produce desired properties in polymethylmethacrylates, polyacrylonitrile, styrene–butadiene resins, polyisobutylene, and natural rubber. Surface halogenation of the diene polymers natural rubber and polyisobutylene resulted in increased adhesion to polar surfaces.

Treatment of a large number of thermoplastics with alkyl or alkenyl halosilanes has yielded materials with improved heat, stain, and scratch resistance. The use of silane coupling agents in a number of systems is particularly instructive. These systems include glass- or carbon-fiber-reinforced composites, phenolic-bonded sand, grinding wheel composites of aluminum oxide, metal-filled resins for the tool and die industry, etc. In fiber-reinforced composites where the interfacial area between the matrix resin and fiber reinforcement is very high, adequate fiber–matrix adhesion is necessary for the efficient transfer of the applied stress across the fiber–matrix interface required for the production of high-performance composite systems.[11] Silane coupling agents are characterized by terminal groups specifically designed to effect chemical linkage between matrix resin and the fiber reinforcement. A typical example is γ-aminopropyltriethoxy silane, $(C_2H_5O)_3$-Si-$(CH_2)_3$-NH_2, with terminal amino and ethoxy groups. When the glass fiber is treated with the silane coupling agent, the ethoxy groups are hydrolyzed to form the silanol group, –Si–$(OH)_3$, which then condenses with the silanols on the glass fiber to form –Si–O–Si– covalent bonds. On incorporation of the silane-treated glass fibers into the matrix resin, the unreacted terminal amino groups react with the epoxy matrix resin. Consequently, the silane coupling agent forms a molecular bridge between the matrix resin and the fiber reinforcement (Figure 5.11). A number of

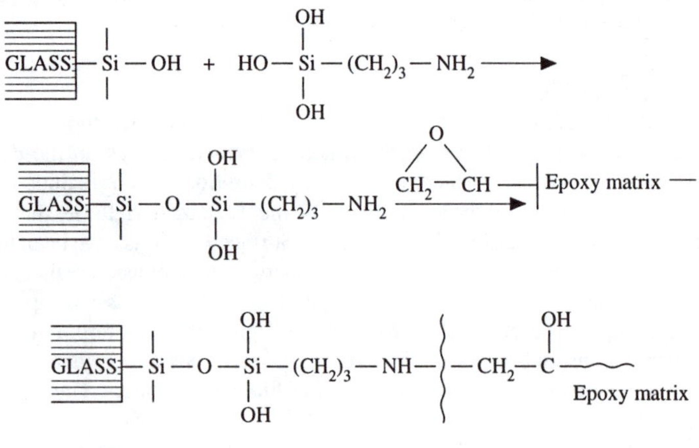

GLASS–Si–OH + HO–Si(OH)(OH)–$(CH_2)_3$–NH_2 ⟶

GLASS–Si–O–Si(OH)(OH)–$(CH_2)_3$–NH_2 + CH_2–CH(–O–) – Epoxy matrix ⟶

GLASS–Si–O–Si(OH)(OH)–$(CH_2)_3$–NH–CH_2–C(OH)~ Epoxy matrix

Figure 5.11 Chemical linkage between matrix resin and glass–fiber reinforcement through a silane coupling agent.

silane coupling agents are available with reactive groups such as amine, epoxy, or vinyl, which can react with epoxy, polyester, and other matrix resins during polymer cure. The resulting polymer materials exhibit improved mechanical properties and durability.

Another illustration of the use of surface modification in the production of improved composite systems is the surface modification of aramid fibers.[12] Poly(*p*-phenylene terephthalamide) (PPTA) (aramid) fibers have excellent mechanical properties and good thermal stability. These properties combined with their low density make aramid fibers prime candidates for reinforcement in high-performance, lightweight composite materials. However, the usefulness and effectiveness of these fibers in such composite systems depend on the adhesion between the fiber reinforcement and matrix resin. Unfortunately, the adhesion between PPTA fibers and most matrices is poor because the chemical inertness and smooth surface quality of fibers preclude chemical bonding and mechanical anchorage. One of the ways of overcoming this problem involves chemical modification of the surface of the fibers by introducing functional groups capable of reacting with the matrix resin.

The surface treatment procedure consists of two stages. In the first step, PPTA fibers are treated with sulfolane/oxalychloride to produce an intermediate (I) in Equation 5.13.

$$\left[-\overset{\text{H}}{\text{N}}-\text{C}_6\text{H}_4-\overset{\text{H}}{\text{N}}-\overset{\text{O}}{\overset{\|}{\text{C}}}-\text{C}_6\text{H}_4-\overset{\text{O}}{\overset{\|}{\text{C}}}-\right]_n + 2n\text{Cl}-\overset{\text{O}}{\overset{\|}{\text{C}}}-\overset{\text{O}}{\overset{\|}{\text{C}}}-\text{Cl} \longrightarrow \left[-\underset{\text{C(=O)C(=O)Cl}}{\text{N}}-\text{C}_6\text{H}_4-\underset{\text{C(=O)C(=O)Cl}}{\text{N}}-\overset{\text{O}}{\overset{\|}{\text{C}}}-\text{C}_6\text{H}_4-\overset{\text{O}}{\overset{\|}{\text{C}}}-\right]_n + 2\text{HCl} \tag{5.13}$$

I

This highly reactive intermediate can react with a number of compounds to introduce different functionalities on the fiber surface. The resulting functional groups, in the case of reaction with ethylenediamine and glycidol, are shown in Equations 5.14 and 5.15, respectively.

$$-\underset{\text{C(=O)}-\text{C(=O)}-\text{Cl}}{\text{N}}-\overset{\text{O}}{\overset{\|}{\text{C}}}- + \text{H}_2\text{N}-(\text{CH}_2)_2-\text{NH}_2 \longrightarrow -\underset{\text{C(=O)}-\text{C(=O)}-\text{N(H)}-(\text{CH}_2)_2-\text{NH}_2}{\text{N}}-\overset{\text{O}}{\overset{\|}{\text{C}}}- + \text{HCl} + \tag{5.14}$$

$$-N(-C(=O)-C(=O)-Cl)-\overset{O}{\overset{\|}{C}}- \; + \; \overset{O}{\overbrace{H_2C-CH}}-CH_2OH \longrightarrow -N(-C(=O)-C(=O)-O-CH_2-\overset{O}{\overbrace{CH-CH_2}})-\overset{O}{\overset{\|}{C}}- \; + \; HCl \qquad (5.15)$$

The introduction of these functionalities on the surface of the aramid fiber resulted in substantial improvement in the fiber–epoxy resin matrix adhesion (about 70%) without a noticeable change in the tensile strength of the fibers. This suggests first that these were possibly covalent and/or hydrogen bonds formed between the fibers and the epoxy resin and that the modification was confined to the surface of the fibers.

Other miscellaneous reactions of polymer have been used to modify the properties of polymers. Polytetrafluoroethylene and polytrifluoroethylene are extremely inert polymers. However, when films of these polymers are treated with solutions of Li, Na, Ca, Ba, or Mg in liquid ammonia, they can be bonded with conventional adhesives to metals and other materials and can be used as backing for pressure-sensitive adhesive tapes. The bonds between active hydrogen-containing polymers, such as natural rubber, and diisocyanate have been enhanced by treatment of the polymer with aliphatic diazo compounds. The weathering properties of poly(vinyl fluoride) have been increased quite considerably by reacting the surface of the polymer with a mixture of diisocyanate and a benzoyl compound, thus resulting in a UV-absorbing surface. Poly(ethylene terephthalate) (PET) fabrics have been freed of the objectionable static charge by reacting PET with 2% NaOH or by reduction with aluminum hydride and subsequently treating with various diisocyanate–polyglycol reaction products.

IV. FUNCTIONAL POLYMERS

There is an ever increasing demand for polymers for specific end-use properties such as enhanced resistance to fire or environmental attack or, in some cases, enhanced degradability. Development of new polymeric materials to meet this challenge has involved the use of a multidisciplinary approach involving chemistry, physics, and engineering. Every decade since the middle of 20th century seems to have witnessed an additional dimension in the thrust of development research in polymer science: the 1950s saw serious advances in polymer chemistry, the 1960s in polymer physics, the 1970s in polymer engineering, and the 1980s in functional polymers.[13] We are currently witnessing the emergence of high-performance polymeric materials such as alloys and blends and advanced composites (so-called polymer abc) being developed through the application of polymer physics and engineering. In this discussion, we focus attention on functional polymers.

Functional polymers may be considered in broad terms as those polymers whose efficiency and characteristics are based on a functional group. A specific functional group is usually carefully designed and located at a proper place on the polymer chain. The functional groups may be dispersed along the polymer main chain (including chain ends) or attached to the main chain either directly or via spacer groups. The main objective of the introduction of special functional groups into the polymer backbone or side chains is to give the polymers special features. Functional groups are, therefore, typically chemical units that are chemically reactive, biologically active, electroactive, mesogenic (liquid crystals), photo-active, and, more commonly, ionic, polar, or optically active.[13] There are two general techniques used for the preparation of functional monomers:[14]

- Polymerization or copolymerization of functional monomers
- Chemical modification of preformed polymers

The use of functional monomers permits ready control of the content and sometimes the distribution of units along the polymer chain; this procedure gives more latitude concerning the physical properties of the final product. On the other hand, chemical modification of an existing polymer, when possible, enables the choice of molecular weight and the dispersity of the polymer. It also allows the synthesis of polymers inaccessible by direct route.[15]

Cross-linking, vulcanization, and grafting are some of the polymer reactions that take place on functional or reactive groups located in the polymer main chain. We have already discussed some of these reactions in the preceding sections. We now discuss other examples of modification of polymers based on reactive monomers, oligomers, or polymers.

A. POLYURETHANES

As we shall see in Chapter 15, polyurethane is a polymer of choice for a wide variety of biomedical applications. Polyurethane is used extensively in the construction of devices such as vascular prostheses, membranes, catheters, plastic surgery, heart valves, and artificial organs. Polyurethanes are also used in drug delivery systems such as the sustained and controlled delivery of pharmaceutical agents, for example, caffeine and prostaglandin.[34]

The major reason for the successful application of polyurethanes as biomaterials is their biocompatibility and formulation versatility. The properties of polyurethanes can be controlled over a wide range by a careful choice of the initial reactants, their ratio, and reaction conditions. Polyurethane elastomers that are of particular interest in medical technology provide a good illustration of the use of functional oligomers/polymers with terminal reactive groups. Bifunctional end-terminated oligomers or low-molecular-weight polymers are made by capping polyols — or polyols of polyether, polyester, or polyalkene — with either aromatic or aliphatic diisocyanate. Typical diisocyanates and polyols used in the synthesis of these polyurethane intermediates are shown in Tables 5.7 and 5.8, respectively. The isocyanate-terminated oligomers (prepolymers or telechelic polymers) are then coupled by reacting with low-molecular-weight diol or diamine chain extenders. A number of aliphatic and aromatic diols and diamines used as chain extenders or cross-linking agents in this synthesis are shown in Table 5.8. This two-step polymerization process (Equations 5.16–5.18), which is usually carried out in solution, results in polyurethane elastomers with good physical characteristics.

Polyurethane elastomers are thermoplastic copolymers of $(AB)_n$ type consisting of an alternating block of relatively long, flexible "soft segment" and another block of highly polar, rather stiff chains or "hard segment." The soft segment is derived from a hydroxyl-terminated aliphatic polyester, polyether, or polyalkene (MW 500 to 5000), while the hard segment is formed from the reaction of diisocyanates with low-molecular-weight diol or diamine chain extenders. The unique physical and mechanical properties of polyurethane elastomers are determined by their two-phase domain structure. These properties can be regulated by exploiting the structural variations possible in R, R′, and R″. Therefore, the ultimate properties of polyurethane-based materials can be controlled largely by the proper choice of the starting monomers used in their synthesis. For example, caprolactone and/or polycarbonate-based polyesterurethanes are more stable to hydrolysis than those based on polyethylene adipate or oxalate.[17,18] Polyetherurethanes exhibit better hydrolytic stability than polyesterurethanes. For hydrophobic urethane elastomers, propylene oxide is used alone or mixed with a small amount of ethylene oxide. Elastomers with high resistance to light and thermal and hydrolytic degradation are prepared from hydroxy-terminated homo- and copolymers of butadiene and isobutylene, copolymers of butadiene and styrene,[19-21] or aliphatic diisocyanates.[22] Polyurethanes produced with aromatic chain extenders are generally stiffer than those prepared with aliphatic extenders.

Typical medical-grade polyurethanes that have been developed and used successfully in various extracorporeal and intracorporeal devices include Biomer (Ethicon Inc.); Cardiothane 51 (Kronton Inc.); and Pellethane (Upjohn/Dow Chemical). For example, resistance to tear and fatigue, as exemplified by adequate modulus and strength, is essential for polymers to be used in reconstructive surgery of soft tissue and cardiovascular surgery. Cardiothane is a block copolymer consisting of 90% poly(dimethylsiloxane) (Figure 5.13).[16] It has been used as intra-aortic balloons, catheters, artificial hearts, and blood tubings. This polymer, with its reactive (cross-linkable) acetoxy groups, combines the good mechanical characteristics of polyurethane with the surface properties of silicone. It further illustrates how the introduction of reactive groups in the polymer backbone can be used to produce a polymer with specific end-use properties.

The chemical and/or biological modification of polyurethane surfaces, such as grafting of hydrogels like acrylamide or poly(hydroxyethyl methacrylate) enhances blood compatibility. Biocompatibility and blood compatibility can be improved by treating the polymer surface with a solution of albumin or gelatin followed by cross-linking with gluteraldehyde or formaldehyde.[23]

Example 5.6: The tendency for hard segment–soft segment phase separation in urethane copolymers increases in the order: polybutadiene and/or polyisoprene soft segment >> polyether soft segment > polyester soft segment. Explain.

Solution: Polyurethane elastomers are segmented copolymers composed of alternating soft and hard blocks. Hydrogen bonding in polyurethanes occurs between the NH hydrogens in the hard segment and the carbonyl oxygens in the hard segment or between the carbonyl and ether oxygens in the soft segment (Table E5.6).

Table E5.6 Possible Hydrogen Bonding in Polyurethane

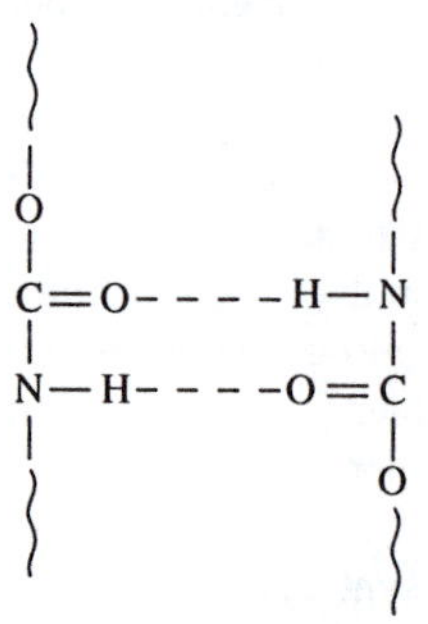

```
O
|
C=O----H—N
|        |
N—H----O=C
|        |
         O
```

Urethane-Urethane

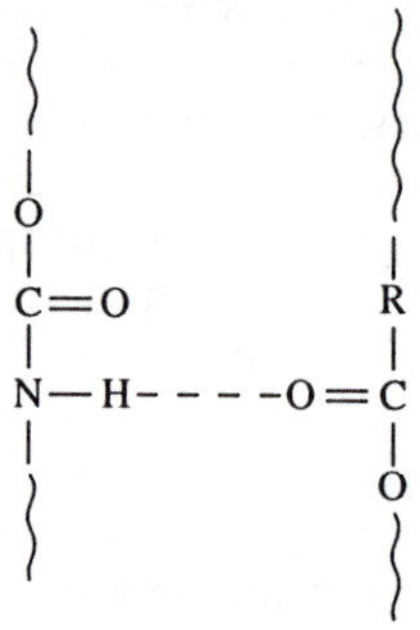

```
O
|        R
C=O      |
|        |
N—H----O=C
|        |
         O
```

Urethane-Ester

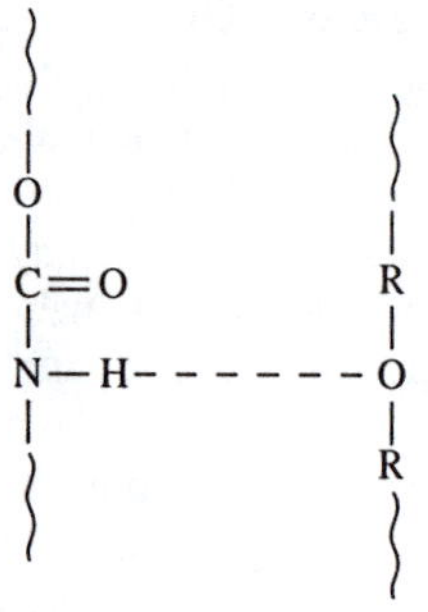

```
O
|        R
C=O      |
|        |
N—H------O
|        |
         R
```

Urethane-Ether

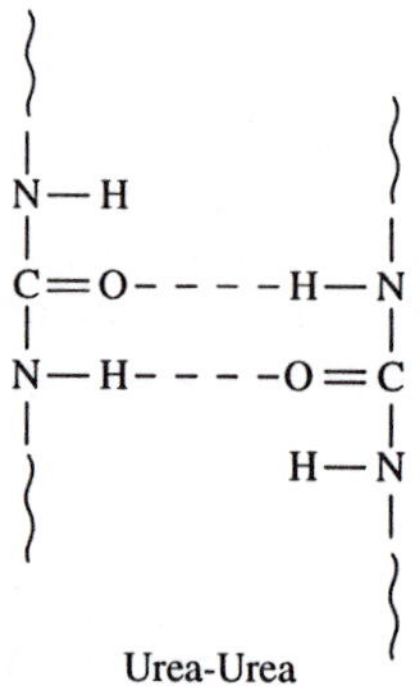

```
N—H
|
C=O----H—N
|        |
N—H----O=C
|        |
       H—N
```

Urea-Urea

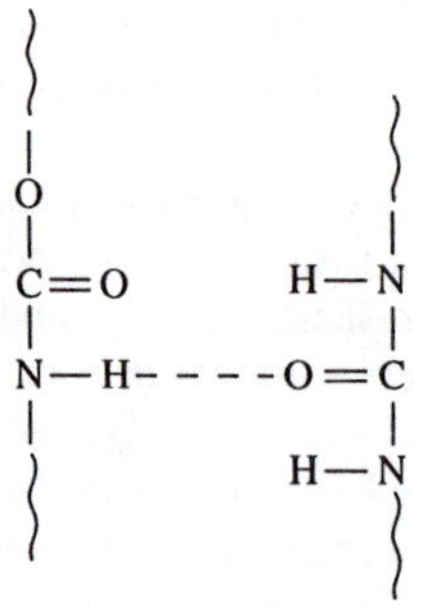

```
O
|
C=O    H—N
|        |
N—H----O=C
|        |
       H—N
```

Urethane-Urea

```
N—H
|        R
C=O      |
|        |
N—H------O
|        |
         R
```

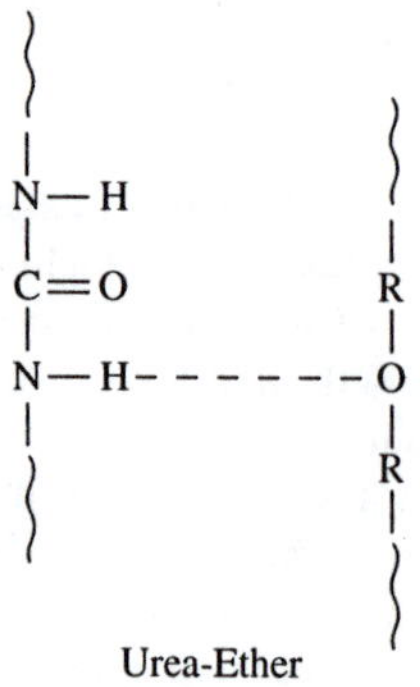

Urea-Ether

Table E5.6 (continued) Possible Hydrogen Bonding in Polyurethane

```
 }
 |          }
 N—H        |
 |          R
 C=O        |
 |          |
 N—H- - - -O=C
 |          |
 }          O
            |
            }
  Urea-Ester
```

The degree of microphase separation, therefore, depends on the relative magnitudes of intra- and interphase H-bonding. The strength and degree of hard segment–soft segment H-bonding are greater for the carbonyl group than the ether oxygen due to its relative polarities. Consequently, hard segment–soft segment compatibility is higher with polyester soft segment than polyether soft segment. With polybutadiene and/or polyisoprene soft segments, the potential for hard segment–soft segment H-bonding is nonexistent; only van der Waals forces operate between the two phases. Consequently, the tendency for phase separation is pronounced.

B. POLYMER-BOUND STABILIZERS[13,24]

Traditionally, stabilizers for polymers such as antioxidants, ultraviolet stabilizers, and flame retardants are low-molecular-weight compounds. However, several problems are associated with the use of these stabilizers. These include their loss from the finished product by evaporation or leaching during fabrication. Also, compatibility at higher levels of incorporation of the stabilizers may be problematic. Another problem is the potential danger of toxicity. Low-molecular-weight chlorine, bromine, or phosphorous compounds commonly used as flame retardants, as well as phenolic derivatives that are frequently employed as antioxidants or ultraviolet stabilizers, are potentially toxic. Oligomeric or, possibly, polymer-bound stabilizers have the potential for eliminating these problems. These are prepared by synthesizing suitable functional monomers followed by homo- and/or copolymerization.

1. Antioxidants

2,6-ditertiarybutyl-1,4-vinyl phenol or 4-isopropenyl phenol polymerizes readily with isoprene, butadiene, styrene, and methyl methacrylate (Equation 5.19).
The resulting copolymers are good antioxidants for their parent polymers at copolymer compositions of only 10 to 15 mol% of polymerizable antioxidant.

2. Flame Retardants

Flame retardants are usually halogen-containing materials. 2,4,6-Tribromophenyl, pentabromophenyl, and 2,3-dibromopropyl derivatives of acrylate and methacrylate esters can be readily polymerized or copolymerized with styrene, methyl methacrylate, and acrylonitrile to produce polymers with improved flame retardancy (Equation 5.20).

Table 5.7 Typical Polyols Used in Syntheses of Polyurethane Elastomers

Polyols	
$HO-(CH_2CH_2-O-)_n\ H$	Polyethylene oxide (PEO)
$HO-(CH_2CH_2CH_2CH_2-O-)_n\ H$	Polytetramethylene oxide (PTMO)
$HO-(CH_2CH(CH_3)-O-)_n\ H$	Polypropylene oxide (PPO)
$HO-[C-(CH_3)_2-CH_2]_n-OH$	Polyisobutylene (PIB)
$HO-[(CH_2)_4-O-OC(CH_2)CO-O-]_n\ H$	Polyethylene adipate (PEA)
$HO-[(CH_2)_3-CO-O-]_n\ H$	Polycapropactone (PCL)
$HO-(CH_2)_4-[Si(CH_3)_2-O-]_n-Si(CH_3)_4-OH$	Polydimethylsiloxane, hydroxybutyl terminate (PDMS)–OH
$HO-CH_2-CH=CH-CH_2-CH-CH_2CH(CH=CH_2)=CH-CH_2-OH$	Polybutadiene (PBD)

From Gogolewski, S., *Colloid Polym. Sci.*, 267, 757, 1989. With permission.

Table 5.8 Typical Chain Extenders Used in Syntheses of Polyurethane Elastomers

Chain Extenders	
$HO-(CH_2)_4-OH$	1,4-Butanediol (BD)
$HO-(CH_2)_6-OH$	Hexanediol (HD)
$HO-(CH_2)_2-OH$	Ethylene diol (ED)
$HO-(CH_2)_2-O-(CH_2)_2-OH$	Diethylene diol (DED)
$H_2N-(CH_2)_2-NH_2$	Ethylenediamine (EDA)
$H_2N-CH_2-CH(CH_3)-NH_2$	Propylenediamine (PDA)

From Gogolewski, S., *Colloid Polym. Sci.*, 267, 757, 1989. With permission.

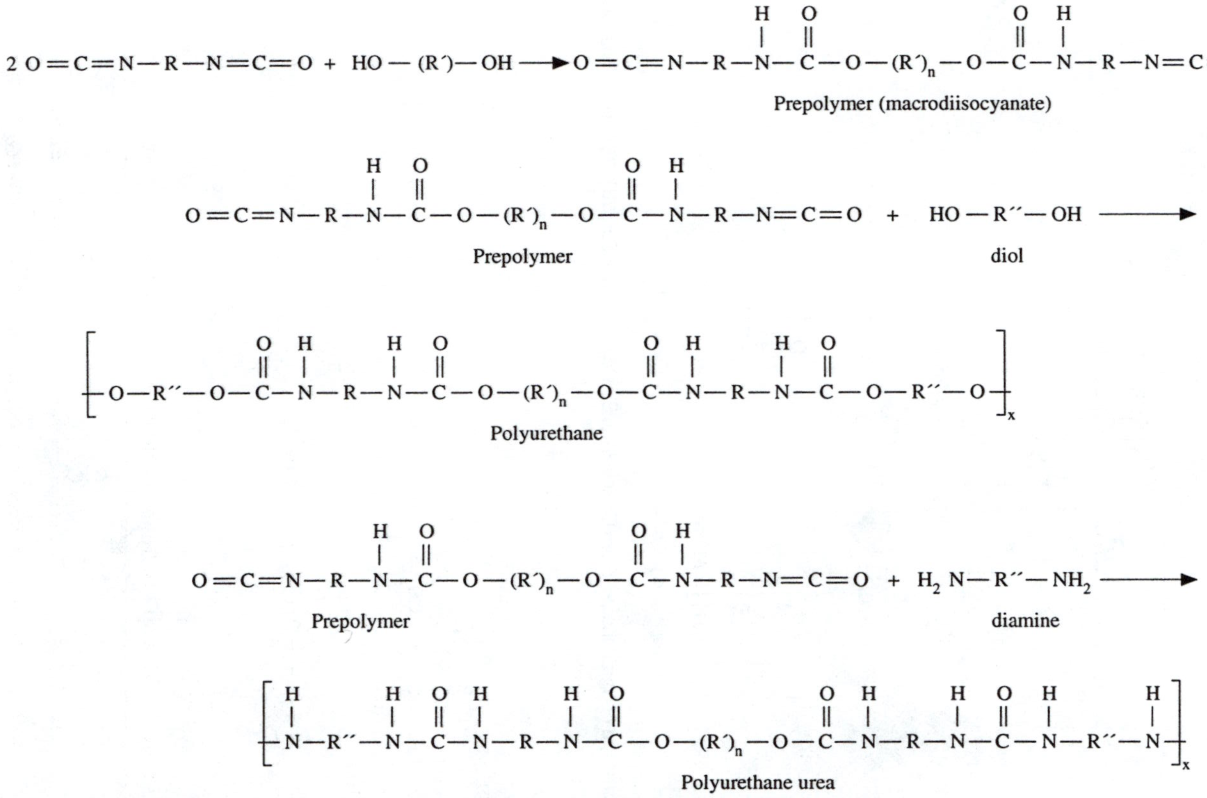

$$2\,O{=}C{=}N{-}R{-}N{=}C{=}O + HO{-}(R')_{}{-}OH \longrightarrow O{=}C{=}N{-}R{-}NH{-}C({=}O){-}O{-}(R')_n{-}O{-}C({=}O){-}NH{-}R{-}N{=}C{=}O \quad (5.16)$$

Prepolymer (macrodiisocyanate)

$$O{=}C{=}N{-}R{-}NH{-}C({=}O){-}O{-}(R')_n{-}O{-}C({=}O){-}NH{-}R{-}N{=}C{=}O + HO{-}R''{-}OH \longrightarrow$$

Prepolymer diol

$$\left[{-}O{-}R''{-}O{-}C({=}O){-}NH{-}R{-}NH{-}C({=}O){-}O{-}(R')_n{-}O{-}C({=}O){-}NH{-}R{-}NH{-}C({=}O){-}O{-}R''{-}O{-}\right]_x \quad (5.17)$$

Polyurethane

$$O{=}C{=}N{-}R{-}NH{-}C({=}O){-}O{-}(R')_n{-}O{-}C({=}O){-}NH{-}R{-}N{=}C{=}O + H_2N{-}R''{-}NH_2 \longrightarrow$$

Prepolymer diamine

$$\left[{-}NH{-}R''{-}NH{-}C({=}O){-}NH{-}R{-}NH{-}C({=}O){-}O{-}(R')_n{-}O{-}C({=}O){-}NH{-}R{-}NH{-}C({=}O){-}NH{-}R''{-}NH{-}\right]_x \quad (5.18)$$

Polyurethane urea

Figure 5.12 Equations representing the two-step process for producing polyurethane copolymers.

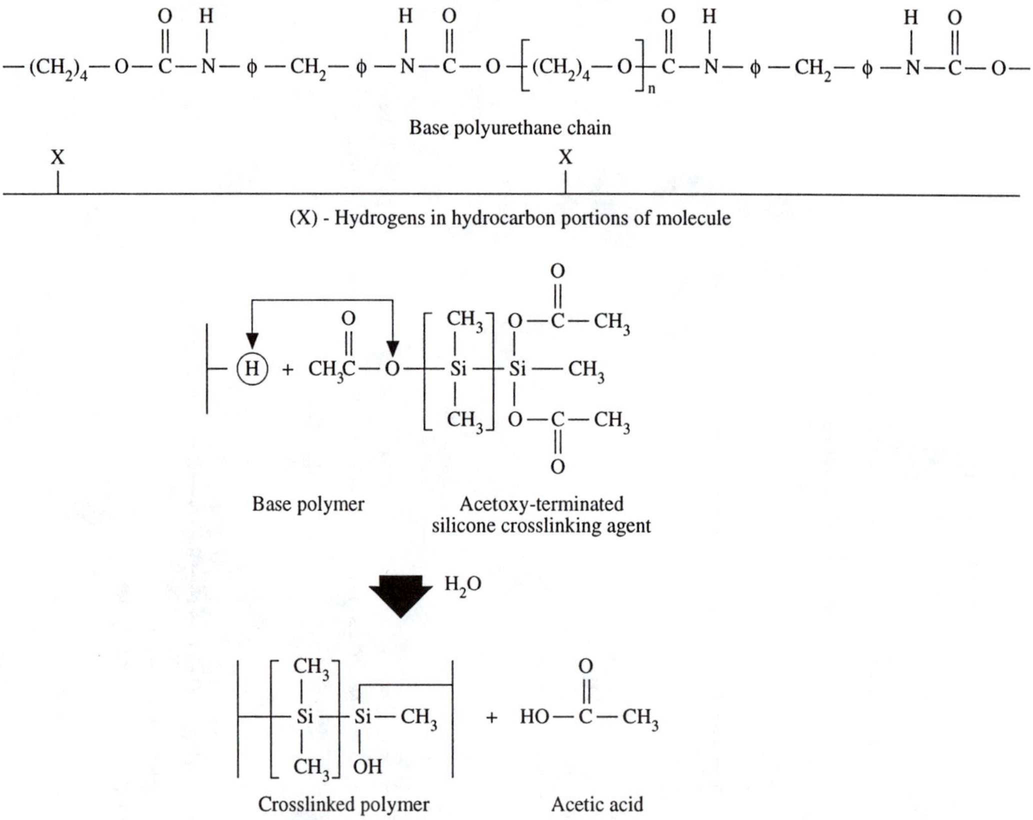

$$-(CH_2)_4-O-\overset{O}{\overset{\|}{C}}-\overset{H}{\overset{|}{N}}-\phi-CH_2-\phi-\overset{H}{\overset{|}{N}}-\overset{O}{\overset{\|}{C}}-O-\left[(CH_2)_4-O\right]_n-\overset{O}{\overset{\|}{C}}-\overset{H}{\overset{|}{N}}-\phi-CH_2-\phi-\overset{H}{\overset{|}{N}}-\overset{O}{\overset{\|}{C}}-O-$$

Base polyurethane chain

(X) - Hydrogens in hydrocarbon portions of molecule

$$\vdash \text{(H)} + CH_3\overset{O}{\overset{\|}{C}}-O-\left[\underset{CH_3}{\overset{CH_3}{Si}}\right]-\underset{O-\overset{}{C}(=O)-CH_3}{\overset{O-C(=O)-CH_3}{Si}}-CH_3$$

Base polymer — Acetoxy-terminated silicone crosslinking agent

$\downarrow H_2O$

$$\left[\underset{CH_3}{\overset{CH_3}{Si}}\right]-\underset{OH}{Si}-CH_3 + HO-\overset{O}{\overset{\|}{C}}-CH_3$$

Crosslinked polymer — Acetic acid

Figure 5.13 Structure of Cardiothane. (From Gogolewski, S., *Colloid Polym. Sci.*, 267, 757, 1989. With permission.)

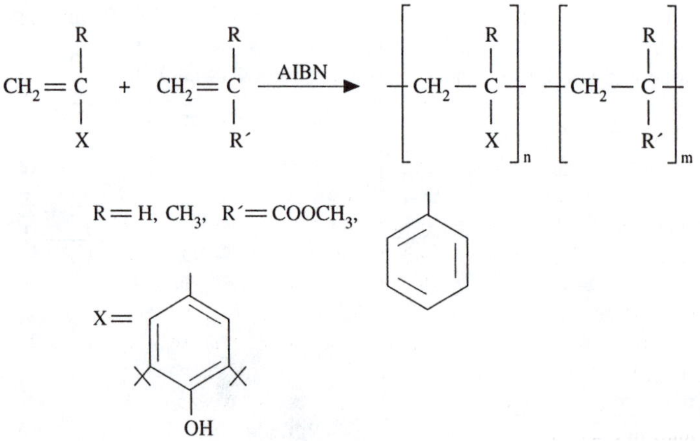

$$CH_2=\underset{X}{\overset{R}{C}} + CH_2=\underset{R'}{\overset{R}{C}} \xrightarrow{AIBN} \left[CH_2-\underset{X}{\overset{R}{C}}\right]_n\left[CH_2-\underset{R'}{\overset{R}{C}}\right]_m \quad (5.19)$$

$R=H, CH_3$, $R'=COOCH_3$, (phenyl)

X = (3,5-di-tert-butyl-4-hydroxyphenyl; OH)

$$CH_2{=}\underset{COOX}{\overset{R}{C}} + CH_2{=}\underset{M}{\overset{R}{C}} \longrightarrow \left[CH_2-\underset{COOX}{\overset{R}{C}}\right]_n \left[CH_2-\underset{M}{\overset{R}{C}}\right]_m \tag{5.20}$$

$R{=}H, CH_3$

$X{=} -CH_2CH_2Br-CH_2Br,$ (tribromophenyl), (pentabromophenyl)

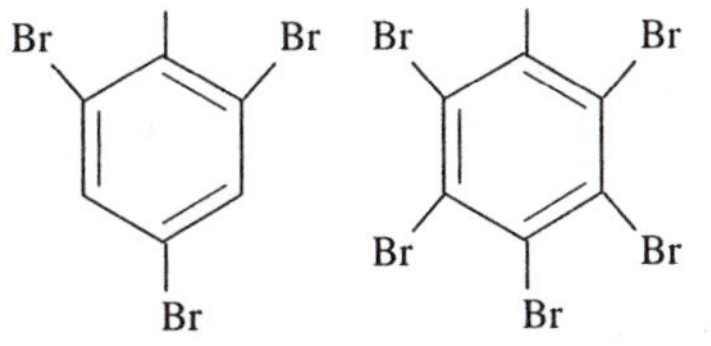

$M{=} -COOCH_3, -CN;$ (phenyl)

Polymerizable groups
$-CR{=}CH_2$
$-OCO-CR{=}CH_2$
$-OCH_2CHOHCH_2OCOCR{=}CH_2$
$-CH_2NHCOCH{=}CH_2$
$R{=}H, CH_3$

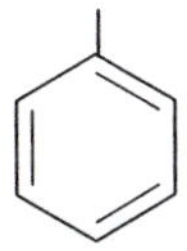

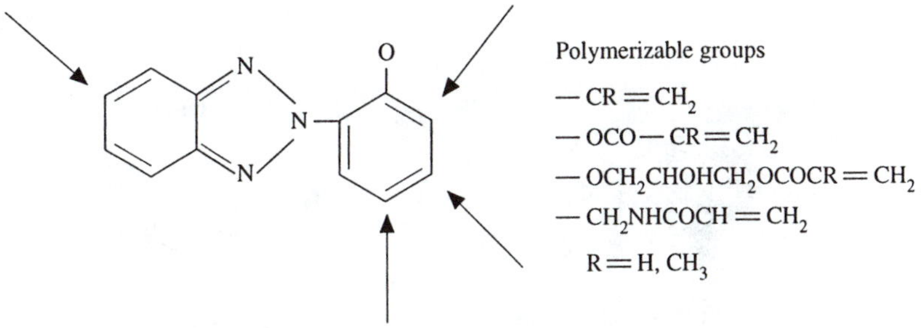

Figure 5.14 Polymerizable 2(2-hydroxyphenyl) 2H-benzotriazole derivatives.

Incorporation of only about 5% of these copolymers in polystyrene increased the limited oxygen index by 30%.[13] This is substantially lower than the 10 to 12% of the usual flame retardant materials required to protect polystyrene and similar polymers from being highly inflammable.

3. Ultraviolet Stabilizers

Upon exposure of polymeric materials to sunlight, ultraviolet stabilizers tend to evaporate from their surfaces where, incidentally, the stabilizing action is particularly required. 2(2-Hydroxyphenyl) 2H-benzotriazole derivatives (Figure 5.14) have been recognized to be the most effective compounds for protecting polymeric materials from ultraviolet and photodegradation. Small quantities (a few percentage points) of photostabilizers have been introduced into polyesters, polycarbonates, and polyamide using hydroxy-, acetoxy-, carboxyl-, and carbomethoxy-derivatives of this compound. These polymer-bound stabilizers are substantially more effective than the lower molecular weight UV stabilizers. The modified polyester has been synthesized directly. For example, poly(ethylene sebacate) was made from dimethyl sebacate and ethylene glycol with 1 to 3% of ethylene glycol replaced by 2(2,4-dihydroxyphenyl) 2H-benzotriazole (Equation 5.21). Similarly, ultraviolet-stabilized polycarbonate was made using bisphenol A and diphenyl carbonate (Equation 5.22).

$$\text{HO–Bzt–}C_6H_3(OH)_2 \;\text{or}\; \text{HO–Bzt–}C_6H_2(OH)_2\text{–Bzt–OH} \;+\; CH_3O\text{–}\overset{O}{\overset{\|}{C}}\text{–R–}\overset{O}{\overset{\|}{C}}\text{–}OCH_3 \;+\; \text{HO–R}'\text{–OH} \tag{5.21}$$

$$\longrightarrow \sim\text{R}'\text{–O–}\overset{O}{\overset{\|}{C}}\text{–R–}\overset{O}{\overset{\|}{C}}\text{–O–Bzt–}C_6H_3(OH)\text{–O–}\overset{O}{\overset{\|}{C}}\text{–R–}\overset{O}{\overset{\|}{C}}\text{–O}\sim$$

or

$$\sim\text{R}'\text{–O–}\overset{O}{\overset{\|}{C}}\text{–R–}\overset{O}{\overset{\|}{C}}\text{–O–Bzt–}C_6H_2(OH)_2\text{–Bzt–O–}\overset{O}{\overset{\|}{C}}\text{–R–}\overset{O}{\overset{\|}{C}}\text{–O}\sim$$

Bzt = (benzotriazol-2-yl, N–); R = $-(CH_2)_8-$, (1,3-phenylene), (1,4-phenylene);

R′ = $-(CH_2)_n-$, n = 4, 6, 10

$$HO{-}Bzt{-}C_6H_3(OH)_2 \;\text{or}\; (HO{-}Bzt)_2C_6H_2(OH)_2 + Ph{-}O{-}\overset{O}{\overset{\|}{C}}{-}O{-}Ph + HO{-}R'{-}OH \longrightarrow \tag{5.22}$$

$$\sim R'{-}O{-}\overset{O}{\overset{\|}{C}}{-}O{-}Bzt{-}C_6H_3(OH){-}O{-}\overset{O}{\overset{\|}{C}}{-}O\sim \quad \text{or} \quad \sim R'{-}O{-}\overset{O}{\overset{\|}{C}}{-}O{-}Bzt{-}C_6H_2(OH)_2{-}Bzt{-}O{-}\overset{O}{\overset{\|}{C}}{-}O\sim$$

$$Bzt = \text{benzotriazol-2-yl}\;(-C_6H_3N_2N{-});\quad R' = -C_6H_4{-}C(CH_3)_2{-}C_6H_4-$$

The discussion so far involves cases of functional polymers where the reactive/functional groups are attached directly to the backbone. The reactivity of a functional group directly attached to the polymer backbone depends on the proximity, flexibility, and polarity of the polymer main chain. The interference from the polymer main chain can be eliminated, or at least reduced, through functional groups attached as side groups or through the use of spacer groups inserted between the reactive groups and the polymer main chain. For example, in addition to the techniques discussed above, the ultraviolet stabilizer (2-(2-hydroxyphenyl) 2H-benzotrizole can be incorporated in the polymer structure using functional monomers or polymers with reactive side groups: the ring-opening reaction of the oxirane ring of glycidyl methacrylate followed by copolymerization with methyl methacrylate produces UV-stabilized PMMA (Equation 5.23).[13]

(5.23)

R = H, OMe, Cl
R´ = H, CH_3
R´´ = COOMe, C_6H_5

PMMA containing a small amount of glycidyl methacrylate units, when allowed to react with tetraalkyl ammonium salts, will undergo ring-opening reaction with 2(2,4-dihydroxyphenyl) 2H-benzotriazole (Equation 5.21).

R = H, CH_3

(5.24)

Natural polymers very frequently link functional groups via spacer groups. For example, the amino acids lysine and glutamic acid have some of their functional groups linked by spacer groups (Figure 5.15). In synthetic polymer, spacer groups might be stiff groups like the phenyl group in *p*-substituted polystyrenes used in ion-exchange resins or, as is usually the case, flexible groups such as methylene or

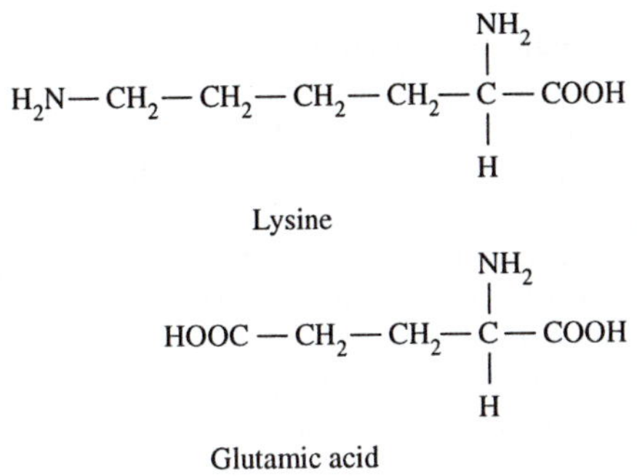

$$H_2N-CH_2-CH_2-CH_2-CH_2-\underset{\displaystyle H}{\overset{\displaystyle NH_2}{\overset{|}{\underset{|}{C}}}}-COOH$$

Lysine

$$HOOC-CH_2-CH_2-\underset{\displaystyle H}{\overset{\displaystyle NH_2}{\overset{|}{\underset{|}{C}}}}-COOH$$

Glutamic acid

Figure 5.15 Spacer groups in amino acids.

fluorocarbon groups. Fluorocarbon ether groups have been used successfully to separate sulfonate or carboxylate groups from the polytetrafluoroethylene main chain to produce fluorocarbon polymer membrane of unusual properties.[24] Polymers have also been synthesized with spacer groups that have polyethylene, polyoxyethylene, and even the much stiffer polymethacrylate backbone. Polymethacrylates have been particularly useful as polymers with functional groups that can be separated effectively from the polymer main chain. Polymers and copolymers based on hydroxyethyl methacrylate or glycidyl methacrylate, for example, have been used in biomedical applications.

C. POLYMERS IN DRUG ADMINISTRATION[28,30]

The medical field is no longer imaginable without synthetic polymers. They are used both as implant and surgical materials and as drugs or carriers for drugs. Drugs are hardly ever administered to patients in an unformulated state. Polymers have long been employed as excipient for adjusting the consistency and release properties of simple solid, cream, and liquid formulations. Polymers that have typically been used in drug delivery include various cellulose derivatives, polyacrylates, poly(vinyl pyrrolidone), polyoxyethylene, poly(vinyl alcohol), and poly(vinyl acetate). Poly(vinyl pyrrolidone), poly(2-hydroxypropyl methacrylamide), and polyoxazoline are examples of synthetic polymers that have been used as plasma expanders.[28] We discuss a few examples to illustrate the use of functional polymers in drug administration.

The traditional approach in pharmaceutical research is to focus effort on the discovery of new compounds with biological activity that can be used in the treatment of diseases. The main problem associated with this approach is that the drug may be distributed to a variety of sites in the body where it may be inactive, harmful, and/or toxic. Both the therapeutic and harmful/toxic effects of drugs depend on their concentrations at the various sites in the body.[28] There is normally an optimal therapeutic range within which the therapeutic effects of the drug outweigh its harmful effects. Outside this range, the drug may either be inactive or its harmful effects may predominate (Figure 5.16).

Two distinct approaches are currently used to improve drug action through its mode of delivery. These are controlled drug release and site-directed or targeted drug delivery.

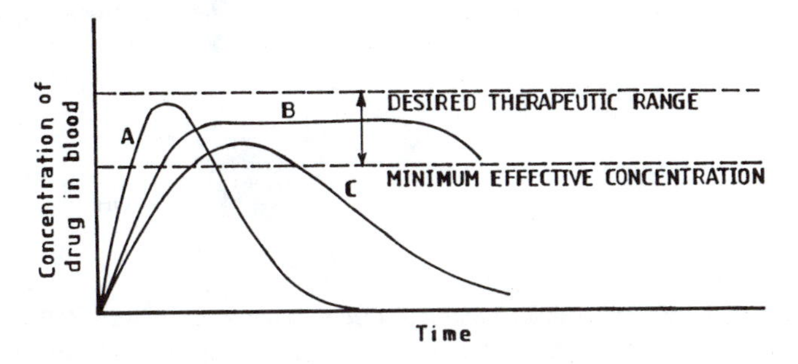

Figure 5.16 A schematic plot of different drug administration routes. (A) Single dose preparation; (B) sustained release preparation; (C) prolonged release preparation.

1. Controlled Drug Release

Most drugs are low-molecular-weight compounds, which when administered conventionally are often rapidly excreted from the body. Intermittent doses result in a sawtooth pattern of the drug in circulation due to the rise and fall of the concentration of the drug. Consequently, large and repeated doses are required to sustain a therapeutic effect. The aim of controlled drug release, therefore, is to eliminate or at least reduce the danger of overdose, with risk of side effects or subtherapeutic blood levels, by producing and maintaining an optimal therapeutic concentration of the drug in the body. Most controlled-release delivery systems rely on a polymer to regulate the flow of the therapeutic agent out of the device. A number of techniques are used to achieve this objective, including encapsulation, dispersion in hydrophobic vehicles or porous polymers, binding of the drug to macromolecules, and formation of drug–carrier complexes. We illustrate a couple of these techniques.

a. Tablets with Prolonged and Sustained Drug Release[29]

The overall release profile of a drug can be controlled by the physical shape of the device or by the chemical nature of the polymer matrix. Tablets with a sustained and prolonged delivery have been prepared using polymers and terpolymers based on acrylic and methacrylic acids as coatings for some dosage forms. The behavior of these polymeric materials in the various parts of the gastrointestinal tract varies depending on the chemical structure and molar mass and on the ratio of basic, acidic, and hydrophobic groups. Thus the location and rate of release of the active substance can be controlled.

Figure 5.17 shows a typical cationic copolymer. It is basic, swells in the pH range of 5 to 8 but is soluble at pH 2.5. The anionic polymers Eudragit L, S (Figure 5.18) are based on methacrylic acid and methacrylic acid methyl ester. They are resistant to gastric juice, but soluble in the small intestine and insoluble and impermeable to water below pH 6 to 7 range. They consequently provide full protection against gastric acid and ensure drug release in the small intestine.

Another variation of these acrylic copolymers are acrylic terpolymers based on methacrylic acid with a small amount of quaternary ammonium group. These terpolymers are water insoluble in the entire physiologic field. They, however, swell and are permeable to water and drugs. Their permeability is independent of pH and can be controlled by appropriate structural modification. The polymeric films behave essentially as semipermeable membranes with a domain microstructure: the predominant hydrophobic parts of the macromolecule cause insolubility in water, while the hydrophobic quaternary groups permit the passage of water and drug molecules.

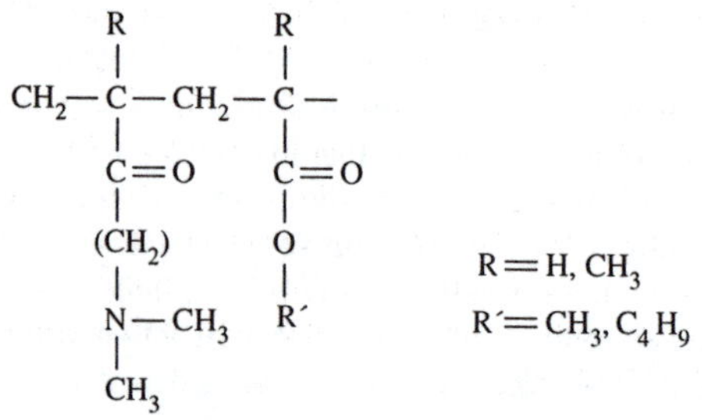

$$-CH_2-C(R)(C{=}O-(CH_2)-N(CH_3)-CH_3)-CH_2-C(R)(C{=}O-O-R')-$$

$R{=}H, CH_3$

$R'{=}CH_3, C_4H_9$

Figure 5.17 Eudragit E-cationic polymer. (From Kalal, J., *Makromol. Chem. Macromol. Symp.*, 12, 259, 1987.)

$$-CH_2-C(CH_3)(C{=}O-OH)-CH_2-C(C{=}O-O-CH_3)-$$

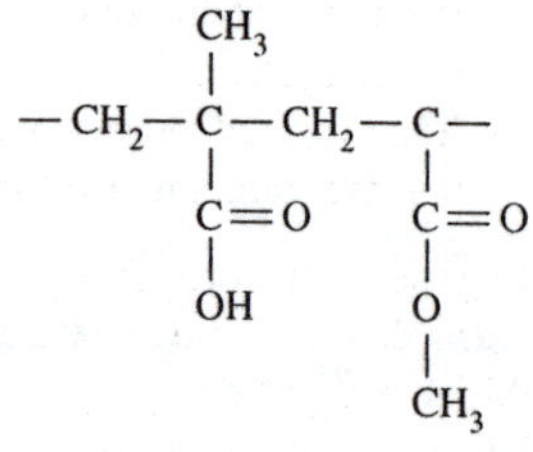

Figure 5.18 Eudragit anionic polymers: (L) $-COOH/-OCH_3 = 1/1$; (S) $-COOH/-OCH_3 = 1/2$. (From Kalal, J., *Makromol. Chem. Macromol. Symp.*, 12, 259, 1987.)

b. Degradable Polymers[13]

Polymers with functionalities in the main chain very often have aromatic groups that consequently cause the polymers to become infusible, intractable, and nonfabricable.[30,31] For example, the copolycarbonates or bithionol and glycols where the glycol units are small are essentially intractable. However, by using an appropriate spacer group, the hydrophilicity of the polymer can be modulated. This is achieved by using bithionol — an antibacterial agent — and a bisphenol as a comonomer in an alternating copolycarbonate with a poly(ethylene oxide) glycol with a DP of at least 10 (Equation 5.24). The resulting polycarbonate has useful physical characteristics. In addition, it is degradable under physiologic conditions for controlled release of the bactericide bithionol and of the carbon dioxide and poly(ethylene oxide) glycol.

$$\mathrm{Cl{-}C({=}O){-}O{-}C_6H_3(Cl){-}S{-}C_6H_3(Cl){-}O{-}C({=}O){-}Cl} + \mathrm{HO{+}CH_2CH_2{-}O{+}_m H} \longrightarrow \left[\mathrm{{-}C({=}O){-}O{-}C_6H_3(Cl){-}S{-}C_6H_3(Cl){-}O{-}C({=}O){-}O{+}CH_2CH_2{-}O{+}_m}\right]_n \qquad (5.25)$$

Polycarbonate of bithionol with poly(ethyleneoxide) glycol as spacer group

2. Site-Directed (Targeted) Drug Delivery

The basic principle in site-directed or targeted drug delivery is that the carrier will recognize the disease-related target or biochemical site of action, interact with it, and concentrate delivery of the drug at the site. The drug is thus rendered inactive elsewhere in the body. The polymer in targeted drug delivery systems acts merely as a carrier with no intrinsic pharmacologic activity or therapeutic effect. In other words, the macromolecule is chemically modified with molecules of the drug and other moieties to overcome some undesirable characteristics of the drug, but can be transformed by a chemical or enzymatic process to release the pharmacologically active drug at the appropriate time or location (Figure 5.19). To satisfy this requirement the macromolecule must be designed to accommodate features that may even be in conflict. Some of these features are[28]

- It must provide overall solubility to the targeting system, and after the drug release the carrier itself must be both water soluble and capable of being totally excreted by the body after the required period of time.
- It may need to contain groups available for the chemical attachment of drugs via biodegradable linkages.
- It must contain structural features necessary for interaction specifically with the discrete features of the biosystem.
- It is expected to prevent nonspecific interactions between the components of the biological system and polymer targeting system.

In general, the properties of a biosystem and a synthetic polymer as well as the nature of the biological medium dictate the degree and type of interaction between a biostructure and a polymer. The biocompatibility of synthetic polymers depends on their chemical nature, physical state, and macroscopic form, which can be modified by functionalization of the polymer skeleton. Many biopolymers, such as proteins and nucleic acids, are natural polyelectrolytes. Similarly, the outer cell membrane of living cells has charged groups. The biological medium is an electrolyte with an aqueous phase. Therefore, electrostatic

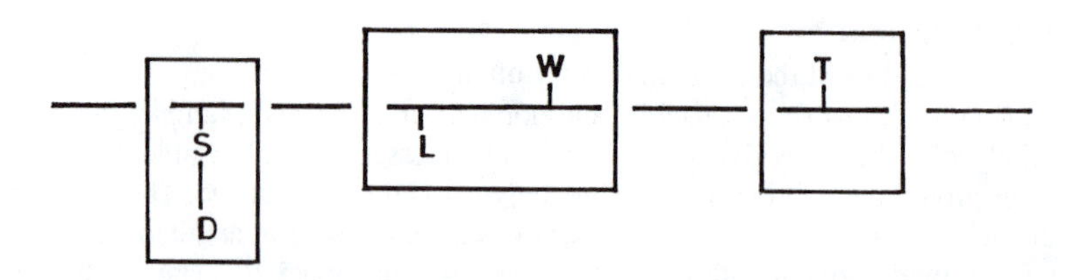

Figure 5.19 Generalized structure of soluble polymeric drug carriers: (D) drug; (S) spacer group; (L) and (W) groups that determine the solubility of macromolecule in nonaqueous and aqueous environment, respectively; (T) a group specifically recognized by an organ/tissue/cell of the body. (From Petrak, K., *Br. Polym. J.*, 22, 213, 1990. With permission.)

forces play a dominant role in the reversible interaction of biopolymer with synthetic polymers, while covalent bonding serves the irreversible immobilization of biostructures.[32]

Most polymer carriers per se do not possess the ability to recognize the target site. Consequently, biological macromolecules such as polysaccharides, antibodies, and toxin fragments are introduced into the polymer carrier as the directional component and to ensure the required selectivity. The functional groups of the polymer are usually modified or activated to effect coupling of the biomolecule (Table 5.9).

Sometimes spacer groups are used to increase the distance between the polymer carrier and the biostructure so as to reduce interference from the polymer main chain and maintain proper functional properties. While there are currently no efficient site-directed drug delivery systems available for practical clinical applications, the feasibility of the general concept has been amply demonstrated. Examples are prodrugs based on *N*-(2-hydroxypropyl)-methacrylamide copolymers containing oligopeptide side chains terminating in an anticancer drug, Adriamycin (Figure 5.20). The bond between the drug and the carrier is stable in the bloodstream, but is cleared intracellularly on exposure to appropriate enzymes.[29]

Table 5.9 Chemical Modification of Synthetic Polymers for Covalent Linkage of Biomacromolecules and Affine Ligands

Functional Group of the Polymer	Chemical Modification (activation) of the Polymer	Binding Group of the Biomacromolcule and Ligand
–OH	Cyanogen bromide	$-NH_2$
	Triazines	$-NH_2$
	Periodate oxidation	$-NH_2$
	Benzoquinones	$-NH_2$
	Epoxides	$-NH_2$, –COOH, –OH,–SH
	Silanization	$-NH_2$, –COOH, $-C_6H_5$-R
	Carboxylating reagents	$-NH_2$
–COOH, $-NH_2$ Carbodiimides	$-NH_2$, –COOH	
	Acylation	$-NH_2$, –COOH
	Active esters	$-NH_2$, –COOH
–CHO		$-NH_2$
$-C_6H_4$–R	Diazotation reagents	$-C_6H_4$–R, histidine, tryptophan
–SH	Thiol-disulfide exchange	–SH

From Kalal, J., *Makromol. Chem. Macromol. Symp.*, 12, 259, 1987.

V. PROBLEMS

5.1. When two homopolymers with the same melt index are blended together by the addition of a compatibilizing agent, the resulting blend has a higher melt index. Explain.

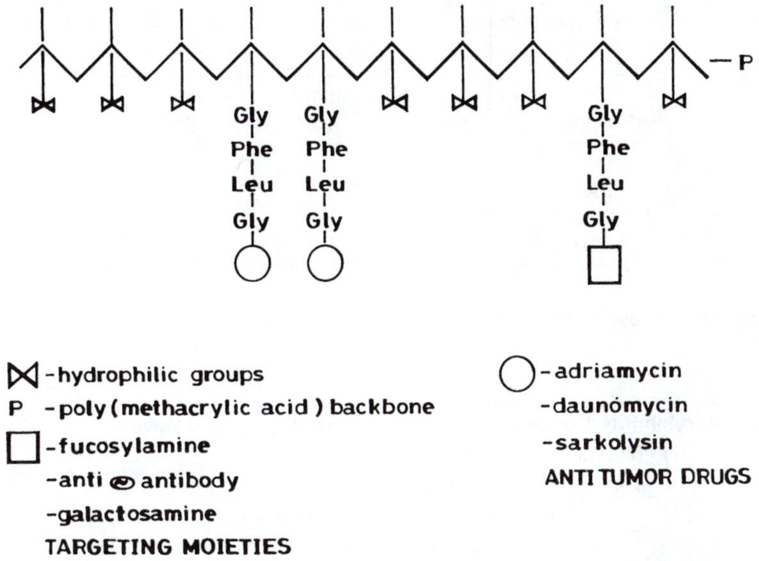

Figure 5.20 Schematic diagram of *N*-(2-hydroxypropyl)-methacrylamide copolymers containing oligopeptide side chains terminating in anticancer drugs or targeting moieties. (From Kalal, J., *Makromol. Chem. Macromol. Symp.*, 12, 259, 1987.)

5.2. Dextrins are degradative products of starch obtained by heating starch in the presence or absence of hydrolytic agents. Typical conditions for the white and yellow (canary) dextrins are shown in the following table.

Manufacturing Condition	White Dextrins	Canary Dextrins
Usual catalyst	HCl	HCl
Temperature, °C	79–121	149–190
Time, h	3–7	6–20

Envelope front seal adhesives are made from solutions of dextrins in water. They require high solid contents, usually 60 to 70%. Which of the two dextrin types (white or yellow) would be more suitable as adhesives for envelope seals?

5.3. Indicate how you would increase:

a. The moisture resistance of nylon 6
b. The toughness and crack resistance of polyethylene (HDPE)
c. Impact resistance of polystyrene

5.4. Explain the observed trends in the data for the melt index and the peel strength to aluminum of polyethylene (PE) and ethylene–acrylic acid copolymer (EAA) shown in the following table.

Polymer	Acrylic acid (mol%)	Melt Index (g)	Peel Strength (lb/in.)
PE	0	1.5	<1
EAA	4.6	83	56
EAA	6.0	58	81
EAA	8.2	52	100

5.5. Nitrile rubbers possess outstanding oil-resistant properties. Explain the following variation in properties of nitrile rubber.

Acrylonitrile Content in Copolymer (%)	Properties of Nitrile Rubber
2–5	Rubbery with poor oil resistance
15	Rubbery with fair oil resistance
25	Rubbery with medium oil resistance
35–40	Leathery wit high oil resistance
50–60	Tough, leathery plastic with high resistance to aromatics

5.6. Suggest a possible method to effect the following changes through copolymerization. Briefly explain the basis of your suggestion.

i. Increase:

(a) the rigidity of $\left[O-(CH_2)_{20}-CO\right]_n$ (Str. 11)

(b) the chemical resistance of $\left[CH_2-CH\right]_n$ with a phenyl group on CH (Str. 12)

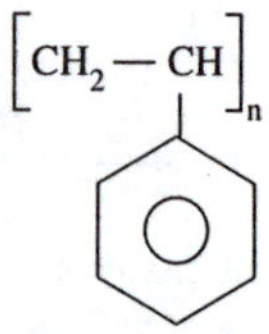

(c) the flexibility of $\left[\overset{O}{\overset{\|}{C}}-C_6H_4-\overset{H}{\overset{|}{N}}\right]_n$ (Str. 13)

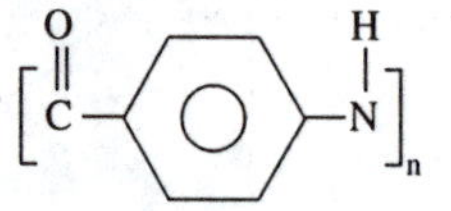

(d) adhesion to polar substrates of $\left[CH_2-CH_2\right]_n$ (Str. 14)

(e) the temperature resistance of $\left[\overset{H}{\overset{|}{N}}-(CH_2)_6-\overset{H}{\overset{|}{N}}-\overset{O}{\overset{\|}{C}}-(CH_2)-\overset{O}{\overset{\|}{C}}\right]_n$ (Str. 15)

ii. Obtain cross-linked polymers from:

(a) 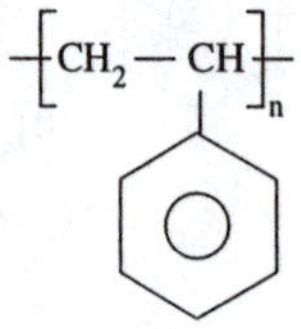 $\left[CH_2-CH\right]_n$ with a phenyl group on CH (Str. 16)

(b) 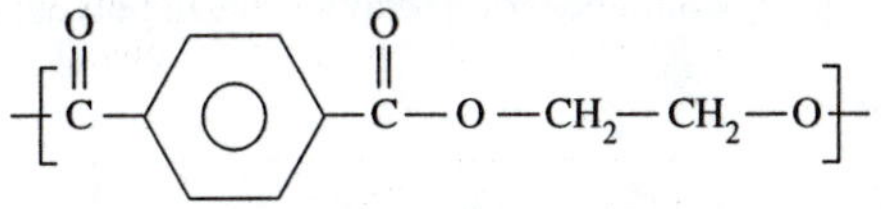 $\left[-\overset{\overset{O}{\|}}{C}-C_6H_4-\overset{\overset{O}{\|}}{C}-O-CH_2-CH_2-O- \right]$ (Str. 17)

5.7. An unsaturated polyester based on maleic anhydride is to be used in an application requiring good creep resistance. Which of the available unsaturated monomers (methyl methacrylate or vinyl acetate) would you choose for curing the polyester? Explain your choice.

5.8. The nature of cross-linked urea–formaldehyde (UF) resins is thought to be

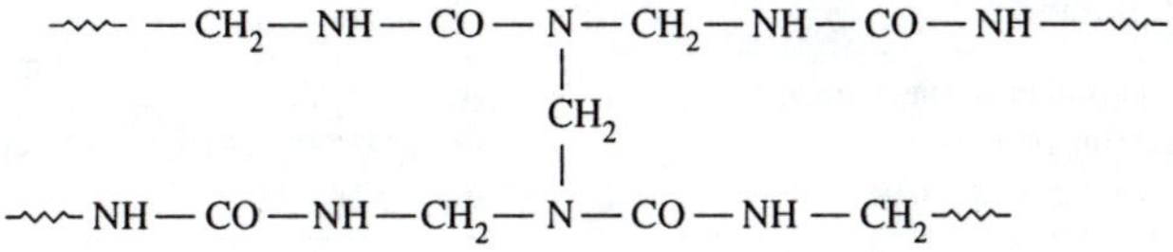

$$\sim\sim -CH_2-NH-CO-N(-CH_2-)-CH_2-NH-CO-NH-\sim\sim$$
$$\sim\sim NH-CO-NH-CH_2-N(-CH_2-)-CO-NH-CH_2\sim\sim$$

(Str. 18)

Films from cured UF resins are very brittle. Arrange the following components in their increasing ability to reduce the brittleness of UF films through copolymerization with urea and formaldehyde.

(a) $H_2N-CO-NH-(CH_2)_{12}-NH-CO-NH_2$ (Str. 19)

(b) 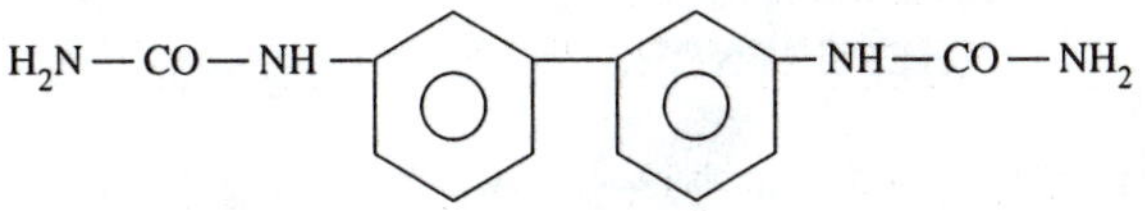 $H_2N-CO-NH-C_6H_4-C_6H_4-NH-CO-NH_2$ (Str. 20)

(c)  $H_2N-CO-NH-C_6H_4-NH-CO-NH_2$ (Str. 21)

(d) $H_2N-CO-NH-(CH_2)_6-NH-CO-NH_2$ (Str. 22)

5.9. Polyimides are formed by the reaction:

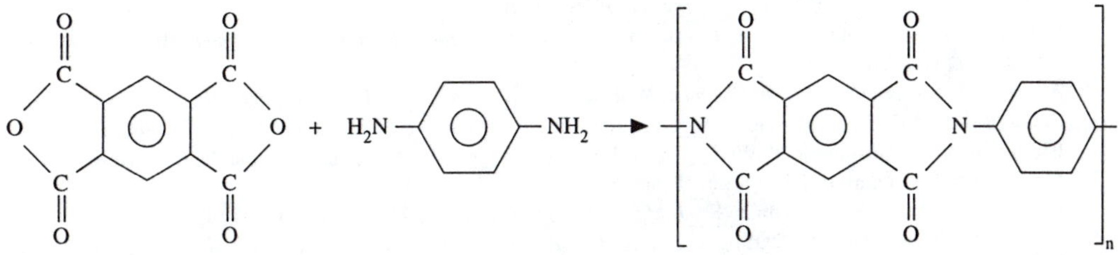

pyromellitic dianhydride + $H_2N-C_6H_4-NH_2$ $\rightarrow$ $\left[-N(CO)_2C_6H_2(CO)_2N-C_6H_4- \right]_n$

(Str. 23)

Polyimides are insoluble, infusible polymers that cannot be used as adhesives. Which of the following two compounds would you copolymerize with the starting monomers to produce a useful adhesive? Comment on the temperature resistance of the resulting adhesive.

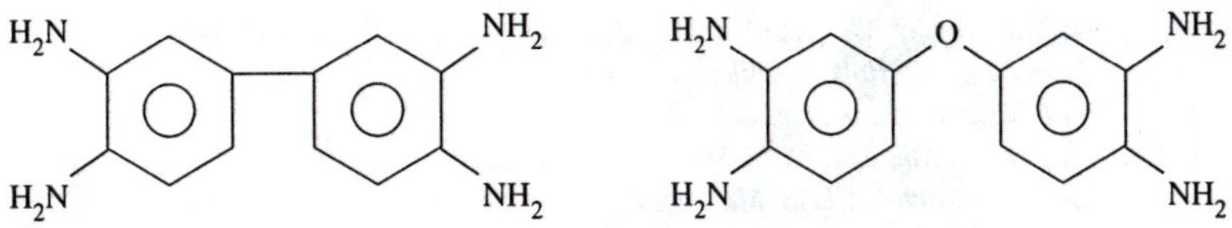

(A) $(H_2N)_2C_6H_3-C_6H_3(NH_2)_2$

(B) $(H_2N)_2C_6H_3-O-C_6H_3(NH_2)_2$

(Str. 24)

5.10. How much sulfur, in parts per hundred resin (phr), would be required to vulcanize polybutadiene rubber of molecular weight 5400 assuming:

a. Disulfide cross-links
b. Decasulfide cross-links

5.11. UF resins are susceptible to hydrolytic degradation. Rank the following monomers in order of the expected increase in the hydrolytic stability of UF resins modified by the incorporation of these monomers through copolymerization of their urea derivatives. Explain the basis of your ranking order.

Monomer	Structure
1. Hexamethylenediamine	$H_2N–(CH_2)_6–NH_2$
2. Trimethylene oxidediamine	$H_2N–CH_2CH_2O–CH_2CH_2OCH_2CH_2-NH_2$
3. Dodecanediamine	$H_2N–(CH_2)_{12}–NH_2$

5.12. The stiffness of the polymer backbone chain has been found to be responsible for the minimum length of spacer groups required to make the reactivity of the functional group free from the interference of the polymer main chain. For poly(ethylene oxide) three to four methylene groups are required, whereas spacer groups of one or two methylene are necessary for polyethylene. Explain this observation.

REFERENCES

1. Stricharczuk, P.T. and Wright, D.E., Styrene-butadiene rubber adhesives, in *Handbook of Adhesives,* 2nd ed., Skeist, I., Ed., Van Nostrand Reinhold, New York, 1977.
2. Wake, W.C., *Adhesion and the Formulation of Adhesives,* Applied Science Publishers, London, 1976.
3. Ebewele, R.O., Myers, G.E., River, B.H., and Koutsky, J.A., *J. Appl. Polym. Sci.* 42, 2997, 1991.
4. Ebewele, R.O., Myers, G.E., River, B.H., and Koutsky, J.A., *J. Appl. Polym. Sci.* 43, 1483, 1991.
5. Jarowenko, W., Starch-based adhesives, in *Handbook of Adhesives,* 2nd ed., Skeist, I., Ed., Van Nostrand Reinhold, New York, 1977.
6. Fettes, E.M., Ed., *Chemical Reactions of Polymers,* Interscience, New York, 1964.
7. Mark, H., Gaylord, N.G., and Bikales, N.M., Eds., *Encyclopedia of Polymer Science and Technology,* Interscience, New York, 1976.
8. Ceresa, R.J., Block and graft copolymers, in *Encyclopedia of Polymer Science and Technology,* Mark, H., Gaylord, N.G., and Bikales, N.M., Eds., Interscience, New York, 1976.
9. Gaylord, N.G., and Aug, F.S., Graft copolymerization, in *Chemical Reactions of Polymers,* Fettes, E.M., Ed., Interscience, New York, 1964.
10. Angier, D.J., Surface reactions, in *Chemical Reactions of Polymers,* Fettes, E.M., Ed., Interscience, New York, 1964.
11. Subramanian, R.U., The adhesive system, in *Adhesive Bonding of Wood and Other Structural Materials,* Vol. 3, Blomguist, R.F., Christiansen, A.W., Gillespie, R.H., and Myers, G.E., Eds., Pennsylvania State University, 1983.
12. Merx, F.P.M. and Lemstra, P.J., *Polym. Commun.,* 31(7), 252, 1990.
13. Vogl, O., Jaycox, G.D., and Hatada, K., *J. Macromol. Sci. Chem.,* A27(13 and 14), 1781, 1990.
14. Prechet, J.M.J., Deratani, A., Darling, G., Lecavalier, P., and Li, N.H., New reactive polymers containing nitrogen functionalities from asymmetric synthesis to supported catalysis, *Makromol. Chem. Macromol. Symp.* 1, 91, 1986.
15. Soutif, J. and Brosse, J., Chemical modification of polymerizations. I. Applications and synthetic strategies, *React. Polym.* 12, 3, 1990.
16. Gogolewski, S., *Colloid Polym. Sci.,* 267, 757, 1989.
17. Frisch, K.C., Advances in Urethanes, seminar materials, Technomic, Westport, CT, 1985.
18. Speckhard, T.A., Verstrate, G., Gibson, P.E., and Cooper, S.L., *Polym. Eng. Sci.,* 23, 337, 1983.
19. Rausch, K.W., McClellan, T.R., Jr., d'Ancicco, V.V., and Sayigh, A.A.R., *Rubber Age,* 99, 78, 1967.
20. Verdol, J.A., Carrow, D.J., Ryan, P.W., and Kuncl, K.L., *Rubber Age,* 98, 62, 570, 1966.
21. Chang, V.S.C. and Kennedy, J.P., *Polym. Bull.,* 8, 69, 1982.
22. Mennicken, G.J., *Oil Colour Chem. Assoc.,* 49, 639, 1966.
23. Imai, Y. and Nose, K., *Trans. Am. Soc. Artif. Intern. Organs,* 17, 6, 1971.
24. Vogl, O. and Sustic, A., *Makromol. Chem. Macromol. Symp.,* 12, 351, 1987.
25. Thijs, L., Gupta, S.G., and Neckers, D.C., *J. Org. Chem.,* 44, 4123, 1979.
26. Gupta, S.N., Thijs, L., and Neckers, D.C., *J. Polym. Sci. Polym. Chem. Ed.,* 19, 855, 1981.
27. Gupta, S.N., Thijs, L., and Neckers, D.C., *Macromolecules,* 13, 1037, 1980
28. Petrak, K., *Br. Polym. J.,* 22, 213, 1990.

29. Kalal, J., *Makromol. Chem. Macromol. Symp.*, 12, 259, 1987.
30. Cho, B.W., Jin, J.I., and Lenz, R.W., *Makromol. Chem. Rapid Commun.*, 3, 23, 1982.
31. Jin, J.I., Antoum, S., Ober, C., and Lenz, R.W., *Br. Polym. J.*, 12, 132, 1982.
32. Scheler, W., *Makromol. Chem. Macromol. Symp.*, 12, 1, 1987.
33. Greek, B.F., *Chem. Eng. News*, p. 25, March 21, 1988.
34. Batyrbekov, E.O., Moshkeich, S.A., Rukhina, L.B., Bogin, R.A., and Zhubanov, B.A., *Br. Polym. J.*, 23(3), 273, 1990.

PART II: POLYMER PREPARATION AND PROCESSING METHODS

Chapter 6

Condensation (Step-Reaction) Polymerization

I. INTRODUCTION

Condensation polymerization is chemically the same as a condensation reaction that produces a small organic molecule. However, as we saw in Chapter 2, in condensation polymerization (i.e., production of a macromolecule) the functionality of reactants must be at least 2. Recall that functionality was defined as the average number of reacting groups per reacting molecule. To derive expressions that describe the physical phenomena occurring during condensation polymerization (polycondensation) — a tool vital to process design and product control — three approaches have been traditionally adopted: kinetic, stoichiometric, and statistical. Various degrees of success have been achieved by each approach. We treat each approach in the succeeding sections. Before then, we briefly discuss the overall mechanism of polycondensation reactions.

II. MECHANISM OF CONDENSATION POLYMERIZATION

The mechanism of polycondensation reactions is thought to parallel that of the low-molecular-weight analogs. As a result of their macromolecular nature, polymers would be expected to have retarded mobility. It was therefore predicted, purely on theoretical arguments, that the chemical reactivity of polymers should be low.

- The collision rate of polymer molecules should be small due to their low kinetic velocity. This should be accentuated by the high viscosity of the liquid medium consisting of polymer molecules.
- Shielding of the reactive group within the coiling chain of its molecule should impose steric restrictions on the functional group. This would lead to a reduction in the reactivity of the reactants.

Flory[1] has shown from empirical data that for a homologous series the velocity constant measured under comparable conditions approaches an asymptotic limit as the chain length increases. He therefore proposed the equal reactivity principle, which states in essence that "the intrinsic reactivity of all functional groups is constant, independent of the molecular size." This principle is in apparent contradiction to the theoretical prediction of low chemical reactivity for macromolecules discussed above. Flory used the following arguments to explain the equality of reactivity between macromolecules and their low-molecular-weight analogs.

1. Low diffusivity of macromolecules — Undoubtedly, large molecules diffuse slowly. But the gross mobility of functional groups must not be confused with the overall diffusion rate of the molecule as a whole. The terminal group, though attached to a sluggish moving polymer molecule, can diffuse through a considerable region due to the constant conformational changes of polymer segments in its vicinity. The mobility of the functional group is much greater than would be indicated by the macroscopic viscosity. Thus it was concluded that the actual collision frequency had little relation to the molecular mobility or macroscopic viscosity.

   A pair of neighboring functional groups may collide repeatedly before they either diffuse apart or react. The lower the diffusion rate, the longer these groups remain in the same vicinity. Consequently, the number of collisions will be greater and hence the probability of fruitful collisions will be greater. However, by the same token, once reactants diffuse apart, it will take a proportionately longer time before two reactants meet to engage in another series of collisions. Flory therefore concluded that for a sufficiently long time span, the decreased mobility resulting from larger molecular size and/or high viscosity will alter the time distribution of collisions experienced by a given functional group but not the average number of collisions.
2. Steric factor — Flory expected steric-factor-related reduction in chemical reactivity due to shielding of functional groups by coiling of their chain to occur only in dilute solution where sufficient space is

available for molecules to coil independently. In concentrated systems, however, polymer chains intertwine extensively. Functional groups show no preference for their own chains. The chains act as diluents whose effects can be accounted for by writing rate expressions in terms of concentration of functional groups and not in terms of molecules or mole fractions of molecules.

Example 6.1: The esterification of a monobasic acid by alcohol proceeds according to the following reaction.

$$H(CH_2)_{10}COOH + C_2H_5OH \xrightarrow{HCl} H(CH_2)_{10}COOC_2H_5 + H_2O \qquad \text{(Str. 1)}$$

The rate of ester formation is given by the equation:

$$\frac{d[\text{ester group}]}{dt} = k\,[COOH]\,[H^+]$$

For carboxyl group and HCl concentrations of 10^3 and 10^{-2} g-eq/l, respectively, the rate of esterification was found to be 7.6×10^{-3} g-eq/l/s. For the esterification of the dibasic acid $(CH_2)_{1000}(COOH)_2$, how much ester is formed in a 8-h day shift using a 100-l reactor if the carboxyl group concentration is 10^2 g-eq/l and the acid (HCl) concentration is 10^{-3} g-eq/l?

Solution: Notes:

1. The reactivity of one carboxyl group in the dibasic acid is unaffected by esterification of the other. Consequently, the same rate equation holds for the esterification of both monobasic and dibasic acids.
2. The rate equation is applicable to the esterifications of all polymethylene dibasic acids for which n is unity or greater.

From the above two observations, it follows that esterification rate constant for homologous series of dibasic acid is the same as those for monobasic acids.

$$\text{Rate} = k\left(10^3\,\frac{g\cdot eq}{\text{liter}}\right)\left(10^{-2}\,\frac{g\cdot eq}{\text{liter}}\right) = 7.6\times 10^{-3}\left(\frac{g\cdot eq}{1\cdot s}\right)$$

$$k = 7.6\times 10^{-4}\left(\frac{g\cdot eq}{\ell}\right)^{-1}\cdot\left(\frac{1}{s}\right)$$

$$\text{Rate (dibasic acid)} = \left(7.6\times 10^{-4}\,\frac{\ell}{g\cdot eq}\,\frac{1}{s}\right)\left(10^2\,\frac{g\cdot eq}{\ell}\right)\left(10^{-3}\,\frac{g\cdot eq}{\ell}\right)$$

$$\left(7.6\times 10^{-5}\,\frac{g\cdot eq}{\ell\cdot s}\right)(100\ \ell)\,(8\times 3600\ s)$$

$$= 7.6\times 10^{-5}\times 100\times 28800\ g\cdot eq$$

$$= 7.6\times 28.8\ g\cdot eq$$

Concentrations are written as equivalents of functional groups. $(CH_2)_{1000}(COOC_2H_5)_2$: gram equivalent of functional group = 73. Ester produced in 8 h = $\frac{7.6 \times 28.8 \times 146}{10^3}$ = 32 kg.

III. KINETICS OF CONDENSATION POLYMERIZATION

Flory's equal reactivity principle, which has been validated on mechanistic and experimental grounds, has greatly simplified an otherwise complicated kinetic analysis of condensation polymerization. This principle in effect amounts kinetically to the proposition that all steps in a condensation polymerization have equal rate constants. We now consider two cases.

Case 1: Polymerization without Added Strong Acid: Consider esterification — the formation of a polyester from a glycol and a dibasic acid. The progress of reaction is easily followed by titrating the unreacted carboxyl groups in samples removed from the reaction mixture. This polyesterification and other simple esterifications are acid-catalyzed. In the absence of an added strong acid, a second molecule of the acid being esterified acts as the

$$\underset{\text{adipic acid}}{HO-\overset{\overset{\displaystyle O}{\|}}{C}-(CH_2)_4-\overset{\overset{\displaystyle O}{\|}}{C}-OH} + \underset{\text{diethylene glycol}}{HO-CH_2CH_2-O-CH_2CH_2-OH} \longrightarrow$$

$$\left[-\overset{\overset{\displaystyle O}{\|}}{C}(CH_2)_4\overset{\overset{\displaystyle O}{\|}}{C}-O-CH_2CH_2-O-CH_2CH_2-O-\right]_n + 2nH_2O \qquad \text{(Str. 2)}$$

catalyst. The rate of polyesterification process can therefore be written:

$$\frac{-d[COOH]}{dt} = k[OH][COOH]^2 \qquad (6.1)$$

where concentrations (written in square brackets) are expressed as equivalents of the functional groups. As written here Equation 6.1 assumes that the rate constant k is independent of molecular size of reacting species and is the same for all functional groups.

If the concentration, C, of the unreacted carboxyl and hydroxyl groups at time t are equal, Equation 6.1 may be rewritten as:

$$\frac{-dc}{dt} = kc^3 \qquad (6.2)$$

On integration, this yields the third-order reaction expression

$$2kt = \frac{1}{c^2} - \text{constant} \qquad (6.3)$$

Now, let us introduce the *extent of reaction,* p, defined as the fraction of the functional group that has reacted at time t. That is,

$$p = \frac{C_o - C}{C_o} \qquad (6.4)$$

where C_o = initial concentration of one of the reactants. From Equation 6.4,

$$C = C_o(1-p) \qquad (6.5)$$

Hence, on substitution, Equation 6.3 becomes

$$2C_o^2kt = \frac{1}{(1-p)^2} - 1 \qquad (6.6)$$

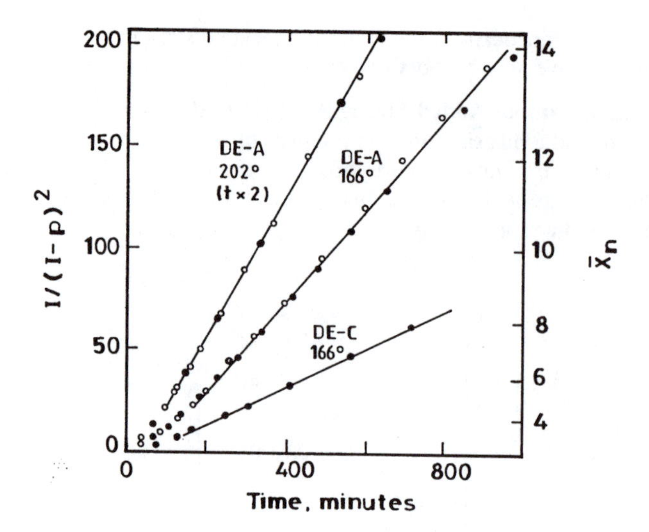

Figure 6.1 Reaction of diethylene glycol with adipic acid (DE-A) and of diethylene glycol with caproic acid (DE-C). Time values at 202°C have been multiplied by two. (From Flory, P.J., *J. Am. Chem. Soc.*, 61, 3334, 1939; 62, 2261, 1940. Copyright 1940 American Chemical Society. With permission.)

A plot of $\frac{1}{(1-p)^2}$ against time should be linear. Figure 6.1 verifies this third-order kinetic expression. Note from Figure 6.1 that when $\frac{1}{(1-p)^2} = 25$ (i.e., when $\frac{1}{1-p} = \bar{x}_n = 5$) the plot is not linear. However, this behavior is not unique to polyesterification. A similar behavior is exhibited by the nonpolymer esterification of diethylene glycol and caproic acid. The similarity in behavior of both mono- and polyesterification provides a direct verification of the nondependence of reactivity on molecular size. The quantity $\frac{1}{1-p}$ is called the degree of polymerization, DP, which as we saw earlier corresponds to the average number of monomer molecules in the chain.

Case II: Polymerization with Added Strong Acid: The kinetic expression can be greatly simplified if the polyesterification is carried out in the presence of a small amount of strong acid, e.g., *p*-toluene sulfonic acid. With the catalyst concentration kept constant throughout the process, the rate expression becomes

$$\frac{-dc}{dt} = k'c^2 \tag{6.7}$$

where k′ = k [catalyst] and where the alcohol and carboxylic acid concentrations are kept constant. Integrating this second-order rate equation and inserting the extent of reaction, p, we have:

$$C_o k't = \frac{1}{1-p} - 1 \tag{6.8}$$

Figure 6.2 shows the linear relation between $\frac{1}{1-p}$ and time.

IV. STOICHIOMETRY IN LINEAR SYSTEMS

As noted earlier, linear polymers are obtained from condensation polymerization when the functionality of the reactants is 2. Two cases may be considered. Some commercial examples of step-growth polymerization are illustrated in Figure 6.3. Reactions A and B are two routes (esterification and ester interchange) for preparing the same compound [poly(ethylene terephthalate), PETP]. Reaction C is

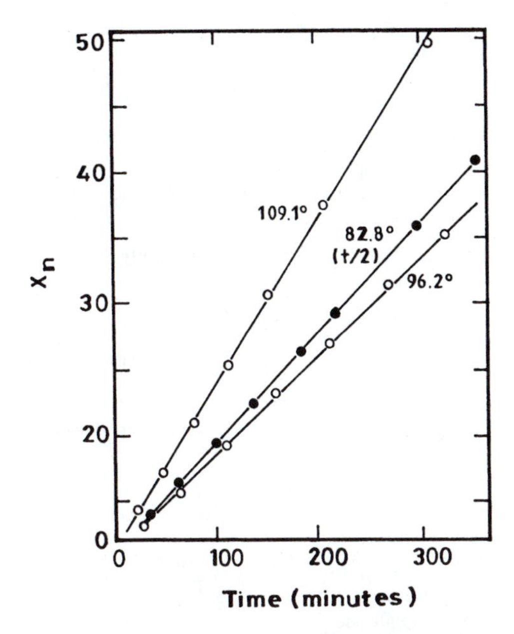

Figure 6.2 Reaction of decamethylene glycol with adipic acid at the temperature indicated, catalyzed by 0.10 eq% of *p*-toluenesulfonic acid. The time scale for the results at 82.8°C is to be multiplied by two. (From Flory, P.J., *J. Am. Chem. Soc.*, 61, 3334, 1939; 62, 2261, 1940. Copyright 1940 American Chemical Society. With permission.)

preparation of nylon 6,6 from adipic acid and hexamethylenediamine. Note that in each of these three reactions, each of the two monomers is bifunctional and contains the same functionality at each end, that is, A–A or B–B. Alternatively, a polymer can also be formed by intramolecular reaction. An example is the formation of an aliphatic polyester by the self-condensation of ω-hydroxycarboxylic acid (reaction D, Figure 6.3). Since the end functional groups of the acid are different, this polyesterification reaction is an example of an A-B step-growth condensation polymerization.

We now proceed to discuss the stoichiometry of these two types of linear system represented by:

$$\begin{aligned} &\text{I: } \text{A–B} \rightarrow (\text{A–B–A–B–A–B}) + \text{small molecule} \\ &\text{II: } \text{A–A} + \text{B–B} \rightarrow \text{A}(\text{B–A–B–A–B–A})\text{–B} + \text{simple molecule} \end{aligned} \quad \text{(Str. 3)}$$

Suppose we start with the bifunctional monomers exclusively such that the number of A groups is precisely equivalent to that of B groups. Obviously, the requirement that A and B groups be present in equivalent amounts is automatically satisfied by A–B type polycondensation provided the monomers are pure and no side reactions occur. For type A–A/B–B polymerization we require, in addition, that the reactants must be present in equivalent proportions. At any stage of the polymerization process, each molecule will be terminated on either side by an unreacted functional group. Therefore, the number of molecules present can be determined by measuring the number of structural units per molecule, i.e., the number-average degree of polymerization, $\overline{X}_n$, is given by the ratio of the initial number of molecules to the final number of molecules, i.e.,

$$\begin{aligned} \overline{X}_n &= \frac{\text{original number of molecules}}{\text{final number of molecules}} \\ &= \frac{C_o}{C_o(1-p)} = \frac{1}{1-p} \end{aligned} \quad (6.9)$$

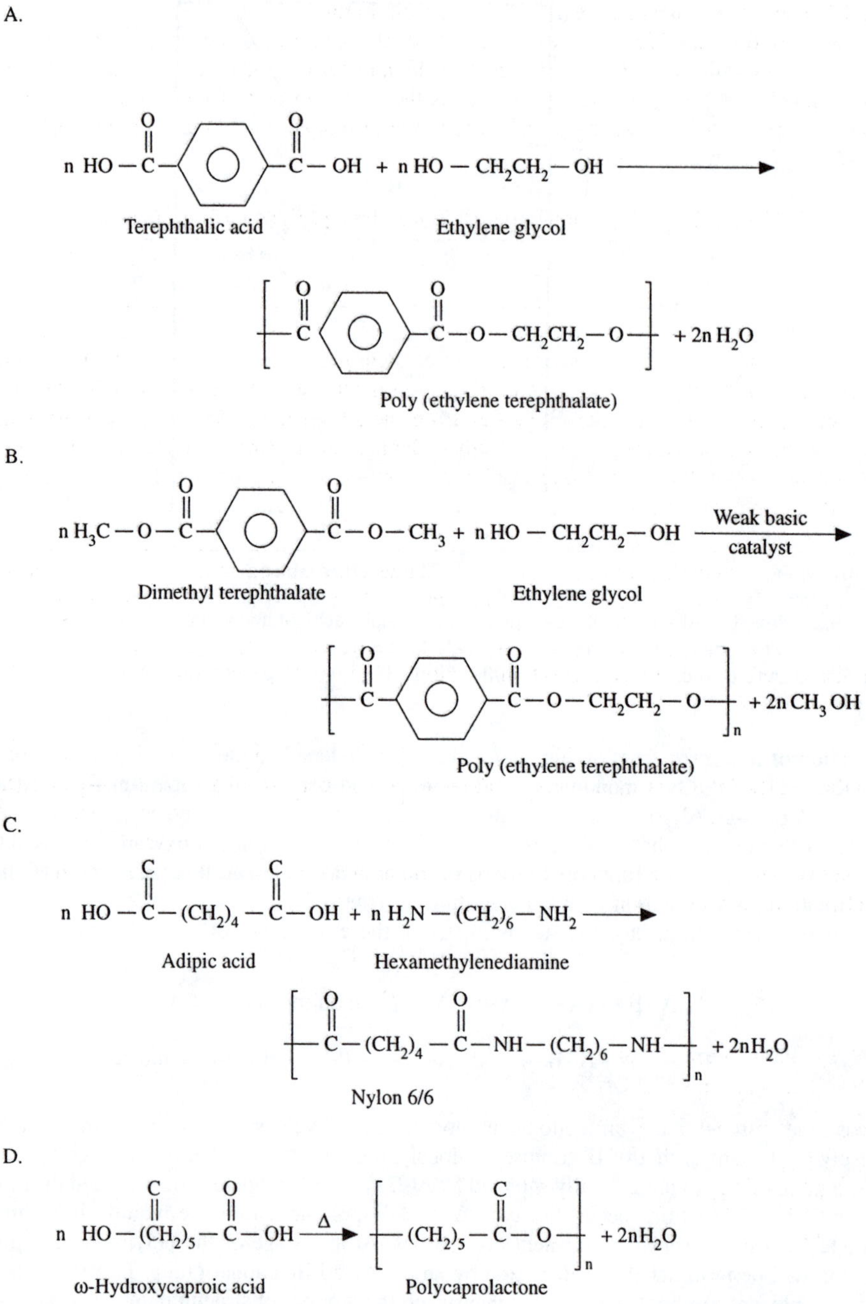

A.

$$n\ HO-\overset{O}{\overset{\|}{C}}-C_6H_4-\overset{O}{\overset{\|}{C}}-OH + n\ HO-CH_2CH_2-OH \longrightarrow$$

Terephthalic acid   Ethylene glycol

$$\left[-\overset{O}{\overset{\|}{C}}-C_6H_4-\overset{O}{\overset{\|}{C}}-O-CH_2CH_2-O-\right] + 2n\ H_2O$$

Poly (ethylene terephthalate)

B.

$$n\ H_3C-O-\overset{O}{\overset{\|}{C}}-C_6H_4-\overset{O}{\overset{\|}{C}}-O-CH_3 + n\ HO-CH_2CH_2-OH \xrightarrow[\text{catalyst}]{\text{Weak basic}}$$

Dimethyl terephthalate   Ethylene glycol

$$\left[-\overset{O}{\overset{\|}{C}}-C_6H_4-\overset{O}{\overset{\|}{C}}-O-CH_2CH_2-O-\right]_n + 2n\ CH_3OH$$

Poly (ethylene terephthalate)

C.

$$n\ HO-\overset{C}{\overset{\|}{C}}-(CH_2)_4-\overset{C}{\overset{\|}{C}}-OH + n\ H_2N-(CH_2)_6-NH_2 \longrightarrow$$

Adipic acid   Hexamethylenediamine

$$\left[-\overset{O}{\overset{\|}{C}}-(CH_2)_4-\overset{O}{\overset{\|}{C}}-NH-(CH_2)_6-NH-\right]_n + 2nH_2O$$

Nylon 6/6

D.

$$n\ HO-\overset{C}{\overset{\|}{(CH_2)_5}}-\overset{O}{\overset{\|}{C}}-OH \xrightarrow{\Delta} \left[-(CH_2)_5-\overset{C}{\overset{\|}{C}}-O-\right]_n + 2nH_2O$$

ω-Hydroxycaproic acid   Polycaprolactone

Figure 6.3 Classical condensation reactions; are (A) esterification, (B) ester interchange, (C) amidization, and (D) intramolecular reaction. Two routes, A and B, can be taken to prepare PET. (From Fried, J.R., *Plast. Eng.*, 38(10), 27, 1982. With permission.)

From the definition of $\overline{X}_n$, it follows that the number-average molecular weight will be given by the expression:

$$\overline{M}_n = \overline{X}_n M_o = \frac{M_o}{1-p} \tag{6.10}$$

where M_o = average molecular weight of the structural unit.

To avoid confusion, it is necessary to reemphasize the definition of structural unit — the residue from a glycol or from a dibasic acid. This means that the number of structural units equals the total number of bifunctional monomers initially present. Thus the repeating unit of a chain derived from A–A/B–B type monomers consists of two structural units, while the repeat unit in the case of type A–B monomers is the same as the structural unit.

Table 6.1 High Molecular Weight Achieved Only by High Conversion

P	0	0.5	0.8	0.9	0.95	0.99	0.999	0.9999	1.0
$\overline{M}_n$	1	2	5	10	20	100	1,000	10,000	∞

A closer look at Equation 6.9 reveals that in condensation polymerization, a high molecular weight product is obtained only when the extent of reaction is almost 100% as shown in Table 6.1. Polymers, by definition, derive their unique properties from their high molecular weights. Thus, commercially useful condensation polymers are obtained only under the rather stringent condition of almost quantitative reaction. This means that in addition to high conversion, a step-growth polymerization requires high yield and high monomer(s) purity.

Example 6.2: For condensation polymerization between a dibasic acid and a glycol, show that M_o is the average molecular weight of the structural units where the structural unit is the residue from each monomer.

Solution: Consider the polycondensation reaction

$$\text{nHOOC}-\text{CH}_2-\text{COOH} + \text{nHO}-\text{CH}_2-\text{HO} \longrightarrow \left[-\overset{\text{O}}{\overset{\|}{\text{C}}}-\text{CH}_2-\overset{\text{O}}{\overset{\|}{\text{C}}}-\text{O}-\text{CH}_2-\text{O}-\right]_n + 2\text{nH}_2\text{O} \quad \text{(Str. 4)}$$

Assume that we start with 10 moles of each monomer and suppose that the extent of reaction at a time t equals p = 0.5. By definition

$$p = \frac{N_i - N}{N_i}$$

N_i = initial number of moles
N = number of moles at time t

At time t,

$$\begin{aligned} N &= N_i (1 - p) \\ &= 20 (1 - 0.5) \\ &= 10 \end{aligned}$$

Now

$$\overline{X}_n = \frac{1}{1-P} = \frac{1}{1-0.5} = 2$$

$$\overline{M}_n = \overline{X}_n M_o = \frac{M_o}{1-P} = 2M_o$$

M_o = molecular weight of structural unit

Three cases have to be considered.

Case 1:

M_o = molecular weight of repeating unit
= 116 kg/kg mol
$\overline{M}_n = 2M_o = 232$ kg/kg mol
Mass of polymer = 232 × 10 = 2320 kg

Case 2:

M_o = average molecular weight of starting monomers
$= \frac{104 + 48}{2} = 76$
$M_n = 2M_o = 152$
Mass of polymer produced at t = 152 × 10 = 1520 kg

Case 3:

M_o = average molecular weight of structural units where structural unit equals residue from each monomer
$= \frac{116}{2} = 58$
$\overline{M}_n = 2 \times 58 = 116$
Mass of polymer produced = M_n × 10 = 1160 kg

To establish which of the above definitions of M_o is correct, we need to carry out a simple mass balance:

Initial mass = sum of masses of reactants = 104 × 10 + 48 × 10 = 1520 kg
Mass of product = mass of polymer + mass of H_2O split off
Mass of water split off = $2nH_2O$ = 2 × 10 × 18 = 360

Case 1: mass of product = 2320 + 360 = 2680 kg
Case 2: mass of product = 1520 + 360 = 1880 kg
Case 3: mass of product = 1160 + 360 = 1520 kg

From mass balance, it is obvious that only Case 3 gives a correct mass balance.

Example 6.3: A 21.3 g sample of poly(hexamethylene adipamide) is found to contain 2.50×10^{-3} mol of carboxyl groups by both titration with base and infrared spectroscopy. From these data calculate

a. The number-average molecular weight
b. The extent of reaction

What assumption is made in your calculations?

$$-\overset{\overset{\displaystyle H}{|}}{N}-(CH_2)_6-\overset{\overset{\displaystyle H}{|}}{N}-\overset{\overset{\displaystyle O}{\|}}{C}-(CH_2)_4-\overset{\overset{\displaystyle O}{\|}}{C}- \quad (MW = 226) \qquad \text{(Str. 5)}$$

$$\overline{M}_n = \frac{\Sigma W_i}{\Sigma N_i} = \frac{W}{N} = \frac{21.3}{2.5 \times 10^{-3}} = 8520 \text{ g/g mol}$$

$$\overline{X}_n = \frac{1}{1-p} = \frac{\overline{M}_n}{M_o}$$

where M_o = average molecular weight of residue = $\frac{226}{2}$ = 113.

$$\frac{1}{1-p} = \frac{8520}{113} = 75.40$$

$$1 - p = 0.013$$

$$p = 98.7\%$$

Assumption is that each polymer molecule contains one –COOH group, i.e.,

$$H_2N\left[(CH_2)_6-\overset{\overset{\large H}{|}}{N}-\overset{\overset{\large O}{\|}}{C}-(CH_2)_4\right]COOH \quad \text{(Str. 6)}$$

V. MOLECULAR WEIGHT CONTROL

From the above discussion, it is obvious that in addition to high conversion, a step-growth polymerization requires high yield and high monomer purity to obtain a polymer of high molecular weight. High yield means the absence of any side reactions that could deactivate the polymerization process. For example, any side reactions that would lead to monofunctional units that are incapable of further reaction would limit the formation of high molecular weight of A–A/B–B or A–B type polymers. In general, depression of molecular weight can be brought about by:

- Nonequivalence of reactants
- Monofunctional ingredients introduced as impurities formed by side reaction
- Unbalance in stoichiometric proportions

Suppose an excess of a functional group is obtained by the addition of reactant designated B+B. In this case the two types of polymerizations discussed above become:

Case I: A–B + little B+B
Case II: A–A + B– + little B+B

Then let

N_A = total number of A groups initially present
N_B = total number of B groups initially present

$$r = \frac{N_A}{N_B} < 1 \quad \text{i.e., } N_A < N_B \tag{6.11}$$

P_A = fraction of A groups that have reacted at a given stage of the reaction

$$\text{Total number of units} = \frac{N_A + N_B}{2}$$

$$= \frac{N_A}{2}\left[1 + \frac{1}{r}\right] \tag{6.12}$$

At any stage of the reaction, possible types of chains are

A–BA–BA–BA–B (type IA)
B–AB–AB–AB–AB–B (type IB)

Number of molecules of type IA = N_A (1 – P_A), and number of molecules of type IB = $\frac{N_B - N_A}{2}$ (i.e., number of molecules due to excess reactant):

Total number of molecules at any stage of the reaction:

$$N_A(1-p_A)+\frac{N_A}{2}\left[\frac{1}{r}-1\right]$$

$$=\frac{N_A}{2}r\left[2r(1-P_A)+(r-r)\right]$$

Now

$$\overline{X}_n=\frac{\text{original no. of molecules}}{\text{final no. of molecules}} \tag{6.13}$$

$$\overline{X}_n=\frac{1+r}{2r(1-P_A)+1-r} \tag{6.14}$$

For r = 1 (for stoichiometric amounts of A and B)

$$\overline{X}_n=\frac{1}{1-p} \tag{6.15}$$

For p = 1, i.e., degree of polymerization is maximum.

$$\overline{X}_n=\frac{1+r}{1-r} \tag{6.16}$$

By playing around with Equation 6.14 we see how the degree of polymerization and hence the molecular weight of the product is influenced by a proper control of the purity of reactants and prevention of extraneous reactions. In practice the molecular weight of nylons can be stabilized by the deliberate addition of a predetermined amount of a monofunctional monomer like acetic acid.

Note that Equation 6.14 can be used to evaluate the degree of polymerization in a system containing bifunctional reactants and a small amount of monofunctional species provided r is defined as

$$r=\frac{N_A}{N_A+2N_{B+}} \tag{6.17}$$

VI. MOLECULAR WEIGHT DISTRIBUTION IN LINEAR CONDENSATION SYSTEMS

From Flory's equal reactivity principle, it follows that each functional group during condensation polymerization has an equal chance of reacting irrespective of the molecular size of the group to which it is attached. Therefore, the probability that a given functional group has reacted should simply be equal to the extent of reaction of that functional group. For type A–B polycondensations or for stoichiometric quantities in type A–A/B–B polymerization, this probability is equal to $P_A = P_B$. Obviously, since the reaction of functional groups is a random event, various molecular sizes of product will be present in the reaction mixture at any particular stage of the reaction. The question therefore is the probability of finding a molecule of size, say, X. Consider the following molecule of A–B type polycondensation:

$$ARB_1-ARB_2-ARB_3\ldots\ldots\ldots\ldots ARB-ARB_{x-1}-ARB_x \tag{Str. 7}$$

The probability that the first functional group B has undergone condensation with A is p, the extent of reaction. By similar reasoning, the probability that the second B has also reacted is also p and so on. A molecule containing X units must have undergone X – 1 reactions. The probability that this number of reactions has occurred is simply the product of individual reaction probabilities, i.e., p^{x-1}. The probability of finding an unreacted end group is 1 – p. Therefore the total probability, P_x, that a given polymer molecule contains X units is

$$P_x = p^{x-1}(1-p). \tag{6.18}$$

Evidently this P_x must be equal to the mole fraction, n_x, of X-mers in the reaction system of the extent of reaction p.

$$P_x = n_x = \frac{N_x}{N} \tag{6.19}$$

where N is the total number of molecules of all sizes in the system and N_x is the number of X-mers. But N is related to the initial number of monomers, N_o, by the relation:

$$N = N_o(1-p) \tag{6.20}$$

On substitution the number of X-mers in terms of the initial amount of monomers and the extent of reaction, Equation 6.19 becomes:

$$N_x = N_o p^{x-1}(1-p)^2 \tag{6.21}$$

The weight fraction, W_x, of X-mers is obtained by dividing the weight of X-mers by the weight of all the molecules, i.e.,

$$W_x = \frac{xN_x}{N_o} \tag{6.22}$$

or substitution, in Equation 6.21, we obtain

$$W_x = xp^{x-1}(1-p)^2. \tag{6.23}$$

The distributions given in Equations 6.21 and 6.23 are known as the *most probable distributions* and are shown graphically in Figures 6.4 and 6.5.

From Equation 6.21 it is observed that on a mole fraction basis, low-molecular-weight chains are most abundant even at high extents of reaction. However, on a weight fraction basis (Equation 6.23), W_x passes through a maximum near the number average value of x. Also, low-molecular-weight chains are less significant.

VII. MOLECULAR WEIGHT AVERAGES

The most probable distributions derived above lead to expressions for molecular weight averages $\overline{M}_n$ and $\overline{M}_w$. From Chapter 3, $\overline{M}_n$ is defined by the relation:

$$\overline{M}_n = \frac{\sum_{x-1}^{\infty} M_x N_x}{\sum_{x=1}^{\infty} N_x} = \frac{\sum_{x=1}^{\infty} (M_o X) N_x}{\sum_{x=1}^{\infty} N_x} \tag{6.24}$$

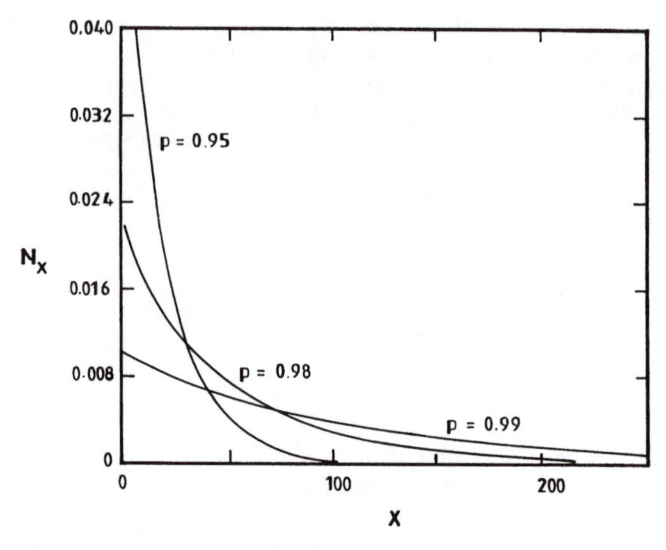

Figure 6.4 Mole fraction distribution of chain molecules in a linear condensation polymer for several extents of reaction. (From Flory, P.J., *Chem. Rev.,* 39, 137, 1946. With permission.)

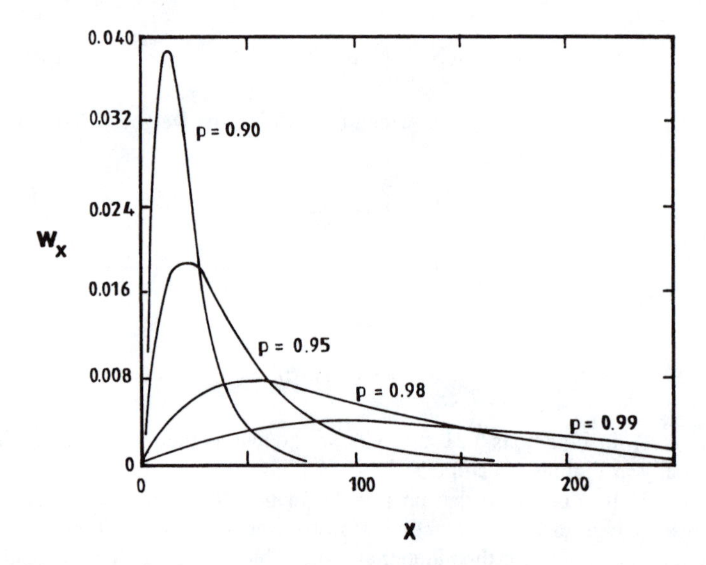

Figure 6.5 Weight fraction distribution of chain molecules in linear condensation polymers for several extents of reaction. (From Flory, P.J., *Chem. Rev.,* 39, 137, 1946. With permission.)

Here M_n = molecular weight of an X-mer. Thus,

$$\overline{M}_n = M_o \Sigma x(1-p)p^{x-1} \tag{6.25}$$

It can be shown that

$$\overline{X}_n = \Sigma x(1-p)p^{x-1} \tag{6.26}$$

$$\overline{X}_n = \frac{1}{1-p} \tag{6.27}$$

and

$$\overline{M}_n = \frac{M_o}{1-p} \tag{6.28}$$

The weight-average molecular weight is defined by:

$$\overline{M}_w = \frac{\sum_{x-1}^{\infty} M_x w_x}{\sum_{x=1}^{\infty} w_x} = \frac{\sum_{x=1}^{\infty} M_o (X)^2 N_x}{\sum_{x=1}^{\infty} M_o (X) N_x} \tag{6.29}$$

where w_x = weight fraction of X-mers

$$= \frac{w_x}{w} \quad \frac{\text{weight of X-mers}}{\text{weight of all molecules}}$$

From where

$$\overline{X}_w = \frac{1+p}{1-p} \tag{6.30}$$

and

$$\overline{M}_w = M_o \overline{X}_w \tag{6.31}$$

The ratio

$$\frac{\overline{X}_n}{\overline{X}_n} = 1 + p \tag{6.32}$$

is a quantitative measure of polydispersity or the spread of the molecular weight curve as

$$p \to 1, \frac{\overline{X}_w}{\overline{X}_n} \to 2 \tag{6.33}$$

Example 6.4: a. Polyester fibers for "lace" material can, in principle, be produced from ω = hydroxycaproic acid. If the initial 100 moles of the hydroxyacid are reduced to 2 moles after 10 h reaction time, calculate:

1. The number average molecular weight $\overline{M}_n$
2. The weight average molecular weight $\overline{M}_w$
3. The probability that the reaction mixture contains tetramers
4. The weight fraction of these tetramers.

b. As a result of extraneous reactions of the hydroxyl groups, a 5% excess of the carboxylic acid is present in the reaction mixture. Calculate the number-average molecular weight for the same extent of reaction in a.

Solutions: a. The polymerization of ω-hydroxycaproic acid proceeds according to the following equation:

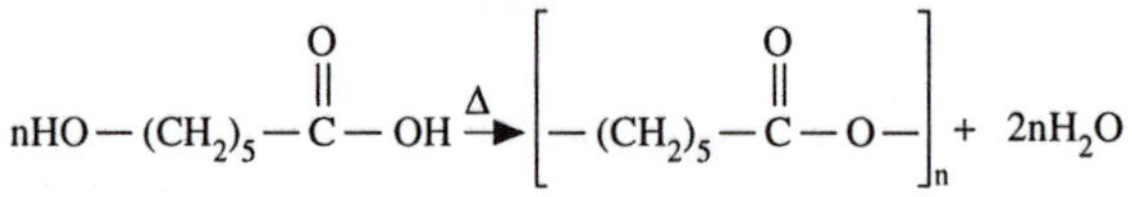

$$nHO-(CH_2)_5-\overset{\overset{O}{\|}}{C}-OH \xrightarrow{\Delta} \left[-(CH_2)_5-\overset{\overset{O}{\|}}{C}-O-\right]_n + 2nH_2O \qquad \text{(Str. 8)}$$

$$\text{Extent of reaction } p = \frac{C_o - C}{C_o} = \frac{100-2}{100} = 0.98$$

1. $\overline{M}_n = \overline{X}_n M_o + \frac{M_o}{1-p} = \frac{114}{0.02} = 5700$

2. $\overline{M}_w = X_w M_o = M_o\left[\frac{1+p}{1-p}\right] = 114\left[\frac{1.98}{0.02}\right] = 11,286$

3. $P_x = p^{x-1}(1-p)$

$= (0.98)^{4-1}(1-0.98) = 0.019$

4. $W_x = X(p^{x-1})(1-p)$

$= 4(0.98)^3(0.02)^2 = 1.51\times10^{-3}$

b. $\overline{X}_n = \frac{1+r}{2r(1-p)+1-r}$

$r = \frac{100}{105} = 0.95$

$\overline{X}_n = \frac{1+0.95}{2(0.95)(1-0.98)+1-0.95} = 22.16$

$\overline{M}_n = \overline{X}_n M_o = 2526$

Notice the depression in $\overline{M}_n$ as a result of stoichiometric imbalance.

VIII. RING FORMATION VS. CHAIN POLYMERIZATION

We observed earlier that polymer formation requires that the reacting monomer(s) must be at least bifunctional. It is pertinent to point out at this stage that polyfunctionality of reactants is a necessary but not sufficient condition for the formation of a polymer. Bifunctional monomers whether of the A–A/B–B or A–B types may react intramolecularly to produce cyclic products. For example, hydroxy acids when heated may yield lactams or linear polyamides.

C=O
R
O
lactam

HORCOOH

$H-\left[-O-R-\overset{O}{\overset{\|}{C}}-\right]_x OH$

polyamide

(Str. 9)

The prime factor that determines the type of product is the size of the ring that is obtained by cyclization. If the ring size is less than five atoms or more than seven, polymer formation is almost entirely favored. If a five-membered ring can be formed, this occurs exclusively. If a ring of six or seven atoms can form, either the ring or linear polymer or both will be the product. Larger rings can be formed under special conditions.

The difficulty of forming rings with less than five members is due to the strain imposed by the valence angles of the ring atom. Five-membered rings are strain-free while all larger rings can be strainless if nonplanar forms are possible.

IX. THREE-DIMENSIONAL NETWORK STEP-REACTION POLYMERS

As recalled from our earlier discussion, we stated that if one of the reactants in step-growth polymerization has a functionality greater than 2, then the formation of a branched or a three-dimensional or cross-linked polymer is potentially possible. An example is the reaction between a dibasic acid and glycerol, which has two primary and one secondary hydroxyl groups (functionality of 3). A monomer that possesses such a functionality is referred to as a *branch unit.* Every reaction of such a molecule introduces a *branch point* for the development of the three-dimensional network. The portion of a polymer molecule lying between two branch points or a branch point and a chain end is called a *chain section* or *segment.*

As polymerization proceeds many branch points are formed. Reaction between large molecules considerably increases the number of reaction groups per polymer chain. The size of the polymer molecules increases rapidly, culminating in the formation of a three-dimensional network polymer of infinite molecular weight. During this process, the viscosity of the reaction medium increases gradually at first and experiences a sudden and enormous increase just before the formation of the three-dimensional network (Figure 6.6). The reaction medium loses fluidity, and bubbles cease to rise through it. *Gelation* is said to have occurred at this point, which is referred to as the *gel point.*

The sudden onset of gelation does not indicate that all the reactants have become bound together in the resulting three-dimensional network. Gelation marks the division of the reaction medium into an insoluble material called the *gel* and a portion still soluble in an appropriate solvent referred to as the *sol.* If polymerization is continued beyond the gel point, the gel fraction increases at the expense of the sol. As would be expected, at gelation there is a considerable depletion of individual molecules, and consequently the number-average molecular weight becomes very low. On the other hand, the weight-average molecular weight becomes infinite.

X. PREDICTION OF THE GEL POINT

We want to be able to define under which conditions a three-dimensional network will be formed, that is, the point at which gelation takes place during a reaction. To do this, let us introduce the term branching coefficient, α, which is defined as the probability that a given functional group of a branch unit is connected via a chain of bifunctional units to another branch unit. Let us illustrate this through the following example.

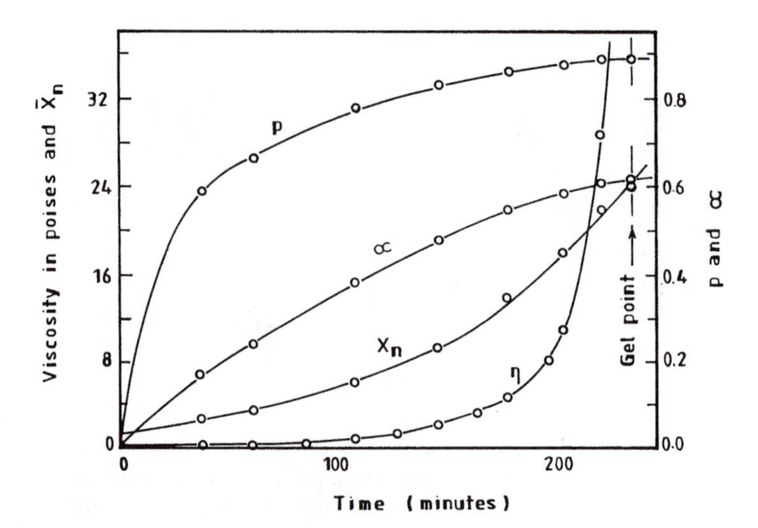

Figure 6.6 The course of typical three-dimensional polyesterification. (From Flory, P.J., *J. Am. Chem. Soc.*, 63, 3083, 1941. Copyright 1941 American Chemical Society.)

$$(A)_2R{-}A + B{-}R'{-}B + A{-}R''{-}A \longrightarrow (A)_2R{-}A^*\left[B{-}R'{-}B{-}A{-}R''{-}A\right]_i{-}B{-}R'{-}B{-}A{-}R(A)_2$$

(Str. 10)

where i varies from 0 to ∞.

In considering this reaction, assumptions are made:

- The principle of equal reactivity of functional groups holds through the condensation. This means that the reactivity of a given A or B group is independent of the size or structure to which the group is attached. In practice this is not strictly true, for it is known, for example, that in glycerol the secondary hydroxyl is less reactive than either of the two primary groups.
- Intramolecular reactions between A and B are forbidden. This, again, is not entirely correct.

We will, however, neglect for the moment errors that will be introduced into our calculations as a result of these assumptions. With these assumptions in mind, we note that the probability that A* has reacted is P_A, the fraction (or extent of reaction) of A groups that have reacted. Similarly, the probability that B** has reacted is P_B.

Now let us define another fractional parameter ρ, as the ratio of A groups both reacted and unreacted on the branch units to all or the total A groups in the mixture. In this case, the probability that a B group has reacted with an A on a branch unit is $P^B\rho$ while the probability that a B group has reacted with an A group on a bifunctional unit (A–A) is $P_B(1 - \rho)$. It follows that the probability that the chain segment shown is formed is given by $P_A|P_B(1 - \rho)\,P_A|^i\,P_B\,\rho$. Summing overall values of i yields

$$\alpha = \frac{P_A P_B \rho}{1 - P_A P_B (1 - \rho)} \tag{6.34}$$

If we define $r = \dfrac{N_A}{N_B} = \dfrac{\text{A groups initially present}}{\text{B groups initially present}}$,

then $P_A = rP_A$.

To eliminate either P_A or P_B, we substitute this relation in Equation 6.34 to obtain:

$$\alpha = \frac{rP_A^2\rho}{1 - rP_A^2(1-\rho)} \tag{6.35}$$

or

$$\alpha = \frac{P_B^2\rho}{r - P_B^2(1-\rho)} \tag{6.36}$$

Consider special cases of interest.

1. When A and B groups are present in equivalent quantities: $r = 1$, $P_A = P_B = P$, and

$$\alpha = \frac{P^2\rho}{1 - P^2(1-\rho)} \tag{6.37}$$

2. If there are no bifunctional A–R–A units, $\rho = 1$ and

$$\alpha = rP_A^2 = P_B^2/r \tag{6.38}$$

3. If the above two conditions apply, $r = \rho = 1$ and

$$\alpha = p^2 \tag{6.39}$$

4. If there are only branch units, then the probability that a functional group on a branch units leads to another branch unit is simply the extent of reaction

$$\alpha = p \tag{6.40}$$

Note that the above equations are completely general and independent of the functionality of the branch unit.

The outstanding problem is to deduce the critical value of α at which the formation of an infinite network becomes possible. To tackle this problem, we consider a branch unit which is trifunctional. In this case, each chain that terminates in a branch unit is succeeded by two more chains. These two chains will, in turn, generate four additional chains and the propagation of chains will continue in like manner. In general therefore for trifunctional branch units, chain sections emanating from n chains will generate 2n chains under the conditions described. Recall, however, that the probability that chain segments originating from branch units will terminate in other branch units is α. Therefore the expected number of branch units from n branch units is $2n\alpha$. Now, it is easy to visualize that the condition for continued expansion of the network is that the number of chains emanating from n chains must be greater than n. That is, the succeeding generation of chains must outnumber the preceding generation. For a trifunctional unit, therefore, if $\alpha < 1/2$ (i.e., $2n\alpha < n$), an infinite network cannot be generated since there is less than an even chance that each chain will terminate in a branch unit. On the hand, if $\alpha > 1/2$ (i.e., $2n\alpha > n$), generation of an infinite network is possible. It is obvious that for a trifunctional branch unit, the critical value of α is $\alpha = 1/2$. This argument can be extended to systems containing polyfunctional branch units with the general result that

$$\alpha_c = \frac{1}{f-1} \tag{6.41}$$

where f is the functionality of the branch unit. If there is more than one type of branch unit present, (f – 1) must be replaced by an appropriate average weighted according to the number of functional groups attached to the various branched units and the molar amount of each present.

XI. MORPHOLOGY OF CROSS-LINKED POLYMERS

As we saw in Section 3, the properties of polymeric materials are dictated by their morphologies. Considerable knowledge has been accumulated about the morphology of semicrystalline polymers over the last four to five decades. Unfortunately, the same cannot be said for cross-linked polymers or thermosets. From the discussion in Section IX for example, in the absence of intramolecular reactions, polyfunctional condensation reactions where different functional groups are chemically equivalent result in a giant molecule with an infinite cross-linked network. For a long time, this concept of an infinitely large molecule with a homogeneous cross-linked network was used to describe the morphology of cross-linked polymers. However, the presence of inhomogeneities in the cross-link density of polymer networks has been established empirically through electron microscopy, thermomechanical studies, and swelling experiments. Today, it is generally recognized that the model that most adequately describes the morphology of thermosets is that of an inhomogeneously cross-linked network consisting of regions of higher cross-linked density (referred to as nodules) immersed in a less cross-linked matrix.[5,6]

XII. PROBLEMS

6.1. United Nigeria Textile (Ltd) UNTL Kaduna is making nylon 6,6 from hexamethylenediamine and adipic acid. One batch is made on every 8-h shift. In each batch equimolar reactants are used and conversion is usually 98.0%. At the end of the run the bulk product is extruded and chopped into pellets.

a. Calculate the number-average molecular weight.
b. The afternoon shift operator dumped in too much adipic acid. From his records you calculate that the mole ratio was 2% excess adipic acid. If the batch went to the usual conversion, what was its number-average molecular weight?
c. The night shift operator weighed things correctly but fell asleep and let the reaction run too long — 99% conversion. What will be the M_n of this batch?
d. How should the UNTL engineer mix these batches to obtain the $\overline{M}_n$ of the usual product. What will be $\overline{M}_w$?

6.2. A polyester system has a number-average molecular weight $\overline{M}_n$ of 6000 and a polydispersity of 2. The system is fractionated into two samples of the $\overline{M}_n$ (2000 and 10,000, respectively). Equimolar proportions of these are mixed.

a. What are the $\overline{M}_n$ and $\overline{M}_w$ for the new (mixed) system?
b. How will the melt viscosity of the new system (fractionated and mixed) change in relation to the unfractionated (original) polymer?

6.3. A polymer chemist prepared nylon from the following amino acid (ω-amino caproic acid): [$NH_2-(CH_2)_5-COOH$]. Due to improper purification of the reactant, side reactions occurred leading to a 5% stoichiometric imbalance (i.e., 5% excess of one of the functional groups). Calculate:

a. The number-average molecular weight if conversion was 98%.
b. The number-average molecular weight for the maximum degree of polymerization.

6.4. Consider a polyesterification reaction.

a. Suppose 1 mol each of dicarboxylic acid and glycol is used. What is the degree of polymerization when the extent of reaction is 0.5, 0.99, and 1.0?
b. Suppose 101 mol glycol is reached with 100 mol of the dicarboxylic acid. What is the maximum degree of polymerization?
c. Suppose the dicarboxylic acid contains 2 mol% monoacid impurity. What is the maximum degree of polymerization?

6.5. The preparation of poly(ethylene terephthalate) from terephthalic acid and ethylene glycol is stopped at 99% conversion. Calculate:

a. The number-average degree of polymerization.
b. The weight-average degree of polymerization.
c. The probability that the reaction mixture contains trimers.
d. The weight-fraction of the trimers.

6.6. The preparation of nylon 12 is shown by the following equation.

$$H_2N-(CH_2)_{11}-COOH \longrightarrow \left[\overset{H}{\overset{|}{N}}-(CH_2)_{11}-\overset{O}{\overset{\|}{C}}\right]_n \qquad \text{(Str. 11)}$$

Calculate the weight fraction of n-mers in the reaction mixture if the conversion is

a. 10%
b. 90%

for n = 1 and 100. Comment on your answer.

6.7. What are the number-average molecular weights for the following nylons if 95% of the functional groups have reacted?

$$-\left[-\overset{O}{\overset{\|}{C}}-(CH_2)_5-\overset{H}{\overset{|}{N}}-\right]-_{100} \text{ nylon 6} \qquad \text{(Str. 12)}$$

$$-\left[-\overset{O}{\overset{\|}{C}}-(CH_2)_4-\overset{O}{\overset{\|}{C}}-\overset{H}{\overset{|}{N}}-(CH_2)_{12}-\overset{H}{\overset{|}{N}}-\right]_{100} \text{ nylon 6,12} \qquad \text{(Str. 13)}$$

What is the weight fraction of monomers in the mixture?

6.8. Calculate (1) the mole fraction and (2) the weight fraction of the following polymer in the reaction medium if 98% of the functional groups has undergone conversion.

$$\left[CO-(CH_2)_5-CO-NH-(CH_2)_9-NH\right]_{100} \qquad \text{(Str. 14)}$$

What is the weight fraction of the monomers in the mixture?

REFERENCES

1. Flory, P.J., *J. Am Chem. Soc.*, 61, 3334, 1939; 62, 2261, 1940.
2. Fried, J.R., *Plast. Eng.*, 38(10), 27, 1982.
3. Flory, P.J., *Chem. Rev.*, 39, 137, 1946.
4. Flory, P.J., *J. Am. Chem. Soc.*, 63, 3083, 1941.
5. Kreibich, U.T. and Schmid, R., *J. Polym. Sci. Symp.*, 53, 177, 1975.
6. Mijovic, J. and Lin, K., *J. Appl. Polym. Sci.*, 32, 3211, 1986.
7. Mijovic, J. and Koutsky, J.A., *Polymer*, 20(9), 1095, 1979.
8. Flory, P.J., *Principles of Polymer Chemistry*, Cornell University Press, Ithaca, NY, 1953.
9. Billmeyer, F.W., Jr., *Textbook of Polymer Science*, 2nd ed., Interscience, New York, 1971.
10. Kaufman, H.S. and Falcetta, J.J., *Introduction to Polymer Science and Technology*, Interscience, New York, 1977.
11. Rudin, A., *The Elements of Polymer Science and Engineering*, Academic Press, New York, 1982.
12. Williams, D.J., *Polymer Science and Engineering*, Prentice-Hall, Englewood Cliffs, NJ, 1971.

Chapter 7

Chain-Reaction (Addition) Polymerization

I. INTRODUCTION

Unsaturated monomers are converted to polymers through chain reactions. In chain polymerization processes, the active center is retained at the end of a growing polymer chain and monomers are added to this center in rapid succession. The rate of addition of monomers to the active center relative to the overall conversion of the monomer to polymer is quite fast. This means that high-molecular-weight polymers are generated even while most of the monomers remain intact. Also, polymers formed at the initial stage of the reaction are little affected by the extent of reaction and are of comparable molecular weight with those formed later in the reaction. Thus in a partially polymerized mixture, only high-molecular-weight polymers and unchanged monomers are present with virtually no intermediates between these ends of the molecular weight spectrum.

II. VINYL MONOMERS

From elementary organic chemistry, we know that the positions and hence reactivities of the electrons in unsaturated molecules are influenced by the nature, number, and spatial arrangement of the substituents on the double bond. As a result of these influences, the double bond reacts well with a free radical for compounds of the types $CH_2 = CHY$ and $CH_2 = CXY$. These compounds constitute the so-called vinyl monomers where X and Y may be halogen, alkyl, ester, phenyl, or other groups. It must, however, be noted that not all vinyl monomers produce high polymers. In symmetrically disubstituted double bonds (e.g., 1, 2 disubstituted ethylenes) and sterically hindered compounds of the type $CH_2 = CXY$, polymerization, if it occurs at all, proceeds slowly.

The active centers that are responsible for the conversion of monomer to polymer are usually present in very low concentrations and may be free radicals, carbanions, carbonium ions, or coordination complexes among the growing chain, the catalyst surface, and the adding monomer. Based on the nature of the active center, chain-reaction polymerization may be subdivided into free-radical, ionic (anionic or cationic), or coordination polymerization. Whether or not the polymerization of a particular monomer proceeds through one or more of these active centers depends partly on the chemical nature of the substituent group(s). Monomers with an electron-withdrawing group can proceed by an anionic pathway; those with an electron-donating group by a cationic pathway. Some monomers with a resonance-stabilized substituent group — for example, styrene — can be polymerized by more than one mechanism. Table 7.1 summarizes the various polymerization mechanisms for several monomers.

III. MECHANISM OF CHAIN POLYMERIZATION

Irrespective of the character of the active center, chain-reaction polymerization, like all chain reactions, consists of three fundamental steps. These are *initiation,* which involves the acquisition of the active site by the monomer; *propagation*, which is the rapid and progressive addition of monomer to the growing chain without a change in the number of active centers; and *termination,* which involves the destruction of the growth activity of the chain leaving the polymer molecule(s). In addition to the above three processes, there is the possibility of another process known as *chain transfer* during which the growth activity is transferred from an active chain to a previously inactive species. We now discuss each of these steps in greater detail.

A. INITIATION

Initiation, as noted above, involves the generation of active species, which can be through free-radical, ionic, or coordination mechanism. This mechanism of chain initiation is the essential distinguishing feature between different polymerization processes. Quite expectedly, therefore, addition polymerization of vinyl

Table 7.1 Initiation Mechanisms Suitable for the Polymerization of Olefin Monomers

Olefin Monomer	Monomer Structure	Free Radical	Anionic	Cationic	Coordination
Ethylene	$CH_2{=}CH_2$	+	–	+	+
Propene	$CH_2{=}CHMe$	–	–	–	+
Butene-1	$CH_2{=}CHEt$	–	–	–	+
Isobutene	$CH_2{=}Cme_2$	–	–	+	–
Butadiene-1,3	$CH_2{=}CH–CH{=}CH_2$	+	+	–	+
Isoprene	$CH_2{=}C(Me)–CH–CH_2$	+	+	–	+
Styrene	$CH_2–CHPh$	+	+	+	+
Vinyl chloride	$CH_2{=}CHCl$	+	–	–	+
Vinylidene chloride	$CH_2{=}CCl_2$	+	+	–	–
Vinyl fluoride	$CH_2{=}CHF$	+	–	–	–
Tetrafluoroethylene	$CF_2{=}CF_2$	+	–	–	+
Vinyl ethers	$CH_2{=}CHOR$	–	–	+	+
Vinyl esters	$CH_2{=}CHOCOR$	+	–	–	–
Acrylic esters	$CH_2{=}CHCOOR$	+	+	–	+
Methacrylic esters	$CH_2{=}C(Me)COOR$	+	+	–	+
Arcrylonitrile	$CH_2{=}CHCN$	+	+	–	+

Note: +, monomer can be polymerized to high-molecular-weight polymer by this form of initiation; –, no polymerization reaction occurs or only low-molecular-weight polymers or oligomers are obtained with this type of initiator.

From Lenz, R.W., *Organic Chemistry of High Polymers,* Interscience, New York, 1967.

monomers differs from system to system, and a proper choice of the nature of the system vis-a-vis the monomer being polymerized is crucial to the degree of success of polymerization of vinyl monomers.

1. Generation of Free Radicals

Free radicals can be generated by a number of ways, including thermal or photochemical decomposition of organic peroxides, hydroperoxides, or azo or diazo compounds. Other methods of generation of free radicals include dissociation of covalent bonds by high-energy irradiation and oxidation–reduction (redox) reactions. The active species produced by these processes are referred to as initiators. These species are frequently but erroneously also called catalysts. Initiators are consumed in the reaction while catalysts are regenerated after the reaction. Compounds usually used for free-radical generation include

- Benzoyl peroxide

$$C_6H_5–C(=O)–O–O–C(=O)–C_6H_5 \longrightarrow 2\,C_6H_5–C(=O)–O^{\bullet} \longrightarrow 2\,C_6H_5^{\bullet} + 2CO_2 \qquad \text{(Str. 1)}$$

- Azo-*bis*-isobutyronitrile (AIBN)

$$(CH_3)_2–C(CN)–N{=}N–C(CN)(CH_3)_2 \longrightarrow (CH_3)_2\,\dot{C}(CN) + N_2$$

- Potassium persulfate

$$K_2S_2O_8 \longrightarrow 2K^+ + 2SO_4^{-}\cdot$$

- *t*-Butylhydroperoxide

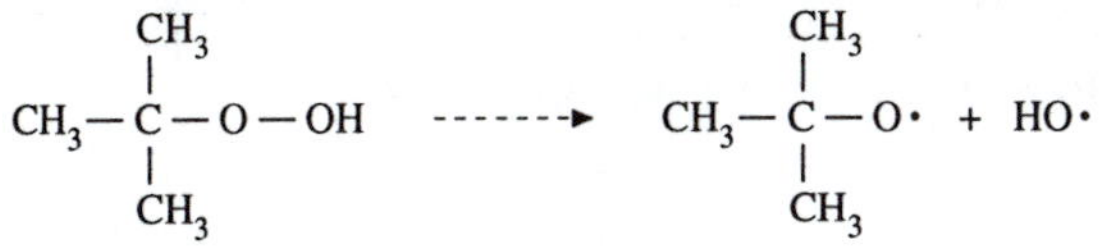

$$CH_3-\overset{\overset{CH_3}{|}}{\underset{\underset{CH_3}{|}}{C}}-O-OH \dashrightarrow CH_3-\overset{\overset{CH_3}{|}}{\underset{\underset{CH_3}{|}}{C}}-O\cdot + HO\cdot \qquad \text{(Str. 2)}$$

Peroxide-type initiators in aqueous system can be decomposed in redox reactions, particularly with reducing agents like ferrous ions, to produce free radicals.

$$H_2O_2 + Fe^{2+} \rightarrow Fe^{3+} + OH^- + HO\cdot$$
$$S_2O_8^{2-} + Fe^{2+} \rightarrow Fe^{3+} + SO_4^{2-} + SO_4^- \cdot \qquad \text{(Str. 3)}$$

The initiation step in free-radical polymerization in the presence of thermal initiators is a two-step process: first is the homolysis of the relatively weak covalent bond in the initiator, designated here by R:R

$$R{:}R \rightarrow 2R\cdot \qquad (7.1)$$

Second, in the presence of a vinyl monomer, the free radical thus generated adds to one of the electrons of one of the bonds constituting the double bonds in the monomer. The remaining electron now becomes the new free radicals:

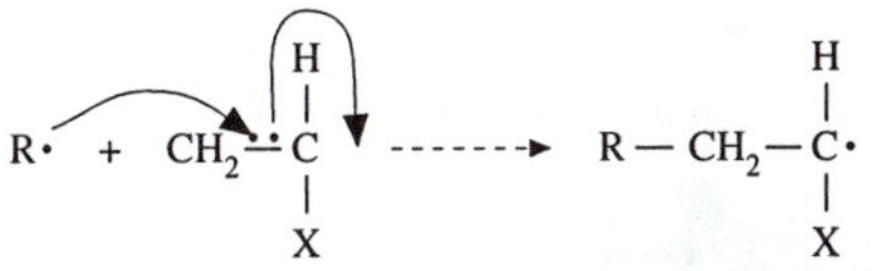

$$R\cdot + CH_2 \overset{..}{=} \overset{\overset{H}{|}}{\underset{\underset{X}{|}}{C}} \dashrightarrow R-CH_2-\overset{\overset{H}{|}}{\underset{\underset{X}{|}}{C}}\cdot \qquad (7.2)$$

The first step is the rate-limiting step in the initiation process. That the above scheme represents that initiation mechanism, and is indeed partly evidence for the free-radical mechanism in addition polymerization of vinyl monomers, has been established by the presence of initiator fragments found to be covalently bound to polymer of molecules. The efficiency with which radicals initiate chains can be estimated by comparing the amount of initiator decomposed with the number of polymer chains formed.

B. PROPAGATION

The free radical generated in the initiation step adds monomers in very rapid succession. Again, this consists of an attack of the free radical on one of the carbon atoms in the double bond. One of the electrons from the double bond pairs up with the free-radical electron forming a bond between the initiator fragment and the attacked carbon atom. The remaining electron shifts to the "unattacked" carbon atom of the double bond, which consequently becomes the free radical. In this way the active center is transferred uniquely to the newly added monomer. This process then continues until termination sets in.

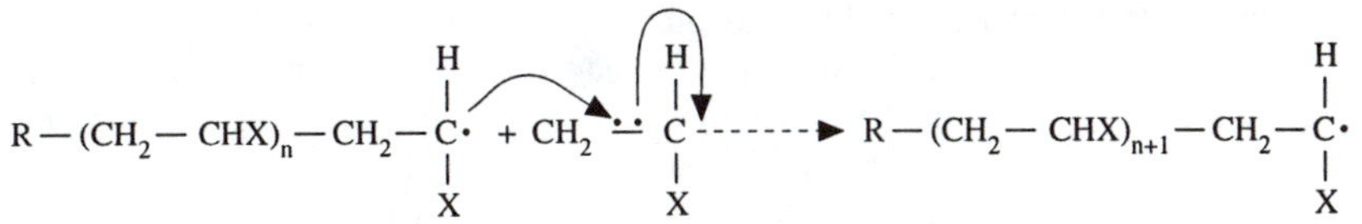

$$R-(CH_2-CHX)_n-CH_2-\overset{\overset{H}{|}}{\underset{\underset{X}{|}}{C}}\cdot + CH_2 \overset{..}{=} \overset{\overset{H}{|}}{\underset{\underset{X}{|}}{C}} \dashrightarrow R-(CH_2-CHX)_{n+1}-CH_2-\overset{\overset{H}{|}}{\underset{\underset{X}{|}}{C}}\cdot \qquad (7.3)$$

For diene monomers, the propagation mechanism is exactly the same as that of vinyl monomers. But the presence of two double bonds means either of two possible attacks can occur. Consequently, propagation of dienes may proceed in one of two ways: 1,4 (*trans* or *cis*) and 1,2:

1,2 addition

$$-CH_2-\underset{H}{\overset{X}{C}}\cdot + CH_2=\overset{X}{C}-C=CH_2 \longrightarrow -CH_2-\underset{H}{\overset{X}{C}}-CH_2-\underset{\underset{CH_2}{\parallel}{CH}}{\overset{X}{C}}\cdot \quad (7.4)$$

1,4 addition

$$-CH_2-\underset{H}{\overset{X}{C}}\cdot + CH_2=\overset{X}{C}-CH=CH_2 \longrightarrow -CH_2-\underset{H}{\overset{X}{C}}-CH_2-\overset{X}{C}=CH-\underset{H}{\overset{H}{C}}\cdot \quad (7.5)$$

In 1,2 addition, the unsaturation is part of the pendant group. In 1,4 addition, however, the unsaturation is part of the backbone. Two configurations are possible — *trans* or *cis*. These considerations have important consequences for the properties and hence end use of the resulting polymers.

C. TERMINATION

The propagation step would theoretically have continued until the consumption of all available monomers but for the tendency of pairs of free radicals to react and annihilate their mutual activities. The termination steps can occur by either of two mechanisms: combination (coupling) or disproportionation.

1. Combination or coupling — Two growing polymer chains yield a single polymer molecule terminated at each end by an initiator fragment.

$$R\sim\sim CH_2-\underset{X}{\overset{H}{C}}\cdot + \cdot\underset{X}{\overset{H}{C}}-CH_2\sim\sim R' \dashrightarrow R\sim\sim CH_2-\underset{X}{\overset{H}{C}}-\underset{X}{\overset{H}{C}}-CH_2\sim\sim R' \quad (7.6)$$

2. Disproportionation — This involves hydrogen transfer with the formation of two polymer molecules, one with a saturated end and the other with an unsaturated terminal olefin and each with an initiator fragment.

$$R\sim\sim CH_2-\underset{X}{\overset{H}{C}}\cdot + \cdot\underset{X}{\overset{H}{C}}-CH_2-\sim\sim R' \longrightarrow R\sim\sim CH_2-\underset{X}{\overset{H}{C}}-H + \underset{X}{\overset{H}{C}}=CH\sim\sim R' \quad (7.7)$$

D. CHAIN TRANSFER

In addition to the three fundamental processes described above for chain-growth (addition) polymerization, another important process, chain transfer, may occur. Chain transfer involves the transfer of radical reactivity to another species, which may be a monomer, a polymer, a solvent, an initiator, or an impurity. This creates a new species that is capable of further propagation but terminates the growth of the original chain. The transfer reaction involves the transfer of an atom between the radical and the molecule. For a saturated molecule like a solvent an atom is transferred to the radical:

$$R\sim\sim CH_2-\underset{X}{\overset{H}{C}}\cdot + CCl_4 \dashrightarrow R\sim\sim CH_2-\underset{X}{\overset{H}{C}}-Cl + CCl_3\cdot \quad (7.8)$$

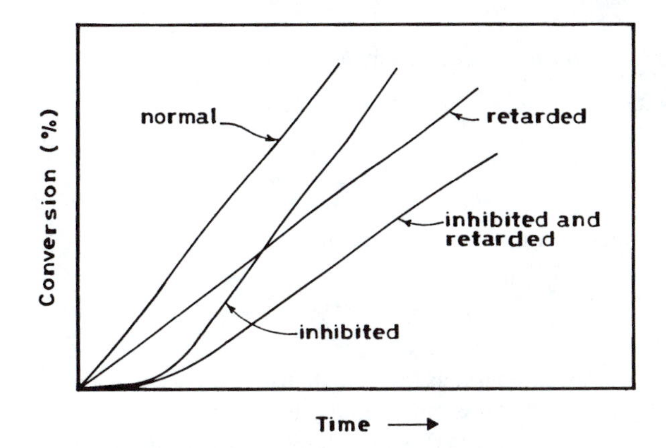

Figure 7.1 The influence of retarders and inhibitors on conversion.

On the other hand, if the molecule is unsaturated, like a monomer, the atom can be transferred to either the monomer or the chain radical:

$$R\sim\sim CH_2-\underset{X}{\overset{H}{C}}\cdot + CH_2{=}CHX \begin{cases} \nearrow\ R\sim\sim CH_2-CH_2X + CH_2{=}\overset{X}{C}\cdot \\ \searrow\ R\sim\sim CH{=}\underset{X}{\overset{H}{C}} + CH_3-\underset{X}{\overset{H}{C}}\cdot \end{cases} \tag{7.9}$$

It can be seen from this scheme that in chain transfer, radicals are neither created nor destroyed. Consequently, the overall polymerization rate is unaffected by chain-transfer processes. However, as we shall see presently, it does limit the obtainable molecular weight.

Chain transfer generally involves the cleavage of the weakest bond in the molecule. Consequently, molecules with active hydrogens play important roles in chain polymerization. A material deliberately added to the system to control molecular weight by chain transfer is called a *chain-transfer agent.* A substance that, if added to the monomer, reacts with chain radicals to produce either nonradical products or radicals with such low reactivity as to be incapable of adding monomer (and thus reducing the polymerization rate) is called a *retarder.* If a retarder is so effective as to completely suppress the rate and degree of polymerization, it is referred to as an *inhibitor.* Thus the distinction between the retarder and inhibitor is merely that of degree. If an inhibitor added to a system is consumed by the radicals generated within the system, an induction period during which no polymerization takes place ensues. Thereafter, normal polymerization occurs. In some cases, added substances first inhibit the polymerization process and then allow polymerization to occur at slower than normal rates. These phenomena are illustrated in Figure 7.1.

Retarders and inhibitors generally operate by chain-transfer mechanism. These substances may be impurities in the system or are formed by side reactions during polymerization. In some cases, they are added deliberately to the system to prevent inadvertent polymerization during transportation or storage. A typical example is hydroquinone, normally added to styrene must be subsequently stripped of the inhibitor before polymerization.

IV. STEADY-STATE KINETICS OF FREE-RADICAL POLYMERIZATION

The overall mechanism for the conversion of a monomer to a polymer via free-radical initiation may be described by rate equations according to the following scheme.

A. INITIATION

As noted earlier, this is a two-step process involving:

1. The decomposition of the initiator into primary radicals

$$I \xrightarrow{k_d} 2R\cdot \tag{7.10}$$

2. The addition of a monomer to the primary free radical

$$R\cdot + M \xrightarrow{k_a} RM_i\cdot \tag{7.11}$$

The constants k_d and k_a are the rate constants for initiator dissociation and monomer addition, respectively. Since initiator dissociation (eqn. 7.10) is much slower than monomer addition (eqn. 7.11), the first step of the initiation step (initiator dissociation) is the rate-limiting step. Some of the initiator radicals may undergo side (secondary) reactions, such as combination with another radical, that preclude monomer addition. Therefore only a fraction, f (an efficiency factor), of the initial initiator concentration is effective in the polymerization process. Also, decomposition of each initiator molecule produces a pair of free radicals, either or both of which can initiate polymerization. Based on these observations, the rate expression for initiation may be written as:

$$R_i = \frac{d[M.]}{dt} = 2\,fk_d[I] \tag{7.12}$$

where [I] represents the initiator concentration.

B. PROPAGATION

The successive addition of monomers during propagation may be represented as follows:

$$\begin{aligned} RM_1\cdot + M &\xrightarrow{k_p} RM_2\cdot \\ RM_2\cdot + M &\xrightarrow{k_p} RM_3\cdot \end{aligned} \tag{7.13}$$

In general,

$$RM_x\cdot + M \xrightarrow{k_p} RM_{x+1}^{\cdot} \tag{7.14}$$

The above scheme is based on the assumption that radical reactivity is independent of chain length. Essentially, this means that all the propagation steps have the same rate constants k_p. In addition, propagation is a fast process. For example, under typical reaction conditions, a polymer of molecular weight of about 10^7 may be produced in 0.1 s. It may therefore be assumed that the number of monomer molecules reacting in the second initiation step is insignificant compared with that consumed in the propagation step. Thus the rate of polymerization equals essentially the rate of consumption of monomers in the propagation step. The rate expression for polymerization rate can therefore be written thus:

$$R_p = -\frac{d[M]}{dt} = k_p[M^*][M] \tag{7.15}$$

where $[M^*] = \sum RM_x\bullet$ i.e., the sum of the concentrations of all chain radicals of type RM_x.

C. TERMINATION

Chain growth may be terminated at any point during polymerization by either or both of two mechanisms:

- Combination (coupling)

$$M_x\cdot + M_y\cdot \xrightarrow{k_{tc}} M_{x+y} \tag{7.16}$$

- Disproportionation

$$M_x\cdot + M_y\cdot \xrightarrow{k_{td}} M_x + M_y \tag{7.17}$$

If there is no need to distinguish between the two types of termination, which in any case are kinetically equivalent, then termination may be represented as:

$$M_x\cdot + M_y\cdot \xrightarrow{k_t} P \tag{7.18}$$

where $k_t(k_{tc} + k_{td})$, k_{tc}, and k_{td} are the rate constants for overall termination process, termination by coupling, and termination by disproportionation, respectively. The termination rate is given by:

$$R_t = \frac{-d[M\cdot]}{dt} = 2k_t[M\cdot]^2 \tag{7.19}$$

The factor of 2 arises from the fact that at each incidence of termination reaction, two radicals disappear.

Over the course of polymerization (at steady state), the total radical concentration remains constant. This means that radicals are being produced and destroyed at equal rates (i.e., $R_i = R_t$). It follows from Equations 7.12 and 7.19 that

$$[M\cdot] = \left(\frac{fk_d}{k_t}\right)^{1/2}[I]^{1/2} \tag{7.20}$$

Since the overall polymerization rate is essentially the rate of monomer consumption during propagation, substitution of Equation 7.20 into Equation 7.15 yields:

$$R_p = k_p\left(\frac{fk_d}{k_t}\right)^{1/2}[I]^{1/2}[M] \tag{7.21}$$

Note that Equation 7.21 predicts that rate of polymer formation in free-radical polymerization is first order in monomer concentration and half order in initiator concentration. This assumes, of course, that the initiator efficiency is independent of monomer concentration. This is not strictly valid. In fact, in practice Equation 7.21 is valid only at the initial stage of reaction; its validity beyond 10 to 15% requires experimental verification. Abundant experimental evidence has confirmed the predicted proportionality between the rate of polymerization and the square root of initiator concentration at low extents of reaction (Figure 7.2). If the initiator efficiency, f, is independent of the monomer concentration, then Equation 7.21 predicts that the quantity $R_p/[I]^{1/2}[M]$ should be constant. In several instances, this ratio has indeed been found to show only a small decrease even over a wide range of dilution, indicating an initiator efficiency that is independent of dilution. This confirmation of first-order kinetics with respect to the monomer concentration suggests an efficiency of utilization of primary radicals, f, near unity. Even where the kinetics indicate a decrease in f with dilution, the decreases have been invariably small. For undiluted monomers, efficiencies near unity are not impossible.

Example 7.1: The following are data for the polymerization of styrene in benzene at 60°C with benzoyl peroxide as the initiator.

$[M] = 3.34 \times 10^3$ mol/m³
$[I] = 4.0$ mol/m³
$k_p^2/k_t = 0.95 \times 10^{-6}$ m³/mol-s

If the spontaneous decomposition rate of benzoyl peroxide is 3.2×10^{-6} m³/mol-s⁻¹, calculate the initial rate of polymerization.

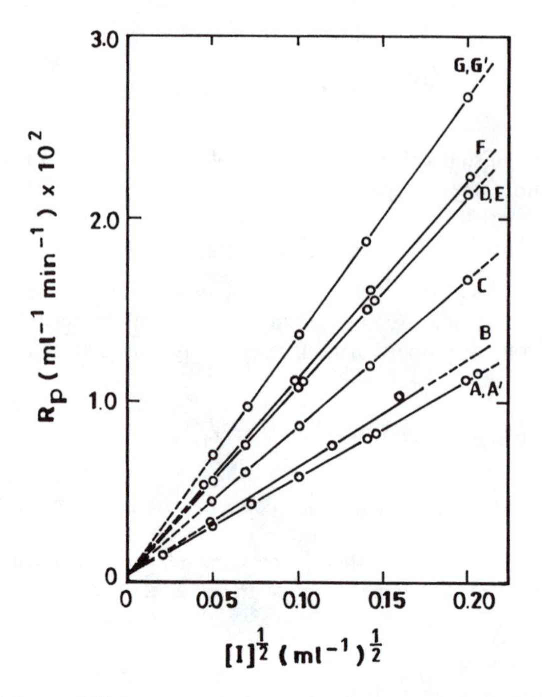

Figure 7.2 Relationship between initiator concentration and polymerization rate for different initiators: A, bis(p-chlorobenzoyl)peroxide; B, benzoyl peroxide; C, acetyl peroxide in dimethylphthalate; D, lauroyl peroxide; E, myristoyl peroxide; F, caprylyl peroxide; G, bis(2,4 dichlorobenzoyl) peroxide. [A.I. Lowell and J.R. Price, *J. Poly. Sci.*, 43, 1 (1996). With permission.]

Solution: For the initial rate of polymerization, Equation 7.21 is valid. Assuming the initiator efficiency f = 1, then

$$R_p = k_p\left(k_d/k_t\right)^{1/2}[I]^{1/2}[M]$$

$$R_p^2 = k_d\left(k_p^2/k_t\right)[I][M]^2$$

$$\left(3.2\times10^{-6}\frac{m^3}{mol\cdot s}\right)\left(0.95\times10^{-6}\frac{m^3}{mol\cdot s}\right)\left(4.0\frac{mol}{m^3}\left(3.34\times10^3\frac{mol}{m^3}\right)^2\right.$$

$$R_p = 11.65\times10^{-3}\ mol/m^3 - s$$

$$R_p = 0.012\ mol/m^3 - s$$

Example 7.2: The data for the bulk polymerization of styrene at 60°C with benzoyl peroxide as initiator are

$[M] = 8.35 \times 10^3$ mol/m^3
$[I] = 4.0$ mol/m^3
$k_p^2/k_d = 1.2 \times 10^{-6}$ m^3/mol-s

If the initial rate of polymerization of styrene is 0.026 mol/m^3-s and the spontaneous decomposition of benzoyl peroxide in styrene is 2.8×10^{-6} s^{-1}, what is the efficiency of the initiator?

Solution:

$$R_p = k_p\left(\frac{fk_d}{k_t}\right)^{1/2}[I]^{1/2}[M]$$

$$R_p^2 = fk_d\left(\frac{k_p^2}{k_t}\right)[I][M]^2$$

$$fk_d = \frac{(0.026)^2}{1.26 \times 4 \times (8.35)^2}$$

$$= 1.92 \times 10^{-6}$$

$$f = \frac{1.92 \times 10^{-6}}{2.8 \times 10^{-6}} = 0.7$$

V. AUTOACCELERATION (TROMMSDORFF EFFECT)

Since the initiator concentration remains fairly unchanged in the course of vinyl polymerization, if the initiator efficiency is independent of monomer concentration, first-order kinetics with respect to the monomer is expected. This is indeed observed over a wide extent of reaction for the polymerization of styrene in toluene solution with benzoyl peroxide as initiator (Figure 7.3). The polymerization of certain monomers, either undiluted or in concentrated solution, shows a marked deviation from such first-order kinetics. At a certain stage in the polymerization process, there is a considerable increase in both the reaction rate and the molecular weight. This observation is referred to as *autoacceleration* or *gel effect* and is illustrated in Figure 7.4 for polymerization of methyl methacrylate at various concentrations of the monomer in benzene.

Observe that for monomer concentrations of up to 40%, plots show that first-order kinetics is followed. However, at higher initial monomer concentrations, a sharp increase in rate is observed at an advanced stage of polymerization. At the same time, high-molecular-weight polymers are produced. Autoacceleration is particularly pronounced with methyl methacrylate, methyl acrylate, and acrylic acid. It occurs independent of an initiator and is observed even under isothermal conditions. In fact, where there is no effective dissipation of heat, autoacceleration results in a large increase in temperature.

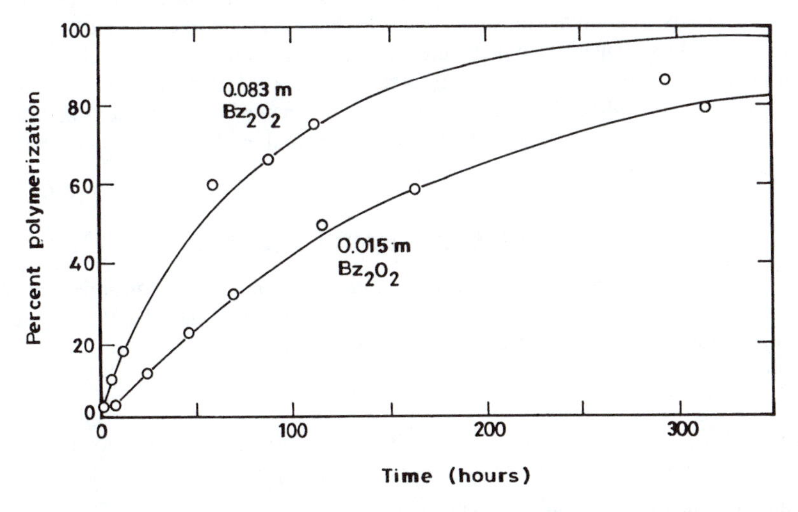

Figure 7.3 Polymerization of 40 percent styrene in toluene at 50°C in the presence of the amounts of benzoyl peroxide shown. (G.V. Schulz and E. Husemann, *Z. Physik Chem.*, B39, 246, 1938. With permission.)

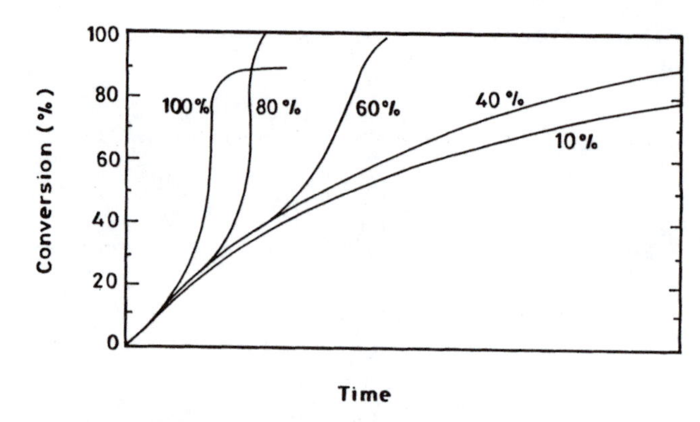

Figure 7.4 The course of polymerization of methyl methacrylate at 50°C in the presence of benzoyl peroxide at different concentrations of monomer in benzene. (G.V. Schulz and G. Harboth, *Macromol. Chem.*, 1, 106, 1947.)

To understand the physical phenomena taking place during autoacceleration, let us look more carefully at the rate equation:

$$R_p \propto k_p \left(\frac{fk_d}{k_t} \right)^{1/2}$$

Since autoacceleration is not a function of the initiator, an increase in the factor fk_d does not provide an explanation for the observation. Consequently the phenomenon of autoacceleration could be due to an increase in the propagation rate constant k_p and/or a decrease in the termination rate constant k_t. Trommsdorff attributed autoacceleration to a decrease in the termination rate. As the concentration of polymer molecules generated builds up, the viscosity of the medium increases. This reduces the overall diffusion rate of growing radical-bearing polymer molecules even though the intrinsic reactivity of the radical remains unaffected. As a result, the bimolecular annihilation of radical reactivity becomes diffusion-controlled. This reduction in termination rate, in turn, increases the concentration of active radicals, and the rate of local consumption of monomers increases proportionately. The overall rate of propagation, however, remains relatively unaffected since the diffusion of monomers is hardly affected by the high medium viscosity. This reduction in termination rate with little or no change in the propagation rate leads to a rapid rise in the molecular weight.

A further reference to Figure 7.4 shows that autoacceleration does not result in complete polymerization of monomers. Also, the higher the dilution, the higher the extent of conversion. Pure poly(methyl methacrylate) has a glass transition temperature of about 90°C. It therefore follows that when the percentage of polymers formed reaches a certain high value, the reaction mixture is transformed into a glass at temperatures below 90°C. Propagation becomes monomer diffusion-controlled, and there is a virtual cessation of polymerization. There is ample experimental evidence to support the above explanation for the phenomenon of autoacceleration.

VI. KINETIC CHAIN LENGTH

The kinetic chain length, ν, is defined as the average number of monomers consumed by each primary radical. Obviously from this definition, the magnitude of the kinetic chain length will depend on the rate of the propagation relative to the termination rate, i.e.,

$$\nu = \frac{R_p}{R_t} \tag{7.22}$$

Since at steady state the rate of initiation equals the rate of termination, Equation 7.22 can also be written as

$$\nu = \frac{R_p}{R_t} = \frac{R_p}{R_i} \tag{7.23}$$

It follows on substitution for R_p and R_t from Equations 7.15 and 7.19, respectively, that

$$\nu = \left(\frac{k_p}{2k_t}\right)\frac{[M]}{[M\cdot]} \tag{7.24}$$

Substituting for [M·] from Equation 7.15 yields:

$$\nu = \left(\frac{k_p^{\,2}}{2k_t}\right)\frac{[M]^2}{R_p} \tag{7.25}$$

Equations 7.24 and 7.25 are quite general and do not depend on the nature of initiation. Notice the inverse relation between the kinetic chain length and the radical concentration. For reactions initiated by the decomposition of an initiator,

$$[M\cdot] = \left(\frac{dk_d}{k_t}\right)^{1/2}[I]^{1/2}$$

Thus

$$\nu = \frac{k_p}{2(fk_d\,k_t)^{1/2}}\frac{[M]}{[I]^{1/2}} \tag{7.26}$$

The kinetic chain length should be related to the number-average degree of polymerization. The degree of complexity of this relation will depend on the existence or otherwise of side (chain-transfer) reactions. We now consider first the case where there are no chain-transfer reactions. By definition

$$\overline{X}_n = \frac{M_o - M_t}{P_t} \tag{7.27}$$

where M_o = initial number of molecules of monomer present
M_t = number of monomer molecules at time t
P_t = number of polymer molecules at time t

On differentiating with respect to time, we obtain:

$$P(t)\frac{d\overline{X}_n(t)}{dt} + \overline{X}_n(t)\frac{dP(t)}{dt} = -\frac{dM(t)}{dt} \tag{7.28}$$

For a short interval, the instantaneous number-average degree of polymerization X_{ni} is constant, hence the above expression becomes

$$\overline{X}_{ni} = -\frac{dM(t)/dt}{dP(t)/dt} \tag{7.29}$$

Recall that the rate of consumption of monomers by an active center (disappearance of monomers) is, by definition, the rate of propagation, R_p. Now, in termination by combination, two growing chains undergo mutual annihilation to produce a single inactive polymer molecule, whereas for termination by disproportionation, a biomolecular annihilation of active polymer chains results in two polymers.

Consequently, $dP(t)/dt = ½ R_t$ for termination by combination, while $dP(t)/dt = R_t$ for termination by disproportionation. From these arguments, and since at steady state $R_t = R_i$,

$$\overline{X}_{ni} = 2R_p/R_t = 2R_p/R_i \quad \text{for termination by combination} \tag{7.30}$$

while

$$X_{ni} = R_p/R_t = R_p/R_i \quad \text{for termination by disproportionation} \tag{7.31}$$

But kinetic chain length is given by Equation 7.23. Hence

$$\overline{X}_{ni} = 2\nu \quad \text{for termination by combination} \tag{7.32}$$

and

$$\overline{X}_{ni} = \nu \quad \text{for termination by disproportionation} \tag{7.33}$$

Although experimental verifications are rare, the above expressions are valid for many systems, unless of course chain-transfer reactions occur. We consider the case of the occurrence of chain transfer in the next section.

Example 7.3: For pure styrene polymerized at 60°C, the value of the ratio k_p^2/k_t is 0.0012 1/mol-s. The corresponding value for pure vinyl acetate polymerized at 60°C is 0.125 l/mol-s.

a. Estimate their relative kinetic length chain lengths.
b. Calculate the kinetic chain length for polystyrene if the rate of polymerization is 10^{-4} mol/l-s and monomer concentration is 8.35 mol/l.

Solution:

a. $\nu = (k_p/2k_t)^2 [M]^2/R_p$

Assume the rates of polymerization and monomer concentrations are the same for styrene(s), and vinyl acetate (VA).

$$\nu_{VA}/\nu_s = (k_p^2/k_t)_{VA}/(k_p^2/k_t)_s = 0.125/0.0012 = 104$$

b. $\nu_s = (0.0012/2)(8.35)^2/10^{-4} = 400$

VII. CHAIN-TRANSFER REACTIONS

Very frequently, a discrepancy exists between the value of the number-average degree of polymerization, and the predicted value of the kinetic chain length ν. The number of polymer molecules is generally found to be more than would be expected to be produced by the primary radicals (i.e., X_n is always less than either 2 ν or ν). Obviously, other components of the reaction medium initiate chain growth. The existence of such polymer-producing extraneous reactions is attributable to chain-transfer reactions that consist essentially of the following sequence of steps:

$$\begin{gathered} M_x\cdot + TL \rightarrow M_xL + T\cdot \\ T\cdot + M \rightarrow TM\cdot, \text{ etc.} \end{gathered} \tag{7.34}$$

where TL = transfer agent, which may be a solvent, monomer, initiator, or polymer
L = a labile group such as hydrogen or chlorine atoms, etc.

Transfer to a polymer does not alter the number of polymer molecules produced, therefore we neglect this type of transfer reaction in the subsequent discussion.

In the presence of transfer reactions, the expression of the kinetic chain length has to be modified to include all possible sources of polymer molecules. Thus

$$\nu = \frac{R_p}{\text{Rate of All Reactions Leading to Polymer Molecule Production}} \quad (7.35)$$

From our discussion in the preceding section, termination mechanisms for generating polymer chains and assuming all transfer species are as reactive as the original (i.e., no inhibition), the expression for ν becomes:

$$\nu = \frac{R_p}{fk_d[I] + k_{tr,M}[M][M\cdot] + k_{tr,s}[S][M\cdot] = k_{tr,I}[I][M\cdot]} \quad (7.36)$$

where $k_{tr,M}$ $k_{tr,S}$ and $k_{tr,I}$ are velocity constants for chain transfer to monomer M, solvent S, and initiator I, respectively. To estimate the tendency for transfer quantitatively, we define transfer constants:

$$C_M = \frac{k_{tr,M}}{k_p};\; C_S = \frac{k_{tr,S}}{k_p};\; C_I = \frac{k_{tr,I}}{k_p} \quad (7.37)$$

Since $\overline{X}_n = \nu$ and substituting for [I] and [M·] Equation 7.36 becomes:

$$\frac{1}{\overline{X}_n} = \frac{k_t}{k_p^2}\frac{R_p}{[M]^2} + C_M + C_s\frac{[S]}{[M]} + C_I\left(\frac{k_t}{k_p^2 fk_d}\right)\frac{R_p^2}{[M]^3} \quad (7.38)$$

$$\frac{1}{\overline{X}_n} = C_M + C_s\frac{[S]}{[M]} + \frac{k_t}{k_p^2}\frac{R_p}{[M]^2} + C_I\left(\frac{k_t}{k_p^2 fk_d}\right)\frac{R_p^2}{[M]^3} \quad (7.39)$$

The complexity of this equation can be greatly reduced if, for example, polymerization of undiluted monomer is done (absence of a solvent) or conditions are chosen so as to minimize or even eliminate transfer to the initiator. In some cases substances exist whose transfer constants are equal to or exceed unity. These substances are useful in controlling the molecular weight of the polymer. With the addition of small quantities of such substances, known as *regulators* or *modifiers,* the molecular weight can be depressed to a desired level. In the preparation of synthetic rubbers from butadiene or diolefins, aliphatic mercaptans are frequently added to reduce the polymer chain length to a desired range required for subsequent processing. The effect of chain transfer on chain length and polymer structure is summarized in Table 7.2.

Table 7.2 Effect of Chain Transfer on Chain Length and Structure

Chain Transfer To	R_p	$\overline{M}_n$	$\overline{M}_w$	Molecular Architecture
Small molecule resulting in active radical	None	Decreases	Decreases	None
Small molecule resulting in retardation or inhibition	Decreases	May decrease or increase	May decrease or increase	None
Polymer (intermolecular)	None	None	Increases	Produces long branches
Polymer (intramolecular)	None	None	Increases	Produces short branches

Example 7.4: For the polymerization of pure (undiluted) styrene with benzoyl peroxide at 60°C, the number-average degree of polymerization, X_n, is given by the expression:

$$X_n = 0.6 \times 10^{-4} + 12.0\, R_p + 4.2 \times 10^4\, R_p^2$$

where R_p = rate of polymerization (mol/l-s).

a. What is the transfer constant to the monomer, C_M?
b. Calculate the transfer constant to the initiator if the monomer concentration [M] is 10 mol/l and $fk_d\, k_p^2/k_t = 2.29 \times 10^{-9}$.

Solution: a. Comparing the above equation with Equation 7.39, it is obvious that $C_M = 0.6 \times 10^{-4}$ and

$$C_I \left(\frac{k_t}{k_p^2 fk_d} \right) \frac{R_p^2}{[M]^3} = 4.2 \times 10^4\, R_p^2$$

$$C_I = 4.2 \times 10^4\, [M]^3 \times 2.29 \times 10^{-9}$$

$$= 0.0962$$

A. TRANSFER TO UNDILUTED MONOMER

For bulk polymerization (i.e., polymerization in the absence of a solvent) Equation 7.39 reduces to:

$$\frac{1}{\overline{X}_n} = C_M + \left(\frac{k_t}{k_p^2} \right) \frac{R_p}{[M]^2} + C_I \left(\frac{k_t}{k_p^2 fk_d} \right) \frac{R_p^2}{[M]^3} \tag{7.40}$$

This expression is quadratic in R_p, and the predicted behavior has indeed been observed in polymerization of styrene at 60° with benzoyl peroxide as the initiator. The first term on the right-hand side of the equation represents the contribution of transfer to the monomer; it is constant and independent of the rate of polymerization. The second term corresponds to normal termination (i.e., ½ ν, transfer reactions), while the third term, which represents transfer to the initiator, increases with increasing rates since high rates require high concentrations of initiator.

B. TRANSFER TO SOLVENT

In the presence of a solvent and with a proper choice of reaction conditions so as to minimize other types of transfer, the general expression for transfer (i.e., Equation 7.39) becomes

$$1/X_n = 1/X_{no} + C_S\, [S]/[M] \tag{7.41}$$

where $1/X_{no}$ combines the terms for normal polymerization with transfer to monomer and represents the reciprocal of the degree of polymerization in the absence of a solvent. Experimentally, in addition to the above precautions, the ratio $R_p/[M]^2$ is held constant while the concentration of the solvent is varied. The transfer constants to solvents for the peroxide initiation of the polymerization of some monomers are listed in Table 7.3.

Example 7.5: The transfer constant to the solvent for the polymerization of styrene in benzene at 100°C is 0.184×10^{-4}. How much dilution is required to halve the molecular weight given that $1/X_{no} = 2.5 \times 10^{-4}$?

Solution: For the molecular weight to be halved, then $1/X_n$ becomes $2/X_{no}$

$$\left(\frac{2}{\overline{X}_{no}} \right) - \frac{1}{X_{no}} = 0.184 \times 10^{-4}\, [S]/[M]$$

$$\left(\frac{1}{\overline{X}_{no}}\right)\frac{10^4}{0.184}=\frac{[S]}{[M]}$$

$$[S]=13.6\,[M]$$

Dilution factor is about 14.

Table 7.3 Transfer Constants to Solvents for the Peroxide Initiation of Monomer Polymerization at 60°C

	Transfer Constant: $C_S \times 10^4$		
Solvent	**Methyl Methacrylate**	**Styrene**	**Vinyl Acetate**
Acetone	0.195	4.1	25.6 (70°C)
Benzene	0.83	0.028	2.4
Carbon tetrachloride	2.40	87.0	2023.0 (70°C)
Chloroform	1.77	3.4	554.0 (70°C)
Toluene	0.202	0.105	20.75

From Schulz, G.V. and Harborth, H., *Makromol. Chem.*, 1, 106, 1947. With permission.

VIII. TEMPERATURE DEPENDENCE OF DEGREE OF POLYMERIZATION

The dependence of the polymerization rate and the number-average degree of polymerization on temperature can be demonstrated if the respective relation is expressed in the form of the Arrhenius equation.

$$k_b = A\exp\left(\frac{-E_b}{RT}\right) \tag{7.42}$$

$$\frac{d\ln R_p}{dT}=\frac{\left(E_p-\frac{E_t}{2}\right)+\frac{E_d}{2}}{RT^2} \tag{7.43}$$

$$\frac{d\ln \overline{X}_n}{dT}=\frac{\left(E_p-\frac{E_t}{2}\right)-\frac{E_d}{2}}{RT^2} \tag{7.44}$$

For radical polymerizations the activation energy of decomposition is of the order of 30 kcal/mol while $(E_p - E_t/2)$ is about 4 to 7 kcal/mol. Thus the temperature coefficients are, respectively,

$$dR_p/dT > 0 \tag{7.45}$$

$$dX_n/dT < 0 \tag{7.46}$$

These results predict that the rate of polymerization increases with increasing temperature while the molecular weight decreases. The same conclusion can be drawn for thermal polymerizations since $(E_p - E_t/2) - E_i/2$ is a negative quantity.

Example 7.6: The energies of activation for the polymerization of styrene with di-tertiary-butyl peroxide as initiator are

E_d = 33.5 kcal/mol
E_p = 7.0 kcal/mol
E_t = 3.0 kcal/mol

Calculate the relative (a) rates of propagation and (b) degree of polymerization (X_η) if the polymerization temperature is changed from 50°C to 60°C.

Solution:

a. $$\frac{d \ln R_p}{dT} = \frac{E}{RT^2}$$

where $E = E_p - \frac{E_t}{2} + \frac{E_d}{2}$.

$$\int_{R_{p/T_o}}^{R_{p/T}} d \ln R_p = \int_{T_o}^{T} \frac{E}{RT^2} dT$$

$$\ln \frac{R_{pT}}{R_{pT_o}} = \frac{E}{R}\left[\frac{1}{T_o} - \frac{1}{T}\right]$$

$$T_o = 50°C = 323 \text{ K}; \; T = 333 \text{ K}$$

$$E = 7 - \frac{3}{2} + \frac{33.5}{2} = 22.25 \text{ kcal/mol}$$

$$\frac{E}{R} = \frac{22.25 \times 10^3 \text{ cal/mol}}{1.99 \text{ cal/mol} - \text{K}}$$

$$\frac{1}{T} - \frac{1}{T_o} = 0.0001 \text{ K}^{-1}$$

$$\ln \frac{R_{p60}}{R_{p50}} = 1.12$$

$$R_{p60}/R_{p50} = 1^{1.12} = 3.06$$

$$\ln \overline{X}_{n60}/\overline{X}_{n50} = \frac{E}{RT}\left[\frac{1}{T_o} - \frac{1}{T}\right]; \; E - E_p - \frac{E_t}{2} - \frac{E_d}{2}$$

$$= -11.25 \text{ kcal/mol}$$

$$E/RT = \left(-\frac{11.25 \times 10^3 \text{ cal/mol}}{1.99 \text{ cal/mol} - \text{K}}\right)\left(0.001 \text{ K}^{-1}\right)$$

$$= -0.57$$

$$\frac{\overline{X}_{\eta 60}}{\overline{X}_{\eta 50}} = 1^{-0.57} = 0.57$$

IX. IONIC AND COORDINATION CHAIN POLYMERIZATION

A. NONRADICAL CHAIN POLYMERIZATION

In addition to the radical chain polymerization mechanisms discussed above, chain-reaction polymerization can also occur through other mechanisms. These include cationic polymerization in which the chain carriers are carbonium ions; anionic polymerization where the carriers are carbanions; and coordination polymerization, which is thought to involve the formation of a coordination compound between the catalyst, monomer, and growing chain. The polymerization mechanisms of these systems are complex and not as clearly understood as the mechanism of radical polymerization. This is because the reactions are generally heterogeneous, involving usually solid inorganic catalysts and organic monomers. In addition, ionic polymerizations are characterized by extremely high reaction rates. High-molecular-weight polymers are generated so fast that it is frequently neither possible to establish nor maintain uniform reaction conditions, thus making it difficult to obtain kinetic data or reproducible results. Two essential differences between free-radical and ionic polymerizations are apparent. First, in ionic polymerization, initiation involves the formation of an ion pair through the transfer of an ion or electron to or from the monomer. This contrasts with the generation and addition of a radical to the monomer in free-radical initiation reactions. Second, termination in ionic polymerization involves the unimolecular reaction of a chain with its counterion or a transfer reaction with the remnant species unable to undergo propagation. In contrast to radical chain polymerization, this termination process in ionic polymerization is strictly unimolecular — bimolecular annihilation of growth activity between two growing chains does not occur. The types of chain polymerization suitable for common monomers are shown in Table 7.1.

While many monomers can polymerize by more than one mechanism, it is evident that the polymerization mechanism best suited for each monomer is related to the polarity of the monomer and the Lewis acid–base strength of the ion formed. Monomers in which electron-donating groups are attached to the carbon atoms with the double bond (e.g., isobutylene) are capable of forming stable carbonium ions (i.e., they behave as Lewis bases). Such monomers are readily converted to polymers by cationic catalysts (Lewis acids). On the other hand, monomers with electron-withdrawing substituent (e.g., acrylonitrile) form stable carbanions and polymerize with anionic catalysts. Free-radical polymerization falls between these structural requirements, being favored by conjugation in the monomer and moderate electron withdrawal from the double bond. The structural requirements for coordination polymerization are less clearly delineated, and many monomers undergo coordination polymerization as well as ionic and radical polymerizations.

B. CATIONIC POLYMERIZATION

Typical catalysts that are effective for cationic polymerization include $AlCl_3$, $AlBr_3$, BF_3, $TiCl_4$, $SnCl_4$, and sometimes H_2SO_4. With the exception of H_2SO_4, these compounds are all Lewis acids with strong electron-acceptor capability. To be effective, these catalysts generally require the presence of a Lewis base such as water, alcohol, or acetic acid as a cocatalyst. As indicated in Table 7.1, monomers that polymerize readily with these catalysts include isobutylene, styrene, α-methylstyrene and vinyl alkyl ethers. All of these monomers have electron-donating substituents, which should enhance the electron-sharing ability of the double bonds in these monomers with electrophilic reagents.

Cationic polymerizations proceed at high rates at low temperatures. For example, the polymerization at −100°C of isobutylene with BF_3 or $AlCl_3$ as catalysts yields, within a few seconds, a polymer with molecular weight as high as 10^6. Both the rate of polymerization and the molecular weight of the polymer decrease with increasing temperature. The molecular weights of polyisobutylene obtained at room temperature and above are, however, lower than those obtained through radical polymerization.

1. Mechanism

Based on available experimental evidence, the most likely mechanism for cationic polymerization involves carbonium ion chain carrier. For example, the polymerization of isobutylene with BF_3 as the catalyst can be represented thus:

First, the catalyst and cocatalyst (e.g., water) form a complex:

$$BF_3 + H_2O \longrightarrow H^+(BF_3OH)^- \tag{7.47}$$

The complex then donates a proton to an isobutylene molecule to form a carbonium ion:

$$H^+(BF_3OH)^- + CH_2{=}\overset{\displaystyle CH_3}{\underset{\displaystyle CH_3}{C}} \dashrightarrow CH_3{-}\overset{\displaystyle CH_3}{\underset{\displaystyle CH_3}{C^+}} + (BF_3OH)^- \tag{7.48}$$

The carbonium ion reacts with a monomer molecule in the propagation step.

$$CH_3-\underset{CH_3}{\overset{CH_3}{|}}{C^+}\,(BF_3OH)^- \xrightarrow{M} (CH_3)_3C-CH_2-\underset{CH_3}{\overset{CH_3}{|}}{C^+}\,(BFOH)^- \qquad (7.49)$$

In general, the propagation reaction can be written as

$$CH_3-\underset{CH_3}{\overset{CH_3}{C}}-\left[CH_2-\underset{CH_3}{\overset{CH_3}{C}}-\right]_n^+ + (BF_3OH)^- \xrightarrow{M} (CH_3)_3-C-\left[CH_2-\underset{CH_3}{\overset{CH_3}{C}}-\right]_{n+1}^+ + (BF_3OH)^- \qquad (7.50)$$

Since cationic polymerization is generally carried out in hydrocarbon solvents that have low dielectric constants, separation of the ions would require a large amount of energy. Consequently, the anion and cation remain in close proximity as an ion pair. It is therefore to be expected that the growth rate and subsequent reactions (e.g., termination and chain transfer) are affected by the nature of the ion pair.

Termination occurs either by rearrangement of the ion pair to yield a polymer molecule with an unsaturated terminal unit and the original complex or through transfer to a monomer.

$$\sim\sim CH_2-\underset{CH_3}{\overset{CH_3}{C^+}}\,(BF_3OH)^- \longrightarrow \sim\sim CH_2=\underset{CH_3}{\overset{CH_3}{C}} + H^+\,(BF_3OH)^- \qquad (7.51)$$

$$\sim\sim CH_2-\underset{CH_3}{\overset{CH_3}{C^+}}\,(BF_3OH)^- \xrightarrow{M} \sim\sim CH_2-\underset{CH_3}{\overset{CH_2}{\overset{\|}{C}}} + CH_3-\underset{CH_3}{\overset{CH_3}{C^+}} + (BF_3OH)^- \qquad (7.52)$$

Unlike in free-radical polymerization, the catalyst is not attached to the resulting polymer molecule, and in principle many polymer molecules can be produced by each catalyst molecule.

2. Kinetics

For the purpose of establishing the kinetics of generalized cationic polymerization, let A represent the catalyst and RH the cocatalyst, M the monomer, and the catalyst–cocatalyst complex $H^+ AR^-$. Then the individual reaction steps can be represented as follows:

$$\begin{aligned}
&A + RH \overset{K}{\rightleftharpoons} H^+ AR^- \\
&H^+ AR^- + M \xrightarrow{k_i} HM^+ AR^- \\
&HM_n^+ AR^- + M \xrightarrow{k_p} HM_{n+1}^+ AR^- \\
&HM_n^+ AR^- \xrightarrow{k_t} M_n + H^+ AR^- \\
&HM_n^+ AR^- + M \xrightarrow{k_{tr}} M_n + HM^+ AR^-
\end{aligned} \qquad (7.53)$$

The rate of initiation R_i is given by

$$R_i = k_i\left[H^+ AR^-\right][M] = k_i K[A][RH][M] \qquad (7.54)$$

As usual, the square brackets denote concentration. If the complex H^+AR^- is readily converted in the second step of Equation 7.48 (i.e., if the complex formation, step 1 Equation 7.48) is the rate-limiting step, then the rate of initiation is independent of the monomer concentration. Since AR^- remains in the close vicinity of the growing center, the termination step is first order

$$R_t = K_t\left[M^+\right] \tag{7.55}$$

where $[M^+]$ is the concentration of all the chain carriers $[HM_n^+ AR^-]$. The retention of the terminating agent AR^- in the vicinity of the chain carrier is responsible for the primary difference between the kinetics of cationic polymerization and that of free-radical polymerization. Assuming that steady state holds, then $R_i = R_t$ and

$$\left[M^+\right] = \frac{Kk_i}{k_t}[A][RH][M] \tag{7.56}$$

The overall rate of polymerization, R_p is given by

$$R_p = k_p\left[M^+\right][M] = K\frac{k_i k_p}{k_t}[A][RH][M]^2 \tag{7.57}$$

The number-average degree of polymerization, assuming predominance of termination over chain transfer, is

$$\overline{X}_n = \frac{R_p}{R_i} = \frac{k_p}{k_t}\frac{\left[M^+\right][M]}{\left[M^+\right]} = \frac{k_p}{k_t}[M] \tag{7.58}$$

If, on the other hand, chain transfer dominates, then

$$\overline{X}_n = \frac{R_p}{R_{tr}} = \frac{k_p\left[M^+\right][M]}{k_t\left[M^+\right][M]} = \frac{k_p}{k_{tr}} \tag{7.59}$$

In this case, the average degree of polymerization is independent of both the concentration of the monomer and the concentration of the catalyst. Available kinetic data tend to support the above mechanism.

Example 7.7: Explain why in the cationic polymerization of isobutylene, liquid ethylene or propylene at their boiling points are normally added to the reaction medium as a diluent. How will an increase in the dielectric constant of the reaction medium affect the rate and degree of polymerization?

Solution: Both the rate of polymerization and the molecular weight decrease with increasing temperature in cationic polymerization. These liquids help to prevent excessive temperature increases because part of the heat of polymerization is dissipated through the heat of evaporation of the liquids. In other words, the liquids act essentially as internal refrigerants.

Using the general relation between the rate of reaction and activation energy ($k = Ae^{-E/RT}$), we note that a decrease in the activation energy, E, increases the rate of reaction while an increase in E has the opposite effect. An increase in dielectric constant increases the rate of initiation, k_i, by reducing the energy required for charge separation. On the other hand, an increase in the dielectric constant decreases k_t by increasing the energy required for the rearrangement and combination of the ion pair. Since both R_p and X_n are directly proportional to k_i/k_t, an increase in dielectric constant increases both quantities.

C. ANIONIC POLYMERIZATION

Monomers with electronegative substituents polymerize readily in the presence of active centers bearing whole or partial negative charges. For example, a high-molecular-weight polymer is formed when methacrylonitrile is added to a solution of sodium in liquid ammonia at –75°C. Typical electron-withdrawing substituents that permit the anionic polymerization of a monomer include –CN, –COOR, $-C_6H_5$, and $-CH{=}CH_2$. The electronegative group pulls electrons from the double bond and consequently renders the monomer susceptible to attack by an electron donor. Catalysts for anionic polymerization include Grignard reagents, organosodium compounds, alkali metal amides, alkoxide, and hydroxides.

1. Mechanism

Propagation in anionic polymerization proceeds according to the following reactions:

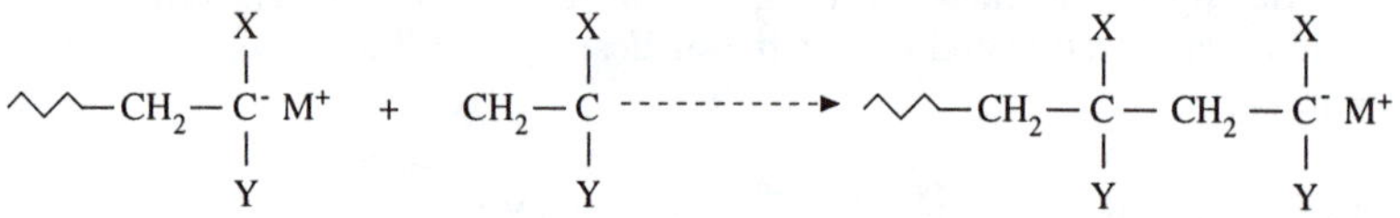

$$\sim\!\!\sim CH_2-\overset{X}{\underset{Y}{C^-}}\ M^+ + CH_2-\overset{X}{\underset{Y}{C}} \dashrightarrow \sim\!\!\sim CH_2-\overset{X}{\underset{Y}{C}}-CH_2-\overset{X}{\underset{Y}{C^-}}\ M^+ \quad (7.60)$$

Here, M^+ represents a counterion that accompanies the growing chain. In most cases, M^+ is an alkali metal ion, whereas X and Y are either electron-withdrawing groups or unsaturated groups capable of resonance stabilization of the negative charge.

Initiation may occur in two ways: a direct attack of a base on the monomer to form a carbanion (Equation 7.61) or by transfer of an electron from a donor molecule to the monomer to form an anion radical (Equation 7.62).

$$M^+B^- + CH_2=\overset{X}{\underset{Y}{C}} \dashrightarrow B-CH_2-\overset{X}{\underset{Y}{C^-}}\ M^+ \quad (7.61)$$

$$M^O + CH_2=\overset{X}{\underset{Y}{C}} \dashrightarrow \cdot CH_2-\overset{X}{\underset{Y}{C^-}}\ M^+ \quad (7.62)$$

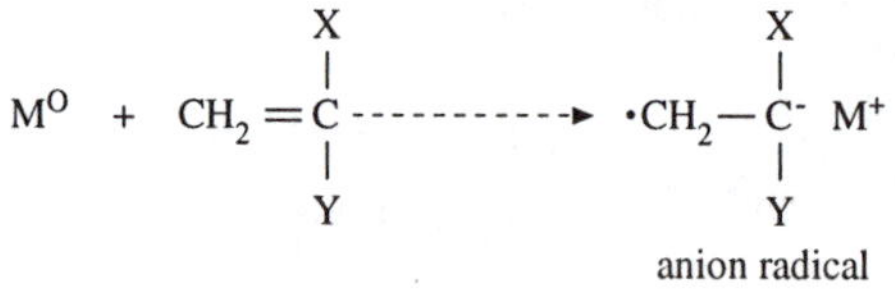

anion radical

M^+B^- may be a metal amide, alkoxide, alkyl, aryl, and hydroxide depending on the nature of the monomer. The effectiveness of the catalyst in the initiation process depends on its basicity and the acidity of the monomer. For example, in the anionic polymerization of styrene, the ability to initiate reaction decreases in the order ⬡$-CH_2^- -NH_2^- > NH_2^- \gg OH^-$. Indeed, OH^- will not initiate anionic polymerization of styrene. Where the anion is polyvalent, such as *tris*(sodium ethoxy) amine $N(CH_2CH_2O^-\ Na^+)_3$, an equivalent number of growing chains (in this case, three) can be initiated simultaneously.

The donor molecular in Equation 7.58 represents, in general, an alkali metal. In this case, transfer results in the formation of a positively charged alkali metal counterion and an anion radical. Pairs of anion radicals combine to form a dianion.

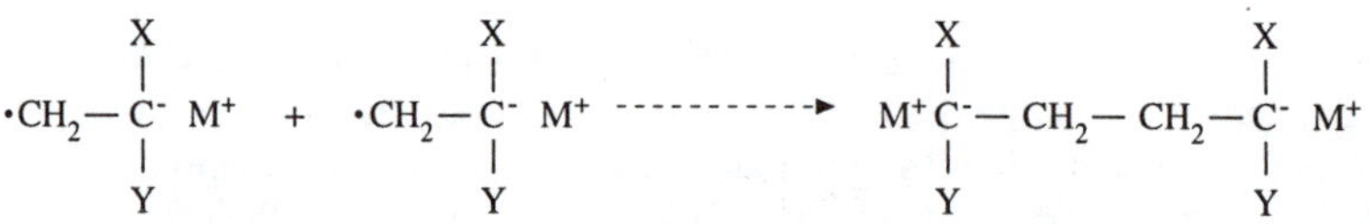

$$\cdot CH_2-\overset{X}{\underset{Y}{C^-}}\ M^+ + \cdot CH_2-\overset{X}{\underset{Y}{C^-}}\ M^+ \dashrightarrow M^+\overset{X}{\underset{Y}{C^-}}-CH_2-CH_2-\overset{X}{\underset{Y}{C^-}}\ M^+ \quad (7.63)$$

In carefully controlled systems (pure reactants and inert solvents), anionic polymerizations do not exhibit termination reactions. As we shall see shortly, such systems are referred to as living polymers; however, because of the reactivity of carbanions with oxygen, carbon dioxide, and protonic compounds, termination occurs according to Equation 7.64.

$$\sim\!\!\sim\overset{|}{\underset{|}{C^-}} + O_2 \dashrightarrow \sim\!\!\sim\overset{|}{\underset{|}{C}}OO^- \quad (a)$$

$$\sim\!\!\sim\overset{|}{\underset{|}{C^-}} + CO_2 \dashrightarrow \sim\!\!\sim\overset{|}{\underset{|}{C}}C(=O)O^- \quad (b)$$

$$\sim\!\!\sim\overset{|}{\underset{|}{C^-}} + H_2O \dashrightarrow \sim\!\!\sim\overset{|}{\underset{|}{C}}H + OH^- \quad (c)$$

$$\sim\!\!\sim C^- + RH \dashrightarrow \sim\!\!\sim\overset{|}{\underset{|}{C}}H + R^- \quad (d)$$

$$\sim\!\!\sim C^- + RC(=O)OH \dashrightarrow \sim\!\!\sim\overset{|}{\underset{|}{C}}H + RC(=O)O^- \quad (e) \qquad (7.64)$$

The terminal groups in Equations 7.64a and 7.64b cannot propagate and, consequently, effectively terminate polymer growth. Equations 7.64c–7.64e involve proton transfer from the solvent to the growing chain resulting in a dead polymer. This is exemplified by the sodium-catalyzed polymerization of butadiene in toluene:

$$\sim\!\!\sim CH_2-CH{=}CH-CH_2^-\ Na^+ \xrightarrow[\text{(toluene)}]{CH_3C_6H_5} \sim\!\!\sim CH_2-CH{=}CH-CH_3 + C_6H_5CH_2^-\ Na^+$$

Other possible termination reactions include (1) the transfer of a hydride ion leaving a residual terminal unsaturation (this, however, is a high-energy process and therefore unlikely); (2) isomerization of the carbanion to give an inactive anion; and (3) irreversible reaction of the carbanion with the solvent or monomer. In general, termination by transfer to the solvent predominates in anionic polymerization.

2. Kinetics

Available kinetic data for the polymerization of styrene by potassium amide in liquid ammonia support the following steps in the mechanism of anionic polymerization.

$$\begin{aligned} &KNH_2 \overset{K}{\rightleftharpoons} K^+ + NH_2^- \\ &NH_2^- + M \xrightarrow{k_i} NH_2M^- \\ &NH_2M_n^- + M \xrightarrow{k_p} NH_2M_{n+1}^- \\ &NH_2M_n^- + NH_3 \xrightarrow{k_{tr}} NH_2M_nH + NH_2^- \end{aligned} \qquad (7.65)$$

Considering the relatively high dielectric constant of the liquid ammonia medium, the counterion K^+ can be neglected. Assuming steady-state kinetics:

$$R_i = k_i[NH_2^-][M] \qquad (7.66)$$

$$R_t = k_{tr}[NH_2-M_n^-][NH_3] \qquad (7.67)$$

Thus from Equations 7.61 and 7.62

$$[NH_2-M_n^-] = \frac{k_i}{k_{tr}}\frac{[NH_2^-][M]^2}{[NH_3]} \qquad (7.68)$$

The rate of polymerization becomes

$$R_p = k_p\left[NH_2 - M_n^-\right][M] \qquad (7.69)$$

$$= k_p \frac{k_i}{k_{tr}} \frac{\left[NH_2^-\right][M]^2}{\left[NH_3\right]}$$

Given the predominance of transfer reactions, the degree of polymerization $\overline{X}_n$ is given by

$$\overline{X}_n = \frac{R_p}{R_t} = \frac{k_p}{k_{tr}} \frac{[M]}{\left[NH_3\right]} \qquad (7.70)$$

Example 7.8: Polymerization of styrene in liquid ammonia gave high yield of low-molecular-weight polystyrene, whereas methacrylonitrile gave high conversion of high-molecular-weight polymethacrylonitrile. Explain.

Solution: Since the reaction is carried out in liquid ammonia, proton transfer from the solvent (ammonia) is the possible termination reaction. The relevant carbanions are

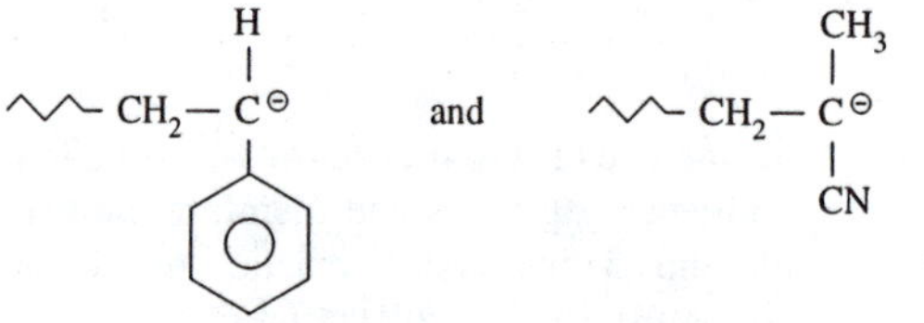

$$\sim\!\!\sim CH_2-\overset{\displaystyle H}{\underset{\displaystyle C_6H_5}{C^{\ominus}}} \quad \text{and} \quad \sim\!\!\sim CH_2-\overset{\displaystyle CH_3}{\underset{\displaystyle CN}{C^{\ominus}}} \qquad \text{(Str. 4)}$$

Since the –CN group is more electronegative than the phenyl group, the styrene carbanion will be more basic than the methacrylonitrile carbanion and as such more susceptible to proton transfer from ammonia. Transfer reactions generally lead to low-molecular-weight polymers.

D. LIVING POLYMERS

The absence of the termination step in anionic polymerizations with carefully purified reactants in inert reactions media results, as indicated above, in living polymers. In such systems, the growing species remain dormant in the absence of monomers but resume their growth activity with a fresh monomer supply. With adequate mixing, the monomer supplied to the system is distributed among the growing centers (living polymers). As a result, the number-average degree of polymerization is simply the ratio of the number of moles of monomer added to the total number of living polymers. That is,

$$\overline{X}_n = \frac{[\text{monomer}]}{[\text{catalyst}]}$$

Unlike other chain reactions where the growing chains undergo spontaneous termination, the reactive end groups in living polymers may be annihilated ("killed") by a choice of suitable reactants by the experimenter at a desired stage of the polymerization process. Thus living polymers offer fascinating potentials for the development of novel polymer systems. We discuss a few examples.

Spontaneous termination is governed by the laws of probability. Therefore, in polymerization involving spontaneous termination, the resulting polymers have a distribution of molecular weights since the lifetimes of the growing centers vary. But with living polymers where termination is lacking, a polymer with a nearly uniform molecular weight (monodisperse) can be obtained if the system is devoid of impurities and is well mixed.

The ability of living polymers to resume growth with the addition of fresh monomer provides an excellent opportunity for the preparation of block copolymers. For example, if a living polymer with one active end from monomer A can initiate the polymerization of monomer B, then an A–AB–B type copolymer can be obtained (e.g., styrene–isoprene copolymer). If, however, both ends of polymer A are active, a copolymer of the type B–BA–AB–B results. Examples are the thermoplastic rubbers polystyrene-polyisoprene-polystyrene and poly(ethylene oxide)-polystyrene-poly(ethylene oxide). In principle, for fixed amounts of two monomers that are capable of mutual formation of living polymers, a series of polymers with constant composition and molecular weight but of desired structural pattern can be produced by varying the fraction and order of addition of each monomer.

The potential versatility of living polymers in chemical synthesis is further demonstrated by the possibility of formation of polymers with complex shapes by employing polyfunctional initiators or terminating monodisperse living polymers with polyfunctional linking agents. For example, star-shaped poly(ethylene oxide) can be prepared with the trifunctional initiator trisodium salt of triethanol amine, $N(CH_2CH_2O^- Na^+)_3$. Another possible area of utilizing living polymers is in the introduction of specific end groups by terminating the living polymer with an appropriate agent. For example, termination of living polystyrene with CO_2 introduces terminal carboxylic groups, while reaction with ethylene oxide introduces hydroxy end groups. The utilization of these two approaches (synthesis of block copolymers or functional-ended polymers) provides the synthetic polymer chemist with a powerful tool for producing polymers with fascinating architectural features and properties (Table 7.4).

Table 7.4. Architectural forms of polymers available by living polymerization techniques.[6]

	Polymer	Application
1	 ~~~~OH *Functional ended*	Dispersing agents Synthesis of macromonomers
2	HO~~~~OH  *α,ω difunctional*	Elastomer synthesis Chain extension Cross-linking agents
3	 *AB Block*	Dispersing agents Compatibilizers for polymer blending
4	 *ABA Block*	Thermoplastic elastomers
5	 *Graft*	Elastomers Adhesives
6	 *Comb*	Elastomers Adhesives
7	 *Star*	Rheology control Strengthening agents
8	 *Ladder*	High-temperature plastics Membranes Elastomers
9	 *Cyclic*	Rheology control
10	 *Amphiphilic network*	Biocompatible polymers

E. COORDINATION POLYMERIZATION

A major development in polymer chemistry was the development in 1953 of new catalysts leading to the formation of polymer with exceptional structural regularity. The first catalysts were described by Ziegler[7] for the low-pressure polymerization of ethylene. These were modified by Natta[8] and his associates and used for the highly stereospecific polymerization of α-olefins, diolefins, and other monomers. As indicated in Chapter 1, Ziegler and Natta were awarded the Nobel prize in chemistry in 1963 for their work in this area. These catalysts are usually referred to as Ziegler–Natta catalysts, and since polymerization processes utilizing these catalysts result in stereoregular structures, they are sometimes called stereospecific or stereoregular polymerization. However, the term *coordination polymerization* is used here to reflect the mechanism which, as we shall see presently, is believed to govern the reaction involving these catalysts.

1. Mechanisms

The metals that are more frequently found as components of Ziegler–Natta catalysts are some light elements of groups I–III of the periodic table (e.g., Li, Be, Mg, Al), present as organometallic compounds and halides, or other derivatives of transition metals of groups IV–VIII (e.g., Ti, V, Cr, Mo, Rh, Rn, Co, and Ni). A typical example is the product(s) of the reaction between triethylaluminum and titanium tetrachloride. The composition of the product is not well defined but is believed to be either an alkylated metal halide (monometallic I) or a bimetallic complex involving a bridge between the two metals (II).

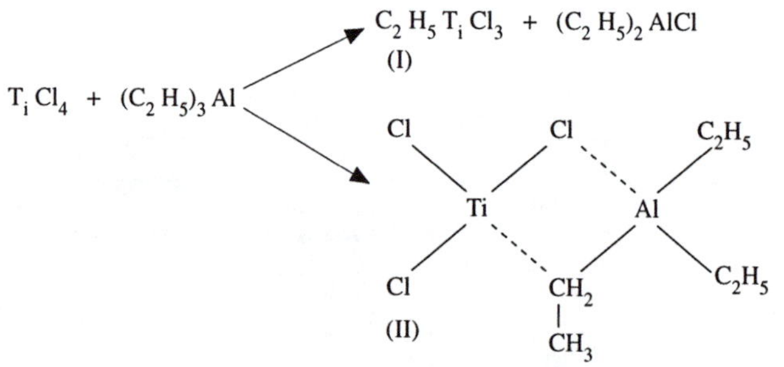

$$T_iCl_4 + (C_2H_5)_3Al \rightarrow C_2H_5T_iCl_3 + (C_2H_5)_2AlCl \quad \text{(I)}$$

$$T_iCl_4 + (C_2H_5)_3Al \rightarrow \text{(II): } Cl_2Ti(\mu\text{-}Cl)(\mu\text{-}CH_2CH_3)Al(C_2H_5)_2 \qquad \text{(Str. 5)}$$

While coordination polymerization may be anionic or cationic, relatively fewer examples of cationic coordination polymerization leading to the formation of stereoregular polymers currently exist. These are limited almost exclusively to the polymerization of monomers containing heteroatoms with lone electron pairs such as oxygen or nitrogen. In any case, the growth reaction in coordination polymerization is considered to be controlled by the counterion of the catalysts, which first coordinates and orients the incoming monomer and then inserts the polarized double bond of the monomer in the polarized bond between the counterion and the end of the growing chain. Coordination involves the overlap of the electrons of the monomer with a vacant sp orbital in the case of groups I–III metals or a vacant d orbital in the case of transition metals. The proposed propagation mechanisms for both the monometallic and bimetallic catalyst are shown below.

Monomettalic catalyst

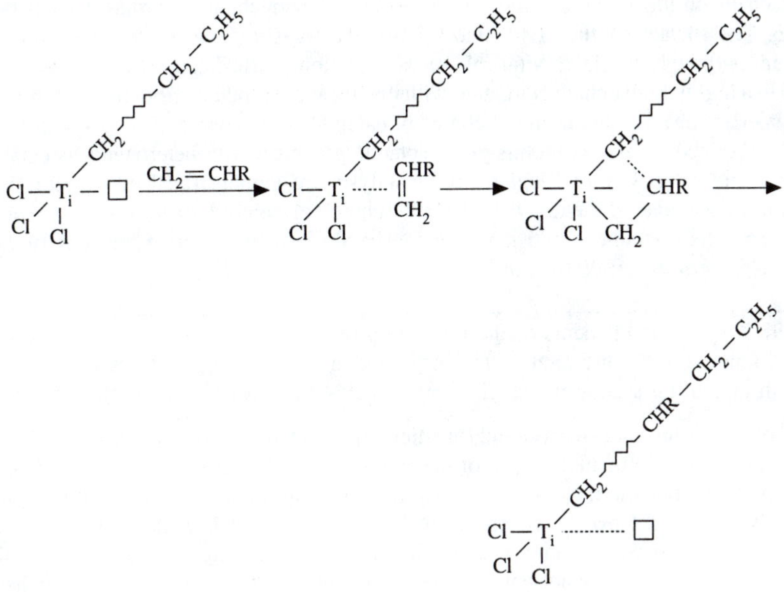

$Cl-Ti(Cl)_2(-CH_2\sim\sim CH_2-C_2H_5)-\square \xrightarrow{CH_2=CHR} Cl-Ti(Cl)_2(-CH_2\sim\sim CH_2-C_2H_5)\cdots(CHR=CH_2) \longrightarrow Cl-Ti(Cl)_2(-CH_2\sim\sim CH_2-C_2H_5)(CHR)(CH_2) \longrightarrow$ (Str. 6)

$Cl-Ti(Cl)_2(-CH_2\sim\sim CHR-CH_2-C_2H_5)\cdots\square$

Bimetallic catalyst

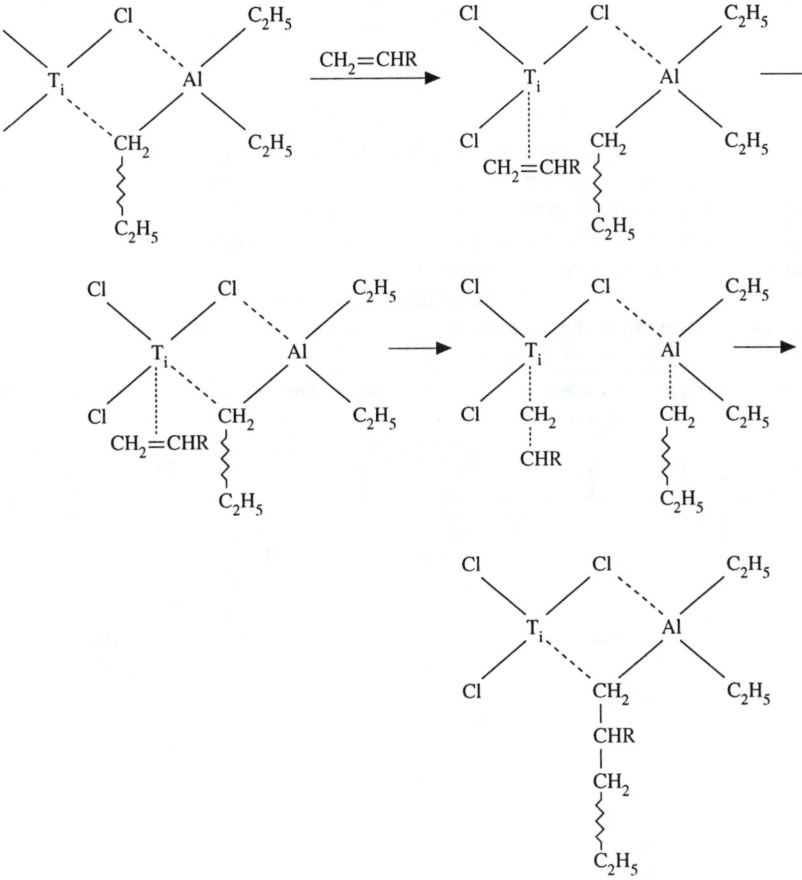

Ti, Cl, Al, C_2H_5, C_2H_5, CH_2, C_2H_5 $\xrightarrow{CH_2=CHR}$ Cl, Ti, Cl, Cl, Al, C_2H_5, $CH_2=CHR$, CH_2, C_2H_5, C_2H_5 $\longrightarrow$ (Str. 7)

Cl, Cl, Ti, Al, C_2H_5, Cl, $CH_2=CHR$, CH_2, C_2H_5, C_2H_5 $\longrightarrow$ Cl, Cl, Ti, Al, C_2H_5, Cl, CH_2, CHR, CH_2, C_2H_5, C_2H_5 $\longrightarrow$

Cl, Cl, Ti, Al, C_2H_5, Cl, CH_2, CHR, CH_2, C_2H_5, C_2H_5

The dominant feature of the coordination polymerization mechanism is the presence of forces that orient and insert each incoming monomer into the growing polymer chain according to a particular steric configuration. The surface of the crystalline salts of the transition metals that constitute part of the catalyst system is thought to play a vital role in this function. With Ziegler–Natta catalysts, ethylene is polymerized to a highly linear chain compared with the branched products from the high-pressure process (radical polymerization). By a convenient choice of catalyst system, isotactic and syndiotactic polypropylene can be obtained. Higher α-olefins yield isotactic polymers with heterogeneous catalyst systems. The versatility and selectivity of Ziegler–Natta catalysts are demonstrated even more sharply in the polymerization of conjugated dienes. By a suitable choice of catalyst system and reaction conditions, conjugated dienes like butadiene and isoprene can be made to polymerize into any of their isomers almost exclusively: *trans*-1,4; *cis*-1,4; or isotactic or syndiotactic 1,2.

Example 7.9: Among the elements of the first group of the periodic table, only organometallic compounds of lithium show a considerable tendency to give stereospecific catalysts. Also, this tendency decreases with increasing atomic radius of the elements of this group. Explain this observation.

Solution: Two conditions that are essential for anionic coordination polymerization are (1) the monomer must form a complex with the metal atom of the catalyst before its insertion in the metal–carbon bond, and (2) the bond between the metal of the catalyst and the carbon atom of the polymer chain must be at least partially polarized. Now, the electropositive nature of the metals of the first group of the periodic table increases with increasing atomic radius. This means that among these elements lithium has the highest electron-withdrawing power (generates the highest electric field) and consequently has the highest ability to coordinate with the double bond of the monomer (condition 1). The ionization of the bond between the metal (M) and the polymer (P) may be represented as follows:

$$\underset{\text{(I)}}{\underset{\text{covalent}}{M\text{———}P}} \rightleftharpoons \underset{\text{(II)}}{\underset{\text{polarized}}{M^{\delta+}\text{———}P^{\delta-}}} \rightleftharpoons \underset{\text{(III)}}{\underset{\text{free ions}}{M^{+} + P^{-}}} \qquad \text{(Str. 8)}$$

The degree of ionization of this bond depends partly but significantly on the electropositivity of the metal. Only the polarized metal–polymer bond of type II leads to the formation of sterically regular polymers. When a metal is very electropositive, it can be assumed that the growing polymer anion is essentially a free ion (III) and as such the positive counterion has negligible influence on the coordination, orientation, and insertion of the incoming monomer into the polymer chain and hence on the structural regularity of the polymer. Consequently, the tendency for stereospecificity decreases with increasing electropositivity (i.e., increasing atomic radius).

Example 7.10: Explain the observed variations in the stereoregularity of polymers when the metal of the catalyst or the reaction medium is varied.

Solution: The growth reaction in coordination polymerization first involves coordination of the monomer to the counterion of the catalyst and the insertion of the polarized double bond of the monomer in the polarized bond between the counterion and the end group of the growing chain. The nature of the coordinating metal in the catalyst determines the particular orientation imposed on the absorbed monomer. Consequently, stereoregularity of polymer may vary with different metals. The ability of the counterion to determine the orientation imposed on the incoming monomer is influenced by the extent of polarization of the bond between the counterion and the end group of the growing chain. The extent of polarization of this bond is itself a function of the solvating power of the reaction medium. Therefore, variation of the reaction medium causes a variation in polymer stereoregularity.

X. PROBLEMS

7.1. The following data were obtained by Arnett[12] for the polymerization of methyl methacrylate in benzene at 77°C with azo-*bis*-isobutyronitrile initiator. Assuming that the initiator efficiency is independent of monomer concentration, are the data consistent with the model for the rate of polymerization by free-radical mechanism?

[M] (kmol/m³)	[I] (mol/m³)	–d[M]/dt (mol/m³ · sec)
9.04	0.235	0.193
8.63	0.206	0.170
7.19	0.255	0.165
6.13	0.228	0.129
4.96	0.313	0.122
4.75	0.192	0.0937
4.22	0.230	0.0867
4.17	0.581	0.130
3.26	0.245	0.0715
2.07	0.211	0.415

From Arnett, L.M., *J. Am. Chem. Soc.*,74, 2027, 1952. With permission.

7.2. A steady-state free-radical styrene polymerization process is being controlled such that the rate of polymerization is constant at 1.79 g of monomer/ml-min. The initiator concentration is 6.6×10^{-6} mol/ml.

a. What must be done to maintain the constant rate of polymerization?
b. If the rate constant for the first-order decomposition of the initiator, k_d, is 3.25×10^{-4} min^{-1}, what is the rate of free radical generation per second per milliliter? What is X_n?
c. What percentage of the original initiator concentration remains after a reaction time of 3 h?

7.3. Consider the isothermal solution polymerization of styrene at 60°C in the following formulation:

100 g styrene
400 g benzene
0.5 g benzoyl peroxide

Assume that the initiator is 100% efficient and has a half-life of 44 h. At 60°C, k_p = 145 l/mol-s, k_t = 0.130 l/mol-s. All ingredients have unit density.

a. Derive the rate expression for this polymerization reaction.
b. Calculate the rate of propagation at 50% conversion.
c. How long will it take to reach this conversion?

7.4. For the polymerization of pure (undiluted) styrene with benzoyl peroxide at 60°C, the number-average degree of polymerization, X_n, is given by the general expression:

$$\frac{1}{\overline{X}_n} = 0.60 \times 10^{-4} + 8.4 \times 10^2 \frac{R_p}{[M]^2} + 2.4 \times 10^7 \frac{R_p^2}{[M]^3}$$

For a rate of polymerization of 10^{-4} mol/l-s and monomer concentration of 8.35 mol/l calculate:

a. The value of the transfer constant to the monomer
b. Number-average degree of polymerization, X_n
c. X_n assuming there is no transfer (normal termination)

d. The kinetic chain length
e. The transfer constant to the initiator if $fk_dk_p^2/k_t = 2.29 \times 10^{-9}$
f. The efficiency of initiation of polymerization of styrene in benzoyl peroxide if $k_d = 3.2 \times 10^{-6}\ s^{-1}$

7.5. The bulk polymerization of styrene at 100°C with benzoyl peroxide as the initiator resulted in a polymer of molecular weight 4.16×10^5. End-use tests showed that this product would be adequate provided the variation in molecular weight did not exceed 20%. However, to ensure better temperature control of the reactor, it was decided that the polymerization should be carried out in a solvent at a dilution factor of 2. The following solvents are available:

Solvent	Transfer Constant, C_S, for styrene at 100°C
Cyclohexane	0.16×10^{-4}
Carbon tetrachloride	180.0×10^{-4}

Assuming that only transfer to the solvent is possible, show that cyclohexane is the better of the two solvents for the end use under calculation.

7.6. The transfer constants to the solvent for the polymerization of styrene, methyl methacrylate, and vinyl acetate in toluene at 80°C are given below:

Chain radical	$C_S \times 10^4$
Styrene	0.31
Methyl methacrylate	0.52
Vinyl acetate	92.00

Find the ratio of the dilution factors for the three monomers if the molecular weight of the resulting polymers is each reduced to one-fourth that from the solvent-free polymerization. Assume that the degree of polymerization X_n in the absence of the solvent for the monomers is the same [$X_{no} = 5.0 \times 10^3$].

7.7. Explain why molecular weight ($\overline{X}_n$) increases with decreasing reaction temperature in cationic polymerization. Assume that termination predominates over transfer.

7.8. The rate constant for propagation of (polystyrene)$^-$ Na^+ in tetrahydrofuran for 25°C was found to be 400 l-mol^{-1} s^{-1}. The rate constants for proton transfer to the anion from water and ethanol were 4000 l-mol^{-1} s^{-1} and 4 l-mol^{-1} s^{-1}, respectively. Which of these impurities (water or ethanol) is more likely to inhibit high polymer formation in sodium-catalyzed anionic polymerization of styrene in tetrahydrofuran?

7.9. If anionic coordination is carried out in ethers or amines, sterically regular structures are generally not obtained. Explain this observation.

7.10. One mole of styrene monomer and 1.0×10 mol of azo-*bis*-isobutyronitrile initiator are dissolved in 1 l benzene. Estimate the molecular weight of the resulting polymer if each initiator fragment starts a chain and

a. All chains start at the same time. Termination is by disproportionation (all chains are of equal length).
b. Same as in case (a) but termination is by coupling.
c. Same as in case (a) but 6.0×10^{-4} mol mercaptan are added and each mercaptan acts as a chain transfer agent once.

7.11. The following data are obtained for the polymerization of a new monomer. Determine

a. The time for 50% conversion in run D
b. The activation energy of the polymerization

Run	Temperature (°C)	Conversion (%)	Time (min)	Initial Monomer Conc. (mol/l)	Initial Initiator Conc. (mol/l)
A	60	50	50	1.0	0.0025
B	80	75	700	0.50	0.0010
C	60	40	60	0.80	0.0010
D	60	50	—	0.25	0.0100

7.12. Sketch, on the same plot, molecular weight conversion curves for

a. Free-radical polymerization
b. Living polymerization
c. Condensation polymerization

REFERENCES

1. Lenz, R.W., *Organic Chemistry of High Polymers,* Interscience, New York, 1967.
2. Lowell, A.I. and Price, J.R., *J. Polym. Sci.,* 43, 1, 1960.
3. Schulz, G.V. and Husemann, E., *Z. Phys. Chem. Abt. B.,* 39, 246, 1938.
4. Schulz, G.V. and Harborth, H., *Macromol. Chem.,* 1, 106, 1947.
5. Young, L.J., in *Polymer Handbook,* 2nd ed., Brandrup, J. and Immergut, E.H., Eds., John Wiley & Sons, New York, 1975.
6. Webster, O.W., *Science,* 251, 887, 1991.
7. Ziegler, K., Holzkamp, E., Breil, H., and Martin, H., *Agnew. Chem.,* 67, 541, 1955.
8. Natta, G., Pino, P., Corradini, P., Danusso, F., Mantica, E., Mazzanti, G., and Moraghio, G., *J. Am. Chem. Soc.,* 77, 1708, 1955.
9. Flory, P.J., *Principles of Polymer Chemistry,* Cornell University Press, Ithaca, NY, 1953.
10. Odian, G., *Principles of Polymerization,* McGraw-Hill, New York, 1970.
11. Billmeyer, F.W., Jr.: *Textbook of Polymer Science,* 2nd ed., Interscience, New York, 1971.
12. Arnett, L.M., *J. Am. Chem. Soc.,* 74, 2027, 1952.

Chapter 8

Copolymerization

I. INTRODUCTION

As indicated in Chapter 1, the polymerization of organic compounds was first reported about the mid-19th century. However, it was not until about 1910 that the simultaneous polymerization of two or more monomers (or copolymerization) was investigated when it was discovered that copolymers of olefins and dienes produced better elastomers than either polyolefins or polydienes alone. The pioneering work of Staudinger in the 1930s and the development of synthetic rubber to meet wartime needs opened the field of copolymerization.

Copolymers constitute the vast majority of commercially important polymers. Compositions of copolymers may vary from only a small percentage of one component to comparable proportions of both monomers. Such a wide variation in composition permits the production of polymer products with vastly different properties for a variety of end uses. The minor constituent of the copolymer may, for example, be a diene introduced into the polymer structure to provide sites for such polymerization reaction as vulcanization; it may also be a trifunctional monomer incorporated into the polymer to ensure cross-linking, or possibly it may be a monomer containing carboxyl groups to enhance product solubility, dyeability, or some other desired property. Copolymerization reactions may involve two or more monomers; however, our discussion here is limited to the case of two monomers.

II. THE COPOLYMER EQUATION

Some observations are relevant to the consideration of copolymerization kinetics are

- The number of reactions involved in copolymerization of two or more monomers increases geometrically with the number of monomers. Consequently, the propagation step in the copolymerization of two monomers involves four reactions.
- The number of radicals to be considered equals the number of monomers. The terminal monomer unit in a growing chain determines almost exclusively the reaction characteristics; the nature of the preceding monomers has no significant influence on the reaction path.
- There are two radicals in the copolymerization of two monomers. Consequently, three termination steps need to be considered.
- The composition and structure of the resulting copolymer are determined by the relative rates of the different chain propagation reactions.

By designating the two monomers as M_1 and M_2 and their corresponding chain radicals as $M_1\cdot$ and $M_2\cdot$, the four propagation reactions and the associated rate equations in the copolymerization of two monomers may be written as follows:

Reaction	Rate equation
$M_1\cdot + M_1 \rightarrow M_1\cdot$	$k_{11}[M_1\cdot][M_1]$
$M_1\cdot + M_2 \rightarrow M_2\cdot$	$k_{12}[M_1\cdot][M_2]$
$M_2\cdot + M_1 \rightarrow M_1\cdot$	$k_{21}[M_2\cdot][M_1]$
$M_2\cdot + M_2 \rightarrow M_2\cdot$	$k_{22}[M_2\cdot][M_2]$

(8.1)

Here the first subscript in the rate constant refers to the reacting radical, while the second subscript designates the monomer. Now, it is reasonable to assume that at steady state, the concentrations of $M_1\cdot$ and $M_2\cdot$ remain constant. This implies that the rates of generation and consumption of these radicals are equal. It follows therefore that the rate of conversion of $M_1\cdot$ to $M_2\cdot$ necessarily equals that of conversion of $M_2\cdot$ to $M_1\cdot$. Thus from Equation 8.1

$$k_{21}[M_2\cdot][M_1] = k_{12}[M_1\cdot][M_2] \quad (8.2)$$

The rates of disappearance of monomers M_1 and M_2 are given by

$$\frac{-d[M_1]}{dt} = k_{11}[M_1\cdot][M_1] + k_{21}[M_2\cdot][M_1] \quad (8.3)$$

$$\frac{-d[M_2]}{dt} = k_{12}[M_1\cdot][M_2] + k_{22}[M_2\cdot][M_1] \quad (8.4)$$

By using Equation 8.2, one of the radicals can be eliminated. By dividing Equation 8.3 by Equation 8.4 we obtain:

$$\frac{d[M_1]}{d[M_2]} = \frac{[M_1]}{[M_2]} \frac{r_1 [M_2] + [M_2]}{[M_1] + r_2 [M_2]} \quad (8.5)$$

where r_1 and r_2 are monomer reactivity ratios defined by

$$r_1 = k_{11}/k_{12} \quad \text{and} \quad r_2 = k_{22}/k_{21} \quad (8.6)$$

Equation 8.5 is the copolymer equation. Let F_1 and F_2 represent the mole fractions of monomers M_1 and M_2 in the increment of polymer formed at any instant during the polymerization process, then

$$F_1 = 1 - F_2 = d[M_1]/d([M_1] + [M_2]) \quad (8.7)$$

Similarly, representing the mole functions of unreacted M_1 and M_2 in the monomer feed by f_1 and f_2, then

$$f_1 = 1 - f_2 = \frac{[M_1]}{[M_1] + [M_2]} \quad (8.8)$$

Substitution of Equations 8.7 and 8.8 in Equation 8.5 yields:

$$F_1 = \frac{r_1 f_1^2 + f_1 f_2}{r_1 f_1^2 + 2f_1 f_2 + r_2 f_2^2} \quad (8.9)$$

III. TYPES OF COPOLYMERIZATION

By definition, r_1 and r_2 represent the relative preference of a given radical that is adding its own monomer to the other monomer. The physical significance of Equation 8.9 can be illustrated by considering the product of the reactivity ratios,

$$r_1 r_2 = \frac{k_{11} k_{22}}{k_{12} k_{21}} \quad (8.10)$$

The quantity r_1r_2 represents the ratio of the product of the rate constants for the reaction of a radical with its own kind of monomer to the product of the rate constants for the cross-reactions. Copolymerization may therefore be classified into three categories depending on whether the quantity r_1r_2 is unity, less than unity, or greater than unity.

A. IDEAL COPOLYMERIZATION ($r_1r_2 = 1$)

$$r_1 r_2 = 1, \quad \text{then}$$
$$r_1 = 1/r_2 \quad \text{or} \quad k_{11}/k_{12} = k_{21}/k_{22} \tag{8.11}$$

In this case the copolymer equation reduces to

$$\frac{d[M_1]}{d[M_2]} = \frac{r_1[M_1]}{[M_2]} \tag{8.12}$$

or

$$F_1 = \frac{r_1 f_1}{f_1(r_1 - 1) + 1} = \frac{r_1 f_1}{r_1 f_1 + f_2} \tag{8.13}$$

It is evident that for ideal copolymerization, each radical displays the same preference for adding one monomer over the other. Also, the end group on the growing chain does not influence the rate of addition. For the ideal copolymer, the probability of the occurrence of an M_1 unit immediately following an M_2 unit is the same as locating an M_1 unit after another M_1 unit. Therefore, the sequence of monomer units in an ideal copolymer is necessarily random.

The relative amounts of the monomer units in the chain are determined by the reactivities of the monomer and the feed composition. To illustrate this, we note that the requirement that $r_1r_2 = 1$ can be satisfied under two conditions:

Case 1: $r_1 > 1$ and $r_2 < 1$ or $r_1 < 1$ and $r_2 > 1$. In this case, one of the monomers is more reactive than the other toward the propagating species. Consequently, the copolymer will contain a greater proportion of the more reactive monomer in the random sequence of monomer units. An important practical consequence of ideal copolymerization is that increasing difficulty is experienced in the production of copolymers with significant quantities of both monomers as the difference in reactivities of the two monomers increases.

Case 2: $r_1 = r_2 = 1$. Under these conditions, the growing radicals cannot distinguish between the two monomers. The composition of the copolymer is the same as that of the feed and as we said above, the monomers are arranged randomly along the chain. The copolymer equation becomes:

$$F_1 = \frac{f_1 f.}{f_1 + f_2} = f_1 \tag{8.14}$$

B. ALTERNATING COPOLYMERIZATION ($r_1 = r_2 = 0$)

When $r_1 = r_2 = 0$ (or $r_1r_2 = 0$), each radical reacts exclusively with the other monomer; that is neither radical can regenerate itself. Consequently, the monomer units are arranged alternately along the chain irrespective of the feed composition. In this case the copolymer reduces to:

$$\frac{d[M_1]}{d[M_2]} = 1 \tag{8.15}$$

or

$$F_1 = 0.5$$

Table 8.1 Reactivity Ratios of Some Monomers

Monomer 1	Monomer 2	r_1	r_2	T (°C)
Acrylonitrile	1,3-Butadiene	0.02	0.30	40
	Methyl methacrylate	0.15	1.22	80
	Styrene	0.04	0.40	60
	Vinyl acetate	4.2	0.05	50
	Vinyl chloride	2.7	0.04	60
1,3-Butadiene	Methyl methacrylate	0.75	0.25	90
	Styrene	1.35	0.58	50
	Vinyl chloride	8.8	0.035	50
Methyl methacrylate	Styrene	0.46	0.52	60
	Vinyl acetate	20	0.015	60
	Vinyl chloride	10	0.1	68
Styrene	Vinyl acetate	55	0.01	60
	Vinyl chloride	17	0.02	60
Vinyl acetate	Vinyl chloride	0.23	1.68	60

From Young, L.J., *Polymer Handbook,* 2nd ed., Brandrup, J. and Immergut, H.H., Eds., John Wiley & Sons, New York, 1975. With permission.

Polymerization continues until one of the monomers is used up and then stops. Perfect alternation occurs when both r_1 and r_2 are zero. As the quantity r_1r_2 approaches zero, there is an increasing tendency toward alternation. This has practical significance because it enhances the possibility of producing polymers with appreciable amounts of both monomers from a wider range of feed compositions.

C. BLOCK COPOLYMERIZATION ($r_1 > 1$, $r_2 > 1$)

If r_1 and r_2 are both greater than unity, then each radical would prefer adding its own monomer. The addition of the same type of monomer would continue successively until there is a chance addition of the other type of monomer and the sequence of this monomer is added repeatedly. Thus the resulting polymer is a block copolymer. In the extreme case of this type of polymerization ($r_1 = r_2 = \infty$) both monomers undergo simultaneous and independent homopolymerization; however, there are no known cases of this type of polymerization. Even though cases exist where r_1r_2 approaches 1 ($r_1r_2 = 1$), there are no established cases where $r_1r_2 > 1$. Indeed, the product r_1r_2 is almost always less than unity. Table 8.1 lists the reactivity ratios for some monomers.

Example 8.1: The reactivity ratios for the copolymerization of methyl methacrylate (1) and vinyl chloride (2) at 68°C are $r_1 = 10$ and $r_2 = 0.1$. To ensure that the copolymer contains an appreciable quantity (>40% in this case) of the vinyl chloride, a chemist decided to carry out the copolymerization reaction with a feed composed of 80% vinyl chloride. Will the chemist achieve his objective?

$$F_1 = \frac{r_1 f_1^2 + f_1 f_2}{r_1 f_1^2 + 2 f_1 f_2 + r_2 f_2^2} = \frac{10(0.2)^2 + (0.2)(0.8)}{10(0.2)^2 + 2(0.2)(0.8) + 0.1(0.8)^2}$$

$$= 0.714$$

$$F_2 = 1 - F_1 = 0.286.*$$

If the difference in the reactivities of the two monomers is large, it is impossible to increase the proportion of the less-reactive monomer in the copolymer simply by increasing its composition in the feed.

IV. POLYMER COMPOSITION VARIATION WITH FEED CONVERSION

Since the reactivity ratios r_1 and r_2 are generally of different magnitudes, there are necessarily differences in the rate of possible growth reactions. Consequently, the composition of the feed, f_1, and that of the

* Although $f_2 = 80\%$, $F_2 = 28.6\%$.

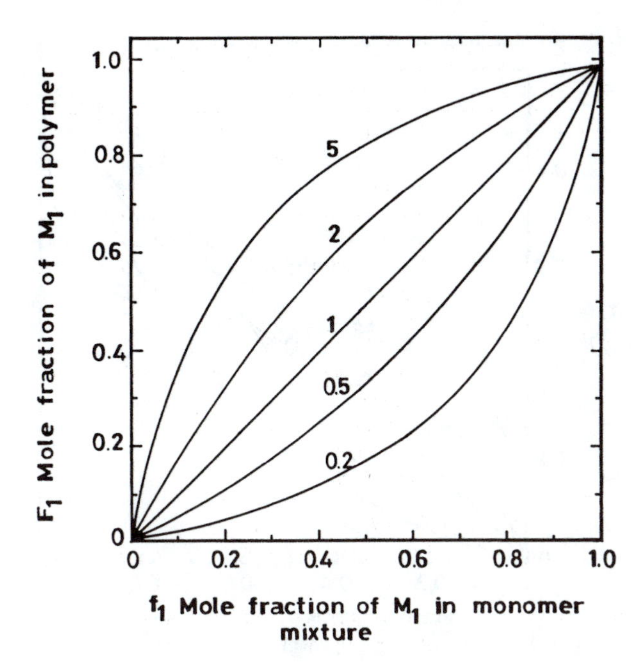

Figure 8.1 Variation of instantaneous composition of copolymer (mole fraction, F_1) with feed composition (mole fraction f_1 for ideal copolymerization with the values of $r_1 = 1/r_2$ indicated).

polymer, F_1, is neither equal nor constant throughout the polymerization. Therefore, the polymer produced over a finite range of conversion consists of a summation of increments of polymer differing in composition. Consider, for example, the copolymerization of styrene (1) and vinyl acetate (2). The reactivity ratios are $r_1 = 55$ and $r_2 = 0.015$. This means that either radical has a much greater preference to add styrene than vinyl acetate. The first polymer formed consists mainly of styrene. This also means a faster depletion of styrene in the feed. As polymerization proceeds, styrene is essentially used up and the last polymer formed consists mostly of vinyl acetate. However, at 100% conversion, the overall polymer composition must reflect the initial composition of the feed. It follows that the copolymers generally have a heterogeneous composition except in very special cases.

Equations 8.5 and 8.9 give the instantaneous polymer composition as a function of feed composition for various reactivity ratios. Figure 8.1 shows a series of curves calculated from these equations for ideal copolymerization. The range of feed composition that gives copolymers containing appreciable amounts of both monomers is small except the monomers have very similar reactivities.

The curves for several nonideal cases are shown in Figure 8.2. These curves illustrate the effect of increasing tendency toward alternation. With increasing alternation, a wider feed composition yields a copolymer containing substantial amounts of each monomer. This tendency is utilized in practice for the preparation of many important copolymers.

In those cases, where r_1 and r_2 are both either less or greater than unity, the curves of Figure 8.2 cross the line $F_1 = f_1$. The points of interception represent the occurrence of azeotropic copolymerization; that is, polymerization proceeds without a change in the composition of either the feed or the copolymer. For azeotropic copolymerization the solution to Equation 8.5 with $d[M_1]/d[M_2] = [M_1]/[M_2]$ gives the critical composition.

$$\frac{[M_1]}{[M_2]} = \frac{1 - r_2}{1 - r_1} \tag{8.16}$$

$$(f_1)_c = \frac{1 - r_2}{2 - r_1 - r_2} \tag{8.17}$$

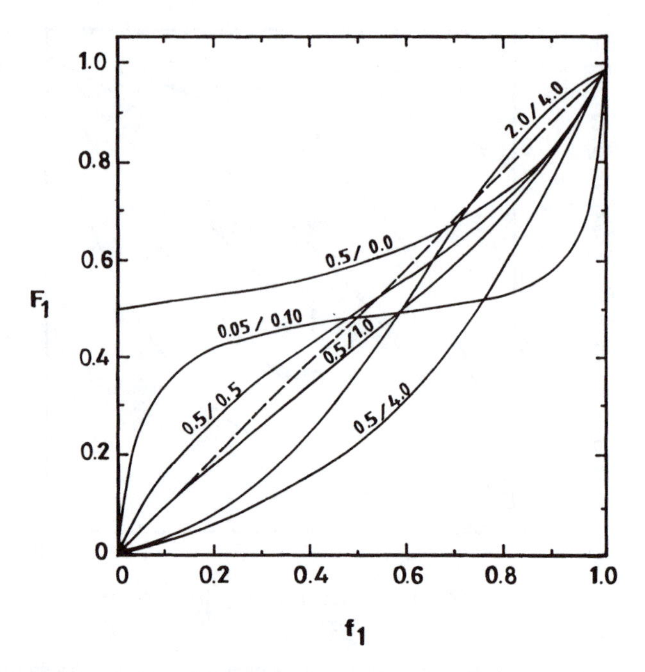

Figure 8.2 Instantaneous composition of copolymer F_1 as a function of monomer composition f_1 for the values of the reactivity ratios r_1/r_2 indicated.

If both r_1 and r_2 are greater than unity or if both are less than unity, then $(f_1)_C$ lies within the acceptable range $0 < f_1 < 1$. If $r_1 > 1$ and $r_2 < 1$ or $r_1 < 1$ and $r_2 > 1$, then there will be no critical feed composition, $(f_1)_C$. As we said earlier, cases where both r_1 and r_2 are greater than unity are not known; whereas, there are numerous cases where r_1 and r_2 are less than unity. When $r_1 \gg 1$ and $r_2 \ll 1$ or $r_1 \ll 1$ and $r_2 \gg 1$, the two monomers have a tendency to polymerize consecutively; the first polymer is composed predominantly of the more reactive monomer; the other monomer polymerizes only after almost all of the more reactive monomer has been exhausted. The case of styrene–vinyl acetate discussed above exemplifies this type of copolymerization.

Our discussion thus far has indicated that during copolymerization, the composition of both the feed and the polymer vary with conversion. To follow this composition drift, it is necessary to integrate the copolymer equation — a problem that is complex. Consider a system that is composed initially of M total moles of the two monomers ($M = M_1 + M_2$) and in which the resulting copolymer is richer in M_1 than the feed ($F_1 > f_1$). When dM moles have been polymerized, the polymer will contain F_1 dM moles of M_1 while the feed content of M_1 will be reduced to $(M - dM)(f_1 - df_1)$ moles. Writing a material balance for M_1:

$$Mf_1 - (M - dM)\left(f_2 - df_1\right) = F_1\, dM \tag{8.18}$$

Expanding and neglecting the second-order differential yields:

$$dM/M = df_1/\left(F_1 - f_2\right) \tag{8.19}$$

On integration, Equation 8.19 becomes

$$\ln M/M_o = \int_{f_{10}}^{f} df_1/\left(F_1 - f_1\right) \tag{8.20}$$

where M_0 and f_{10} are the initial values of M and f_1. Using Equation 8.19, it is possible to calculate F_1 by choosing values for f_1, and the integral may be evaluated graphically or numerically. This gives the relation between feed composition and the degree of conversion $(1 - M/M_0)$. An analytic solution to Equations 8.19 and 8.20 has been obtained. For $r_1 \neq 1$ and $r_2 \neq 1$,

$$\frac{M}{M_0} = \left[\frac{f_1}{(f_1)_0}\right]^{\alpha}\left[\frac{f_2}{(f_2)_0}\right]^{\beta}\left[\frac{(f_1)_0 - \delta}{f_1 - \sigma}\right]^{\gamma} \quad (8.21)$$

where the superscripts are given by:

$$\alpha = \frac{r_2}{1-r_2};\ \beta = \frac{r_1}{1-r_1};\ \gamma = \frac{1-r_1 r_2}{(1-r_1)(1-r_2)}$$
$$\delta = \frac{1-r_2}{2-r_2-r_2} \quad (8.22)$$

The feed composition variation with conversion gives the instantaneous composition of the copolymer as a function of conversion. It is also useful to know the *overall average composition* $\langle F_1 \rangle$ for a given conversion. This can be obtained through a material balance, say, for M_1 in a batch reactor. Moles M_1 in feed = moles of M_1 in polymer + moles of unreacted M_1.

$$(f_1)_0\, M_0 = \langle F_1 \rangle\,(M_0 - M) + f_1\, M \quad (8.23)$$

On rearrangement Equation 8.23 becomes:

$$\langle F \rangle = \frac{f_{10} - f_1\,(M/M_0)}{(1 - M/M_0)} \quad (8.24)$$

Example 8.2: What is the composition of the copolymer formed by the polymerization of an equimolar mixture of butadiene (1) and styrene (2) at 66°C? Which will contain more styrene — the polymer formed first or that formed later in the reaction?

Solution: From Table 8.1, $r_1 = 1.35$ and $r_2 = 0.58$

$$F_1 = \frac{r_1 f_1^2 + f_1 f_2}{r_1 f_1^2 + 2 f_1 f_2 + r_2 f_2^2}$$
$$= \frac{1.35(0.5)^2 + (0.5)^2}{1.35(0.5)^2 + 2(0.5)^2 + 0.58(0.5)^2}$$
$$= 0.60$$

$$k_{11} = 1.35\, k_{12},\ k_{22} = 0.58\, k_{21}$$

The rate of consumption of butadiene by either radical is greater than the rate of addition of styrene. Therefore the polymer formed first will be richer in butadiene while the polymer formed at the later stages of the reaction will be richer in styrene.

Example 8.3: Estimate the feed and copolymer compositions for the azeotropic copolymerization of acrylonitrile and styrene at 60°C.

Solution: For azeotropic copolymerization:

$$f_{1c} = \frac{1 - r_2}{2 - r_1 - r_2}$$

From Table 8.1

$$f_{1c} = \frac{1 - 0.4}{2 - 0.04 - 0.4}$$

$$f_1 = \frac{r_1 f_1^2 + f_1 f_2}{r_1 f_1^2 + 2 f_1 f_2 + r_2 f_2^2}$$

$$= \frac{0.04(0.38)^2 + (0.38)(0.62)}{0.04(0.38)^2 = 2(0.38)(0.62) + 0.4(0.62)^2}$$

$$= 0.38$$

Example 8.4: Plot graphs showing the variation of the instantaneous copolymer composition F_1 with monomer composition for the following systems:

I. Vinyl acetate (1), maleic anhydride (2), 75°C, $r_1 = 0.055$, $r_2 = 0.003$.
II. Styrene (1), vinyl acetate (2), 60°C, $r_1 = 55$, $r_2 = 0.01$.
III. Vinyl chloride (1), methyl methacrylate (2) 68°C, $r_1 = 0.1$, $r_2 = 10$

Comment on the shapes of the curves.

Solution:

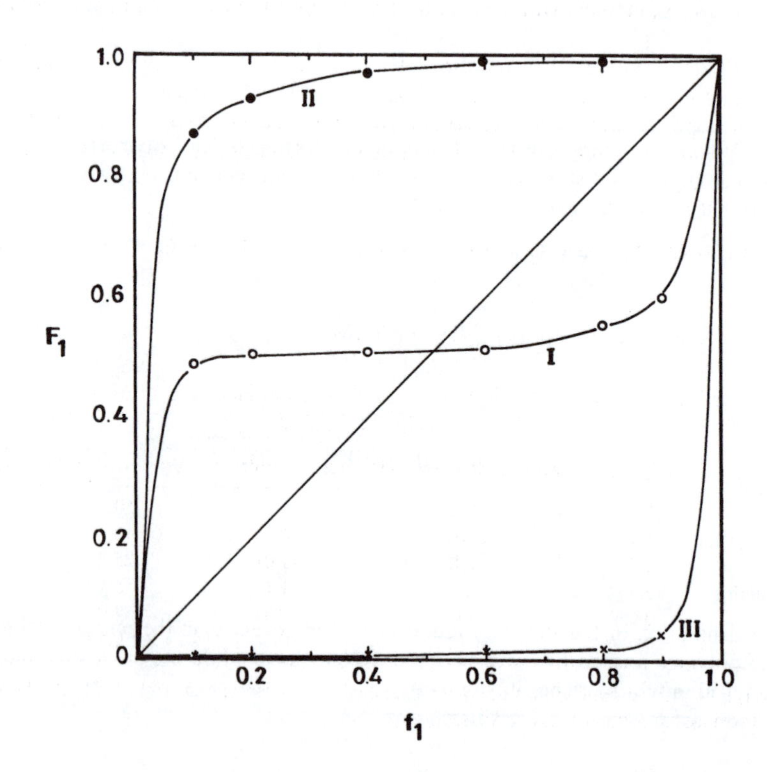

The range of feed composition that gives copolymers with appreciable quantities of both monomers is small for Cases II and III where the differences in reactivity ratios are large. However, in Case I where the reactivity ratios are comparable, copolymers with comparable quantities of both monomers are obtainable.

V. CHEMISTRY OF COPOLYMERIZATION

By definition, monomer reactivity ratios are independent of the initiation and termination steps of copolymerization reaction. In addition, they are virtually independent of the reaction medium, and their dependence on temperature is minimal. The relative reactivities of a series of monomers are determined by the reactivity of the individual monomer and that of the attacking radical. The reactivities of monomers and radicals are themselves dependent on the nature of the substituents on the monomer double bond. The influence of substituents on reactivity is threefold:

- Enhancement of monomer reactivity through activation of the double bond
- Conferment of resonance stabilization on the resulting radical
- Provision of steric hindrance at the reaction site

We now briefly discuss the contribution of these factors to radical copolymerization reactions.

A. MONOMER REACTIVITY

The relative reactivities of monomers to a given radical can be obtained from analysis of the relative reactivity ratios. This can be seen by considering the inverse of the monomer reactivity ratio:

$$1/r_1 = k_{12}/k_{11} \tag{8.25}$$

The inverse of the reactivity ratio (Equation 8.25) is the ratio of the rate of reaction of the given radical with another monomer to its rate of reaction with its own monomer. If the rate of reaction of the reference radical with its own monomer is taken as unity (i.e., $k_{11} = 1$), then the resulting k_{12} values give the relative reactivities of monomers with respect to the reference radical. A list of such values is given in Table 8.2.

Note that a different reference is taken for each column and as such only values in each column can be compared; horizontal comparisons are meaningless. In general, the relative reactivities of monomers decrease from the top to the bottom of each column; the order of decrease is irregular due to the specificity of addition of monomers by radicals. It is evident that the effect of substituents in enhancing monomer reactivity is in the order:

$$-C_6H_5 > -CH{=}CH_2 > -COCH_3 > -CN > -COOR > -Cl > -CH_2Y > -OCOCH_3 > -OR \tag{Str. 1}$$

The effect of a second 1-substituent is generally additive.

Table 8.2 Relative Reactivities of Monomers to Reference Radicals at 60°C

	Reference Radical				
Monomer	**Styrene**	**Methyl Methacrylate**	**Acrylonitrile**	**Vinyl Chloride**	**Vinyl Acetate**
Styrene	(1.0)	2.2	25	50	100
Methyl methacrylate	1.9	(1.0)	6.7	10	67
Acrylonitrile	2.5	0.82	(1.0)	25	20
Vinylidene chloride	5.4	0.39	1.1	5	10
Vinyl chloride	0.059	0.10	0.37	(1.0)	4.4
Vinyl acetate	0.019	0.05	0.24	0.59	(1.0)

From Billmeyer, F.W., Jr., *Textbook of Polymer Science,* 2nd ed., Interscience, New York, 1971. © John Wiley & Sons. Reprinted with permission of John Wiley & Sons.

Table 8.3 Radical–Monomer Propagation Rate Constants at 60°C (l/mol-s)

	Radical					
Monomer	**Butadiene**	**Styrene**	**Methyl Methacrylate**	**Methyl Acrylate**	**Vinyl Chloride**	**Vinyl Acetate**
Butadiene	100	250	2,820	42,000	350,000	—
Styrene	74	143	1,520	14,000	600,000	230,000
Methyl methacrylate	134	278	705	4,100	123,000	150,000
Methyl acrylate	132	206	370	2,090	200,000	23,000
Vinyl chloride	11	8	70	230	12,300	10,000
Vinyl acetate	—	2.6	35	520	7,300	2,300

From Billmeyer, F.W., Jr., *Textbook of Polymer Science,* 2nd ed., Interscience, New York, 1971.

The above order of monomer reactivities corresponds to the order of increased resonance stabilization of the resulting radical by the particular substituent. Substituents with unsaturation leading to a monomer with conjugated bonds are most effective in conferring resonance stabilization on radicals. On the other hand, substituents such as chlorine that have only nonbonding electrons for interaction with the radical show only weak stabilization. In the case of styrene, for example, the resonance stabilization energy is about 20 kcal/mol compared with 1 to 4 kcal/mol for the unconjugated systems. It is important to remember, however, that substituents that stabilize the product radical also stabilize the monomer, albeit to a much lesser extent. Again, using styrene as an example, the monomer is stabilized to about 3 kcal/mol. Therefore a limited compensation is derived from the stabilization of monomers by substituents.

B. RADICAL REACTIVITY

The relative reactivities of radicals to a reference monomer can be obtained from the product of the inverse of the reactivity ratio ($1/r_1 = k_{12}/k_{11}$) and the appropriate propagation rate constants for homopolymerization (k_{11}). Some values are listed in Table 8.3.

The order of enhancement of radical reactivity due to substituents is the reverse of that for the monomers. This should be expected because the enhancement of monomer reactivity by a substituent is due to its stabilization and the consequent decrease of the reactivity of the corresponding radical. The degree of depression of radical reactivity by substituents turns out to be much greater than the extent of enhancement of monomer reactivity. The styrene radical is about 1000 times less reactive than the vinyl acetate radical to a given monomer if the effects of alternation are disregarded, but then the styrene monomer is only about 50 times more reactive than the vinyl acetate to a given radical.

C. STERIC EFFECTS

Monomer–radical reaction rates are also influenced by steric hindrance. The effect of steric hindrance in reducing monomer reactivity can be illustrated by considering the copolymerization reaction rate constants (k_{12}) for di- and tri-substituted ethylene. Table 8.4 lists some of these values.

Table 8.4 Rate Constants (k_{12}) for Radical–Monomer Reactions

	Polymer Radical		
Monomer	**Vinyl acetate**	**Styrene**	**Acrylonitrile**
Vinyl chloride	10,000	8.7	720
Vinylidene chloride	23,000	78	2,200
cis-1,2-dichloroethylene	370	0.60	—
trans-1,2-dichloroethylene	2,300	3.9	—
Trichloroethylene	3,450	8.6	29
Tetrachloroethylene	460	0.70	4.1

From Odian, G., *Principles of Polymerization,* McGraw-Hill, New York, 1970. With permission.

We have already alluded to the fact that the addition of a second substituent to the 1- or α-position increases monomer reactivity. However, when the same substituent is in the 2- or β-position (i.e., 1,2-disubstitution), the reactivity of the monomer decreases 2- to 20-fold. This has been attributed to the resulting steric hindrance between the substituent and the attacking radical. The role of steric hindrance in the reduction of the reactivity of 1,2-disubstituted vinyl monomers can be further illustrated by the fact that while these monomers undergo copolymerization with other monomers, say, styrene, they exhibit extreme reluctance to homopolymerize. Homopolymerization is prevented because of the steric hindrance between a 2-substituent on the attacking radical and the monomer. On the other hand, there is no 2- or β-substituent on the attacking styrene radical; consequently, copolymerization is possible.

Example 8.5: Arrange the following monomers in the possible order of decreasing reactivity with an acrylonitrile radical. What is the basis of your arrangement?

a. Methacrylate
b. Vinyl methyl ether
c. Methyl methacrylate
d. Vinyl acetate

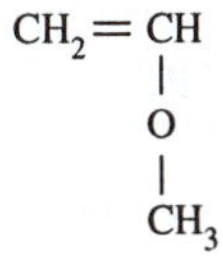

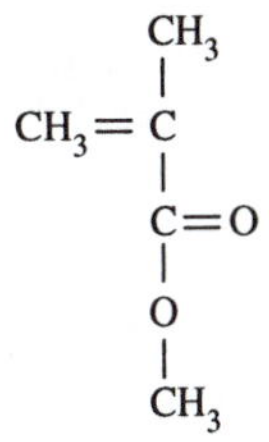

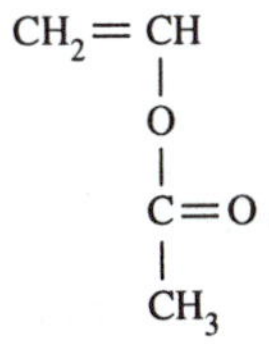

Solution:

Monomer	Structure
a. Methacrylate	$CH_2{=}CH{-}C({=}O){-}O{-}CH_3$
b. Vinyl methylether	$CH_2{=}CH{-}O{-}CH_3$
c. Methyl methacrylate	$CH_3{=}C(CH_3){-}C({=}O){-}O{-}CH_3$
d. Vinyl acetate	$CH_2{=}CH{-}O{-}C({=}O){-}CH_3$

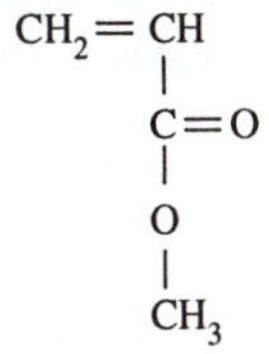

Order of reactivity: c > a > d > b. Reason: (1) increasing stabilization of resulting radical; (2) second substituent in 1-position increases monomer reactivity.

D. ALTERNATION-POLAR EFFECTS

We indicated in our earlier discussion that the relative reactivities of different monomers were not exactly the same for different radicals. Similarly, the order of reactivity of different radicals is influenced by the

Table 8.5 Product of Reactivity Ratios of Monomers Showing Order of Alternating Tendency

Vinyl acetate								
—	Butadiene							
0.55	0.78	Styrene						
0.39	0.31	0.34	Vinyl chloride					
0.30	0.19	0.24	1.0	Methyl methacrylate				
0.60	0.10	0.16	0.96	0.61	Vinylidene chloride			
0.90	0.04	0.11	0.24	0.96	0.08	Methyl acrylate		
0.21	0.006	0.016	0.11	0.18	0.34	0.84	Acrylonitrile	
0.004	—	0.02	0.06	—	0.56	—	—	Diethyl fumarate

reference monomer. This suggests, therefore, that monomer reactivity is a function of radical reactivity and vice versa. A close examination of monomer reactivity data reveals a tendency toward enhanced copolymerization reactivities for certain pairs of monomers. This is evidently a result of some monomer–radical affinity. This tendency toward enhanced reactivity for certain monomer pairs is a general phenomenon in radical copolymerization and is due to the alternating tendency in comonomer pairs. As the quantity r_1r_2 deviates from unity (ideal behavior) and approaches zero, the tendency for alternation increases. It is possible to arrange monomers in order of their r_1r_2 values such that the farther apart any two pairs of monomers are, the greater their tendency toward alternation. Table 8.5 shows such an arrangement. Notice, however, that there some exceptions, presumably due to the predominance of steric factors; for example, vinyl chloride and styrene show a greater alternation tendency than vinyl chloride and vinyl acetate, even though the former monomer pair is closer in Table 8.5 than the latter.

The order of occurrence of monomers in Table 8.5 is obviously a reflection of the polarity of the double bond. Observe that the product r_1r_2 approaches unity only in those cases where the substituents on the monomers are either both electron-donating or electron-withdrawing substituents. Expressed differently, alternation tendency increases if the substituents on both monomers exhibit different electron-donating or electron-attracting characteristics. Alternation tendency is enhanced by an increased difference in the polarity of monomer pairs.

Example 8.6: Which of the following pairs of monomers will most probably form an alternating copolymer?

a. Butadiene (1), Styrene (2), 60°C, $r_1 = 1.89$, $r_2 = 0.78$.
b. Vinyl acetate (1), styrene (2), 60°C, $r_1 = 0.01$, $r_2 = 55$.
c. Maleic anhydride (1), isopropenyl acetate (2), 60°C, $r_1 = 0.002$, $r_2 = 0.032$.

Solution:

Monomer	r_1r_2
(a)	1.4742
(b)	0.5500
(c)	0.000064

Alternating tendency increases generally as r_1r_2 approaches zero. Hence monomer pair (c) has the highest alternating tendency.

VI. THE Q-e SCHEME

The Q-e scheme is an attempt to express free radical copolymerization data on a quantitative basis by separating reactivity ratio data for monomer pairs into parameters characteristic of each monomer. Under this scheme, radical–monomer reaction rate constant k_{12} is written as:

$$k_{12} = P_1 Q_2 \exp(-e_1 e_2) \tag{8.26}$$

Table 8.6 Q and e Values for Monomers

Monomer	e	Q
t-Butyl vinyl ether	−1.58	0.15
Ethyl vinyl ether	−1.17	0.032
Butadiene	−1.05	2.39
Styrene	−0.80	1.00
Vinyl acetate	−0.22	0.026
Vinyl chloride	0.20	0.44
Vinylidene chloride	0.36	0.22
Methyl methacrylate	0.40	0.74
Methyl acrylate	0.60	0.42
Methyl vinyl ketone	0.68	0.69
Acrylonitrile	1.20	0.60
Diethyl fumarate	1.25	0.61
Maleic anhydride	2.25	0.23

From Alfrey, T., Jr. and Price, C.C., *J. Polym. Sci.*, 2, 101, 1947. With permission.

where P_1 represents the reactivity of radical $M_1\cdot$ and Q_2 the reactivity of monomer M_2; e_1 and e_2 represent the degrees of polarity of the radical and monomer, respectively. Both P and Q are determined by the resonance characteristics of the radical and monomer. It is assumed that both the monomer and its radical have the same e value. Consequently,

$$k_{11} = P_1 Q_1 \exp\left(-e_1^2\right) \tag{8.27}$$

Therefore

$$r_1 = k_{11}/k_{12} = Q_1/Q_2 \exp\left[-e_1\left(e_1 - e_2\right)\right] \tag{8.28}$$

Similarly

$$r_2 = k_{22}/k_{21} = Q_2/Q_1 \exp\left[-e_2\left(e_2 - e_1\right)\right] \tag{8.29}$$

Thus, according to the Q-e scheme, by assigning values to Q and e, it should be possible to evaluate r_1 and r_2 for any monomer pair. A selected list of Q and e values is shown in Table 8.6. Negative values of e indicate electron-rich monomers, while positive e values indicate electron-poor monomers.

The Q-e scheme is subject to criticisms. First, there seems to be no justification for assuming the same e values for the monomer and the radical derived from it. Second, the Q and e values for a particular monomer are not unique; they vary with the monomer to which the monomer is paired (Table 8.7). In spite of these flaws, however, the Q-e scheme provides a semi-empirical basis for correlating the effect of structure on monomer reactivity.

Table 8.7 Variation in Q and e Values

Monomer (M_1)	Comonomer	e_1	Q_1
Acrylonitrile	Styrene	1.20	0.44
	Vinyl acetate	0.90	0.37
	Vinyl acetate	1.0	0.67
	Vinyl chloride	1.3	0.37
	Vinyl chloride	1.6	0.37
Vinyl chloride	Styrene	0.2	0.024
	Methyl acrylate	0.0	0.035

From Odian, G., *Principles of Polymerization,* McGraw-Hill, New York, 1970. With permission.

VII. PROBLEMS

8.1. Styrene can undergo copolymerization reaction with either methyl methacrylate or vinyl chloride. It has been suggested that the resulting copolymers become brittle for styrene composition greater than 70%. Explain why even with a feed composition as low as 25% styrene, styrene–vinyl chloride copolymer is brittle, while styrene–methyl methacrylate copolymer is not.

8.2. Discuss the composition and structure of the polymer product obtained from the reaction of acrylonitrile (1) and 1,3-butadiene (2) in the molar ratios 25:75, 50:50, and 75:25. $r_1 = 0.02$, $r_2 = 0.3$ at 40°C.

8.3. Methyl methacrylate (1) and vinyl chloride (2) form an ideal copolymerization at 68°C. What is the composition of this copolymer for a feed composition $f_1 = 0.75$ and $r_2 = 0.1$?

8.4. For a particular application, it has been established that an alternating copolymer is most suitable. Given the following monomers — butadiene, styrene, acrylonitrile, and vinyl chloride — which monomer pair would be best studied for the application?

8.5. Arrange the following monomers in *increasing* order of reactivity with a vinyl acetate radical. Explain the basis of your arrangement.

a. Acrylonitrile
b. Styrene
c. α-Methyl styrene
d. 2-Methyl styrene
e. Vinyl ethyl ether

8.6. Maleic anhydride does not homopolymerize, but will readily form alternating copolymer with styrene. Explain.

8.7. The following are the Q and e values for monomers A, B, C, and D:

Monomer	Q	e
A	2.39	−1.05
B	0.15	−1.58
C	0.69	0.68
D	0.42	0.60

Monomers A and B; A and C; and C and D are subjected, respectively, to copolymerization reactions. Which of these reactions will most likely result in an alternating polymer? What is the composition of the copolymer formed at low conversion from equimolar mixtures of the pairs of monomers?

8.8. Describe the probable proportions and sequences of monomers entering a copolymer chain at the beginning of polymerization of a 1:1 molar mixture of each of the following pairs:

a. Vinyl acetate (1) and isopropenyl acetate (2); $r_1 = r_2 = 1$
b. Butadiene (1) and styrene (2); $r_1 = 1.89$, $r_2 = 0.78$
c. Maleic anhydride (1) and stilbene (2); $r_1 = r_2 = 0.01$

d. Maleic anhydride (1) and isopropenyl acetate (2); $r_1 = 0.002$, $r_2 = 0.032$

In each case, would the composition of polymer formed toward the end of the reaction (where the monomers are nearly consumed) be much different, and if so how?

REFERENCES

1. Young, L.J., Copolymerization reactivity ratios, in *Polymer Handbook,* 2nd ed., Brandrup, J. and Immergut, H.H., Eds., John Wiley & Sons, New York, 1975.
2. Billmeyer, F.W., Jr., *Textbook of Polymer Science,* 2nd ed., Interscience, New York, 1971.
3. Odian, G., *Principles of Polymerization,* McGraw-Hill, New York, 1970.
4. Young, L.J., Tabulation of Q-e values, in *Polymer Handbook,* 2nd ed., Brandrup, J. and Immergut, E.H., Eds., John Wiley & Sons, New York, 1975.
5. Alfrey, T., Jr. and Price, C.C., *J. Polym. Sci.,* 2, 101, 1947.
6. Mayo, F.R. and Walling, C., *Chem. Rev.,* 46, 191, 1950.
7. Kruse, R.K., *J. Polym. Sci. Part B,* 5, 437, 1967.
8. Tidwell, P.W. and Mortimer, G.A., *J. Macromol. Sci. Rev.,* C4, 281, 1970.
9. Williams, D.J., *Polymer Science and Engineering,* Prentice-Hall, Englewood Cliffs, NJ, 1971.
10. Flory, P.J., *Principles of Polymer Chemistry,* Cornell University Press, Ithaca, NY, 1953.

Chapter 9

Polymer Additives and Reinforcements

I. INTRODUCTION

In Chapter 5, we discussed the techniques for the modification of polymer properties based on structural modification of polymers either during or post polymerization. Even though structural modification of polymers often leads to significant property modification, very few polymers are used technologically in their chemically pure form; it is generally necessary to modify their behavior by the incorporation of additives. Additives are usually required to impart stability against the degradative effects of various kinds of aging processes and enhance product quality and performance. Thus many commercial polymers must incorporate thermal and light stabilizers, antioxidants, and flame retardants.[1] In addition to these additives that influence essentially the chemical interaction of polymers with the environment, other additives are usually employed to reduce costs, improve aesthetic qualities, or modify the processing, mechanical, and physical behavior of a polymer. Such additives include plasticizers, lubricants, impact modifiers, antistatic agents, pigments, and dyes. These additives are normally used in relatively small quantities; however, nonreinforcing fillers are employed in large quantities to reduce overall formulation costs provided this does not result in significant or undesirable reduction in product quality or performance. In some cases, a given polymer may still not meet the requirements of a specific application even with the incorporation of additives. In such cases, the desired objective may be achieved through alloy formation or blending of two or more polymers. In this chapter, we discuss the upgrading of the performance of polymers through the use of additives and reinforcements.

II. PLASTICIZERS

Many commercial polymers such as cellulosics, acrylics, and vinyls have glass transition temperatures, T_g, above room temperature. They are therefore hard, brittle, glasslike solids at ambient temperatures. To extend the utility of these polymers, it is usually necessary to reduce the T_g to below the anticipated end-use temperature. The principal function of a plasticizer is to reduce the T_g of a polymer so as to enhance its flexibility over expected temperatures of application. For example, unplasticized PVC is a rigid, hard solid used in such applications as credit cards, plastic pipes, and home siding. Addition of plasticizers such as phthalate esters reduces the modulus and converts the polymer into a leathery material used in the manufacture of upholstery, electrical insulation, and similar items.

Plasticizers are usually high boiling organic liquids or low melting solids. They are also sometimes moderate-molecular-weight polymers. Like ordinary solvents, plasticizers act through a varying degree of solvating action on the polymer. The plasticizer molecules are inserted between the polymer molecules thereby pushing them apart. This reduces the intensity of the intermolecular cohesive forces. The plasticizer may also depend on polar intermolecular attraction between the plasticizer and polymer molecules, which effectively nullifies dipole–dipole interactions between polymer molecules. As a result, plasticization is difficult to achieve in nonpolar polymers like polyolefins and highly crystalline polymers.

Polymer plasticization can be achieved either through internal or external incorporation of the plasticizer into the polymer. Internal plasticization involves copolymerization of the monomers of the desired polymer and that of the plasticizer so that the plasticizer is an integral part of the polymer chain. In this case, the plasticizer is usually a polymer with a low T_g. The most widely used internal plasticizer monomers are vinyl acetate and vinylidene chloride. External plasticizers are those incorporated into the resin as an external additive. Typical low-molecular-weight external plasticizers for PVC are esters formed from the reaction of acids or acid anhydrides with alcohols. The acids include ortho- and iso- or terephthalic, benzoic, and trimellitic acids, which are cyclic; or adipic, azeleic, sebacic, and phosphoric acids, which are linear. The alcohol may be monohydric such as 2-ethylhexanol, butanol, or isononyl alcohol or polyhydric such as ethylene or propylene glycol. The structures of some plasticizers of PVC are shown in Table 9.1.

Table 9.1 Chemical Structures of Some PVC Plasticizers

Plasticizer Type	Chemical Structure	Example
Phthalate Esters (Dialkylphthalate)	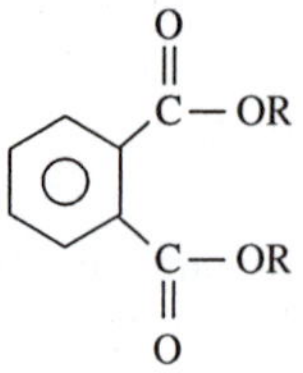 $C_6H_4(COOR)_2$ (benzene ring with two adjacent C(=O)–OR groups)	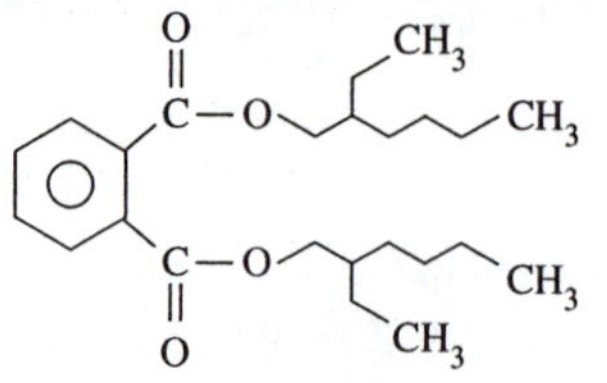 Di-2-ethylhexyl phthalate or Dioctylphthalate (DOP)
Phosphate Esters (Trialkyl-phosphate)	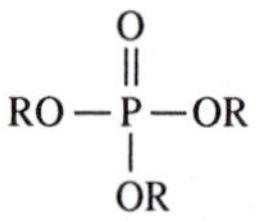 $RO-P(=O)(-OR)-OR$	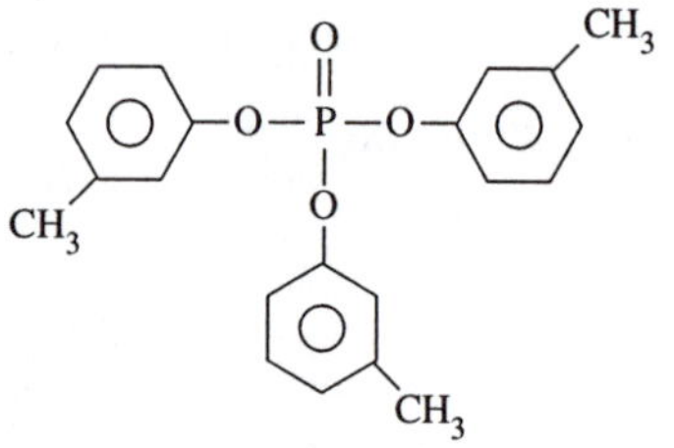 Tricresyl phosphate (TCP)
Adepates, azelates, oleates, sebacates (Aliphatic diester)	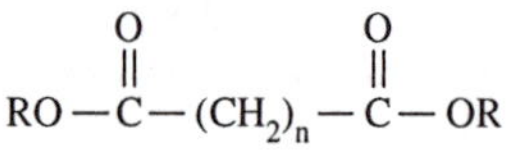 $RO-C(=O)-(CH_2)_n-C(=O)-OR$	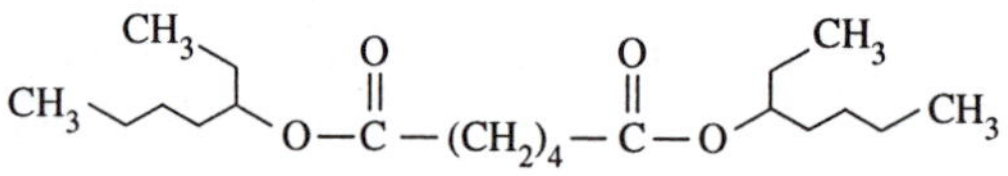 $CH_3 \ldots O-C(=O)-(CH_2)_4-C(=O)-O \ldots CH_3$ Di-2-ethylhexyl adepate (DOA)
Glycol Derivatives	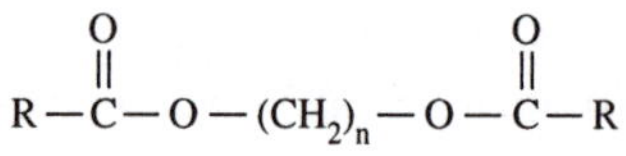 $R-C(=O)-O-(CH_2)_n-O-C(=O)-R$	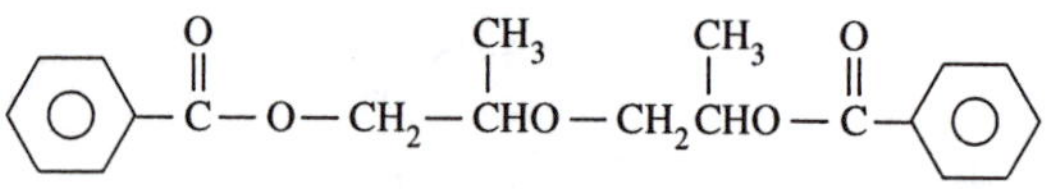 $C_6H_5-C(=O)-O-CH_2-CH(CH_3)O-CH_2CH(CH_3)O-C(=O)-C_6H_5$ Dipropyleneglycol benzoate
Trimellitates (Trialkyltrimellitate)	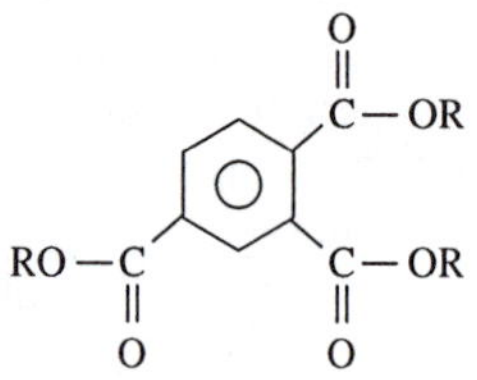 benzene ring with three C(=O)–OR groups ($RO-C(=O)$, $C(=O)-OR$, $C(=O)-OR$)	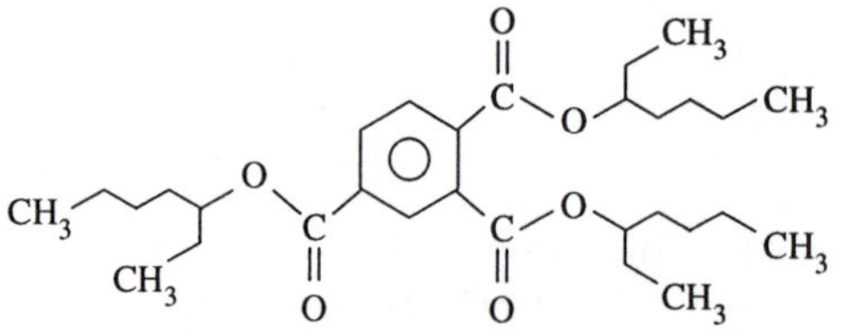 Trisethylhexy trimellitate (TOTM)

Phthalate, terephthalate, adipate, and phosphate esters are generally referred to as monomeric plasticizers. Linear polyesters obtained from the reaction of dibasic acids such as adipic, sebacic, and azelaic acids with a polyol constitute the group of polymeric or permanent plasticizers. They have much higher molecular weights than the monomerics and as such are less volatile when exposed to high temperatures either during processing or in end-use situations, less susceptible to migration, and less extractible.

They impart enhanced durability or permanence on PVC products, hence the name permanent plasticizers. Another group of plasticizers known as epoxy plasticizers is derived from vegetable oils. An example is epoxidized soybean. These plasticizers confer heat and light stability on PVC products, which, however, usually have relatively poor low-temperature properties.

The ideal plasticizer must satisfy three principal requirements. These are compatibility, performance, and efficiency. In addition, it should be odorless, tasteless, nontoxic, nonflammable and heat stable. Compactibility of a plasticizer with the host polymer demands the absence of blooming even with long usage of the plasticized material. Incompatibility can also be manifested by poor physical properties, possibly after some period of usage.

Permanence requires low volatility, extractability, nonmigration, and heat and light stability. Lack of permanence involves long-term diffusion into the environment. The consequent loss of plasticizer gradually enhances brittleness as the T_g of the plasticized polymer increases. Volatility is generally a function of molecular weight. Increasing the molecular weight of the plasticizer by using polymeric plasticizers tends to decrease volatility and hence increase permanence. However, this may lead to a decrease in low-temperature flexibility. Internal plasticization precludes plasticizer migration or volatility since the plasticizer molecules are an integral part of the polymer chain.

The level of plasticizer required to achieve the desired changes in properties is a measure of plasticizer efficiency. Plasticizer efficiency may also be measured on the basis of the magnitude of change induced in a number of physical properties of the polymer such as tensile strength, modulus, or hardness. For example, the actual reduction in T_g of the polymer per unit weight of plasticizer added is also known as the plasticizer efficiency.

It must be emphasized, however, that quite frequently no single plasticizer can satisfy all the above requirements or produce all the desired property enhancements. It is generally necessary to blend several plasticizers and compromise some properties, particularly those that are not critical to the specific application. For example, some applications require that DOP-plasticized PVC remains flexible at low temperatures. This requires further addition of DOP to the formulation to achieve the desired flexibility, but the additional DOP would adversely affect the hardness and performance of the product at ambient temperatures. Instead, aliphatic diesters such as DOA and DBS (dibutyl sebacate), which are more effective in enhancing low-temperature flexibility than DOA, are needed to improve low-temperature performance. However, the use of these esters is accompanied by unacceptable levels of volatility and oil extraction. Consequently, in such applications a combination of phthalate and aliphatic esters is required to produce the desired product even though some compromise in product performance will occur.

Plasticizers, particularly for PVC, constitute one of the largest segments of the additives market. The wide range of applications of flexible PVC depends largely on broad plasticizer technology that imparts the desired flexibility. The characteristic properties and typical applications of plasticized PVC are shown in Table 9.2.

III. FILLERS AND REINFORCEMENTS (COMPOSITES)

The need to meet exacting end-use requirements and at the same time reduce costs is stimulating a broad spectrum of product development involving the use of fillers and reinforcements to upgrade product performance rather than the development of new and usually more expensive resins. For example, because of their advantageous light weight, high strength, fatigue life, and corrosion resistance, structural composites have been used successfully and admirably in aircraft and in numerous industrial and consumer applications in place of conventional materials like metals. Fiber-reinforced materials have moved within a short time from being a curiosity to having a central role in engineering materials development. Polymers, thermoplastics, and thermosets can be reinforced to produce quite frequently a completely new kind of structural materials.

Different types of fillers are employed in resin formulations; the most common are calcium carbonate, talc, silica, wollastonite, clay, calcium sulfate, mica, glass structures, and alumina trihydrate. Fillers serve a number of purposes. Inert materials like wood flour, clay, and talc serve to reduce resin costs and, to a certain extent, improve processability and heat dissipation in thermosetting resins. Both alumina trihydrate and talc improve flame retardance. Mica is used to modify the electrical and heat-insulating properties of a polymer. Parts molded from composites containing phlogopite mica as a reinforcement exhibit little or no warpage on demolding or subjection to elevated temperature. A variety of fillers, e.g., particulate fillers like carbon black, aluminum flakes, and metal or metal-coated fibers may be used to reduce mold shrinkage as well as to produce statically conductive polymers, shielding of electromagnetic interference/radio frequency interference (EMI/FMI).[4] Particulate fillers such as carbon black or silica are used as reinforcing fillers to improve the strength and abrasion resistance of commercial elastomers. Fibers such as asbestos, glass, carbon, cellulosics, and aramid are used principally to improve some mechanical property/properties such as modulus, tensile strength, tear strength, abrasion resistance, notched impact strength, and fatigue strength as well as enhance the heat-deflection temperature.

Fibers are available in a variety of forms. For example, carbon fibers are obtained from the pyrolysis of organic materials such as polyacrylonitrile (PAN) and rayon for long fibers and pitch for short fibers.

Table 9.2 General-Purpose PVC Plasticizers

Abbreviations	Chemical Designations	Advantages	Limitations	Applications
		Low-Molecular-Weight Phthalates		
BBP	Butyl benzyl phthalate	Rapid fusing (high-solvating); low migration into flooring asphaltics	High volatility as compared with the commodity phthalates; costs slightly more than DOP	Flooring: processing aid in calendering and extrusion; expanded-foam formulations
BOP	Butyl octyl phthalate	Rapid fusing	Same as above	Same as above
DHP	Dihexyl phthalate	Same as above	Same as above	Same as above
DIHP	Diisoheptal phthalate	Same as above	Same as above	Same as above
		Commodity Phthalates		
DOP (DEHP)	Di-2-ethylhexyl phthalate	Good property balance; industry standard for general applications; relatively low cost; good weatherability; good compatibility in PVC)	Moderate plasticizer volatility; moderate low-temperature properties (–38°C at 40% concentration in PVC).	General-purpose calendering and extrusion; plastisols; flooring
DINP	Diisononyl phthalate	Low volatility (less than 2% per ASTM D1203); good electrical properties; similar to DOP in cost	Fair weatherability; fair low-temperature properties (similar to that of DOP)	Competes "head on" with DOP
Linear phthalates	C_7,C_9,C_{11}, C_6,C_8,C_{10}	Low volatility (–2%); better low-temperature properties: 7 to 9°C better than DOP and DIMP (to –47°C at 40% plasticizer concentration); very good weathering	Costs slightly more than DOP; slightly poorer electrical properties	Automotive wire and cable jacketing compounds; outdoor applications; plastisols; automotive interiors (except for crash pads)
DIDP	Diisodecyl phthalate	Very low volatility (–1%); good electrical properties; good viscosity stability	Low-temperature properties (–38°C); fair weatherability	Wire and cable; sealants; plastisols; calendering applications
		High-Molecular-Weight Specialties		
DUP	Diundecyl phthalate	Good electrical properties; volatility less than 1% at 48°C to –50°C at 40% concentration; exceptionally good viscosity stability in a plastisol; very low fogging in interior applications (automotive)	Slow processing characteristics; marginal compatibility with PVC; costs about 20 cents per pound more than DOP; high hydrocarbon extraction	Wire and cable up to 90°C; sealants; automotive interiors (crash pads); roofing membranes
DTOP	Ditridecyl phthalate	Less than 1% volatility; good electrical properties	Slow processor; high viscosity; fair low-temperature properties (–38°C); marginal compatibility with PVC	Wire and cable up to 90°C; automotive interiors; sealants
UDP	Undecyl dodecyl phthalate	Volatility less than 1%	Fair low-temperature properties to –38°C; slow processor; marginal compatibility with PVC	Wire and cable up to 90°C; automotive interiors

Table 9.2 (continued) General-Purpose PVC Plasticizers

Abbreviations	Chemical Designations	Advantages	Limitations	Applications
Linear phthalates	C_9,C_{10},C_{11}	Volatility less than 1%; low-temperature properties to 50°C; good outdoor weathering; low viscosity; improved compatibility with PVC	Costs more than DOP	Wire and cable to 90°C; automotive interiors; roofing membranes

Note: Plasticizers for PVC also include special-purpose products in applications such as flooring (benzyl phthalates); stain resistance (monobenzoates or benzyl phthalates); food and film wrap (adipate esters with superior low-temperature and oxygen-permeation properties); flame retardance (phosphate esters); wire and cable (trimellitates with exceptionally low volatility); heat stabilizers as well as plasticizers (epoxy); mar resistance of styrene (polymerics, isophthalates, and terephthalates); and polymeric plasticizers for long-term usage. These materials are more expensive than DOP and have specialty niches in the market.

From Wigotsky, V., *Plast. Eng.*, 40(12), 19, 1984. With permission.

Glass structures, the most widely used reinforcement, are available as roving, mat, hollow or solid spheres, bubbles, long or short fibers, and continuous fibers.[5] The form has a significant influence on properties. The impact strength of glass-mat-reinforced polypropylene is approximately four to five times that obtained with short fibers.[6] The most important form of filler is E-glass, which is used typically to reinforce thermoset polyester and epoxy resins. E-glass is boroaluminosilicate glass having low alkali-metal context and containing small percentages of calcia (CaO) and magnesia (MgO). For special applications such as in the manufacture of aerospace materials, fibers of boron, Kevlar, PBT, and especially carbon are generally preferred.[7] Typical properties of some fiber reinforcements are shown in Table 9.3. Table 9.4 shows the properties of some fiber-filled engineering thermoplastics composites. The effect of mineral fillers on the properties of nylon 6/6 is shown in Table 9.5.

Table 9.3 Properties of Some Reinforcing Fibers

Property	Boron	Kevlar-49	Carbon	Glass		Steel
				E-Glass	S-Glass	
Tensile strength (10^3 psi)	495–500	525	400	500	700	600
Tensile modulus (10^6 psi)	56	20	32	10.5	12.4	59
Density (lb/in^3)	0.086–0.096	0.052	0.063	0.092	0.090	0.282

The mechanical properties of particulate-filled composites are generally isotropic; that is, they are invariant with direction provided there is a good dispersion of the fillers. On the other hand, fiber-filled composites are typically anisotropic. In general, fibers are usually oriented either uniaxially or randomly in a plane. In this case, the composite has maximum modulus and strength values in the direction of fiber orientation. For uniaxially oriented fibers, Young's modulus, measured in the orientation direction (longitudinal modulus, E_L) is given by Equation 9.1:[7]

$$E_L = E_m(1-\phi_f)+E_f\,\phi_f \qquad (9.1)$$

where E_f is the tensile modulus of the fiber, E_m is the modulus of the matrix resin, and ϕ_f is the volume fraction of filler.[7]

The performance of composites is influenced by the adhesive strength of the filler–matrix interface. The presence of water absorbed by the filler surface and of thermal stresses generated by the differential thermal coefficients of linear expansion between the filler and matrix materials reduces interfacial adhesive strength. Polymers have relatively higher linear expansion coefficients than filler materials (60 to 80 × 10^{-6} per °C for graphite). Therefore, in practice, interfacial adhesive strength is enhanced by the use of coupling agents, which are usually low-molecular-weight organofunctional silanes, stereates, or titanates that act as a bridge between the matrix resin and fiber material (Table 9.6). Coupling

Table 9.4 Properties of Some Engineering Thermoplastics Composites

Property	Poly(etheretherketone) (PEEK)			Polyetherimide		Polyethersulfone			Nylon 6/6		
	Unreinforced	30% Glass	30% Graphite	Unreinforced	30% Glass	30% Graphite	30% Carbon Fibers	Unreinforced	30% Glass	30% Carbon Fiber	20% Kevlar-49
Tensile strength (10^3 psi)	13.2	20.3	31.2	15.2	24.5	30.0	30.0	11.9	25.0	36.0	18.2
Flexural modulus (10^5 psi)	5.65	11.6	22.4	4.80	12.0	25.0	25.0	4.8	13.0	27.0	8.8
Elongation at break (%)	150	3	3	60	3	1.4	1.7	40	2.5	2.0	2.3
Izod Impact Strength (ft-lb/in)	—	—	—	—	—	1.6	1.6	1.2	2.5	2.2	3.1
Heat-deflection temperature under load (264 psi) (°F)	298	572	572	392	410	410	420	190	485	490	490

Table 9.5 Effect of Mineral Fillers on Nylon 6/6

Property	Unfilled Resin	Mica	Calcium Carbonate	Wollastonite	Glass Beads	Alumina	Talc
Specific gravity	1.14	1.50	1.48	1.51	1.46	1.45	1.49
Tensile strength (psi)	11,800	15,260	10,480	10,480	9,780	9,200	8,980
Tensile elongation (%)	60	2.7	2.9	3.0	3.2	2.8	2.0
Flexural modulus (10^3 psi)	410	1,540	660	780	615	645	925
Izod impact(notched) (ft-lb/in.)	0.06	0.6	0.5	0.6	0.4	0.5	0.6
Heat-deflection temperature (°F)	170	460	390	430	410	395	445
Mold shrinkage (in./in.)	0.018	0.003	0.012	0.009	0.011	0.008	0.908

Table 9.6 Structures of Representative Coupling Agents

Vinyl silane	$CH_2{=}CH{-}Si{-}(OCH_3)_2$
Epoxy silane	$\overset{O}{CH_2\,\diagup\quad\diagdown\,CH}{-}CH_2{-}OCH_2CH_2CH_2{-}Si\ (OCH_3)_3$
Primary amine silane	$H_2NCH_2{-}CH_2{-}CH_2{-}Si{-}(OC_2H_5)_3$
Titanate	$(CH_2{=}\overset{CH_3}{\overset{\vert}{C}}{-}COO)_3{-}Ti{-}OCH{-}(CH_3)_2$

agents generally have two different terminal groups one of which is designed specifically to combine with the matrix resin and the other with the fiber reinforcement. For example, the amino group in γ-aminopropyltriethoxy silane, $(C_2H_5O)_3$-Si-$(CH_2)_3$-NH_2, can react with the epoxy functionality in an epoxy matrix, while the ethoxy groups on hydrolysis during application on the fibers form the silanol group –Si–$(OH)_3$, which condenses with silanol groups on the glass fiber to form covalent Si–O–Si bonds.

The search for new reinforcements or new applications for older materials is frequently linked to the development of new resins or new grades of the same resin to optimize performance. In broad terms, composites are characterized by two extremes of materials. On one end are the polyester-fiberglass automotive exterior body panels of Fiero or Corvette, representative of the relatively low cost and minimal performance materials. Thermoplastics — including polycarbonate, polypropylene, nylon, ABS, polystyrene, high-density polyethylene, acetal copolymer, SAN, poly(butylene terephthalate), PET, thermoplastic polyurethane, and polysulfone — as well as thermosets such as cross-linkable polyesters, phenolics, and polyurethanes are the matrix resins used in these composites. On the other end of the composite materials spectrum, are the high-cost, high-performance advanced composites typified by the carbon- and boron-fiber-reinforced structural materials used in many aircraft and aerospace applications. Thermosets, notably epoxies, have dominated advanced composite matrix resins. However, thermosetting materials are being challenged because of the increasing interest in high-performance thermoplastic resin matrices. The growing interest in thermoplastic matrix composites stems from their promise of improved durability and toughness and possibilities for increased cost effectiveness in production and maintenance. Thermoplastics have relatively longer shelf-life than thermosets and their reprocessability offers potential for possible repair of design and fabricating faults. Processing of thermoplastics potentially simplifies fabrication and reduces manufacturing cycle times, particularly by eliminating the curing cycles normally required for cross-linking thermoset materials. This also permits access to the wider variety of manufacturing methods for processing thermoplastics. Figure 9.1 shows schematically the simplified requirements for utilizing thermoplastic resin matrices relative to thermosets in composite production.

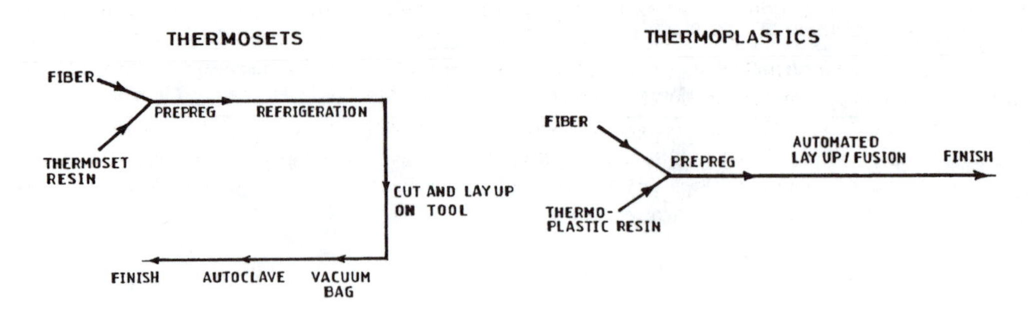

Figure 9.1 Schematic diagram illustrating the simplified requirements for utilizing thermoplastic resin matrices compared with thermoset resin matrices.

Table 9.7 Markets and Typical Applications for Composites

Market	Typical Applications
Aircraft/aerospace/military	Helicopter blades and shell; control surfaces and floor beams (Boeing 767); F-16s horizontal stabilizer, skin, dorsal access panels and leading edge fairing, and vertical tail fin skins and louver fin leading edge and rudder
Appliances/business equipment	Refrigerator compressor bases, room air conditioner parts, interior dishwasher parts, business machine bases and housings
Construction	Reinforced plastic bathtubs and shower fixtures, dimensionally stable reinforced plastic window lineals and doors, traffic signs, pedestrian shelters, telephone booths, and traffic kiosks
Consumer products	Recreational and sporting goods and recreational-vehicle equipment; microwave dinnerware and office furniture
Corrosion-resistant equipment	Underground gasoline tanks and reinforced plastic pipes and fittings; flooring for chemical processing plants and pulp and paper mills; tank linings for the storage of crude oil; piping and fittings for petroleum refineries; caustic and sulfur dioxide scrubbers; mixing tanks for phosphoric acid fertilizer solutions
Electrical/electronic	Fiber-optic cable; satellite dish and antenna, protective coverboards, light poles, electrical conduit, bus bar insulation, and third-rail insulators and covers
Marine	Sailboat, powerboat, minesweeper, military craft clad with reinforced plastics
Land transportation	Exterior body parts, suspensions, chassis, load floors, bumpers, driveshafts, lead springs, under-hood components, and automotive frames

The market for polymer composites has continued to expand in all forms of transportation (aerospace, aircraft, marine, automotive), in the construction industry and numerous other industrial and consumer applications. The increased use of composites in place of conventional materials is driven by their established advantages such as corrosion resistance, high strength-to-weight ratio, and moderate costs as well as the design flexibility offered by novel resin/reinforcement combinations together with new processing and machinery innovations. Table 9.7 summarizes the markets and typical applications for composites.

IV. ALLOYS AND BLENDS

The traditional method of enhancing properties by adding fillers and reinforcements, while still effective in some applications, has been inadequate for coping with the increasing performance requirements of design problems and the changing material specifications. Development of new resin systems to meet demands for high performance materials would undoubtedly take too long and would certainly be too expensive since it would require huge investments in totally unexplored technologies and new plant facilities. An alternative to the development of new polymers is the development of alloys and blends that are a physical combination of two or more polymers to form a new material. The basic objective is to combine the best properties of each component in a single functional material that consequently has properties beyond those available with the individual resin components and that is tailored to meet specific requirements. Another goal is to optimize cost/performance index and improve processability of a high-temperature or heat-sensitive polymer.

Although the terms *alloys* and *blends* are used interchangeably, they differ in the levels of inherent thermodynamic compatibilities and in resulting properties. In general, a necessary though not sufficient condition for thermodynamic compatibility (miscibility) is a negative change in the free energy of mixing, ΔG_m, given by Equation 9.2.

$$\Delta G_m = \Delta H_m - T\,\Delta S_m \tag{9.2}$$

where ΔH_m and ΔS_m are, respectively, the changes in enthalpy and entropy of mixing at temperature T. Generally because of their large size and unfavorable energy requirements, chains from a given polymer prefer to intertwine among themselves than with those of another polymer. Consequently, the magnitude of ΔS_m is usually small. Therefore, for two polymers to be thermodynamically miscible, ΔH_m must be negative, zero, or, at most, slightly positive. If ΔH_m is strongly positive, the components of a physical mixture separate into different phases resulting in a blend. However, some polymers tend to be mutually soluble, at least over a limited concentration range. Such are called alloys. In other words, alloys represent the high end of the compatibility spectrum. Individual polymeric components in alloys are intimately mixed on a molecular level through specific interactions such as donor–acceptor or hydrogen bonding between the polymer chains of the different components.

The composition dependence of a given property, P, of a two-component polymer system may be described by Equation 9.3:[10]

$$p = p_1\,C_1 + p_2\,C_2 + I\,p_1\,p_2 \tag{9.3}$$

where P_1 and P_2 are the values of the property for the isolated components and C_1, C_2 are, respectively, the concentrations of the components of the system. I is the interaction parameter that measures the magnitude of synergism resulting from combining the two components. If I is positive, then the magnitude of the property for the system exceeds that expected for a simple arithmetic averaging of the two component properties. The system in this case is referred to as synergistic. If I is negative, then the mixture has a property value less than that predicted from the weighted arithmetic average. This is known as nonsynergistic. Polymer systems for which I is either zero or nearly zero are called additive blends. They have properties that are essentially arithmetic averages of the properties of their components (Figure 9.2).

Compatible polymer blends form single-phase systems and have a single property, like the glass transition temperature, the value of which is generally the weighted arithmetic average of the values of the components of the blend. At certain compositions, some compatible polymer blends exhibit strong intermolecular attraction and hence have a high level of thermodynamic compatibility. This results in properties superior to those of the individual components alone. Such blends display synergistic properties (e.g., tensile strength and modulus). Only a few commercially available polymers are truly compatible. Some of these are shown in Table 9.8. Examples of the most significant commercial engineering alloys are polystyrene (PS)-modified poly(phenylene oxide) (PPO) and polystyrene (PS)-modified poly(phenylene ether) (PPE).

Incompatible polymer blends consist of a heterogeneous mixture of components and exhibit discrete polymer phases and multiple glass transition temperatures corresponding to each of the components of the blend. A polymer blend with completely incompatible components has limited material utility because the components separate during processing due to lack of interfacial adhesion, which is required for optimum and reproducible polyblend properties. Improvement in adhesion in such blends can be effected by the addition of compatibilizers. Compatibilizing agents provide permanent miscibility or compatibility between otherwise immiscible or partially immiscible polymers creating homogeneous materials that do not separate into their component parts. The most effective compatibilizing agents are generally block and graft copolymers whose polymer blocks or segments are the same as the components of the polyblend. The inherent compatibility between the segments of the compatibilizer and each component

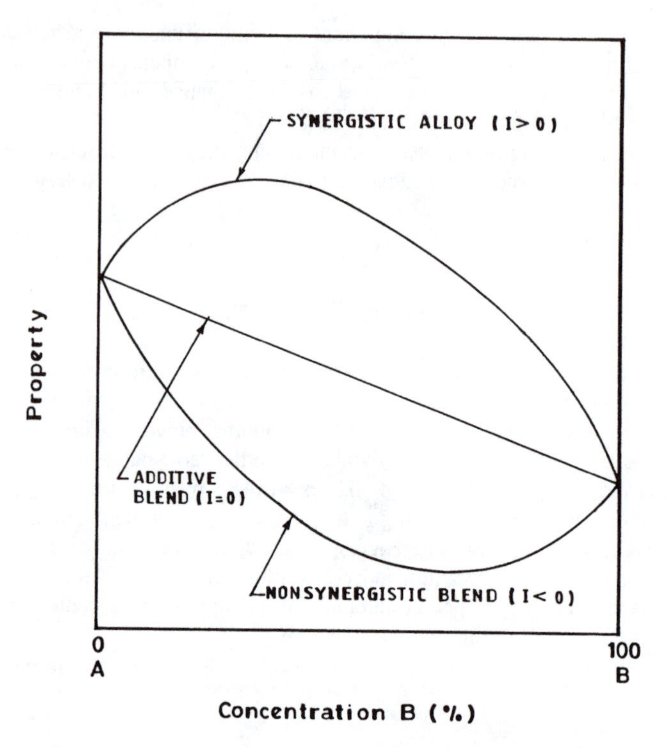

Figure 9.2 Composition dependence of the property of a two-component polymer system: alloy properties better than arithmetic averages; blend properties equal to or less than arithmetic averages.

Table 9.8 Representative Miscible Polymer Blends

Polymer 1	Polymer 2
Polystyrene	Poly(2,6-dimethyl-1,4-phenylene oxide)
	Poly(methyl vinyl ether)
	Tetramethyl BPA polycarbonates
Poly(vinyl chloride)	Polycaprolactone
	Poly(butadiene-*co*-acrylonitrile)
	Chlorinated polyethylene
	Poly(ethylene-*co*-vinyl acetate)
Poly(methyl methacrylate)	Poly(vinylidene fluoride)
	Poly(styrene-*co*-acrylonitrile)
	Poly(vinylidene fluoride)

From Fried, J.R., *Plast. Eng.*, 39(9), 37, 1983. With permission.

of the blend effectively ties the discrete polymer phases together. It is useful to visualize the function of a compatibilizing agent as akin to that of the emulsifier in water–oil emulsified systems.

As we said earlier, the primary objectives in blending are to enhance material processing characteristics and optimize product performance while minimizing costs. A common strategy for achieving these objectives is to combine a crystalline polymer with an amorphous polymer. The aim is to exploit the strengths of each component while deemphasizing their weaknesses. Crystalline materials such as nylon, poly(butylene terephthalate) (PBT), and poly(ethylene terephthalate) (PET) offer excellent chemical resistance, processing ease, and stiffness, but suffer from poor impact strength and limited dimensional stability. On the other hand, amorphous polymers such as polycarbonate (PC) and polysulfone have outstanding impact strength and dimensional stability but poor chemical resistance. Polycarbonate–poly(ethylene terephthalate) blends combine the desirable properties of polycarbonates with the

Table 9.9 Major Properties and Applications of Some Commercially Available Alloy/Blends

Alloy/Blends	Trade Name/ Manufacturer	Major Properties	Applications
PC/SMA	Arloy/Arco Chemical	Toughness and heat resistance, excellent moldability	Switches, power tools, food trays, lighting fixtures, automotive interior trim, connectors
PC/ABS	Bayblend/Mobay	Improve low temperature impact strength, processability and stiffness	Business machine housings and components, industrial and mechanical parts, switches, terminal blocks, food trays
SMA/ABS	Cadon/Monsanto	Improved ductility, and impact and heat resistance	Automotive interior trim, instrument panel, appliance and equipment housings
PC/ABS	Cycoloy/Borg-Warner Chemicals	Improved no-load heat resistance, impact strength	Electroplated parts, grilles, wheel covers, appliances, instrument panels, telecommunications
ABS/Nylon	Elemid/Borg-Warner Chemicals	Chemical resistance, low-load warpage resistance, toughness processing	Power tools, automotive and agricultural components
PPE/Nylon	GTX/General Electric	High heat and chemical resistance, dimensional stability, ductility	Automotive exterior body panels, wheel covers, mirror housings
Polysulfone/ABS	Mindel-A631/Union Carbide	Platability, processability, toughness	FDA- and NSF-recognized for food processing and food service systems
PPE/PS	Noryl/General Electric	Improved processing, combination of heat resistance, excellent dimensional stability and toughness	Computer and business equipment housings, automotive instrument panels, interior trim, connectors, electrical housings, medical components
PC/PBT, PC/PET	Xenoy/General Electric	Balance of chemical resistance, toughness, low-temperature impact strength and high temperature rigidity	Automotive, lawn and garden materials handling, sporting goods, military

From Wigotsky, V., *Plast. Eng.*, 42(7), 19, 1986. With permission.

excellent chemical resistance of PET and have been specified for use in many applications, for instance, as replacements for metal, including automotive, lawn and garden appliances and electrical/electronic, consumer, industrial/mechanical, sporting and recreation, and military equipment.[11] A blend of PPO and nylon is used for fenders and rocker panels of some automobiles, applications demanding chemical-resistant performance under high impact and high heat. Table 9.9 lists some alloys and blends, their characteristic properties, and areas of application.

V. ANTIOXIDANTS AND THERMAL AND UV STABILIZERS

Polymers, during fabrication or storage or in service, may be exposed, sometimes for long periods, to the separate or combined effects of moderate or high temperatures, ultraviolet radiation, and air or other potential oxidants. Under these environmental conditions, polymers are susceptible to thermal, UV, and/or oxidative degradative reactions initiated, in most cases, by the generation of free radicals. Polymer stabilization, therefore, involves incorporation of antioxidants and thermal and UV stabilizers to minimize, if not avoid, such degradative reactions.

A. POLYMER STABILITY

Polymers deteriorate through a complex sequence of chemical reactions resulting from the separate or combined effects of heat, oxygen, and radiation. In addition, polymers may be susceptible to attack and mechanical failure on exposure to water (hydrolysis) or a variety of chemical agents. Molecular weight is changed considerably in most of these reactions by chain scission and/or cross-linking. However, deterioration can also occur without significant change in the size of the polymer molecules.

1. Nonchain-Scission Reactions

Nonchain-scission reactions resulting, for example, from the application of heat, involve elimination of a small molecule — usually a pendant group — leaving the backbone essentially unchanged (Equation 9.4).

$$-CH_2-\underset{R}{\underset{|}{CH}}-CH_2-\underset{R}{\underset{|}{CH}}- \dashrightarrow -CH_2-\underset{R}{\underset{|}{CH}}-CH=CH- + RH \quad (9.4)$$

Vinyl polymers are particularly susceptible to thermal degradation. A typical example is rigid PVC, which is impossible to process under commercially acceptable conditions without the use of thermal stabilizers. Unstabilized PVC undergoes dehydrochlorination near the melt processing temperature. This involves liberation of hydrochloric acid and the formation of conjugated double bonds (polyene formation). The intense coloration of the degradation products is due to polyene formation. A second example of a polymer that undergoes nonchain-scission reaction is poly(vinyl acetate) or PVAc. When heated at elevated temperatures, PVAc can liberate acetic acid, which is followed by polyene formation.

2. Chain-Scission Reactions

The chemical bonds in a polymer backbone may be broken with the generation of free radicals by heat, ionizing irradiation, mechanical stress, and chemical reactions (Equations 9.5).

$$-CH_2-\underset{R}{\underset{|}{CH}}-CH_2-\underset{R}{\underset{|}{CH}} \dashrightarrow CH_2-\overset{H}{\overset{|}{\underset{R}{\underset{|}{C}}}}\cdot + \cdot CH_2-\underset{R}{\underset{|}{CH}}- \quad (9.5)$$

In the thermal degradation of polyethylene and polystyrene, for example, chain scission occurs through the homolytic cleavage of weak bonds in the polymer chain. This results in a complex mixture of low-molecular-weight degradation products. In some cases, particularly 1,1-disubstituted vinyl polymers, the long-chain radical formed from initial chain scission undergoes depolymerization resulting in the reduction of molecular weight and formation of monomers. In a few cases, e.g., PMMA, the initial chain scission occurs at one end of the molecule and the subsequent depolymerization results in a gradual decrease in molecular weight. In other cases, e.g., poly(α-methylstyrene), the initial chain scission occurs at random sites and as such there is an initial rapid reduction in molecular weight.

3. Oxidative Degradation

In the presence of oxygen or ozone, as soon as free radicals form, oxygenation of the radicals gives rise to peroxy radicals, which through a complex series of reactions result in polymer degradation. Oxidative degradation may occur at moderate temperature (thermal oxidation) or under the influence of ultraviolet radiation (photooxidation). Unsaturated polyolefins are particularly susceptible to attack by oxygen or ozone (Equation 9.6).

$$-CH_2-C(CH_3)=CH-CH_2-CH_2-C(CH_3)=CH-CH_2-CH_2-C(CH_3)=CH-CH_2\sim\sim$$

$$\xrightarrow{+O_3,\ +H_2O} -CH_2-C(CH_3)=O + HCOOH + H-\overset{O}{\overset{\|}{C}}-CH_2- \qquad \xrightarrow{O_2,\ \text{light}} -CH_2-C(CH_3)=O + -CH_2-\overset{O}{\overset{\|}{C}}-CH_2- \quad (9.6)$$

4. Hydrolysis and Chemical Degradation

In addition to the separate or combined effects of heat, oxygen, and radiation, polymers may deteriorate due to exposure to water (hydrolysis) or different types of chemical agents. Condensation polymers like nylons, polyesters, and polycarbonates are susceptible to hydrolysis. Structural alteration of some polymers may occur as a result of exposure to different chemical environments. Most thermoplastics in contact with organic liquids and vapors, which ordinarily may not be considered solvents for the polymers, can undergo environmental stress cracking and crazing. This may result in a loss of lifetime performance or mechanical stability and ultimately contribute to premature mechanical failure of the polymer under stress.

B. POLYMER STABILIZERS

The two main classes of antioxidants are the free-radical scavengers, (primary antioxidants, radical or chain terminators) and peroxide decomposers (secondary antioxidants or synergists). Free-radical scavengers, as the name suggests, inhibit oxidation through reaction with chain-propagating radicals, while peroxide decomposers break down peroxides into nonradical and stable products. Commercial antioxidants include organic compounds like hindered phenols and aromatic amines, which act as free-radical scavengers, as well as organic phosphites and thioesters that serve to suppress homolytic breakdown.

Thermal stabilizers may be based on one or a combination of the following classes of compounds: barium/cadmium (Ba-Cd), calcium/zinc (Ca-Zn), organotin, organo-antimony, phosphite chelates, and epoxy plasticizers. Ba/Cd stabilizer systems, which represent the largest share of the PVC stabilizer market, are available as liquids or powders. The liquids normally employ one or more of the following anions combined with the respective metal: nonyl phenate, octoate, benzoate, naphthenate, and neodacanoate. Conventional Ba/Cd powder stabilizer systems have one or more of the anions of the following acids: stearic, palmitic, and lauric. The powder Ba/Cd stabilizers, which act both as thermal stabilizer and lubricant, are normally employed in rigid and semirigid products. Ca/Zn stabilizers represent a small volume of the stabilizer market. The best known Ca/Zn stabilizer systems are associated with FDA-sanctioned products, including blown films, medical and beverage tubing, blood bags, blister packs, and bottles. The organotin compounds are butyl-, methyl-, and octyltin and are available in liquid and solid forms. Most organotin stabilizers are used in rigid extrusion and injection molding processes for such products as PVC pipes, profiles, and bottles as well as fittings, siding, and clear PVC bottles.

UV radiation in the range 290 to 400 nm has potentially degradative effects on polymers since most polymers contain chemical groups that absorb this radiation and undergo chain scission, forming free radicals that initiate the degradative reactions. UV stabilizers are employed to impede or eliminate the process of degradation and, as such, ensure the long-term stability of polymers, particularly during outdoor exposure. Light stabilizers are typically UV absorbers or quenchers. The former preferentially absorbs UV radiation more readily than the polymer, converting the energy into a harmless form. Quenchers exchange energy with the excited polymer molecules by means of an energy transfer mechanism. Other UV stabilizers deactivate the harmful free radicals and hydroperoxides as soon as they are formed. Pigments offer good protection for polymers by absorbing UV radiation. Carbon black, used widely in tire manufacture, absorbs over the entire range of UV and visible radiation, transforming the absorbed energy into less harmful infrared radiation. Pigments and carbon black cannot be used, unfortunately, in applications where transparency is required. In these applications, stabilizers that contribute minimal color or opacity are used. For example, transparent polymers like polycarbonate can be protected against yellowing and embrittlement from UV light (photolysis) by incorporating compounds like benzophenone derivatives (e.g., 2-hydroxybenzophenone) and 2-hydroxybenzotriazoles. These highly conjugated compounds are able to convert absorbed UV radiation to less harmful heat energy without chemical change by forming a transient rearranged quinoid structure that changes back to the initial form. Other UV stabilizers include acrylic and aryl esters, hindered amines, and metal salts. Unlike conventional UV stabilizers, the metallic complexes interact with photoexcited polymer molecules, deactivating them by dissipating excess energy as infrared radiation.

VI. FLAME RETARDANTS[13,14]

Most materials, synthetic or natural, burn on exposure to temperatures that are sufficiently high. The response of polymers to high temperatures depends on their formulation and configuration in the end-use situation. The essential goal of flame retardancy is to preserve life and prevent or at least minimize damage to property. Therefore, the function of flame retardants in a resin formulation is ideally the outright inhibition of ignition where possible. Where this is impossible, a flame retardant should slow down ignition significantly and/or inhibit flame propagation as well as reduce smoke evolution and its effects. The presence of flame retardants also tends to cause substantial changes in the processing and ultimate behavior of commercial resins.

The burning characteristics of polymers are modified by certain compounds — including alumina trihydrates; bromine compounds; chlorinated paraffins and cycloaliphatics; phosphorus compounds, notably phosphate esters; and antimony oxides, which are used basically as synergists with bromine and chlorine compounds. The halogens are most effective in the vapor phase; they act in the flame zone by forming a blanket of halogen vapor that interferes with the propagation of the flame by interrupting the generation of highly reactive free radicals, thus tending to quench the flame. Others such as phosphorus or boron operate in the condensed or solid phase, minimizing the availability of fresh fuel. They form a glaze that limits the heat and mass transfer necessary for flame propagation and/or lowers the melt temperature of the polymer causing it to flow away from the flame. Table 9.10 shows the characteristics and end use of various flame retardants.

Flame retardants may be classified as additives, reactives, intumescents, and nonflame-retardant systems based on their method of incorporation in the resin formulation or their mode of action. Additive flame retardant systems can be further classified essentially as fillers, semiplasticizers, or plasticizers depending on their melting points and area of application. Typical additive flame retardants are halogenated additives used alone or in synergistic combinations with antimony oxide (with PS, PP, polyester, and nylons), phosphate esters, mineral hydrates, boric acid, sodium tetraborate, and ammonium bromide. Additive flame retardants are used with both thermoplastics and thermosets. The use of additive flame retardants involves the addition of very high-melting inorganic materials that will reduce combustibility. This is based on the rationale that the fewer the combustibles, the less burning will occur. Alumina trihydrate, which dominates the flame retardant market in terms of sales volume, serves this function. In addition, it liberates water at high temperatures and this further dampens flame energy. Alumina trihydrate is a typical dual function filler additive (like calcium carbonate and clays), which provides noncombustible filling plus gas or moisture release at elevated temperatures. Alumina trihydrate is used as a primary flame-retardant additive in carpet backing, in electrical parts, and in applications employing glass-fiber-reinforced unsaturated polyester. It is also used in sheet and bulk molding compounds for electrical enclosures in business machine and computer housings, and in laminated circuit boards. Alumina trihydrate offers low-cost flame retardance in epoxy systems and for wire and cable insulation, cross-linked PE and ethylene vinyl acetate copolymers and for flexible polyurethane foams.

Flame retardants that melt or flux at or near the polymer processing temperature are known as semiplasticizing additives, while those that truly flux during processing are referred to as plasticizing additives. These types of flame retardants can be tailor-made for a specific polymer. Plasticizing flame retardant filler additives provide beneficial effects in the processing of rigid polymers and quite often improve the physical properties of the polymer. Commercial examples of plasticizing filler additives include phosphate systems in polyphenylene oxides, phosphates or chlorinated paraffins in vinyl chloride and vinyl acetate polymers, and octabromobiphenyl oxide and brominated aromatic compound in ABS.[13]

In reactive flame retardants, retardancy is built into the polymer chain through appropriate polymerization or postpolymerization techniques. For example, the current additive method of producing flame retardant high-impact polystyrene involves blending decabromobiphenyl oxide and antimony oxide to produce a blend with 8 to 10% combined bromine content. Reactive flame retardants with the same bromine content can be obtained either by brominating the resin or copolymerizing styrene and brominated styrene. Reactive flame retardants include halogenated acids such as tetrabromophthalic anhydride and tetrachlorophthalic anhydrides, halogenated or phosphorated alcohols (e.g., dibromoneopentyl glycol, dibromopropanol) and halogenated phenol bisphenol A. These materials are usually incorporated into the polymer molecular structure through copolymerization with other reactive species.

The use of reactive flame retardants has a number of advantages over that of additive flame retardants. With flame retardants as an integral part of the polymer chain rather than incorporated as an additive,

Table 9.10 Characteristics and Applications of Flame Retardants

Market/ Characteristics	Alumina Trihydrate		Bromine Compounds		Chlorine Compounds	
	Resin	End Use	Resin	End Use	Resin	End Use
Markets						
Electrical parts	Unsaturated, polyesters, epoxies, phenolics, thermoplastic rubbers	Switch gear, standoff insulation, electrical connections, junctions boxes, circuit boards, insulating sheet, potting compounds	Nylon 6/6, 6, PBT PET, polycarbonate, epoxy, polypropylene	Connectors, terminal strips, bobbins, switches, relays, circuit boards	Nylon 6/6, 6 polypropylene PBT	Same as for bromine compounds
Electronic housing and enclosures	Unsaturated polyester	Portable power tools, business machine housings, computer housings	ABS, HIPS, polycarbonate	Business machines, computers, TV monitors, copiers	—	—
Wire and cable	LDPE, XLPE, PVC, EPDM	Automotive wire, appliance wire, insulation and jacketing compounds	XLPE, LDPE, EPDM, thermoplastic rubber	Power, industrial and marine cable; electronic wire	XLPE, LDPE, EPDM, thermoplastic rubbers	Same as for bromine compounds
Appliances	Unsaturated polyester	Home laundry equipment, air conditioning equipment	HIPS, ABS, polypropylene	TV cabinetry, power tool housings, kitchen appliances	Polypropylene	Power tool housings, electrical sockets
Building and construction	Unsaturated polyester, PVC, acrylic, EPDM, polyurethane, flexible polyurethane	Paneling, bathroom tubs, shower stalls, countertops, wall coverings, flooring, roofing compounds, cushioning, mattresses	Unsaturated polyester, rigid polyurethanes foam, expandable polystyrene foam	Building panels, translucent sheet, corrosion-resistant equipment, thermal insulation	Unsaturated polyester	Same as for bromine compounds
Transportation (automotive)	Epoxies, acrylics, PVC	Adhesives sealants, upholstery fabrics	Flexible polyurethane foam, rigid polyurethane foam, unsaturated polyesters	Seating, crash padding, thermal insulation (trucks), marine usage (shipboard components, coatings)	Unsaturated polyester	Seating, panels, sheathing
Characteristics	Require high loading to achieve modest flame retardance. Low cost. Low smoke advantage. Nontoxic. Multifunctional extender. Flame-retardant mechanism: endothermic (heat-absorption) cooling of flame upon release of water of hydration		Highly efficient and versatile for wide range of applications. Requires uses of antimony oxide. Broad range of high-performance products available, as aromatic, cycloaliphatic, and bromine/chlorine paraffins. Vapor-phase flame-retardant mechanism. Flame-retardant elements interrupt chemical reactions of combustion in flame zone.		Relatively low smoke generation. One high-performance additive and one high-performance reactive formulation used. Paraffins have low heat stability and are plasticizing. Flame retardant mechanism similar to bromine.	

Table 9.10 (continued) Characteristics and Applications of Flame Retardants

	Phosphorus Compounds		Antimony Oxide	
	Resin	**End Use**	**Resin**	**End Use**
Markets				
Electrical	Modified polyphenylene oxide (PPO)	Connectors, terminal strips, etc.	Nylon 6/6, 6, PBT, PET	Connectors, switches, circuit boards, terminal strips
Electronic housing and enclosures	Modified PPO	Business machines	ABS, HIPS	Business machines
Wire and cable	PVC	Communications cable	PVC, XLPE, LDPE, EPDM	PVC building wire, similar uses as for bromine and chlorine compounds
Appliances	Modified PPO	Kitchen appliances, TV cabinetry	HIPS, ABS, polypropylene	TV cabinetry, power tools, kitchen appliances
Building and construction	Rigid polyurethane foam, PVC, unsaturated polyesters	Thermal insulation, wall coverings, flooring, sanitary ware, laminates	Unsaturated polyester, rigid PVC, flexible PVC	Building panels, panels, windows, insulation covering
Transportation (automotive)	PVC, flexible polyurethane foam, rigid polyurethane foam	Upholstery, seat cushioning, thermal insulation (trucks)		
Characteristics	Phosphorus promotes char formation to protect substrate, and halogen acts in vapor phase. Good thermal stability. Process with modified PPO up to 550–600°F. Flame-retardant mechanism: condensed phase. Flame retardant induces reactions in host resin that lead to charring and insulation against further burning.		Inert material by itself. Must be used with halogen-type compound. Used in all polymers where halogen compounds are the selected flame-retardant system. As a synergist, enhances efficiency.	

From Wigotsky, V., *Plast. Eng.*, 41(2), 23, 1985. With permission.

the loss of performance of the base polymer due to diluents (other constituents) in the resin formulation is minimized. In addition, migration and/or bloom are completely eliminated while compounding costs are substantially reduced. Leaching of reactive flame retardants by solvents and corrosives is relatively more difficult than leaching additive retardants.

Reactive flame retardants are used mainly with thermoset resins, particularly unsaturated polyesters, both reinforced and unreinforced. A technique for introducing flame retardants into unsaturated polyesters involves the use of compounds with residual unsaturation such as diallyl tetrabromophthalate, which is used as a cross-linking agent alone or with styrene.

Intumescent flame retardance is based on the formation of a char on the surface of the resin on the application of heat, consequently insulating the substrate from further heat and flame. Phosphorous compounds such as inorganic or organic phosphates, nitrogen compounds such as melamine, and polyhydroxy compounds — usually pentaerythritol — are used in intumescing formulations. High loadings, are usually required to achieve the required level of fire retardancy. This tends to degrade the physical properties of the base polymer. Intumescent flame retardant systems are generally limited to polymers with low processing temperature (e.g., PP) because they expand considerably on application of heat.

Nonflame-retardant systems are polymeric systems that inherently have some level of flame retardance and therefore do not require additive or reactive flame retardants. Examples include PVC and its compounds, poly(vinylidene chloride) films and compounds, phenolic foams, amide-imides, polysulfones, and poly(aryl sulfides).

VII. COLORANTS[15,16]

As we said earlier, very few polymers are used technologically in their chemically pure form; it is generally necessary to incorporate various additives and reinforcements to assist processing and achieve desired properties. Unfortunately, these components also often produce a significant amount of undesirable color and opacity in the resin. Each resin itself has its color that may vary from grade to grade or batch to batch. For example, polystyrene crystal is transparent, whereas high-impact polystyrene has a white, somewhat translucent, appearance while the common grade flame-retardant polystyrene is opaque. The color of general-purpose ABS is off-white and opaque. Glass fibers, the most common reinforcements added to nylons and polyesters, darken the color of these resins.[17] The marketability of a polymer product quite frequently depends on its color; therefore the purpose of adding a colorant to a resin is to overcome or mask its undesirable color characteristics and enhance its aesthetic value without seriously compromising its properties and performance.

Colorants are available either as organic pigments and dyes or inorganic pigments. They may be natural or synthetic. By convention, a dye is a colorant that is either applied by a solution process or is soluble in the medium in which it is used, while pigments are generally insoluble in water or in the medium of use. Dyes are generally stronger, brighter, and more transparent than pigments. As a result of the intrinsic solubility, dyes have poor migration fastness and this restricts their use as polymer colorants. Inorganic pigments are largely mixed metal oxides with generally good-to-excellent light fastness and heat stability but variable chemical resistance. Organic pigments and dyes are generally transparent and possess good brightness. The heat stability and light and migration fastness of organic pigments range from poor to very good. Table 9.11 shows some colorants, their characteristics, and their applications.

Colorants are used in polymers either as raw pigments (and dyes), concentrates (solid and liquid), or precolored compounds. Precolored resins, solid and liquid concentrates, are all offsprings of the basic dry pigments. Colorants are available in a variety of forms, including pellets, cubes, granules, powder and liquid, and paste dispersions. Raw pigments are generally supplied as fine particles, which require dust control measures. To optimize color development when raw pigments are used, the size of pigment particles or agglomerates must be reduced and coated with appropriate resin. Most finished colors use multiple pigments. This requires a homogeneous mixing of all the pigments in the formula in high-shear mixing equipment to produce a uniform color. Precise metering into the processing machine is required to produce consistent colors since some components of the pigment system, though present in relatively small quantities, have strong color characteristics. Raw colorants or pigments generally cost less than other forms of colorants, but they can be more difficult to disperse and may result in inconsistent master batches.

Table 9.11 Characteristics and Applications of Some Colorants

Pigment	Characteristics	Applications
	Reds	
Quinacridones (medium red-magenta)	Good to excellent heat stability in 500–525°F range, excellent high fastness, expensive, some grades difficult to disperse	Widely used in automotive applications, propylene fibers
Perylenes (scarlet to violet)	Excellent fastness, and heat stability in 325–500°F range, expensive, some grades difficult to disperse	Extensive use in automobile, finishes, synthetic fibers
Diazol (scarlet to medium red)	Very good heat stability, light fastness, easily dispersable, relatively expensive	Widely used in vinyls and polyolefins
Azo (scarlet to bluish reds)	Heat stability in the 450–500°F range, light fastness range from poor to good, economical, limited light fastness in tint	Polyolafin packaging, toys housewares
Permanent red 2B (scarlet to bluish reds)	Economical light red shades, good heat stability and light fastness in mass tone and near mass tone; tendency to plate out, limited light fastness in tints with white	Widely used in vinyls, polyolefins, rubbers
	Oranges	
Isoindolinonone (medium orange)	Transparent, excellent heat stability and light fastness, relatively expensive	Used for vinyls, polyolefins, and styrenics
Diaryl orange (medium orange)	Heat stability in 450–500°F range, fair light fastness, bright, clear, high tinting strength, moderate price, limited light fastness, some tendency toward bleeding	Used in polyolefins for packaging, toys
	Yellows	
Isoindoline (reddish yellow)	Very good to excellent heat stability (500–575°F), good to very good light fastness, transparent or opaque	Used in PVC and polypropylene and some engineering plastics
Quinophthalone (greenish yellow)	Moderate light fastness, good to excellent heat stability; tint light fastness limited	Used in vinyls and polyolefins
Metal complex (greenish yellow)	Heat stable (550–575°F), light fastness good to excellent	Some engineering resins, PVC and polypropylene
	Greens	
Phthalocyanine (bluish to yellowish green)	Heat stable to 600°F, light fast, easy to disperse	Very broad use in vinyls, polyethylene styrenics, polypropylene and some engineering resins
	Blues	
Indanthrone (reddish blue)	Very good light fastness and heat stable expensive	Light tinting of vinyls, in rigid PVC
	Violets	
Carbazole (reddish violet)	Very high tinting strength, expensive limited light fastness	Polyolefins and vinyls
	Metal Oxides	
Iron oxides (synthetic), red-maroon	Good heat stability, inexpensive, inert, poor tinting strength, dull	Variety of plastics but not in rigid PVC
Zinc ferrite tan	Good heat stability, inert, light fast, good weatherability, more expensive than other iron oxides	Variety of plastics but not in rigid PVC
Iron oxides (natural), siennas	Inexpensive, color uniformity can contain impurities	Limited use in plastics, polyethylene film bags
Chromium oxide green	Good heat stability and light fastness, excellent weatherability and chemical resistance, inexpensive, dull, poor tinting strength	Can be used in most thermoplastics and thermosets
	Mixed metals oxides	
Nickel tinuium yellow	Excellent weatherability, inert, easy to disperse, good chemical resistance, poor tinting strength, weak color	Engineering resins, PVC siding

Table 9.11 (continued) Characteristics and Applications of Some Colorants

Pigment	Characteristics	Applications
Inorganic browns	Heat and light stable, good chemical resistance, good color uniformity, relatively expensive	Most thermoplastics and thermosets
Cadmiums, cadmium yellow sulfide	Excellent heat stability, good alkali resistance, light fast, sensitive to moisture and acids, poor weatherability, toxicity concerns	Wide use in plastics including engineering resins
Chrome yellow	Bright colors, inexpensive, good hiding power, poor heat stability, poor chemical resistance, possibly toxic	Fairly broad use in thermoplastics and thermosets

From Wigotsky, V., *Plastic. Eng.*, 42(10), 21, 1986. With permission.

Color concentrates are intimately mixed dispersions of pigments in a base carrier resin. The pigment content of concentrate is usually in the 2 to 30% range, but higher loadings are being developed to enhance versatility and cost/performance benefits. The pigments and resin are subjected to high stress during processing to promote thorough dispersion of the colorant. The concentrate is blended with the resin material being colored in a predetermined proportion by weight known as the let-down ratio to produce the desired color and opacity in the master batch or end product. To ensure compatibility between the concentrate and the let-down resin, the color concentrate is generally made from the same generic polymer as the let-down resin but with a higher melt index so as to promote ready and even mixing. A uniform blend of concentrate must be fed either continuously by a metering device or by weight on a batch basis into the processing equipment, which must be able to convert the blend of pellets into a uniform melt. Color variation is produced if the melt is nonuniform or the concentrate is not completely incorporated into the resin.

Concentrates are available in solid or liquid forms. Solid color concentrate, the major form in which colorants are manufactured and sold, comes in pellet, cube, granulated, and powder forms. They usually consist of dry pigments, additive components encapsulated in a base resin carrier. Dispersion and color control are excellent, weighing is less critical, and flowability is good for easy feeding while cleanup is easier because of the need to process color with the carrier resin.[15] It is, however, necessary to ensure that the pigment carrier and other components of the concentrate do not compromise the resistance to heat, light, basic physical properties, and rheological compatibility of the resin. Liquid concentrates possess many of the advantages of solid concentrates but are usually more costly. They allow lower pigment loading and can sometimes be used at slightly lower concentrations because they cover more surface. Liquid concentrates require less material handling and floor space for inventory, and their production does not involve a previous heat history. Some resins, however, are unable to absorb a high percentage of liquid.

There is currently a trend to increase the multifunctional nature of color concentrates so that color as well as other desired properties are added to the resin system. This increases the processor's flexibility in meeting user's needs and simplifies user inventory by reducing the need to stock large quantities of tailored resins.

Precolored compounds provide the processor with a single source of color and resin. This eliminates the need for mixing during processing and provides highly accurate color control, particularly in cases where the base resin is subject to color variations arising from lot-to-lot color differences. In complex part designs where resin flow may not be uniform or where equipment is unable to provide uniform mixing, precolored compounds offer the best method of making a uniformly colored part. However, processors sometimes tend to have preference for dry pigments and concentrates for greater flexibility of inventory and color changeover. For example, from an inventory standpoint, it is much simpler to stock a supply of basic natural resin to satisfy diverse, changing requirements with smaller supplies of dry pigment or color concentrate. A processor who prefers to compound his color from pigment must, of course, have the proper equipment and an in-depth knowledge of the process. One of the growth areas for precolored resin is in specialized engineering applications where not only colorants but a variety of other modifiers must be included in the compound. For example, fiber-reinforced resins that may also contain mold-release agents, flame retardants, and other additives are best made from fully compounded precolored resin.

Certain fundamental criteria must be considered in selecting a colorant for a particular application. These include the ability of the colorant to provide the desired color effect and withstand processing conditions and whether or not the fastness will satisfy end-use requirements. Therefore, the initial step in the selection of a colorant is to determine whether it will provide the coloristic properties desired, alone or in combination with others. The performance properties of a colorant are generally of two types — those related to processing and those related to the ultimate end use. The most important processing-related property is the heat stability of the colorant. It must be able to withstand not only the process temperature encountered during manufacture but also, for possibly prolonged periods, the temperature in the end-use situation.

Migration fastness is related to the solubility of the colorant in the polymer. Color migration is manifested through bleeding, blooming, or plate-out. Bleeding is the migration from a colored polymer film to an adjacent uncolored or differently colored material, while blooming involves colorant migration, recrystallization, and formation of a dustlike coating on the surface of the polymer. Plate-out is characterized by the building of a coating on the metallic surfaces of processing equipment. Inadequate light fastness of a colorant is usually manifested in the form of fading or, in the case of some colorants, darkening. The severity and rapidity of color change depend on the chemical structure of the colorants, its concentration in the part, and part thickness.

The compatibility of a colorant is assessed not only on the basis of the ease with which it can be mixed with the base resin to form a homogeneous mass but also on the requirement that it neither degrades nor is degraded by the resin. In relation to product functional properties, incompatibility of a colorant can affect mechanical properties, flame retardancy, weatherability, chemical and ultraviolet resistance, and heat stability of a resin through interaction of the colorant with the resin and its additives. Flame retardancy, for example, may impinge directly on the performance of a colorant. Pressure to produce materials with lower levels of toxic combustion products can involve organic fire retardant additives that interact with the colorant either to negate the effect of the additives or affect the color.

VIII. ANTISTATIC AGENTS (ANTISTATS)

Most synthetic polymers, unless specially treated, are good electrical insulators. They are therefore capable of generating static electricity, which can be potentially costly and dangerous. Static-induced accumulation of dust reduces the attractiveness and thus saleability of products displayed on store shelves. The attraction of a formed part to the charged surfaces of a processing mold prevents proper ejection of the formed part and consequently slows down production. Electrostatic charges can cause problems when textile, films, or powders join up in automatic machinery. Sparks, and possibility explosions or fires, can occur when static electricity is induced from plastics on nearby conductor. Damage of sensitive semiconductors and similar complex microelectronic devices can also occur from either the direct discharge from the conductive skin of personnel or by exposure of such devices to the close approach of a static-charged polymer material.[20]

When two surfaces that are in intimate contact are rubbed together or pulled apart, static electricity is generated. This is due to the transfer of electrons from the surface of the donor material, which consequently becomes positively charged, to the surface of the acceptor material, which then becomes negatively charged. For materials that are nonconductive, these static charges do not flow easily along the surfaces and therefore remain fixed or static. Whether a material behaves as a donor or an acceptor depends on its position in the triboelectric series (Table 9.12). For example, if nylon and propylene are rubbed together, nylon is the donor, while polypropylene is the acceptor.

The generation of static charges is not confined strictly to nonconductors only; conductors also generate static charges, but since they dissipate the charge quickly the level of static charges developed by these materials is difficult to measure. When a conductor and nonconductor are separated or rubbed against each other, the nonconductor develops a measure of static charge. The nonconductor will not lose its charge to the ground. Consequently, the charge is removed by employing other techniques. The use of air-ionizing bars and blowers provides an atmosphere of ionized air capable of neutralizing the charged objects or surfaces. However, this does not provide lasting protection since it does not prevent another charge from forming once the object is removed from the ionized air atmosphere. To ensure an extended removal of static charges from the surface, the nonconductor must be made sufficiently conductive to carry charge to the ground. A layer of water, even a few molecules thick, will do this

Table 9.12 The Triboelectric Series

Negative [–] end		
1. Teflon	9. Rubber	17. Polyester
2. PVC	10. Brass, stainless steel	18. Aluminum
3. Polypropylene	11. Nickel, copper, silver	19. Wool
4. Polyethylene	12. Acetate fiber	20. Nylon
5. Saran	13. Steel (carbon)	21. Human hair
6. Polyurethane foam	14. Wood	22. Glass
7. Polystyrene foam	15. Cotton	23. Acetate
8. Acrylic	16. Paper	
		Positive [+] end

satisfactorily since high conductivity is not required for the removal of static charges. Antistatic agents (antistats) are therefore hygroscopic chemicals that can generate this layer of water by pulling moisture from the atmosphere.

There are essentially two types of antistats that are commonly used in polymers to get rid of static electricity: those that are applied topically and those that are incorporated internally into the polymer. Both improve the conductance of polymer surfaces by absorbing and holding a thin, invisible layer of moisture from the surrounding air onto the polymer surface. Topical coatings are usually applied using wipe, spray, dip, or roller coating techniques. Topically applied antistats are particularly useful where antistat cost control and performance are essential or when the mechanical performance of a polymer is negatively affected by an internal additive.[20] Reapplication of topical antistats may be required, particularly in cases of high-friction end uses. Hence, antistats applied to the surface of parts are used primarily in applications where temporary static protection is desired. The techniques for applying topical antistats necessarily involve wasteful amounts of antistats and leave an undesirable oil-like surface.

Internal antistats are compounded directly into the polymer mix prior to processing. They then migrate slowly and continually through the molecular interstices and the bulk polymer to its surface where they absorb the water necessary to prevent accumulation of static charges. In this case, it is necessary to balance the rate of such migration and the rate of surface removal of the antistat to provide long-term protection for the part.

Major types of organic antistatic agents include quaternary ammonium compounds, amines and their derivatives, phosphate esters, fatty acid polyglycol esters, and polyhydric alcohol derivatives such as glycerine and sorbitol.[20] Selection of the appropriate antistat depends on its compatibility with the polymer, the end use of the part, and the desired level of antistatic activity. Other factors that need to be considered include the effect of antistatic agent on color, transparency, and finish of the polymer part; its possible toxicity; stability during processing; and degree of interference with physical properties and ultimately cost effectiveness.

IX. PROBLEMS

9.1. Explain the following observations:

a. The magnitude of an electrostatic charge increases with increasing intimacy of contact and the speed of separation of the two materials whose surfaces are in contact with each other.
b. Graft copolymers of PVA/EVA are available commercially as flexible films. Applications include outdoor exposed materials such as roofing, pond membranes, and swimming pool covers as well as medical uses such as processing of blood.
c. Code specification of PVC wire and cable used in building has been raised from 60 to 90°C. This has caused a reformulation from the use of DOP and/or DINP to phthalates such as DUP, DTDP, and UDP, which are used in combination with trimellitate ester such as trioctyl trimellitate.
d. Although phthalate esters less volatile than DOP are used in most automotive interior applications, trialky trimellitates are recommended.
e. Plasticizers sanctioned for food-contact applications include soya and linseed oil epoxides and various adipates.

f. Furniture covered with tablecloths and upholstery made with plasticized PVC sometimes get discolored and tacky. Suggest a possible solution to the problem.
g. Polystyrene is completely immiscible with PVC. Styrene–maleic anhydride (SMA) copolymers, on the other hand, show a degree of miscibility with PVC that assures part integrity and toughness.
h.

Blend	Properties
PC/PBT	Toughness, chemical resistance
PC/PET	Toughness, chemical resistance, and high temperature rigidity

i. ABS is more susceptible to oxidation than polystyrene.
j. PVC requires incorporation of thermal stabilizers and antioxidants but not flame retardants.
k. Black pigments are not normally used as colorants for rigid PVC used for home siding and window profiles, while pigments containing heavy metals are usually avoided as colorants for polymers used for toys and food wrapping.

9.2. The front and rear bumper beams on the 1993 Toyota Corolla were made from polypropylene/fiberglass composite. Estimate the longitudinal modulus of these beams assuming a fiberglass (E-glass) composition of 50%.

REFERENCES

1. Kamal, M.R., *Plast. Eng.,* 38(11), 23, 1982.
2. Wigotsky, V., *Plast. Eng.,* 40(12), 19, 1984.
3. Beeler, A.D., and Finney, D.C., Plasticizers, in *Modern Plastics Encyclopedia,* 56(10A), 212, 1979–80.
4. Theberge, J.E., *J. Elastomers Plast.,* 14, 100, 1982.
5. Mock, J.A., *Plast. Eng.,* 39(2), 13, 1983.
6. Kamal, M.R., *Plast. Eng.,* 38(12), 31, 1982.
7. Fried, J.R., *Plast. Eng.,* 39(9), 37, 1983.
8. Wigotsky, V., *Plast. Eng.,* 44(10), 25, 1988.
9. Castagno, J.M., *Plast. Eng.,* 42(4), 41, 1986.
10. Kienzle, S.Y., *Plast. Eng.,* 43(2), 41, 1987.
11. Wigotsky, V., *Plast. Eng.,* 44(11), 25, 1988.
12. Wigotsky, V., *Plast. Eng.,* 42(7), 19, 1986.
13. Sutker, B., *Plast. Eng.,* 39(4), 27, 1983.
14. Wigotsky, V., *Plast. Eng.,* 41(2), 23, 1985.
15. Wigotsky, V., *Plast. Eng.,* 40(9), 21, 1984.
16. Wigotsky, V., *Plast. Eng.,* 42(10), 21, 1986.
17. Gordon, S.B., *Plast. Eng.,* 39(11), 37, 1983.
18. Marvuglio, P., Colorants, in *Modern Plastics Encyclopedia,* 56(10A), 154, 1979–80.
19. Roe, D.M., Color concentrates, in *Modern Plastics Encyclopedia,* 56(10A), 162, 1979–80.
20. Halperin, S.A., Antistatic agents, in *Modern Plastics Encyclopedia,* 56(10A), 153, 1979–80.
21. Walp, L.E., *Plast. Eng.,* 42(8), 41, 1986.

Chapter 10

Polymer Reaction Engineering

I. INTRODUCTION

Several important differences exist between the industrial production of polymers and low-molecular-weight compounds:[6,7]

- Generally, polymers of industrial interest have high molecular weights, usually in the range of 10^4 to 10^7. Also, in contrast to simple compounds, the molecular weight of a polymer does not have a unique value but, rather, shows a definite distribution. The high molecular weight of polymers results in high solution or melt viscosities. For example, in solution polymerization of styrene, the viscosity can increase by over six orders of magnitude as the degree of conversion increases from zero to 60%.
- The formation of a large polymer molecule from small monomer results in a decrease in entropy. It follows therefore from elementary thermodynamic considerations that the driving force in the conversion process is the negative enthalpy gradient. This means, of course, that most polymerization reactions are exothermic. Consequently, heat removal is imperative in polymerization reactions — a problem that is accentuated by the high medium viscosity that leads to low heat transfer coefficients in stirred reactors.
- In industrial formulations, the steady-state concentration of chain carriers in chain and ionic polymerization is usually low. These polymerization reactions are therefore highly sensitive to impurities that could interfere with the chain carriers. Similarly, in step-growth polymerization, a high degree of conversion is required in order to obtain a product of high molecular weight (Chapters 6 and 7). It is therefore necessary to prevent extraneous reactions of reactants and also exclude interference of impurities like monofunctional compounds.
- The quality of a product from a low-molecular-weight compound can be usually improved by such processes as distillation, crystallization, etc. However, if the performance of a product from a polymerization process is inadequate, it is virtually impossible to upgrade its quality by subsequent processing.

Given these possible differences in the production processes between polymers and low-molecular-weight compounds, it is vitally important to choose the most suitable reactor and operating conditions to obtain the required polymer properties from a polymerization reaction. This demands a detailed knowledge of the phenomena that occur in the reactor. This, in turn, requires an accurate model of the polymerization kinetics, the mass and heat transfer characteristics of the polymerization process. Our approach in this chapter is largely qualitative, and our treatment involves a discussion of the various polymerization processes, followed by a brief review of polymerization reactors. It is hoped that this approach will enable the reader to gain insight into the complex problem of selecting a reactor for a specific polymerization reaction.

II. POLYMERIZATION PROCESSES

Polymerization processes may be conveniently classified as homogeneous or heterogeneous. In homogeneous polymerization, as the name suggests, all the reactants, including monomers, initiators, and solvents, are mutually soluble and compatible with the resulting polymer. On the other hand, in heterogeneous systems, the catalyst, the monomer, and the polymer product are mutually insoluble. Homogeneous polymerization comprises bulk (mass) or solution systems while heterogeneous polymerization reactions may be categorized as bulk, solution, suspension precipitation, emulsion, gas phase, and interfacial polymerization.

A. HOMOGENEOUS SYSTEMS

1. Bulk (Mass) Polymerization

In bulk polymerization the reaction mixture consists essentially of the monomer and, in the case of chain growth polymerization, a soluble initiator and possibly modifiers. In the case of homogeneous bulk

Table 10.1 Heats of Polymerization of Some Monomers

Monomer	Heat of Polymerization, Btu/lb[a]
Ethylene	1530–1660 (gas to solid)
Propylene	860
Isobutylene	370
Butadiene	620 (1,4 addition)
Vinyl chloride	650
Vinyl acetate	445 (at 170°F)
Acrylic acid	400 (at 166°F)
Ethyl acrylate	340
Methyl methacrylate	245
Styrene	290
Isoprene	470
Vinylidene chloride	330
Acrylonitrile	620

[a] Value given for liquid monomer converted to a condensed solid polymer at 77°F unless indicated otherwise.

polymerization, the product polymer and monomer are miscible. Since polymerization reactions are generally exothermic, the temperature of polymerization depends on the polymerization system. Mixing and heat transfer become difficult as the viscosity of the reaction mass increases.

Bulk polymerization is widely practiced in the manufacture of step-growth polymers. However, since condensation reactions are not very exothermic and since the reactants are usually of low activity, high temperatures are required for these polymerization processes. Even though medium viscosities remain low throughout most of the course of polymerization, high viscosities are generally experienced at later stages of the reaction. Such high viscosities cause not only problems with the removal of volatile by-products, but also a possible change in the kinetics of the reaction from a chemical-controlled regime to a diffusion-controlled one. To obtain a product of an appropriate molecular weight, therefore, proper cognizance must be given to this problem in the design of the reactor.

Organic systems have low heat capacities and thermal conductivities. Free-radical reactions are highly exothermic in nature. This coupled with the extremely viscous reaction media for these systems prevents effective convective (mixing) heat transfer, leading, as a consequence, to very low overall heat transfer coefficients. The problem of heat removal is accentuated at higher conversions because the rates of polymerization and heat generation are usually enhanced at these stages of reaction. This leads to the development of localized hot spots or runaway reactions, which if uncontrolled may be ultimately disastrous. The occurrence of local hot spots could result in the discoloration and even possible degradation of the polymer product, which usually has a broadened molecular weight distribution due to chain transfer to polymer.

Because of the above heat transfer problems, bulk polymerization of vinyl monomers is restricted to those with relatively low reactivities and enthalpies of polymerization. This is exemplified by the homogeneous bulk polymerization of methyl methacrylate and styrene (Table 10.1). Some polyurethanes and polyesters are examples of step-reaction polymers that can be produced by homogeneous bulk polymerizations. The products of these reactions might be a solid, as in the case with acrylic polymers; a melt, as produced by some continuous polymerization of styrene; or a solution of polymer in monomer, as with certain alkyd-type polyesters.

Sheets, rods, slabs, and other desired shapes of objects are produced from poly(methyl methacrylate) in batch reactors by keeping at least one dimension of the reaction mass thin, thereby facilitating heat transfer. Typically, the monomer containing a small amount of an initiator such as benzoyl peroxide is poured between two glass plates separated by a flexible gasket of poly(vinyl chloride) tubing and held together by spring clips to compensate for shrinkage. Depending upon the thickness, the filled mold is heated from 45 to 90°C for about 12 to 24 h. After cooling, the molds are stripped from the casting and the sheets are annealed at 140 to 150°C. The resulting sheet has good optical properties, but the process

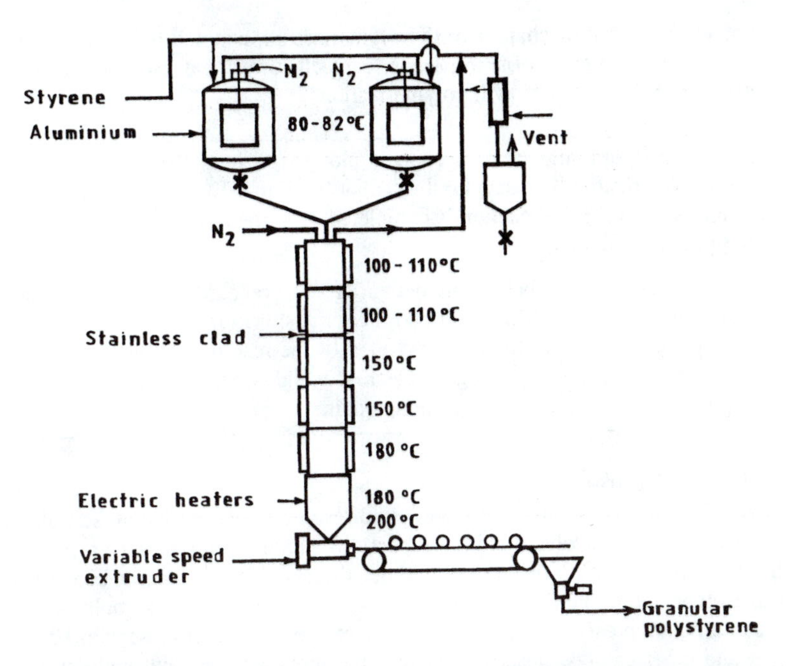

Figure 10.1 Vertical column reactor for the continuous bulk polymerization of styrene. (From Winding, C.C. and Hiatt, G.D., *Polymeric Materials,* McGraw-Hill, New York, 1961. With permission.)

suffers from a number of problems. These include possible bubbles from dissolved gases, long curing times, and large shrinkage (about 21%). These problems may be overcome by using prepolymerized syrup rather than the pure monomer. The syrup is prepared by carefully heating methyl methacrylate monomer containing 0.02 to 0.1% benzoyl peroxide or azo-bis-isobutyronitrile in a well-agitated stainless steel vessel to initiate polymerization. A good syrup is obtained after a prepolymerization cycle of 5 to 10 min at a batch temperature of 90°C.

Another way of circumventing the heat transfer problems is by continuous bulk polymerization. An example is the polymerization of polystyrene, which is carried out in two stages. In the first stage, styrene is polymerized at 80°C to 30 to 35% monomer conversion in a stirred reactor known as a prepolymerizer. The resulting reaction mass — a viscous solution or syrup of polymer in monomer — subsequently passes down a tower with increasing temperature. The increasing temperature helps to keep the viscosity at manageable levels and also enhances conversion, which reaches at least 95% at the exit of the tower (Figure 10.1). By removal of the heat of polymerization at the top of the tower and proper temperature control of the finished polymer at the bottom of the tower, an optimum molecular weight may be achieved and channeling of the polymer may be minimized.

Bulk polymerization is ideally suited for making pure polymeric products, as in the manufacture of optical grade poly(methyl methacrylate) or impact-resistant polystyrene, because of minimal contamination of the product. However, removal of the unreacted monomer is usually necessary, and this can be a difficult process. This may be achieved in vacuum extruders where the molten polymer is extruded under vacuum to suck off the residual monomer.

Example 10.1: Explain why bulk polymerization is generally more suited for step-growth polymerization than for chain-growth polymerization.

Solution: The major problems in bulk polymerization are heat removal and mixing. Both step-growth and chain-growth polymerization are exothermic. But the enthalpy of polymerization for chain-growth reactions is of the order of 15 to 20 kcal/g mol compared with 2 to 6 kcal/g mol for step-growth polymerization. Also, chain-growth polymerization is characterized by high viscosities even at low conversions due to the generation of polymers early in the reaction, whereas viscosities are generally low in step-growth polymerization until the later stages of the reaction (high conversions). Thus the much higher exothermic nature of chain-growth polymerization, coupled with the difficulty of mixing the reaction mixture due to high viscosities, leads to much lower heat transfer efficiencies in chain-growth polymerization than in step-growth polymerization.

Example 10.2: Purified styrene monomer is charged along with initiators into an aluminum prepolymerization vessel. Polymerization is carried out at about 90°C to 30% conversion. The resulting syrup is then poured into molds where the reaction is completed. Comment on the probable molecular weight distribution of the product polymer.

Solution: This process is essentially batch polymerization of styrene. Most of the conversion takes place in the finishing trays. Because of the relatively poor heat transfer in these trays, the polymerizing mass may reach high temperatures at some spots. Consequently, the resulting polystyrene will have a broad molecular weight distribution, with very high molecular weights being produced at low temperatures and lower molecular weights produced at high temperature.

B. Solution Polymerization

In solution polymerization, the monomer, initiator, and resulting polymer are all soluble in the solvent. Solution polymerization may involve a simple process in which a monomer, catalyst, and solvent are stirred together to form a solution that reacts without the need for heating or cooling or any special handling. On the other hand, elaborate equipment may be required. For example, a synthetic rubber process using a coordination catalyst requires rigorous exclusion of air (to less than 10 ppm); moisture; carbon dioxide; and other catalyst deactivators from the monomer, solvent, and any other ingredient with which the catalyst will come in contact before the reaction. In addition, exclusion of air prevents the tendency to form dangerous peroxides. To avoid product contamination and discoloration, materials of construction also need to be selected with the greatest care.

Polymerization is performed in solution either batchwise or continuously. Batch reaction takes place in a variety of ways. The batch may be mixed and held at a constant temperature while running for a given time, or for a time dictated by tests made during the progress of the run. Alternatively, termination is dictated by a predetermined decrease in pressures following monomer consumption. A continuous reaction train, on the other hand, consists of a number of reactors, usually up to about ten, with the earlier ones overflowing into the next and the later ones on level control, with transfer from one to the next by pump.

As the reaction progresses, solution polymerization generally involves a pronounced increase in viscosity and evolution of heat. The viscosity increase demands higher power and stronger design for pumps and agitators. The reactor design depends largely on how the heat evolved is dissipated. Reactors in solution polymerization service use jackets; internal or external coils; evaporative cooling with or without compression of the vapor or simple reflux-cooling facilities, a pumped recirculation loop through external heat exchanger; and combinations of these. A typical reactor has agitation, cooling, and heating facilities; relief, temperature level, and pressure connections; and, frequently, cleanout connections in addition to inlet and outlet fittings.

Solution polymerization has certain advantages over bulk, emulsion, and suspension polymerization techniques. The catalyst is not coated by polymer so that its efficiency is sustained and removal of catalyst residues from the polymer, when required, is simplified. Solution polymerization is one way of reducing the heat transfer problems encountered in bulk polymerization. The solvent acts as an inert diluent, increasing overall heat capacity without contributing to heat generation. By conducting the polymerization at the reflux temperature of the reaction mass, the heat of polymerization can be conveniently and efficiently removed. Furthermore, relative to bulk polymerization, mixing is facilitated because the presence of the solvent reduces the rate of increase of reaction medium viscosity as the reaction progresses.

Solution polymerization, however, has a number of drawbacks. The solubility of polymers is generally limited, particularly at higher molecular weights. Lower solubility requires that vessels be larger for a given production capacity. The use of an inert solvent not only lowers the yield per reactor volume but also reduces the reaction rate and average chain length since these quantities are proportional to monomer concentration. Another disadvantage of solution polymerization is the necessity of selecting an inert solvent to eliminate the possibility of chain transfer to the solvent. The solvent frequently presents hazards of toxicity, fire, explosion, corrosion, and odor problems not associated with the product itself. Also, solvent handling and recovery and separation of the polymer involve additional costs, and removal of unreacted monomer can be difficult. Complete removal of the solvent is difficult in some cases. With certain monomers (e.g., acrylates) solution polymerization leads to a relatively low reaction rate and low-molecular-weight polymers as compared with aqueous emulsion or suspension polymerization. The problem of cleaning equipment and disposal of dirty solvent constitutes another disadvantage of solution polymerization.

Solution polymerization is of limited commercial utility in free-radical polymerization but finds ready applications when the end use of the polymer requires a solution, as in certain adhesives and coating processes [i.e., poly(vinyl acetate) to be converted to poly(vinyl alcohol) and some acrylic ester finishes]. Solution polymerization is used widely in ionic and coordination polymerization. High-density polyethylene, polybutadiene, and butyl rubber are produced this way. Table 10.2 shows the diversity of polymers produced by solution polymerization, while Figure 10.2 is the flow diagram for the solution polymerization of vinyl acetate.

C. HETEROGENEOUS POLYMERIZATION

1. Suspension Polymerization

Suspension polymerization generally involves the dispersion of the monomer, mainly as a liquid in small droplets, into an agitated stabilizing medium usually consisting of water containing small amounts of suspension and dispersion agents. The catalyst or initiator is dissolved in the monomer if the monomer is a liquid or included in the reaction medium if a gaseous monomer is used. The ratio of monomer to dispersing medium ranges from 10 to 40% suspension of monomer or total solids content of polymer at the finish of polymerization. When polymerization is completed, the polymer suspension is sent to a blowdown tank or stripper where any remaining monomer is removed, using a vacuum or antifoaming agent if necessary. Several stripped batches are transferred and blended in a hold tank. The slurry mixture is finally pumped to a continuous basket-type centrifuge or vibrating screen where the polymer is filtered, washed, and dewatered. The wet product, which may still contain as much as 30% water or solvent, is dried in a current of warm air (66 to 149°C) in a dryer. It is then placed in bulk storage or transferred to a hopper for bagging as powdered resin or put through an extruder for forming into a granular pellet product. Figure 10.3, which shows a flow sheet of the suspension polymerization of methyl methacrylate, is typical of many suspension polymerization processes for the production of thermoplastic resins.

For the monomer to be dispersible in the suspension system, it must be immiscible or fairly insoluble in the reaction medium. In some instances, partially polymerized monomers or prepolymers are used to decrease the solubility and also increase the particle size of the monomer. The initiators employed in the polymerization reaction are mainly of the peroxide type and, in some cases, are azo and ionic compounds. Examples include benzoyl, diacetyl, lauroyl, and *t*-butyl-peroxides. Azo-bis-isobutyronitrile (AIBN) is one of the most frequently used azo initiators, while aluminum and antimony alkyls, titanium chloride, and chromium oxides are typical ionic initiators. The amount of catalyst used depends on the reactivity of the monomer and the degree of polymerization, varying from 0.1 to 0.5% of the weight of the monomer.

Apart from the monomer itself, the most important ingredient in suspension polymerization is the suspension agent. Even though it is used in relatively small amounts (0.01 to 0.50% weight of monomer), it is vital to the successful control of the polymerization process and the uniformity of the product obtained. The major problem in suspension polymerization is in the formation and the maintenance of the stability of the thermodynamically unstable droplets as they are slowly transformed from a highly mobile immiscible liquid, through the viscous, sticky stage to the final solid beads (rigid granules) without their coalescence or agglomeration into a conglomerate mass. During the transformation of the liquid monomer droplet to the solid resin, the viscous or sticky phase first appears when 10 to 20% conversion has occurred. This phase persists for up to 75 to 80% conversion before the particles take on a nonsticky solid appearance. The tendency for agglomeration of the particles, which is particularly critical at the stage when the particles become sticky, is prevented by proper agitation and the use of suspending agents. The stabilizing agents are employed in two ways. (1) Surface-acting agents (surfactants) such as fatty acids and some inorganic salt such as magnesium and calcium carbonates, calcium phosphate, titanium and aluminum oxides reduce the surface tension between water and the monomer droplets, thus providing greater stability for this interface. They also reduce the surface viscosity of the

Table 10.2 Typical Solution-Polymerization Processes

Monomer	Product	Solvent	Catalyst	Pressure (Psig)	Temperature (°F)	Isolation
Conjugated diene[a]	Synthetic rubber	Various[b]	Coordination,[c] or alkyllithium	40	50	Steam coagulation, slurrying in hot water, extrusion into hot water, evaporation of solvent on drum, or dryer
Isobutylene + isoprene	Butyl rubber	Methyl chloride	$AlCl_2$	Atm	–140	Slurrying in hot water
Ethylene alone, or with 1-butene	Linear polyethylene,homo- or copolymer	Cyclohexane, pentane, or octane	Chrome oxide on silica alumina base	400–500	275–375	Precipitation with water then steam stripping
Ethylene	Polyethylene	Ethylene	Peroxygenic	500–2,000 atm	210–480	
Vinyl acetate	Polyvinyl acetate	Alcohol, ester, or aromatic	Peroxygenic compound		Precipitation	
Urea-formaldehyde	Resin	Water				
Bisphenol A + phosgene	Polycarbonate resin				to 104	Precipitation
Dimethyl terephthalate + ethylene glycol	Polyester resin	Ethylene glycol	Various[d]		320–570	Distillation of solvent and recovery of the polymer as a melt
Resorcinol + formaldehyde	Latex adhesive for tire cord	Water	NaOH	Atm	Ambient	Not isolated but used as a solution
Melamine + formaldehyde	Laminating resin	Water NH_4OH			Spray dried, or used as a solution	
Acrylamide + acrylonitrile	Resin	Water	Ammonium persulfate	Atm	165–175	
Acrylate	Adhesive coating	Ethyl acetate	Free-radical initiator	Atm	Refluxing temperature	Not isolated but used as a solution
Maleic anhydride + styrene + divinylbenzene	Water-soluble thickener	Acetone or benzene	Benzoyl peroxide	Atm	Refluxing temperature	Neutralized with ammonium hydroxide and used as a solution, or precipitated with petroleum ether
Ethylene + propylene + a diene	EPT rubber	Hydrocarbon	Coordination	200–500	100	Steam coagulation
Epichlorohydrin	Polyepichlorodydrin elastomer	Cyclohexane, or ether	Organo-aluminum compound	Autogenous pressure	–20 to 210	
Phenol + drying oil + hexamethylenetetramine	Thermosetting resin	Ester-alcohol mixture	H_3PO_4	Atm	350, 200, and 185 in stages	Not isolated but used as solution for paint vehicle
Propylene[e]	Polypropylene	Hexane	Coordination	175	150–170	Precipitates as formed
Formaldehyde	Polyoxy-methylene	Hexane	Anionic type[f]	Atm	–60 to 160	Precipitates as formed

Table 10.2 (continued) Typical Solution-Polymerization Processes

[a] For example, 1,3-butadiene or isoprene
[b] Includes hexane, heptane, an olefin, benzene, or a halogenated hydrocarbon. Must be free from moisture, oxygen, and other catalyst deactivators.
[c] $TiCl_4$, an aluminium, alkyl, and cobalt halide are reported to be used to make Ameripol CB *cis*-polybutadiene.
[d] In the transesterification step, inorganic salts, alkali metals or their alkoxides, or Cu, Cr, Pb, or Mn metal are used. In the next step, the catalyst is not disclosed.
[e] Isotactic polymers are not usually formed completely in solution but precipitate in the course of reaction.
[f] Amines, cyclic nitrogen compounds, arisine, stibine, or phosphine.

From Back, A.L., *Chem. Eng.*, p. 65, August 1, 1966.

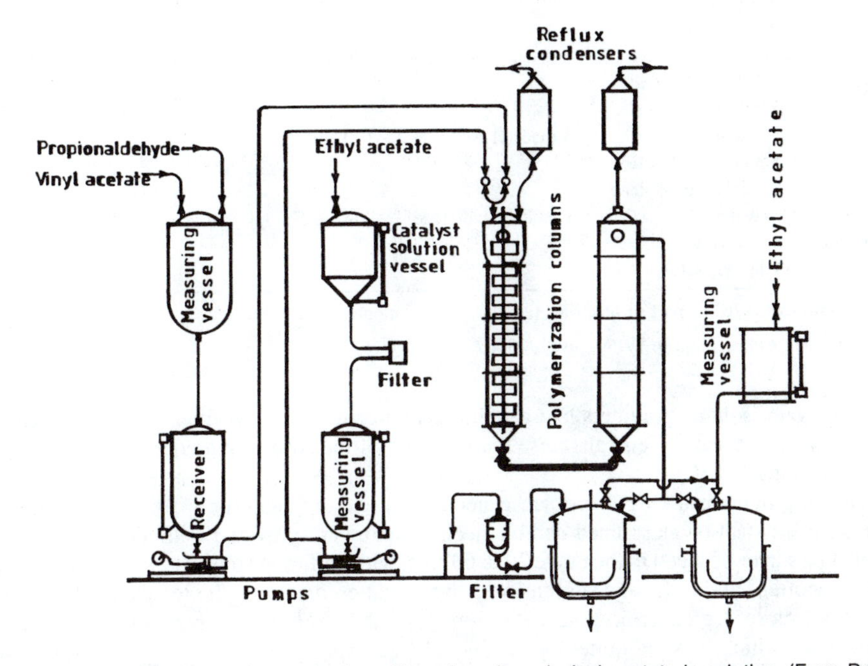

Figure 10.2 Continuous process for production of low viscosity polyvinyl acetate in solution. (From Back, A.L., *Chem. Eng.*, p. 65, August 1, 1966. With permission.)

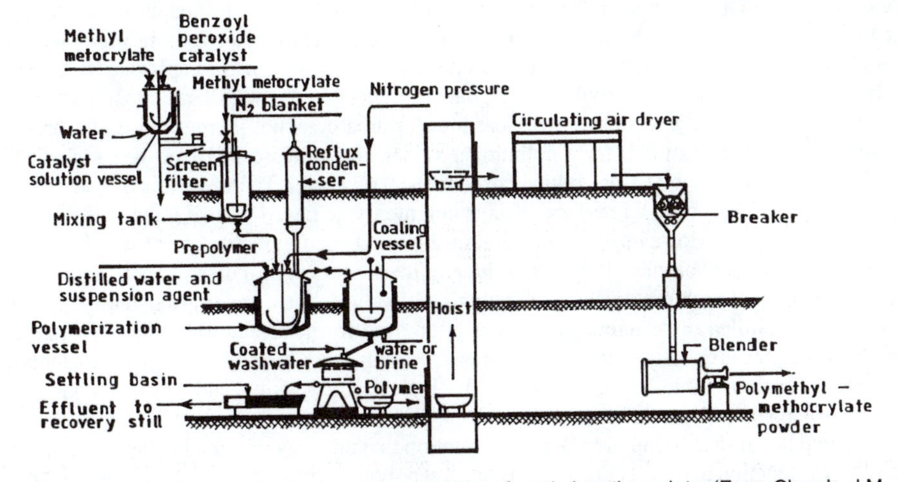

Figure 10.3 Flow sheet for the suspension polymerization of methyl methacrylate. (From Church, J.M., *Chem. Eng.*, p. 79, August 1, 1966.)

Table 10.3 Composition and Reaction Conditions for Some Suspension Polymerization Systems

Polymer	Suspension Medium[a]	Catalyst System[a]	Temperature	Pressure	Time Cycle	Polymer Slurry
Polyethylene (low density)	1,000 water 100 benzene	0.1 oxygen (air), or 0.5 di-*tert*-butyl peroxide	350–400 F	800–1000 atm	1–2 min	10%
Polyethylene (high density)	500 cyclohexane or heptane	0.3 aluminum triethyl or chromium oxide	180–250 F	150–200 psi	1–3 h	20%
Polyvinyl chloride	300 water 0.2 polyvinyl alcohol, or 0.5 gelatin 0.05 emulsifier	0.2 lauroyl peroxide	130–140 F	100–200 psi	10–12 h	18%
Polymethyl-methacrylate	350 water 0.8 polyacrylic acid 0.7 lauryl sulfonate	0.5 benzoyl peroxide	160–180 F	—	6–8 h	23%
Polystyrene	400 water 0.5 methyl cellulose 0.2 calcium phosphate	0.3 diacetyl peroxide	190–200 F	—	3–4 h	20%
Polytetra-fluoroethylene	600 water 0.3 polyvinyl alcohol 0.1 lauryl sulfonate	0.1 isobutylazonitrile	80–140 F	—	2–3 h	12%

[a] Concentrations expressed as part of materials per 100 parts of monomer.

From Church, J.M., *Chem. Eng.*, p. 79, August 1, 1966.

droplets. (2) Water-soluble polymers like gelatin, methyl cellulose, poly(vinyl alcohol), starches, gums, and poly(acrylic acids) and their salts raise the viscosity of the suspending aqueous medium and act as protective coatings.

Suspension polymerization reactors are generally vertical, agitated (or stirred tank) vessels usually made of stainless steel or glass-lined carbon steel. Reactors are provided with agitators with a paddle or anchor-type stirrer of speed in the range 20 to 60 rpm, with baffles in some cases to enhance dispersion. The most important design for the suspension polymerization reactor is the temperature control, which must ensure a close degree of accuracy. Reactors must therefore be capable of removing the heat of polymerization, which may be quite appreciable. For example, the heats of polymerization for vinyl chloride and styrene are 650 Btu/lb and about 300 Btu/lb, respectively (Table 10.1). This heat is released over a relatively short period (5 to 10 h). Consequently, reactors are jacketed and water cooled. The overall heat transfer coefficient for a glass-lined steel unit is 55 to 70 Btu/h · ft² · °F. The difference in the heat transfer coefficients between the two reactor types is due to the additional heat transfer resistance offered by the glass layer. Glass lining, of course, helps to reduce reactor fouling problems, but glass thickness must be minimized so as not to compromise reactor heat transfer capability.

As the reactor size increases, problems are generally encountered with heat transfer surfaces. Even when dimensional similarity is maintained, heat transfer area does not increase in direct proportion to the reactor volume. For example, for a cylindrical vessel, the increase in jacket heat-transfer area for the straight side is proportional to the volume raised to the 0.67 power. Therefore it is frequently necessary to provide additional heat transfer surface. This is commonly achieved by using the baffles required for agitation as cooling aids. However, extreme care must be exercised in the design of supplemental baffles to avoid creation of dead volumes. Polymer buildup in these spots contributes to poor product properties and creates operational problems such as the plugging of valves, lines, and pipes by chunks of polymers from these spots. Chilled or refrigerated water is often used for heat removal in lieu of adding supplementary heat-transfer area. Hot water or low-pressure steam may also be used to heat up reactants to initiate polymerization. Table 10.3 summarizes the typical composition and reaction conditions for a number of suspension polymerization systems.

We indicated in our discussion of bulk polymerization that one way of reducing heat transfer problems is to conduct the reaction in thin sections. In suspension polymerization, this concept is utilized practically in its extreme by dispersing and suspending monomer droplets (0.0001 to 0.50 cm diameter) in an inert nonsolvent, which is almost always water. This is achieved by maintaining an adequate degree of turbulence and an interfacial tension between the monomer droplets and the water. This way, the monomer droplets and the resulting viscous polymer form the discontinuous (dispersed) phase in the continuous aqueous phase. This, combined with the added heat capacity of water, enhances the overall heat transfer efficiency. Kinetically, therefore, each of the suspended particles has the characteristics of a miniature bulk reactor without a mechanism for internal mixing but with a good surface-to-volume ratio for efficient heat removal.

Suspension polymerization, also known as bead, pearl, or granular polymerization because of the physical nature of the product polymer, has numerous attractive features. First, the use of water as the heat exchange medium is more economical than the organic solvents used in most solution polymerizations. Second, with water as the dispersed phase acting as the heat-transfer medium, the removal of the excessive heat of polymerization presents minimal problems, and control of temperature is relatively simple. Another advantage is the quality of the product obtained. Separation and handling of the polymer product are relatively easier than in emulsion and solution polymerizations. The product is far easier to purify than emulsion systems; very little contamination occurs, with only trace amounts of catalyst, suspension, and dispersing agents remaining in the resin. Only a minimum amount of these ingredients are used in the polymerization process and much of these are removed in the subsequent purification steps.

Suspension polymerization is the most widely used process for making plastic resins both in terms of the number of polymer products and in tonnage production. Practically all of the common thermoplastic resins, including some of the newer polymers, are made by this method. Styrene, methyl methacrylate, vinyl chloride, vinylidene chloride, vinyl acetate, the fluorocarbons, and some gaseous monomers, including ethylene, propylene and formaldehyde, may be polymerized by the suspension polymerization process.

Example 10.3: Describe briefly the potential implications with respect to process operation and/or the resulting product of the following problems.

a. Power or equipment failure in solution polymerization
b. Addition of excess amount of initiator in suspension polymerization

Solution:

a. The reaction must be stopped in an appropriate way depending on the nature of the reactants. Failure to stop a highly exothermic polymerization during an extended power loss can be potentially dangerous since agitators and compressors stop functioning and the necessary heat transfer to the coolant is effectively terminated. The reaction may proceed to a violent stage or a completely solidified mass depending on the reactants. A deactivator such as a ketone, an alcohol, or water is effective with a coordination-type initiator. In some cases, e.g., the phenolic type, the reaction is effectively terminated by lowering the temperature and adding more solvent. For a free-radical polymerization reaction, a free-radical acceptor ("shot stop") halts the polymerization. These additives are charged into the reactor where necessary or, preferably, after the batch leaves the reactor. In the case of a train of reactors, the reactor in which the rate of polymerization is highest is dumped into an off-grade vessel and terminated there.
b. An excess of initiator should be avoided because of the undesirable effects it has on process control and product quality. Some of these adverse effects include low-molecular-weight polymer formation resulting from chain termination, difficult control of temperature due to enhanced reaction rate and exothermic heat resulting in gel formation and agglomerization, and decreased thermal stability of the product polymer due to the presence of the excess peroxide causing degradation of the product at processing temperatures.

Example 10.4: Explain the following observation. The use of internal cooling coils or an external circulating loop and exchanger is generally not an appropriate means of providing supplemental heat-transfer area in suspension polymerization.

Solution: Polymer fouling is generally a problem in suspension polymerization reactors. Supplemental heat-transfer area in the form of coils is generally avoided because of the difficulty encountered in cleaning the area between coils and the vessel wall.

The use of an external circulating loop and exchanger requires that a portion of the reaction mass is pumped through an external heat exchanger and cooled and returned to the reactor. In suspension polymerization, a controlled degree of agitation is imperative to prevent agglomeration and maintain the desired polymer particle size. It is difficult to design the equipment and the recirculating loop to avoid zones of too little agitation, where the coalescence of monomer droplets occurs. When the reaction slurry is pumped through an external loop, the pump may impose an order-of-magnitude increase in the shear rate, thereby forcefully agglomerating polymer particles as they pass through a sticky phase of polymerization. Also, some polymer buildup is inevitable and an external circulating cooling loop will present cleaning problems.

2. Emulsion Polymerization

Emulsion polymerization was developed in the U.S. during World War II for the manufacture of GR-S rubber (Government-Rubber-Styrene) or SBR (styrene–butadiene rubber) when the Japanese cut off the supply of rubber from the East. Emulsion polymerization is now widely used commercially for the production of a large variety of polymers. All polymers made by this process are addition polymers rather than condensation polymers and require free-radical initiators. In general, an emulsion polymerization system would consist of the following ingredients: monomer(s), dispersing medium, emulsifying agent, water-soluble initiator, and, possibly, a transfer agent. Water serves as the dispersing medium in which the various components are suspended by the emulsifying agent. The water also acts as a heat transfer medium. Monomers such as styrene, acrylates, methacrylates, vinyl chloride, butadiene, and chloroprene used in emulsion polymerization show only a slight solubility in water.

Reactors for the emulsion polymerization process vary in size from 1000 to 4000 gal depending upon production requirements. Reactors may be glass lined or made of stainless steel. Glass-lined reactors are preferred for the production of acrylic polymer emulsions, while stainless steel is usually preferred for the manufacture of poly(vinyl acetate) because it can be cleaned easily with a boiling solution of dilute caustic. Both types of reactors have been employed in the production of butadiene–styrene copolymers and poly(vinyl chloride). Reactors are necessarily jacketed for heat control purposes. In processes [e.g., poly(vinyl acetate) manufacture] where monomer, catalyst, and surfactant are added to the reactor incrementally and the available jacket heat-transfer area is not initially available for cooling, supplemental cooling is both necessary and attractive. This involves the use of reflux cooling. Here, the heat of polymerization vaporizes unreacted monomer, and the monomer vapor is condensed in a reflux condenser and returned to the reactor. In this case, however, the foaming characteristics of the latex must be determined first since a stable foam carried into the reflux condenser will foul the exchanger surface. Reactors must also be rated to withstand a minimum internal pressure ranging from 50 psi for acrylic, methacrylic and acrylic–styrene vinyl acetate and its copolymers to at least 300 psi for vinyl chloride homopolymers and copolymers. Figure 10.4 shows a flow sheet for a typical emulsion polymerization plant.

In order to understand the quantitative relations governing emulsion polymerization kinetics, it is necessary to give a qualitative description of the process.

a. Distribution of Components

A typical recipe for emulsion polymerization in parts by weight consists of 180 parts of water, 100 parts of monomer, 5 parts of fatty acid soap (emulsifying agent), and 0.5 parts of potassium persulfate (water-soluble initiator). The question, of course, is how these components are distributed within the system. By definition, soaps are sodium or potassium salts of organic acids, for example, sodium stearate:

$$[\underbrace{CH_3\ (CH_2)_{16}}_{R} - \overset{\overset{\displaystyle O}{\|}}{C} - O^-]\ Na^+ \qquad \text{(Str. 1)}$$

When a small amount of soap is added to water, the soap ionizes and the ions move around freely. The soap anion consists of a long oil-soluble portion (R) terminated at one end by the water-soluble portion.

$$\left(- \overset{\overset{\displaystyle O}{\|}}{C} - O^- \right) \qquad \text{(Str. 2)}$$

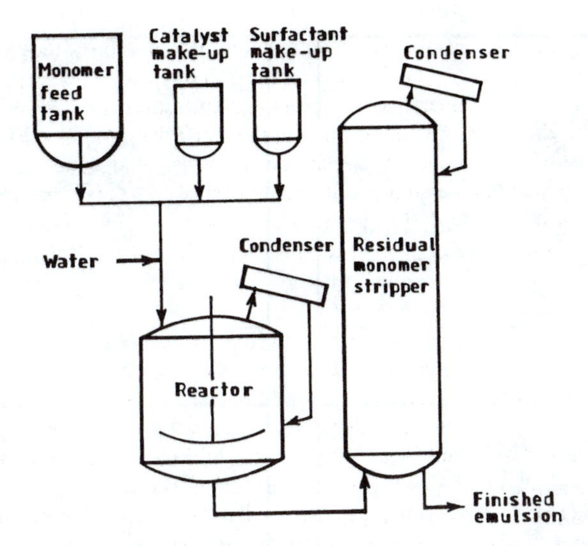

Figure 10.4 Flow sheet of typical emulsion polymerization plant. (From Gellner O., *Chem. Eng.*, p. 74, August 1, 1966.)

In other words, soap anions consist of both hydrophilic and hydrophobic groups. In water containing a partially soluble monomer molecule, the soap anion molecules orient themselves at the water–monomer interfaces with the hydrophilic ends facing the water, while the hydrophobic ends face the monomer phase. Each monomer droplet therefore has a protective coat of negative charge. Consequently, the emulsified monomer droplet is stabilized not only by the reduction of surface tension but also by the repulsive forces between the negative charges on its surface.

Above a critical concentration of the emulsifying agent known as the critical micelle concentration (CMC), only a small fraction of the emulsifying agent is dissolved in the water. The bulk of the emulsifier molecules arrange themselves into colloidal particles called micelles. The micelles remain in dynamic equilibrium with the soap molecules dissolved in water. Arguments persist with regard to the shape of the micellar aggregate, but energy considerations favor a spherical arrangement with the hydrophilic (polar) groups on the surface facing the aqueous phase while the hydrophobic chains are arranged somewhat irregularly at the interior. Each micelle consists of 50 to 100 soap molecules. Proposed rodlike-shaped micelles range in length from 1000 to 3000 Å and have diameters that are approximately twice the length of each soap molecule. The number and size of micelles depend on the relative amounts of the emulsifier and the monomer. Generally, large amounts of emulsifier result in larger numbers of smaller size particles.

The presence of soap or emulsifying agents considerably enhances the solubility of a water insoluble or sparingly soluble monomer. It has been demonstrated by X-ray and light scattering measurements that in the presence of monomers, micelles increase in size, a clear manifestation of the occupation of the hydrophobic interior part of the micelles. Meanwhile, a very small portion of the monomer remains dissolved. However, the bulk of the monomer is dispersed as droplets are stabilized, as discussed above, by the emulsifier. Consequently, when a slightly water-soluble monomer is emulsified in water with the aid of soap and agitation, three phases are present: the aqueous phase with small amounts of dissolved soap and monomer, the emulsified monomer droplets, and the monomer-swollen micelles. The level of agitation dictates the size of the monomer droplets, but they are generally at least 1 μm in diameter. The emulsifier in micelle form and monomer concentrations would typically be in the range of 10^{18} micelles per ml and 10^{10} to 10^{11} droplets per ml, respectively. Figure 10.5 is a schematic representation of the components of the reaction medium at various stages of emulsion polymerization.

b. Locus and Progress of Polymerization

When the water-soluble initiator potassium persulfate is added to an emulsion polymerization system, it undergoes thermal decomposition to form sulfate radical anion:

$$S_2O_8^- \xrightarrow{\text{heat}} 2SO_4^{-\cdot} \qquad \text{(Str. 3)}$$

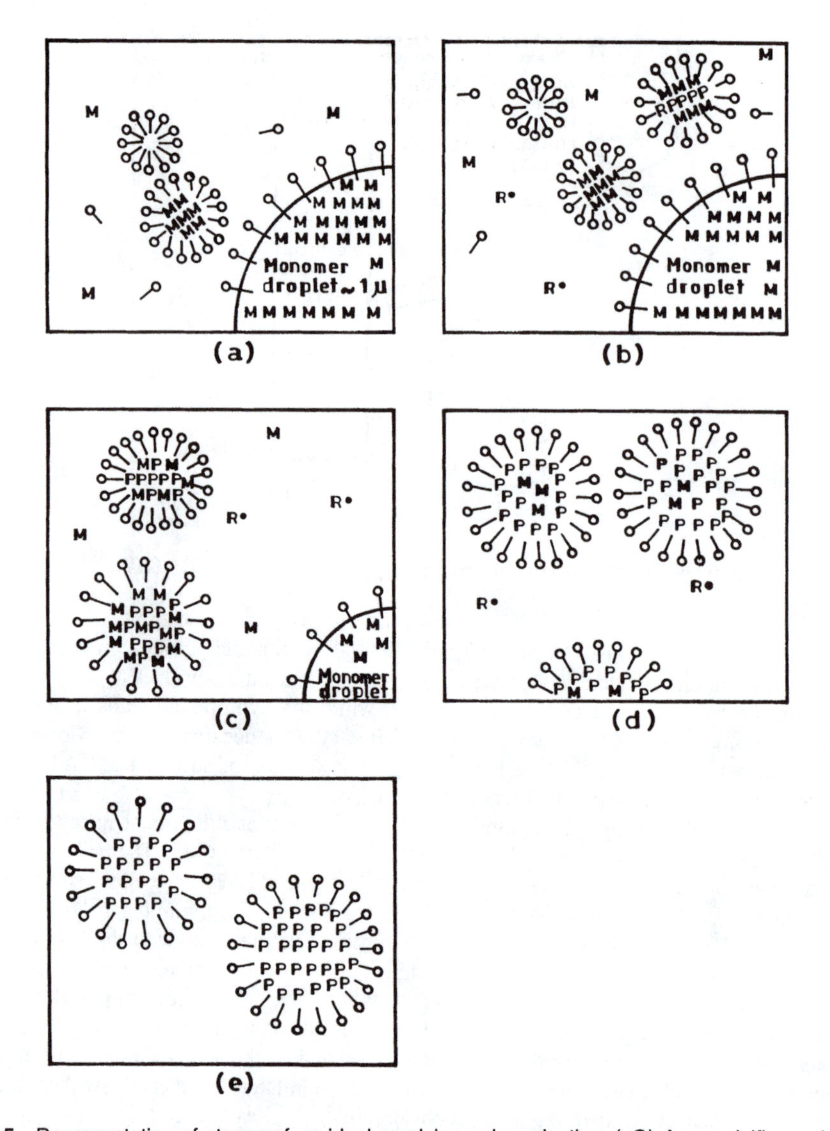

Figure 10.5 Representation of stages of an ideal emulsion polymerization. (–O) An emulsifier molecule; (M), a monomer molecule; (P) a polymer molecule; and (R·) a free radical. (a) Prior to initiation; (b) polymerization stage 1; shortly after initiation; (c) polymerization stage 2; all emulsifier micelles consumed; (d) polymerization stage 3; monomer droplets disappear; and (e) end polymerization.

The water-soluble radical anions react with monomer dissolved in aqueous phase to form soap-type free radicals:

$$SO_4^{-}\cdot + (n+1)M \xrightarrow[50-60°C]{} -SO_4-(CH_2-CX_2)_n-CH_2-CX_2\cdot \qquad \text{(Str. 4)}$$

Given the three phases present in an emulsion polymerization system, the locus of polymerization can conceivably be in the monomer droplets, in the aqueous phase within the micelles, or possibly at an interface. Some polymerization obviously takes place in the aqueous phase but with a limited contribution to the overall polymerization because of the low solubility of the monomer in water. The monomer droplets also do not provide the loci for polymerization because the negatively charged sulfate anions find the soap-stabilized monomer droplets virtually impossible to penetrate. Also, the primary sulfate radical anions are oil insoluble. The absence of polymerization in the monomer droplets has been verified

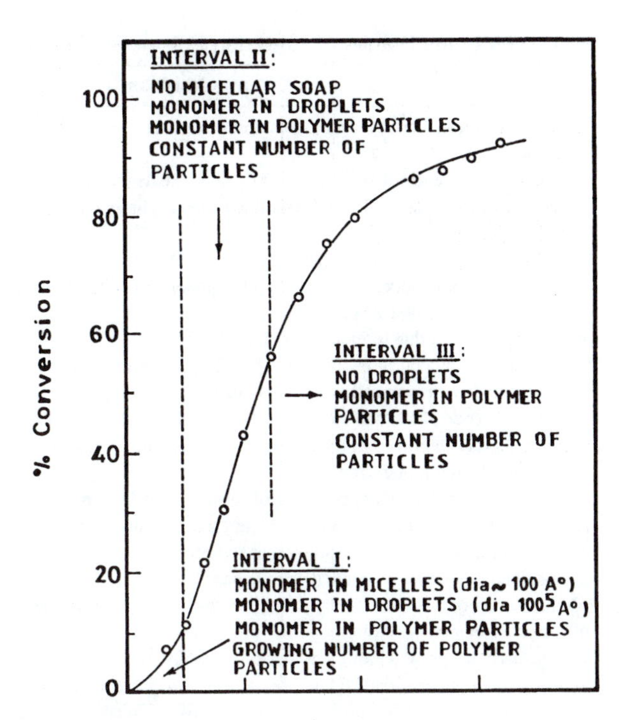

Figure 10.6 Stages in an emulsion polymerization. (From Rudin, A., *The Elements of Polymer Science and Engineering,* Academic Press, New York, 1982. With permission.)

experimentally. Consequently, emulsion polymerization takes place almost exclusively in the micelles. This is because of two reasons: (1) A micelle has a dimension in the range 50 to 100 Å while, as we said earlier, the dimension of the monomer droplet is at least 1 μm (10,000 Å). Since the surface/volume ratio of a sphere is 3/R and even though the total volume of micelles is considerably less than that of the droplets, the micelles present a much greater total surface area. (2) The concentration of the micelles is higher than that of the monomer droplets (10^{18} vs. 10^{11}).

Emulsion polymerization is considered to take place in three stages (Figure 10.6).

- **Stage I:** — Polymerization is initiated in only a small number (about 0.1%) of the micelles present initially. As polymerization proceeds, the active micelles consume the monomers within the micelle. Monomer depletion within the micelle is replenished first from the aqueous phase and subsequently from the monomer droplets. The active micelles grow in size with polymer formation and monomers contained therein, resulting in a new phase: monomer–polymer particles. To preserve their stability, these growing monomer-swollen polymer particles absorb the soap of the parent micelles. As the number of micelles converted to monomer-swollen particles increases and as these particles grow larger than the parent micelles, soap from the surrounding (existing micelles and emulsified monomer droplets) is rapidly absorbed. At about 13 to 20% conversion, the soap concentration decreases to a level below that required to form and sustain micelles (i.e., CMC). Consequently, the inactive micelles (those without a growing polymer) become unstable and disappear. As evidence for this, the initially low surface tension of the aqueous emulsion rises rather suddenly due to the decrease in the soap concentration, and if agitation is stopped at this stage the monomer droplets will coalesce because they are no longer stable. This marks the completion of stage I, which is characterized by a continuous increase in the overall rate of polymerization (Figure 10.6).
- **Stage II** — At the end of stage I, the locus of further polymerization is shifted exclusively to the monomer-swollen polymer particles since the micelles (the site of generation of new polymer particles) have all disappeared. Polymerization proceeds homogeneously in the polymer particles by the maintenance of a constant monomer concentration within the particles through the diffusion of monomers from the monomer droplets, which, in effect, serve as monomer reservoirs. As the polymer particles increase in size, the size of the monomer droplets disappears. Since there is no new particle nucleation

Table 10.4 Characteristics of Various Stages of Emulsion Polymerization

Stage	Characteristics
Prior to initiation	Dispersing medium, usually water, containing small amount of dissolved soap (emulsifier) and monomer.
	Monomer droplets each of size ca. 10,000 Å separate due to stabilization by a coat of emulsifier molecules whose hydrophilic ends face the aqueous phase; concentration of monomer droplets 10^{10}–10^{11} per ml.
	Beyond CMC, emulsifier molecules number 50–100 form spherically shaped micelles each of size 40–50 Å; some micelles are swollen by monomer and have dimensions 50–100 Å; micelle concentration is about 10^{18} per ml.
	Low surface tension due to emulsifier concentration.
Stage I	Polymerization initiated in ca. 0.1% of initially present micelles.
(12–20% conversion)	As monomers in active micelles are consumed, and then replenished by diffusion of monomers from the aqueous phase and subsequently from monomer droplets, the micelles form swollen particles that are stabilized by their absorbing soap molecules from neighboring inactive micelles and emulsified monomer droplets.
	End of stage marked by disappearance of inactive micelles and increase in surface tension; agitation needed to prevent coalescence of monomer droplets.
Stage II	Low concentration of dissolved monomer molecules.
(25–50% conversion)	No dissolved emulsifier or emulsifier micelles.
	Polymerization occurs exclusively in monomer-swollen polymer (latex) particles through diffusion of monomers from monomer droplets.
	Polymer particles grow while monomer droplets decrease in size.
	No new particle nucleation (i.e., number of latex particles is constant) and since monomer concentration within particles is constant, rate of polymerization is constant.
	End of stage marked by disappearance of monomer droplets.
Stage III	No dissolved monomer, dissolved emulsifier, emulsifier micelles, monomer droplets or monomer-swollen micelles.
(50–80% conversion)	Since monomer droplets have disappeared, a supply of monomers from the monomer reservoir (i.e., monomer droplets) is exhausted, hence rate of polymerization drops with depletion of monomers in latex particles.
	At end of polymerization (i.e., 100% conversion) system contains polymer particles, i.e., 400–800 Å dispersed in aqueous phase.

at this stage, the number of polymer particles remains constant. As a result of this and the constant monomer concentration within the particles, this stage is characterized by a constant rate of reaction.

- **Stage III** — At an advanced stage of the polymerization (50 to 80% conversion), the supply of excess monomer becomes exhausted due to the disappearance of the monomer droplets. The polymer particles contain all the unreacted monomers. As the concentration of monomer in the polymer particles decreases, the rate of polymerization decreases steadily and deviates from linearity. The characteristics of various stages of emulsion polymerization are summarized in Table 10.4.

c. Kinetics of Emulsion Polymerization

A number of questions need to be resolved from the qualitative description of emulsion polymerization given in the previous section. For example, it is necessary to consider whether the diffusion of monomers to the polymer particles is high enough to sustain polymerization given the low solubility of monomer in the aqueous phase. It is also important to know the average radical concentration in a polymer particle. Also, the validity of the assumption that only the monomer–polymer particles capture the radicals generated by the initiator needs to be established convincingly. The answers to these questions were provided by Smith and Ewart[3] and this forms the basis for the quantitative treatment of the steady-state portion of emulsion polymerization.

(1) Rate of Emulsion Polymerization

In emulsion polymerization, the rate of generation of free radicals is about 10^{13}/m-s while the number of monomer–polymer particles for typical recipes, N, is in the range 10^{13} to 10^{15} particles/ml of the aqueous phase. Consequently, if all the initiator radicals are captured by the monomer–polymer particles, each particle will acquire, at the most, a radical every 1 to 100 s. It can be shown that if a particle contains two radicals, mutual annihilation of radical activity will occur within a time span of the order of 10^{-13} s. This is much smaller than the interval between radical entry into the polymer particle. It is, therefore, perfectly safe to assume that the entry of a second radical with an active polymer particle results in immediate bimolecular termination. The particle remains dormant until another radical enters about 1 to 100 s later to reactivate polymerization. The activity of the particle will once again be terminated after 1 to 100 s due to the entry of another radical and the cycle is repeated. It follows, therefore, that at any given point in time a particle will either have one or zero radicals. This, in turn, means that a given particle will be active half of the time and dormant the other half of the time. By extension of this argument, at any given instant of time, one half of the polymer particles will contain a single radical and be active while the other half will remain dormant. The polymerization rate is given by:

$$R_p = k_p[M][M\cdot].$$

Therefore, the polymerization rate per cubic centimeter of water is given by

$$R_p = k_p \frac{N}{2}[M] \tag{10.1}$$

where N = number of polymer particles per cubic centimeter of aqueous phase
k_p = homogeneous propagation rate constant
[M] = monomer concentration in the polymer particles

Notice that Equation 10.1 predicts a direct dependence of the polymerization rate on the number of particles but not on the rate of radical generation. The equation holds true, of course, only when the radicals are being produced. Again, since the derivation of the equation is pivoted on the argument that each particle does not contain more than one radical at a given time, it follows that the particle size is not large enough to violate this condition.

Figure 10.7 is an experimental verification of the linear dependence of the rate of polymerization on the number of particles and monomer concentration. The polymerization rate increases with an increase in soap concentration due to the increase in N with soap concentration.

Example 10.5: The rate of diffusion, I, into a sphere of radius r is given by

$$I = D4\pi r\Delta C$$

where D = diffusion coefficient (cm^2/s)
ΔC = concentration difference between the surface of the sphere and the surroundings

In the emulsion polymerization of styrene at 60°C the diffusion coefficient is 10^{-10} cm^2 s^{-1} and the termination rate constant is 3×10^7 l mol^{-1} s^{-1}. Show that the rate of termination of the initiating radicals in the aqueous phase is of the order 10^3 radicals/ml/s. What is the average lifetime of a radical in the aqueous phase? Assume that the concentration of the radicals at the surface of the polymer particles is zero.

Solution: The rate of termination R_t is given by

$$R_t = 2k_t[M\cdot]^2$$

where [M·] = the overall concentration of radicals.

Since the concentration of radicals within and at the surface of the polymer particles is zero then

ΔC = [M·] = the concentration of radicals in the surrounding aqueous phase
[M·] = $I/D4\pi r$

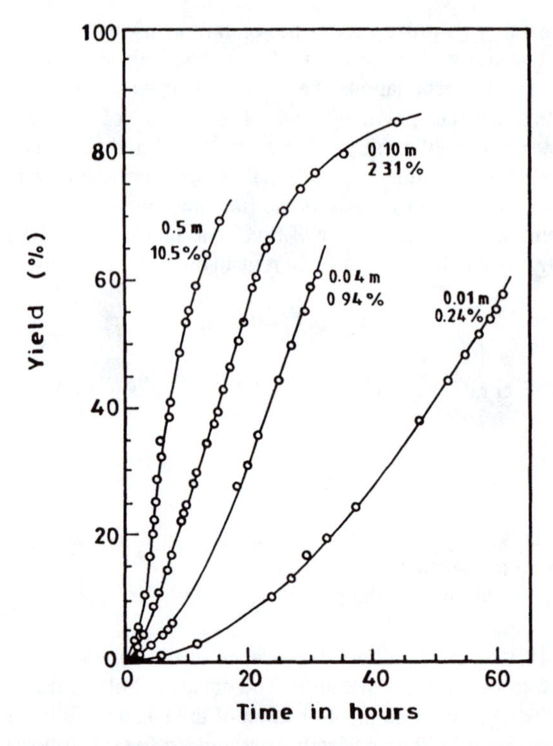

Figure 10.7 Polymerization of isoprene in emulsion at 50°C using 0.3 g of $K_2S_2O_8$ per 100 g of monomer, and with the amounts of soap (potassium chlorate) indicated in weight percent and in molality m. (From Harkins, H.B., *J. Am. Chem. Soc.*, 69, 1428, 1947. With permission.)

For emulsion polymerization, the rate of generation of radicals from the initiator is of the order of 10^{13} radicals/cm^3/s; the number of polymer particles is about 10^{14} particles/cm^3; and the average diameter of polymer particles is 1000 Å. The rate of diffusion of radicals to the polymer surface is

$$I = \frac{\textit{rate of generation of radicals}}{\textit{concentration of polymer particles}}$$

$$= \left(10^{13}\,\frac{radicals}{cm^3\,\mathrm{s}}\right)\left(\frac{1}{10^{14}}\,\frac{cm^3}{particle}\right) = 10^{-1}\ \text{radicals s}^{-1}$$

$$\mathrm{r} = \frac{1}{2} \times 1000\ \text{Å} = 5 \times 10^{-6}\ cm$$

$$[\mathrm{M.}] = \frac{10^{-1}\ radicals\ \mathrm{s}^{-1}}{\left(10^{-5}\ cm^2\ \mathrm{s}^{-1}\right)(4\pi)\left(5 \times 10^{-6}\ cm\right)}$$

$$= 10^{8}\ \text{radicals cm}^{-3}$$

$$\mathrm{k_t} = 3 \times 10^{7}\ \frac{\ell}{mol\text{-}\mathrm{s}} = 3 \times 10^{7}\left(\frac{1000\ cm^3}{\ell}\right)\left(\frac{1\ mol}{6.02 \times 10^{23}\ molecules}\right)\left(\frac{1}{\mathrm{s}}\right)$$

$$= 5 \times 10^{-14}\ \text{cm}^3\ \text{molecules}^{-1}\ \text{s}^{-1}$$

$$\mathrm{R_t} = 2\left(5 \times 10^{-14}\ \text{cm}^3\ \text{molecules}^{-1}\ \text{s}^{-1}\right)\left[10^{8}\ \text{radicals cm}^{3}\right]^2$$

$$= 10^{3}\ \text{radicals cm}^{-3}\ \text{s}^{-1}$$

Average lifetime of a radical from generation to its capture by a polymer particle is

$$\frac{\textit{concentration of radicals}}{\textit{rate of radical generation}}$$

$$= \frac{10^{8} \text{ radicals cm}^{-3}}{10^{13} \text{ radicals cm}^{-3}\text{ s}^{-1}}$$

$$= 10^{-5} \text{ s}$$

d. Degree of Polymerization

Whenever a primary radical enters an inactive polymer particle, polymerization occurs as it would in normal homogeneous polymerization. In this case the rate of polymerization is given by

$$r_p = k_p [M] \tag{10.2}$$

where k_p is the propagation rate constant. The rate of capture of primary radicals is given by

$$r_c = \frac{r_i}{N} \tag{10.3}$$

where r_i is the rate of generation of primary radicals (in radicals per milliliter per second). From our discussion in the previous section, the growth of a polymer is terminated immediately following the entry of another radical. Therefore the rate of termination should be essentially equal to the rate of capture of primary radicals. The degree of polymerization, in the absence of transfer, should then be the ratio of the rate of polymer growth to the rate of capture of primary radicals.

$$\overline{X}_n = \frac{r_p}{r_c} = \frac{k_p[M]}{r_i/N}$$

$$\overline{X}_n = \frac{k_p N[M]}{r_i} \tag{10.4}$$

Both the degree of polymerization and the rate of polymerization show a direct variation with the number of polymer particles N. However, unlike the rate of polymerization, the degree of polymerization varies indirectly with the rate of generation of primary radicals. This is as should be expected intuitively since the greater the rate of radical generation, the greater the frequency of alternation between polymer particle growth and dormancy and therefore the lower the chain length.

e. The Number of Particles

Equations 10.1 and 10.4 show that the number of polymer particles is crucial in determining both the rate and degree of polymerization. The mechanism of polymer particle formation indicates clearly that the number of polymer particles will depend on the emulsifier, its initial concentration (which determines the number of micelles), and the rate of generation of primary radicals. Smith and Ewart[3] have shown that

$$N = k\left(\frac{r_i}{\mu}\right)^{0.4} \left(a_s[E]\right)^{0.6} \tag{10.5}$$

where k is a constant with a value between 0.4 and 0.53; μ_s is the rate of increase in volume of a polymer particle, a_s is the interfacial area occupied by an emulsifier molecule; and [E] is the soap or emulsifier concentration. Note that all the units are in cgs. Equations 10.1, 10.4, and 10.5 establish the quantitative relations between the rate and degree of polymerization and the rate of generation of primary radicals and the emulsifier concentration.

In bulk polymerization, the rate of polymerization depends directly on the rate of initiation. However, the degree of polymerization is inversely related to the rate of initiation. Consequently, an increase in the rate of initiation results in a high rate of polymerization but a decrease in the degree of polymerization. This constitutes a major differences between bulk and emulsion polymerization. In emulsion polymerization, it is possible to increase the rate of polymerization by increasing the concentration of polymer radicals through a high initial emulsifier concentration. If the rate of initiation (generation of primary radicals) is kept constant, the degree of polymerization is increased as well.

f. Deviations from Smith–Ewart Kinetics

The Smith–Ewart kinetic theory of emulsion polymerization is simple and provides a rational and accurate description of the polymerization process for monomers such as styrene, butadiene, and isoprene, which have very limited solubility in water (less than 0.1%). However, there are a number of exceptions. For example, as we indicated earlier, large particles (> 0.1 to 0.5 cm diameter) may and can contain more than one growing chain simultaneously for appreciable lengths of time. Some initiation in, followed by polymer precipitation from the aqueous phase may occur for monomers with appreciable water solubility (1 to 10%), such as vinyl chloride. The characteristic dependence of polymerization rate on emulsifier concentration and hence N may be altered quantitatively by the absorption of emulsifier by these particles. Polymerization may actually be taking place near the outer surface of a growing particle due to chain transfer to the emulsifier.

Emulsion polymerization has a number of unique advantages compared with other polymerization methods. The viscosity of the reaction mass is relatively much less than that of a comparable true solution of polymers in the same molecular weight range. This, coupled with the increased heat capacity due to the presence of water, results in excellent heat transfer and creates a physical state that is much easier to control. Efficient removal of heat of polymerization is one of the factors determining the rate at which polymer may be produced on a commercial scale: efficient heat transfer permits faster rates to be used without overheating the mass and thus avoiding possible polymer degradation. As indicated above, it is possible to obtain both high rates of polymerization and relatively high-molecular-weight polymers through high emulsifier concentration and low initiator concentration. In bulk, solution, or suspension polymerization, rapid polymerization rates can be attained only at the expense of lower-molecular-weight polymers, except in anionic polymerization. In contrast to suspension polymerization, where there is a high risk of agglomeration of polymer particles into an intractable mass, emulsion polymerization is suitable for producing very soft and tacky polymers. Relatively low viscosity with high polymeric solids is advantageous in many applications. The latex product from emulsion polymerization can be used either directly or through master-batching to obtain uniform compounds that find useful applications in coatings, finishes, floor polishes, and paints. Emulsion polymerization, however, has some drawbacks. The large surface area presented by the tiny surfaces of a large number of small particles is ideal for absorption of impurities, thus making the product polymer impure. For example, the presence of water-soluble surface-active agents used in the polymerization process results in some degree of water sensitivity of the polymer itself, while ionic materials such as surfactants and inorganic salts result in poor electrical properties of the final polymer. Only free-radical-type initiators can be used in emulsion polymerization. This precludes the possibility of producing stereoregular polymers by this method.

Example 10.6: From the data given below for the emulsion polymerization styrene in water at 60°C:

a. Calculate the rate of polymerization.
b. Show that the number average degree of polymerization $\overline{X}_n$ is 3.52×10^3
c. Estimate the number of polymer chains in each. Data:

$k_p = 176$ l mol^{-1} s^{-1}
$r_i = 5 \times 10^{12}$ radicals cc^{-1} s^{-1}
$N = 10^{13}$ particles cc^{-1}
$[M] = 10$ *M*
Latex particle size = 0.10 μm
Particle density = 1.2 g/cc

Solution: a. $$R_p = k_p \cdot \frac{N}{2}[M]$$

$$k_p = \frac{176\ l}{mol\text{-s}} = 176\left(\frac{1000\ cc}{\ell}\right)\left(\frac{1}{mol-s}\right) = 1.76\times10^5\ \frac{cc}{mol-s}$$

$$N = 10^{13}\ \frac{particles}{cm^3} = 10^{13}\left(\frac{particle}{cm^3}\right)\left(\frac{1\ mol}{6.023\times10^{23}\ particles}\right)$$

$$= 1.66\times10^{-11}\ \frac{mol}{cm^3}$$

$$[M] = 10\ M = \frac{10\ mol}{\ell} = 10\ \frac{mol}{\ell}\left(\frac{1\ \ell}{1000\,cc}\right) = 10^{-2}\ mol\ cc^{-1}$$

$$R_p = \left(\frac{1.76}{2}\times10^5\ \frac{cm^3}{mol-s}\right)\left(1.66\times10^{-11}\ \frac{mol}{cm^3}\right)\left(10^{-2}\ \frac{mol}{cm^3}\right)$$

$$= 1.46\times10^{-8}\ \text{mol cc}^{-1}\ \text{s}^{-1}$$

b. $$\overline{X_n} = k_p N \frac{[M]}{r_i}$$

$$r_i = 5\times10^{-12}\ \frac{radicals}{cm^3\ \text{s}}\left(\frac{1 mol}{6.023\times10^{23}\ radicals}\right)$$

$$= 8.30\times10^{-12}\ \frac{mol}{cm^3-s}$$

$$\overline{X_n} = \left(1.76\times10^5\ \frac{cm^3}{mol-s}\right)\left(1.66\times10^{-11}\ \frac{mol}{cm^3}\right)\left(10^{-2}\ \frac{mol}{cm^3}\right)\left(\frac{10^{12}}{8.30}\ \frac{cm^3-s}{mol}\right)$$

$$= 3.52\times10^3$$

c. Volume of a particle $= \frac{4}{3}\pi r^3$, $r = 0.05\ \mu m = 0.05\times10^{-4}\ cm$

$$= 4.19\ r^3 = 4.19\times\left(5\times10^{-6}\ \text{cm}\right)^3$$

$$= 5.24\times10^{-16}\ \text{cm}^3$$

Density of each particle $= 1.2$ g/cc

Mass of each particle $= \left(1.2\ \text{g/cm}^3\right)\left(5.24\times10^{-16}\ \text{cm}^3\right)$

$$= 6.29\times10^{-16}\ \text{g}$$

Molecular wt of styrene = 104

Each particle contains

$$6.29 \times 10^{-16}\ \text{g}\left(\frac{1\ g\ mol}{104\ g}\right)\ \text{monomer units}$$

$$= 6.05 \times 10^{-18}\ \text{g mol monomer units}$$

$$= (6.05) \times 10^{-18}\ \text{g-mol monomer units)}$$

$$\left(6.023 \times 10^{23}\ \frac{monomer}{g\text{-}mol\ monomer}\right)$$

$$= 3.64 \times 10^{6}\ \text{monomers}$$

$$\text{Since } \overline{X_n} = 3.5 \times 10^{3} \text{ or } 3.52 \times 10^{3} \text{ monomer per chain}$$

$$\text{Chains per particle} = \left(3.64 \times 10^{6}\ monomers\right)\left(\frac{10^{-3}\ chains}{3.25\ monomer}\right)$$

$$= 103 \times 10^{3}\ \text{chains}$$

---

3. Precipitation Polymerization

Precipitation polymerization, also known as slurry polymerization, involves solution systems in which the monomer is soluble but the polymer is not. It is probably the most important process for the coordination polymerization of olefins. The process involves, essentially, a catalyst preparation step and polymerization at pressures usually less than 50 atm and low temperatures (less than 100°C). The resultant polymer, which is precipitated as fine flocs, forms a slurry consisting of about 20% polymer suspended in the liquid hydrocarbon employed as solvent. The polymer is recovered by stripping off the solvent, washing off the catalyst, and if necessary, extracting any undesirable polymer components. Finally, the polymer is compounded with additives and stabilizers and then granulated.

The suspension of the polymer flocs in the solvent produces a physical system of low viscosity that is easy to stir. However, problems may arise due to settling of the polymer and the formation of deposits on the stirrer and reactor walls. Most industrial transition-metal catalysts are insoluble, and consequently polymerization occurs in a multiphase system and may be controlled by mass transfer. Therefore, the type of catalyst employed exerts a larger influence on parameters and reactor geometry.

4. Interfacial and Solution Polycondensations

Monomers that are very reactive are capable of reacting rapidly at low temperatures to yield polymers that are of higher molecular weight than would be produced in normal bulk polycondensations. The best and most widely used reactants are organic diacid chlorides and compounds containing active hydrogens (Table 10.5):

$$-\overset{\overset{\displaystyle O}{\|}}{C}-Cl\ +\text{-}NH_2\ \xrightarrow{\text{base}}\ -\overset{\overset{\displaystyle O}{\|}}{C}-NH-\ +\ HCl \qquad \text{(Str. 5)}$$

In interfacial polymerization a pair of immiscible liquids is employed, one of which is usually water while the other is a hydrocarbon or chlorinated hydrocarbon such as hexane, xylene, or carbon tetrachloride. The aqueous phase contains the diamine, diol, or other active hydrogen compound and the acid receptor or base (e.g., NaOH). The organic phase, on the other hand, contains the acid chloride. As the name suggests, this type of polymerization occurs interfacially between the two liquids. In contrast to high-temperature polycondensation reactions, these reactions are irreversible because there are no significant reactions between the polymer product and the low-molecular-weight by-product at the low

Table 10.5 Typical Interfacial/Solution Polycondensation Reactions

Active Hydrogen Compound	Acid Halide	Product	
$-NH_2$	$-\overset{O}{\overset{\|}{C}}-Cl$	$-\overset{O}{\overset{\|}{C}}-\overset{H}{\overset{\|}{N}}-$	Polyamide
$-NH_2$	$-\overset{O}{\overset{\|}{C}}-Cl$	$-\overset{H}{\overset{\|}{N}}-\overset{O}{\overset{\|}{C}}-\overset{H}{\overset{\|}{N}}-$	Polyurea
$-NH_2$	$Cl-\overset{O}{\overset{\|}{C}}-O-$	$-\overset{H}{\overset{\|}{N}}-\overset{O}{\overset{\|}{C}}-O-$	Polyurethane
$-OH$	$Cl-\overset{O}{\overset{\|}{C}}-$	$-O-\overset{O}{\overset{\|}{C}}-$	Polyester
$-OH$	$Cl-\overset{O}{\overset{\|}{C}}-Cl$	$-O-\overset{O}{\overset{\|}{C}}-O-$	Polycarbonate

From Williams, D.J., *Polymer Science and Engineering,* Prentice-Hall, Englewood Cliffs, NJ, 1971. With permission.

temperatures employed. Consequently, the molecular weight distribution is a function of the kinetics of the polymerization system; it is not determined statistically as in normal equilibrium polycondensations. The rate of reaction is controlled by the rate of monomer diffusion to the interface. This obviates the necessity to start the reaction with stoichiometric quantities of reactants. Since the reactions are irreversible, high conversions are not necessarily required to obtain high-molecular-weight polymers.

In solution polycondensation, all the reactants are dissolved in a simple, inert solvent. However, for some solution polymerizations, the solvent can facilitate the reaction. For example, a tertiary amine such as pyridine is an acid acceptor in the solution phosgenation in polycarbonate manufacture.

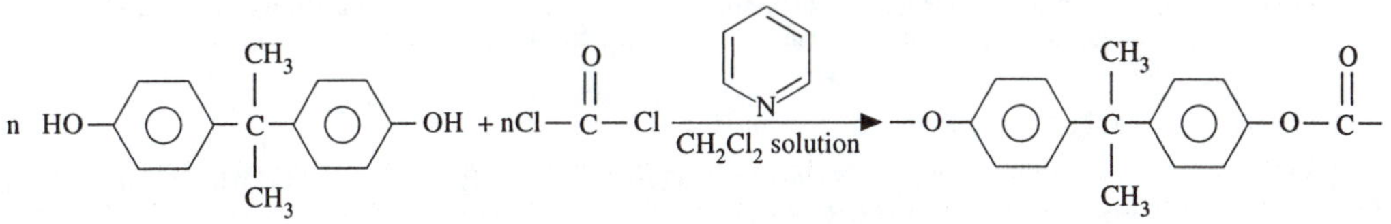

$$n\ HO-C_6H_4-\underset{CH_3}{\overset{CH_3}{\underset{|}{\overset{|}{C}}}}-C_6H_4-OH + nCl-\overset{O}{\overset{\|}{C}}-Cl \xrightarrow[CH_2Cl_2\ \text{solution}]{\text{pyridine}} -O-C_6H_4-\underset{CH_3}{\overset{CH_3}{\underset{|}{\overset{|}{C}}}}-C_6H_4-O-\overset{O}{\overset{\|}{C}}- + 2nHCl \qquad \text{(Str. 6)}$$

Interfacial and solution polycondensations are commercially important. For example, an unstirred interfacial polycondensation reaction is utilized in the production of polyamide fibers. Another important application of interfacial polycondensation is the enhancement of shrink resistance of wool. The wool is immersed first in a solution containing one of the reactants and subsequently in another solution containing the other reactant. The polymer resulting from the interfacial reaction coats the wool and improves its surface properties.

III. POLYMERIZATION REACTORS

The course of a polymerization reaction and hence the properties of the resultant polymer are determined by the nature of the polymerization reaction and the characteristics of the reactor employed in the

Table 10.6 Characteristics, Advantages and Disadvantages of Various Polymerization Types

Polymerization Process	Characteristics	Advantages	Disadvantages
Bulk	Reaction mixture consists essentially of monomer; and initiator in the case of chain reaction polymerization Monomer acts as solvent for polymers	Products relatively pure due to minimum contamination Enhanced yield per reactor volume	Exothermic nature of polymerization reactions (particularly chain reaction polymerizations) makes temperature control of system difficult Product has broad molecular weight distribution Removals of tracers of unreacted monomer difficult
Solution	Solvent miscible with monomer, dissolves polymer	Heat transfer efficiency greatly enhanced resulting in better process control Resulting polymer solution may be directly usable	Necessary to select an inert solvent to avoid possible transfer to solvent Lower yield per rector volume Reduction of reaction rate and average chain length Not particularly suitable for production of dry or relatively pure polymer due to difficulty of complete solvent removal
Suspension	Monomer and polymer insoluble in water, initiator soluble in monomer	Heat removal and temperature control relatively easier Polymer obtained in a form that is convenient and easily handled Resulting polymer suspension or granules may be directly usable	Need to maintain stability of droplets requires continuous and a minimum level of agitation Possibility of polymer contamination by absorption of stabilizer on particle surface Continuous operation of system difficult
Emulsion	Monomer and polymer insoluble in water, initiator soluble in water Emulsifier needed for stabilization of system component particularly at initial stages of polymerization	Physical state of the system enhances heat transfer efficiency Possible to obtain high rates of polymerization and high average chain lengths Narrow molecular weight distribution Latex (emulsion) often directly usable	Difficult to get pure polymer due to contamination from other components of polymerization system Difficult and expensive if solid polymer product is required Presence of water lowers yield per reactor volume
Precipitation	Polymer insoluble in monomer or monomer miscible with precipitant for polymer	Physical state of system permits easy agitation Relatively low temperatures employed	Separation of product difficult and expensive Catalyst systems are special and need careful preparation Molecular weight distribution depends on type of catalyst
Interfacial	Polymerization occurs at interface of two immiscible solvents, usually water and an organic solvent	Polymerization is rapid and occurs at low temperatures High conversions are not necessarily required to obtain high molecular weight Unnecessary to start with stoichiometric quantifiers of reactants	Limited to highly reactive systems Need appropriate choice of solvent to dissolve reactants

reaction. In other words, the reactor is essentially the heart of any polymerization process. The reactor affects the conversion of the monomer to the polymer. The reactor also effectively establishes the ultimate properties of the polymer such as polymer structure, molecular weight, molecular weight distribution, and copolymer composition. To perform its functions satisfactorily, the reactor must remove the heat of polymerization, provide the necessary residence time, provide good temperature control and reactant homogeneity, control the degree of back-mixing in a continuous polymerization, and provide surface exposure. In addition, the reactor must be applicable to mass production and economical to operate.

Therefore, the control of polymer properties requires a careful selection of a reactor appropriate for the particular polymerization process.

We dealt with various polymerization processes in the previous sections. We now consider polymerization reactors. Our treatment of this subject is essentially qualitative; the principal focus is to highlight salient features of each reactor type.

Reactors may be divided into three simple, idealized model categories: batch reactor, tubular or plug flow reactor, and the continuous stirred tank reactor (CSTR).

A. BATCH REACTORS

In the case of the batch reactor, the reactants are charged into the reactor and mixed properly for the duration of the reaction and then the product is discharged. The batch reactor has essentially the following characteristics:

- It is simple and does not need extensive supporting equipment.
- It is ideal for small-scale operations.
- The operation is an unsteady-state operation, with composition varying with time.

Now let us discuss how these features affect the various polymerization reactions and the resultant polymer. We start by considering the general material balance equation for the batch reactor:

$$\begin{array}{c}\text{Rate of}\\ \text{monomer flow}\\ \text{into the reactor}\end{array} = \begin{array}{c}\text{Rate of}\\ \text{monomer flow}\\ \text{out of the reactor}\end{array} + \begin{array}{c}\text{Rate of monomer loss}\\ \text{due to reaction}\\ \text{within the reactor}\end{array} + \begin{array}{c}\text{Rate of monomer}\\ \text{accumulation}\\ \text{within the reactor}\end{array} \quad (10.6)$$

For a batch reactor, the first two terms of Equation 10.6 are equal to zero since by definition nothing flows in or out of the reactor. Consequently, the equation reduces to

$$\begin{array}{c}\text{Rate of monomer loss}\\ \text{due to reaction}\\ \text{within the reactor}\end{array} = \begin{array}{c}\text{Rate of accumulation}\\ \text{monomer}\\ \text{within the reactor}\end{array} \quad (10.7)$$

$$-\frac{dM}{dt} = R_p \quad (10.8)$$

$$\int_o^t dt = -\int_{[M_o]}^{[M]} \frac{dM}{R_p} \quad (10.9)$$

For free radical polymerization,

$$R_p = k_p \left(\frac{f k_d}{k_t}\right)^{1/2} [I]^{1/2}[M].$$

If f is independent of monomer concentration and the initiator concentration remains constant, then the above operation is first order in monomer concentration and may be rewritten:

$$R_p = k[M] \quad (10.10)$$

where $k = k_p \left(\frac{f k_d}{k_t}\right)^{1/2} [I]^{1/2}$. On substitution of Equation 10.10, Equation 10.9 becomes

$$\int_o^t dt = -\int_{[M]_o}^{[M]} \frac{dM}{kM} \quad (10.11)$$

Integrating this equation yields

$$\ell n\frac{[M]}{[M_o]} = -kt \tag{10.12}$$

or

$$[M] = [M]_o e^{-kt} \tag{10.13}$$

$$\text{Percent Conversion} = 100\,\frac{[M]_o - [M]}{[M]_o} = 100\left(1 - e^{-kt}\right) \tag{10.14}$$

The derivation of Equation 10.13 assumes that the composition of the reaction mixture is uniform throughout the reactor at any instant of time and the initiator concentration is constant. Good mixing is an important way to ensure not only uniform concentration but also uniform temperature and prevents the occurrence of localized inhomogeneities. However, the mixing efficiency for bulk polymerizations in batch reactors, while difficult to generalize, usually varies with conversion because the viscosity and density of the reaction mass changes continuously with reaction time. For example, the viscosity of styrene at 150°C increases by an order of three at 60% conversion. The corresponding changes in emulsion and suspension polymerizations are less drastic. An important assumption made in the deviation of Equation 10.10 is that the initiator concentration remains constant. This assumption may not be realistic under industrial conditions beyond a few percentage points of conversion. A more accurate expression is obtained by assuming a first-order decay of the initiator:

$$[I] = [I_o]e^{-k_d t} \tag{10.15}$$

Example 10.7: Calculate the time required for 10% polymerization of pure styrene at 60°C with benzoyl peroxide as the initiator in a batch reactor. Assume that the initiator concentration remains constant. Data:

$f = 1$
$k_p^2/k_t = 0.95 \times 10^{-3}$ l/mol-s
$[I] = 4.0 \times 10^{-3}$ mol/l
$k_d = 1.92 \times 10^{-6}\ s^{-1}$

Solution: Since the initiator remains constant, Equation 10.12 is applicable:

$$\ln\frac{[M]}{[M]_o} = -kt$$

$$k = k_p\left(\frac{f k_d}{k_t}\right)^{1/2}[I]^{1/2}$$

or

$$k^2 = \frac{k_p^2}{k_t}(f k_d)[I]$$

$$= \left(0.95\times10^{-3}\,\frac{l}{mol{-}s}\right)\left(1.92\times10^{-6}\,\frac{1}{s}\right)\left[4.0\times10^{-3}\,\frac{mol}{l}\right]$$

$$= 7.30\times10^{-12}\ s^{-2}$$

$$k = 2.70 \times 10^{-6}\ s$$

$$\ln\frac{[M]}{[M_o]} = -2.70 \times 10^{-6}\, t$$

$$\frac{[M_o]-[M]}{[M_o]} = 0.10, \text{ i.e., } \frac{[M]}{[M]_o} = 0.9$$

$$\ln 0.9 = -2.70 \times 10^{-6}\ t$$

$$t = 8.10\ h$$

In chain growth reactions, temperature control is a major problem in bulk polymerization and, to a lesser extent, in solution polymerization in batch reactors. This is due to the large increase in viscosity of the reaction medium with conversion. The heat transfer to a jacket in a vessel varies approximately inversely with the one-third power of the viscosity. For example, in a stirred batch (tank) reactor with thick walls and unfavorable surface-to-volume ratio, polymerization essentially proceeds adiabatically. A variety of methods are employed for heat removal, including heat transfer to a jacket, the internal cooling loop, and, in the case of a vaporizable constituent, an overhead condenser through reflux. While reactors are almost always jacketed, the use of additional heat removal devices is generally necessary as the size of the reactor increases since the heat transfer area of the reactor increases with reactor volume to the two-thirds power, while the rate of heat generation varies directly with the reactor volume. As we said earlier, the use of external cooling devices, of course, depends on the polymerization. Where the viscosity is relatively low and/or the latex stable, a portion of the reaction mixture is recycled through a heat exchanger. This is impossible in suspension polymerization where a continuous and minimum level of agitation is required to ensure the stability of the particles and avoid the formation of coagulate and wall deposits in the dead volumes. Also, the high viscosity in bulk polymerization precludes the use of external heat exchangers because the poor agitation would lead to wall deposits and a rapid loss of cooling efficiency. The use of internal cooling coils is restricted to low viscosity reactions to avoid poor product quality resulting from improper mixing. The idealized model batch reactor together with its residence time is shown in Figure 10.8a.

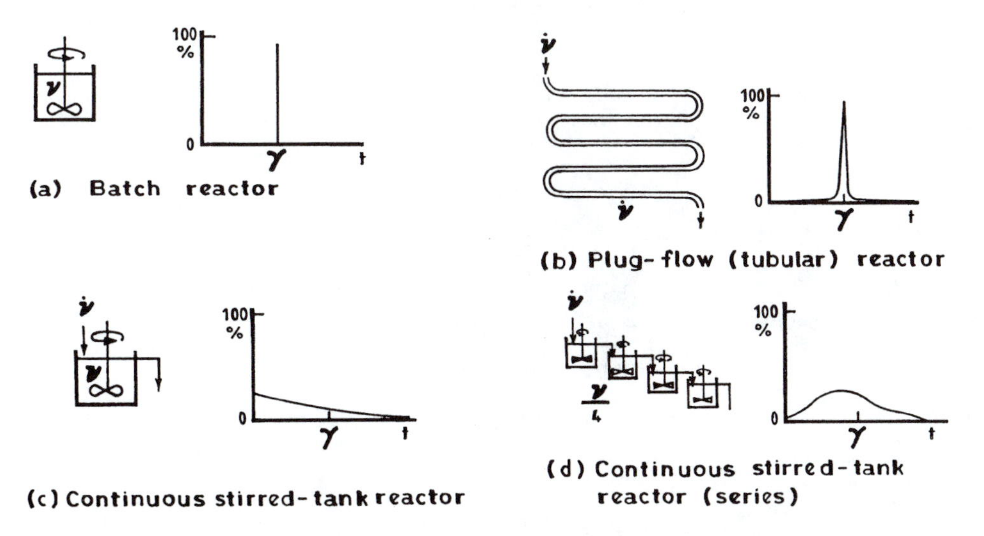

Figure 10.8 Idealized reactors and their associated residence time distribution. (From Gerrens, H., *Chem. Technol.*, p. 380, June 1982.)

B. TUBULAR (PLUG FLOW) REACTOR

The tubular or plug flow reactor, as the name suggests, is essentially a tube or unstirred vessel with very high length/diameter (l/d) ratio. Fluid flow in the ideal plug flow reactor is orderly, with no element of fluid overtaking any other element. The description of a tubular reactor as plug flow connotes that each element moves through the tube as a plug with no axial diffusion along the flow path and no difference in the velocity of any two elements flowing through the reactor (i.e., no lateral or back-mixing). For the ideal plug flow reactor, therefore, the residence time for all elements flowing through the tube is the same.

In principle, the tubular reactor has a favorable surface-to-volume ratio and relatively thin walls — features that should enhance heat-transfer efficiency. In addition, like the batch reactor, it is suited for achieving high conversions. However, tubular reactors have limited application in polymer production. The high viscosities characteristic of polymerization reaction media present temperature control problems. In addition, the material temperature increases from the tube wall to its radius. This broad distribution of temperature leads to a broad molecular weight distribution — a situation that is accentuated in chain-growth polymerizations by the decrease in initiator and monomer concentrations with increasing conversions. Since the reactor walls are cooler that the center of the reactor, there is a tendency to form a slow-moving polymer layer on the walls. This reduces the production capacity and aggravates the heat-transfer problems. A typical idealized-model tubular reactor together with the resistance time distribution and concentration profile for a first-order reaction is shown in Figure 10.8b.

Example 10.8: The production of high-pressure low-density polyethylene is carried out in tubular reactors of typical dimensions 2.5 cm diameter and 1 km long at 250°C and 2500 atm. The conversion per pass is 30% and the flow rate is 40,000 kg/h. Assuming that the polymerization reaction is first order in ethylene concentration, estimate the value of the polymerization rate constant.

Solution: Consider a mass balance about an elemental volume of the reactor.

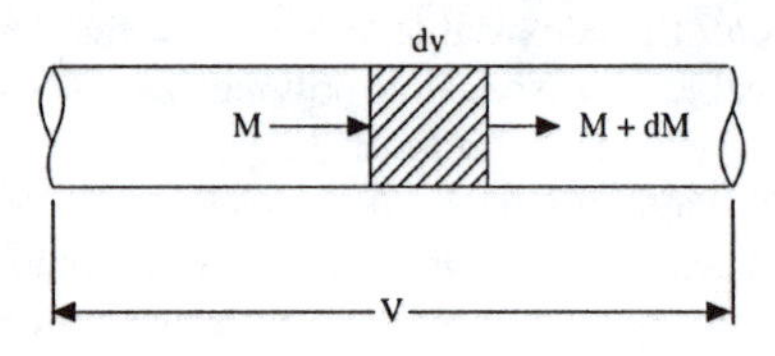

On substitution:

$$vM_o dp = R_p\, dV$$

$$\int_o^V \frac{dV}{vM_o} = \int_o^p \frac{dp}{R_p}$$

Now,

$$p = \frac{M_o - M}{M_o}$$

$$dp = -\frac{dM}{M_o}$$

$$\int_o^V \frac{dV}{vM_o} = \int_{M_o}^M -\frac{dM}{M_o R_p}$$

$$\frac{V}{v} = \int_{M_o}^M -\frac{dM}{R_p}$$

$$R_p = k[M]$$

$$\frac{V}{v} = -\int_{M_o}^M \frac{dM}{kM}$$

$$k\frac{V}{v} = -\int_{M_o}^M \frac{dM}{M} \text{ or } k\tau = -\int_{M_o}^M \frac{dM}{M}$$

where $\tau = \frac{V}{v}$ = mean residence time.

$$\ln\frac{M}{M_o} = -k\tau$$

or

$$M = M_o\, e^{-k\tau}$$

$$k = -\frac{1}{\tau}\ln\frac{M}{M_o}$$

For a conversion of 30%

$$\frac{M_o - M}{M_o} = 0.3$$

or

$$\frac{M}{M_o} = 0.7$$

$$k = \frac{1}{\tau}\ln 0.7$$

$$V = \pi R^2 L = \pi\left(\frac{2.5}{2}\times 10^{-2}\right)^2 \times 1.0 = \pi(1.25)^2 \times 10^{-4}$$

$$= 4.9 \times 10^{-4}\ \mathrm{m}^3$$

Assume, for simplicity, under the conditions of polymerization ethylene behaves as an ideal gas

$$pv = nRT$$

or

$$v = \frac{nRT}{P}$$

$$n = 40{,}000\ \mathrm{kg/h} = \frac{40{,}000}{28}\ kg\ mol/h$$

$$= 1.43 \times 10^3\ \mathrm{kg\ mol/h}$$

$$R = 0.08206\ (\mathrm{l})(\mathrm{atm})/(\mathrm{g\ mol})\,(\mathrm{K})$$

$$= 0.08206\left(1 \times 10^{-3}\frac{m^3}{1}\right)(\mathrm{atm})/\left(\mathrm{kg\ mol} \times 10^3\right)(\mathrm{K})$$

$$= 0.08206\left(\mathrm{m}^3\right)(\mathrm{atm})/(\mathrm{kg\ mol})(\mathrm{K})$$

$$T = 273 + 250\mathrm{K} = 523\mathrm{K}$$

$$v = \left(1.43 \times 10^3\frac{kg\ mol}{h}\right)\left(0.08206\frac{(m^3)(atm)}{(kg\ mol)(K)}\right)(523\ K)\left(\frac{1}{2500\ atm}\right)$$

$$= 24.55\ \mathrm{m}^3/\mathrm{h}$$

Assuming the flow rate is constant along the entire length of the reactor, then

$$\tau = \frac{V}{v} = \frac{4.91 \times 10^{-4}\, m^3}{24.55\, m^3/h} = 2.0 \times 10^{-5}\, h$$

$$k = -\frac{1}{\tau} \ln 0.7 = \frac{1}{\tau} 2.0 \times 10^{-5}\, h \ln 0.7$$

$$k = 1.78 \times 10^4 \text{ h}^{-1}$$

$$= 4.95 \text{ s}^{-1}$$

C. CONTINUOUS STIRRED TANK REACTOR (CSTR)

The ideal continuous stirred tank reactor is a reactor with well-stirred and back-mixed contents. As a result, instant blending of the feed with the reactor contents is assumed to occur. The composition of the contents of the reactor is uniform throughout the reactor. Consequently, the exit stream from the reactor has the same composition and temperature as the reactor contents.

During polymerization with a CSTR, the monomer and the other components of the polymerization recipe are fed continuously into the reactor while the polymerization product mixture is continually withdrawn from the reactor. The application of the CSTR in suitable polymerization processes reduces, to some extent, the heat removal problems encountered in batch and tubular reactors due to the cooling effect from the addition of cold feed and the removal of the heat of reaction with the effluent. Even though the supporting equipment requirements may be relatively substantial, continuous stirred tank reactors are economically attractive for industrial production and consistent product quality.

In a CSTR, each element of monomer feed has an equal chance of being withdrawn from the reactor at any instant regardless of the time it has been in the reactor. Therefore, in a CSTR, unlike in batch and tubular reactors, the residence time is variable. The contents of a well-stirred tank reactor show an exponential distribution of residence times of the type shown in Equation 10.15.

$$R(t) = e^{-t/\tau} \tag{10.16}$$

where R(t) is the residence time distribution and t is the mean residence time, which, as we said previously, is the ratio of the reactor volume to the volumetric flow rate. The residence time distribution influences the degree of mixing of the reaction mixture. This, in turn, determines the uniformity of the composition and temperature of the reactor contents. As a consequence, the polymer product properties are influenced by the residence time distribution. Figure 10.8c shows an ideal CSTR with its residence time distribution and concentration profile. Tables 10.7 and 10.8 summarize polymerization reactions and processes, industrially employed reactors, and resulting polymer products.

Table 10.7 Polymerization Processes and Industrially Employed Reactors

Polymerization Reaction	Polymerization Process	Reactor Batch	Plug Flow	CSTR
Chain	Bulk	X		X
	Solution	X	X	X
	Suspension	X		
	Emulsion	X		X
	Precipitation	X		X
Ionic	Solution	X		
	Precipitation	X	X	X
Step growth	Solution	X	X	
	Interfacial	X		

Table 10.8 Polymerization Reactions and Processes and Some Polymer Products

Reactor	Polymerization Reaction	Polymerization Process	Example
Batch	Free radical	Solution	Paint resins by polymerization and often also copolymerization of acrylates and methacrylates
			Vinlylacetate in methanol with subsequent alcoholysis to poly(vinyl alcohol)
			Polyacrylonitrile spinning solution in dimethylformamide
		Precipitation	Acrylonitrile in water
			Azeotropic copolymerization of styrene and acrylonitrile in methanol
		Suspension (bead)	Poly(vinylchloride)
			Expanded polystyrene
			Poly(methyl methacrylate)
			Poly(vinyl acetate)
		Emulsion	Water soluble paints
			Poly(vinyl acetate)
	Ionic	Solution	Butadiene
			Isoprene
			Ethylene or propylene and their copolymers
			Block copolymers such as butadiene-styrene or styrene-butadiene-styrene
			Polyether polyols block copolymers prepared from reaction of ethylene oxide or propylene oxide with polyhydric alcohols used for polyurethane preparation
			Polymerization of ε-caprolactam
		Precipitation	Polyethylene-propylene copolymer
	Condensation	Solution	Formaldehyde resin (UF, MF, PF)
			Nylon 6,6
			Polyurethanes
		Interfacial	Heat resistant aromatic polyamides and polyimides
Plug Flow	Free radical	Solution	High pressure polymerization of ethylene to give LDPE
	Ionic	Precipitation	Polymerization of isobutene in liquid ethylene with BF_3 catalyst
			Polymerization of trioxane with BF_3 etherate as catalyst to produce polyoxymethylene
	Condensation	Solution	Continuous manufacture of polyurethene foam blocks
			Production of nylon 6,6 on the extruder reactor as the last stage
CSTR	Free radical	Solution	Vinyl acetate esters of acrylic acid acrylonitrile
		Precipitation	Acrylonitrile
		Emulsions	Polymerization of vinyl chloride
			Polyacrylates and polymethacrylates
			Butadiene, isopreal and then copolymers
		Precipitation	Cationic copolymerization of isobutene-isoprene with slurry $AlCl_3$ as the initiator and methyl chloride as diluent
			$AlCl_3$ slurry polymerization of propylene in the presence of transition methyl catalyst and excess monomer as a diluent
			Fluidized bed reactor is used in the gas phase polymerization; a powdered polymer is produced in a gaseous monomer-low pressure polymerization of ethylene (HDPE) and propylene

Example 10.9: Explain the molecular weight distribution shown in the following table.

Reactor	Polymerization Reaction	
	Chain-Growth	Step-Reaction
Batch or plug flow	Wider than Schulz-Flory distribution	Schulz-Flory distribution
Continuous stirred tank reactor	Schulz-Flory distribution	Much wider Schulz-Flory distribution

Solution: In radical chain reactions, the overall rate of polymerization, R_p, and the number-average degree of polymerization, $\overline{X}_n$, are functions of the initiator concentration [I], the monomer concentration [M], and also the temperature via the temperature dependence of the individual rate constants. At constant [M] and [I], the Schulz–Flory MWD is produced. However, if [M] and [I] vary with time, a number of Schulz–Flory distributions overlap and thus a broader MWD is produced. In the ideal CSTR [M] and [I] are constant and the temperature is relatively uniform. Consequently, chain polymerizations in CSTR produce the narrowest possible MWD. In the batch reactor, [M] and [I] vary with time (decrease with conversion) while in the tubular reactor [M] and [I] vary with position in the reactor and the temperature increases with tube radius. These variations cause a shift in $\overline{X}_n$ with conversion and consequently a broadening of MWD.

The molecular weight distribution also depends on the relative lifetime of the growing macromolecule and the mean residence time. In step-growth polymerization, even though monomers disappear very early in the reaction, high conversions are required to generate high $\overline{X}_n$. In general, therefore, the lifetime of a growing macromolecule is much higher than the residence time in step-growth polymerizations. For the CSTR, the composition of the product is not altered by increasing the duration of the polymerization process. In addition, because the residence time in CSTR is variable, there is an equal chance of finding both small and large molecules in the polymerization product. Consequently, there is a relatively broad molecular weight distribution. On the other hand, for step-growth polymerization in batch and tubular reactors, narrow MWD with increasing $\overline{X}_n$ is generated with increasing conversion.

IV. PROBLEMS

10.1. Polyesters and polyamides are prepared according to the following equilibrium reactions.

$$\text{-OH} + \text{-COOH} \overset{K_e}{\rightleftharpoons} -\overset{\overset{\displaystyle O}{\|}}{C}-O- + H_2O \qquad \text{(Str. 7)}$$

$$\text{-NH}_2 + \text{-COOH} \overset{K_a}{\rightleftharpoons} -\overset{\overset{\displaystyle H}{|}}{N}-\overset{\overset{\displaystyle O}{\|}}{C}- + H_2O \qquad \text{(Str. 8)}$$

K_e and K_a are 10 and 400, respectively.

In the manufacture of linear polyesters, the final stages of polymerization are carried out at pressures of about 1 mm Hg and temperatures of 280°C. On the other hand, the reaction for the manufacture of nylon 6,6 is completed at 240 to 280°C and atmospheric pressure. Explain.

10.2. Repeat Example 10.7 in this chapter but assume that the initiator concentrations show a first-order decay.

10.3. Calculate the time required for 80% conversion of pure styrene at 60°C with benzoyl peroxide as the initiator in a batch reactor. Assume that the initiator concentration remains constant. Use the data given in Example 10.7 in this chapter and comment on your result.

10.4. It has been proposed that to avoid possible health problems, the polystyrene used for drinking cups must contain less than 1% monomer. Polystyrene was prepared at 100°C by the thermal polymerization of 10 *M* styrene in toluene in a batch reactor. The shift operator stopped the reaction after 2.5 h. Is the product from this operation suitable for producing drinking cups without further purification?

Data:

$$\frac{k_p^2}{k_t} = 8.5 \times 10^{-3}\ l/mol\text{-s}$$

$$k_i = 4.2 \times 10^{-11}\ l/mol\text{–s}$$

10.5. Estimate the adibatic increase in temperature for the batch polymerization of butadiene.

Data:

$\Delta Hp = -18.2$ kal/g mol

Heat capacity = 29.5 cal/g mol °C

10.6. For the bulk polymerization of methyl methacrylate at 77°C with azo-*bis*-isobutyronitrile as the initiator, the initial rate of polymerization was 1.94×10^{-4} mol/l-s. The concentrations of the monomer and initiator were 9 mol/l and 2.35×10^{-4} mol/l, respectively. To reduce the heat transfer problems, the polymerization was repeated in a solution of benzene by the addition of 4.5 l of benzene with an initiator concentration of 2.11×10^{-4} l/mol. Assuming that the rate constants are the same for both the bulk and solution polymerizations, calculate the rate of solution polymerization of methyl methacrylate. If there is no transfer to the solvent and the initial rate of initiation is 4×10^{-4} mol/l-s, what is the ratio of the number-average degree of polymerization for bulk-to-solution polymerization? Comment on your results.

10.7. It has been found that the polymer from a certain toxic monomer has excellent mechanical properties as a food wrap. To ensure maximum conversion and obtain a relatively pure product, the monomer was bulk polymerized in a batch reactor. Do you agree with the decision to use this polymer as a food wrap?

10.8. A typical emulsion polymerization recipe consists of 180 g water, 100 g monomer, 5 g soap, and 0.5 g potassium persulfate. Estimate the ratio of the total surface area of micelles to that of monomer droplets. Assume that the relative volume of micelle to a droplet is equal to the ratio of their volumes in the polymerization recipe. The density of a micelle is 0.2 g/cc and that of a droplet is 0.8 g/cc.

10.9. The following data were found for the emulsion polymerization of vinyl acetate at 60°C:

$N = 12.04 \times 10^{14}$ particles/ml

$R_i = 1.1 \times 10^{12}$ radicals/ml-s

$k_p = 550$ l/mol-s

$[M] = 5\ M$

Polymer particle density = 1.25 g/cm^3

a. Assuming that the Smith–Ewart kinetics is valid for emulsion polymerization up to 80% conversion, how much time was required to obtain this conversion?
b. What is the average size of the latex polymer particles if, on the average, each particle contains 133 chains?

10.10. A batch reactor of length 2.5 m and diameter 5.046 m is filled with vinyl acetate monomer. Polymerization is carried out isothermally at 50°C. The reactor is jacketed for heat removal. What is the temperature of the coolant?

Data:

Rate of polymerization = 10^{-4} mol/l-s

$\Delta Hp = 21.3$ kcal/mol

Overall heat-transfer coefficient = 0.0135 cal/cm^2 s °K

10.11. In an emulsion polymerization of styrene in a 50 m^3 CSTR, the feed contains 5.845 mol of styrene per liter. For an average conversion of 75% in the reactor and assuming the Smith–Ewart behavior is followed, estimate:

a. The mean residence time
b. The volumetric flow rate

Data:

k_p = 2200 l/mol-s
N = 1.80×10^{14} particles/cm^3

10.12. The bulk polymerization of a polymer in a plug flow reactor had been found to follow first-order kinetics. The flow rate is 3.14×10^{-4} m^3 s^{-1}, while the reaction rate constant and residence time are 10^{-2} s^{-1} and 10^2 s, respectively. Calculate:

a. The conversion
b. The length of the reactor if its diameter is 2×10^{-2} m

10.13. The bulk polymerization of the poly(ethylene terephthalate) or polyisobutylene is to be carried out. For which of these polymerizations would it be more favorable to use a plug flow reactor?

10.14. Explain why a continuous stirred tank reactor (CSTR) is not normally used in industry in the production of nylon 6,6. What modification is needed to make (CSTR) suitable for the preparation of nylon 6,6?

10.15. The specific heats of polymerization of styrene and butene monomers are 160 and 300 kcal/kg, respectively. For which of these monomers would solution polymerization be more appropriate?

10.16. In certain *cis*-butadiene processes, the volume of the diluting solvent is sufficiently high that cooling is achieved by feeding cold diluent solvent to the reactor. For an end-use application, the final polymer concentration was 10%. The specific heat of the solvent is 0.96 Btu/lb°F. If the final reaction temperature is 122°F, at what temperature should the solvent be fed into the reactor so as to remove all of the heat of polymerization?

REFERENCES

1. Winding, C.C. and Hiatt, G.D., *Polymeric Materials,* McGraw-Hill, New York, 1961.
2. Roe, C.P., *Ind. Eng. Chem.*, 69(9), 20, 1968.
3. Smith, W.V. and Ewart, R.H., *J. Chem. Phys.*, 16, 592, 1948.
4. Harkins, H.B., *J. Am. Chem. Soc.*, 69, 1428, 1947.
5. Williams, D.J., *Polymer Science and Engineering,* Prentice-Hall, Englewood Cliffs, NJ, 1971.
6. Gerrens, H., *Chem. Technol.*, p. 380, June 1982.
7. Gerrens, H., *Chem. Technol.*, p. 434, July 1982.
8. Rudin, A., *The Elements of Polymer Science and Engineering,* Academic Press, New York, 1982.
9. Flory, P.J., *Principles of Polymer Chemistry,* Cornell University Press, Ithaca, NY, 1953.
10. Levenspiel, O., *Chemical Research Engineer,* John Wiley & Sons, New York, 1962.
11. Cameron, J.B., Lundeen, A.J., and McCulley, J.H., Jr., *Hydrocarbon Process.*, B9, 50, 1980.
12. Diedrich, B., *Appl. Polym. Symp.*, 26, 1, 1975.
13. Wallis, J.P.A., Ritter, R.A., and Andre, H., *AIChE J.*, 21, 691, 1975.
14. Wohl, M.H., *Chem. Eng.*, p. 60, August, 1966.
15. Back, A.L., *Chem. Eng.*, p. 65, August 1, 1966.
16. Church, J.M., *Chem. Eng.*, p. 79, August 1, 1966.
17. Gellner, O., *Chem. Eng.*, p. 74, August 1, 1966.
18. Schlegel, W.F., *Chem. Eng.*, p. 88, March 20, 1972.

Chapter 11

Unit Operations in Polymer Processing

I. INTRODUCTION

Polymer processing may be divided into two broad areas. The first is the processing of the polymer into some form such as pellets or powder. The second type describes the process of converting polymeric materials into useful articles of desired shapes. Our discussion here is restricted to the second method of polymer processing. The choice of a polymer material for a particular application is often difficult given the large number of polymer families and even larger number of individual polymers within each family. However, with a more accurate and complete specification of end-use requirements and material properties the choice becomes relatively easier. The problem is then generally reduced to the selection of a material with all the essential properties in addition to desirable properties and low unit cost. But then there is usually more than one processing technique for producing a desired item from polymeric materials or, indeed, a given polymer. For example, hollow plastic articles like bottles or toys can be fabricated from a number of materials by blow molding, thermoforming, and rotational molding. The choice of a particular processing technique is determined by part design, choice of material, production requirements, and, ultimately, cost–performance considerations.

The number of polymer processing techniques increases with each passing year as newer methods are invented and older ones modified. This chapter is limited to the most common polymer processing unit operations, but only extrusion and injection molding, the two predominant polymer processing methods, are treated in fairly great detail. Our discussion is restricted to general process descriptions only, with emphasis on the relation between process operating conditions and final product quality. Table 11.1 summarizes some polymer processing operations, their characteristics, and typical applications.

II. EXTRUSION[1]

Extrusion is a processing technique for converting thermoplastic materials in powdered or granular form into a continuous uniform melt, which is shaped into items of uniform cross-sectional area by forcing it through a die. As shown in Table 11.1, extrusion end products include pipes for water, gas, drains, and vents; tubing for garden hose, automobiles, control cable housings, soda straws; profiles for construction, automobile, and appliance industries; film for packaging; insulated wire for homes, automobiles, appliances, telephones and electric power distribution; filaments for brush bristles, rope and twine, fishing line, tennis rackets; parisons for blow molding. Extrusion is perhaps the most important plastics processing method today.

A simplified sketch of the extrusion line is shown in Figure 11.1. It consists of an extruder into which is poured the polymer as granules or pellets and where it is melted and pumped through the die of desired shape. The molten polymer then enters a sizing and cooling trough or rolls where the correct size and shape are developed. From the trough, the product enters the motor-driven, rubber-covered rolls (puller), which essentially pull the molten resin from the die through the sizer into the cutter or coiler where final product handling takes place.

A. THE EXTRUDER

Figure 11.2 is a schematic representation of the various parts of an extruder. It consists essentially of the barrel, which runs from the hopper (through which the polymer is fed into the barrel at the rear) to the die at the front end of the extruder. The screw, which is the moving part of the extruder is designed to pick up, mix, compress, and move the polymer as it changes from solid granules to a viscous melt. The screw turns in the barrel with power supplied by a motor operating through a gear reducer.

The heart of the extruder is the rotating screw (Figure 11.3). The thread of an extruder screw is called a flight, and the axial distance from the edge of one flight to the corresponding edge on the next flight is called the pitch. The pitch is a measure of the coarseness of the thread and is related to the helix

Table 11.1 Some Unit Operations in Polymer Processing and Typical Products

Process	Characteristics	Resin Employed	Typical Products
Extrusion	A process for making indeterminate length of thermoplastics with constant cross-section	Most thermoplastics including PE, PP, PVC, ABS, PS	Pipes and tubing used for soda straws, garden hose, drains and vents, control cable housings, gas and water pipes; profiles, e.g., auto trim, home siding, storm windows; sheets used for window glazing, refrigerator liners, signs, plates, lighting; films for bags, coverings, laminates, packaging; fibers or filaments for brush bristles, rope, fishing line; insulated wire for homes, automobiles, appliances, telephone and electric power distribution; coated paper for milk cartons, meat packaging
Injection molding	Versatile process, most suitable for high speed, low cost molding of intricate plastic parts required in high volume	Virtually all thermoplastics and some thermoset; most common are commodity plastics such as PVC, PE, PP, and PS; others include ABS, nylon, cellulosics, acrylics	Automobile parts, appliance housings, camera cases, knobs, gears, grilles, fan blades, bowls, spoons, lenses, flowers, wastebaskets and garbage cans
Blow molding	Process used for making bottles and other hollow plastic parts having relatively thin walls	Several thermoplastics with PE (particularly ND, PE) having the largest volume; others include PVC, PP, PS, ABS, acrylics, nylons, acrylonitrile, acetates, and PC	Bottles, watering cans, hollow toys, gas tanks for automobiles and trucks
Rotational molding	Economical process for the production of hollow seamless parts with heavy walls and/or complex shapes	PE, (highest volume) PP, PVC together account for almost all plastics used; others include a number of engineering thermoplastics, including ABS, acetal copolymers, nylon (6 and 11), polycarbonate	Hollow balls, squeeze toys, storage and feed tanks, automobile dashboards, door liners and gearshift covers, industrial storage tanks and shipping containers, whirlpool tubs, recreational boats, canoes and camper tops, hobby horses, heater ducts, auto armrest skins, athletic balls, portable toilets
Thermoforming	Process for forming moderately complex shapes that are not readily amenable to injection molding	Almost all thermoplastics but most commonly used include ABS, PP, PS, PVC polyesters; others include acrylics, polycarbonate, cellulosic, nitrile resins	Automobile, airline and mass transportation industries for such uses as auto headliners, fender walls, overhead panels, aircraft canopies; construction industry for exterior and interior paneling, bathtubs, shower stalls; outdoor signs; appliances, e.g., refrigerator liners, freezer panels; packaging trays for meat packing, egg cartons, fast-food disposables and carryouts, blister packages, suitcases, tote boxes, cups, and containers
Compression and transfer molding	Most widely used techniques for molding thermosets	Phenolic (largest volume), urea, melamines, epoxy, rubber, diallyl phthalate, alkyds	Pot handles, electrical connectors, radio cases, television cabinets, bottle closures, buttons, dinnerware, knobs, handles, replacement for metal parts in electrical, automotive, aircraft industries
Casting	Process for converting liquid resins into rigid objects of desired shape	Polyesters, nylons, polyurethanes, silicones, epoxies, phenolics, acrylics	Tooling and metal-forming industries' cast epoxy dies used to produce airplane and missile skins, automobile panels and truck parts; epoxies used by artists and architects for outdoor sculpture, churches, homes and commercial buildings, and encapsulation in electronic industry; cast acrylic sheets used in airplanes, helicopters, schools

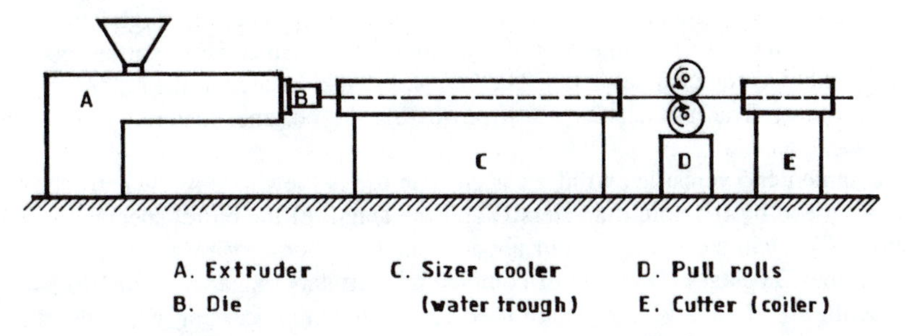

Figure 11.1 Sketch of an extrusion line. (From Richards, P.N., *Introduction to Extrusion,* Society of Plastics Engineers, CT, 1974. With permission.)

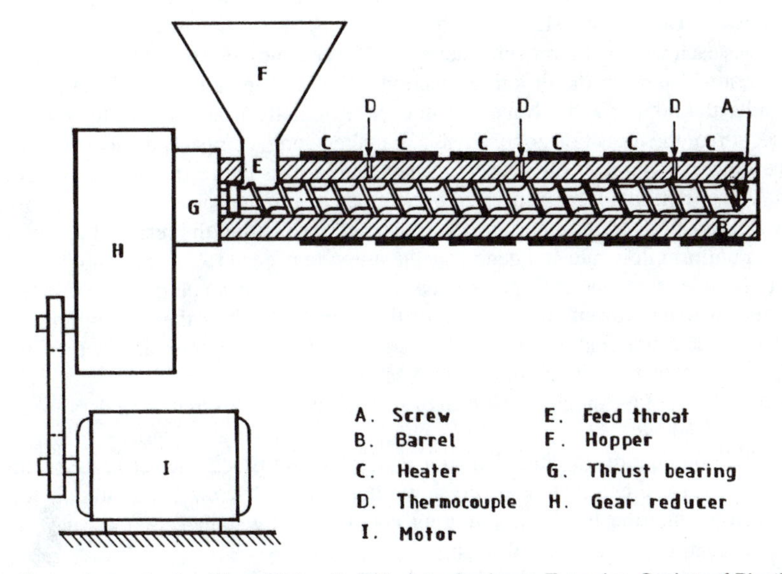

Figure 11.2 Parts of an extruder. (From Richards, P.N., *Introduction to Extrusion,* Society of Plastics Engineers, CT, 1974. With permission.)

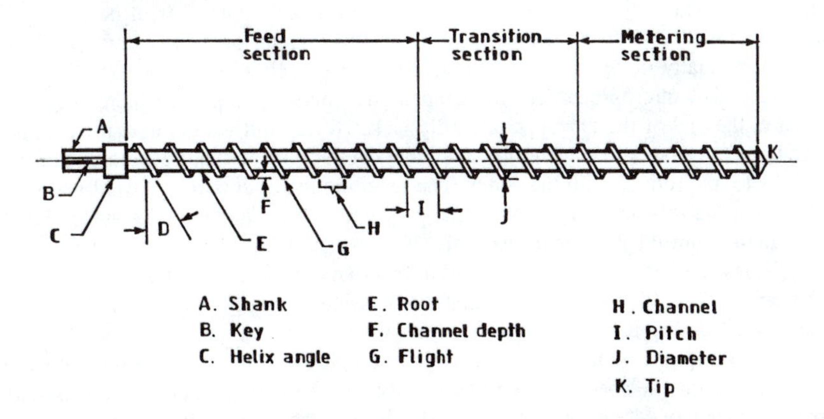

Figure 11.3 Parts of an extruder screw. (From Richards, P.N., *Introduction to Extrusion,* Society of Plastics Engineers, CT, 1974. With permission.)

angle. The polymer is melted and pumped in the open section between the flights called the channel. The bottom of the channel is called the root of the screw. The distance between the root and the top of the flight is referred to as the channel depth. The hub and shank, located at the rear of the screw, fit into the driving mechanism. The hub acts as a seal preventing the polymer material from leaking into the machinery.

The most common and versatile extruder currently in use is the single screw extruder. Extruders are normally characterized by the bore diameters D and the length of the barrel specified as the length-to-diameter ratio, L/D. Diameters range from about 1 in. for laboratory to about 8 in. for specialized production machines. Typical L/D ratios for commercial extruders range from 20/1 to 34/1. In general, the shorter machines (L/D = 20/1 range) are used for elastomer processing while the longer ones are employed for processing thermoplastics. Screws with pitch equal to diameter are referred to as square-pitched screws and have a helix angle of 17.7°. The number of turns of the flight in a square-pitched screw gives the L/D ratio.

The channel depth changes along the screw length; it is deepest at the section under the hopper and shallowest toward the tip. In a few cases where the channel depth changes gradually from one end to the other, the screw is called a constant taper screw. In most cases, however, the screw is divided into three distinct zones each with a different channel depth. These zones are referred to as the feed, transition, and metering sections based on the differing functions. The feed section, 1 to 10 diameters long, picks up the resin under the hopper and advances it into the externally heated barrel to begin melting. The compression section or the transition zone, about 5 diameters long, compacts the loosely packed polymer feed and in the process eliminates air pockets. It melts and forms the resin into a continuous stream of molten material. The frictional force generated between the resin, the barrel wall, and the rotating screw (viscous energy dissipation) is an important energy source for melting the resin. The metering or pump section ensures a uniform flow rate and generates the pressure needed to force the polymer melt through the rest of the extruder and out through the die. Screws are also sometimes described in terms of compression ratio, which is essentially the ratio of the channel depth of the first turn at the rear end of the screw to that of the channel at the last turn. Screws vary from a compression ratio of two for a low-compression screw to four for a high-compression screw.

Various modifications of the single screw design are available. Multiscrew extruders are also in current use for specialized application for which the single screw designs are inefficient. Examples of such applications include production of large-diameter PVC pipe and processing of heat-sensitive materials or resins that must leave the die at relatively low temperatures. Screw configuration for multiscrew extruders may involve intermeshing, corotating, or counterrotating screws. Screws may also be nonintermeshing, of constant depth, or conical in shape.

The barrel of the extruder is normally a long tube. It is made of a special hard steel alloy to provide resistance against wear and corrosion and has thickness sufficient to withstand high internal pressure (about 10,000 to 20,000 psi) without failure. The barrel is equipped with systems for both heat input and extraction. The heaters are arranged in zones and are usually made of cast aluminum. Both halves of the heater are clamped to the barrel to ensure intimate thermal contact. Heat is extracted either by air or a liquid coolant. The simplest design is to mount the coolant or blower under each heater. The heaters, operated manually or by automatic controllers, keep each zone of the extruders at a defined temperature. The use of microprocessors for temperature control has proved increasingly successful. The feed throat at the end of the barrel is generally jacketed for cold water circulation. This keeps the surfaces of the barrel section below the temperature that could cause premature softening of the polymer, which then sticks to and rotates with the screw. The resulting material buildup could seal off the screw channel and prevent the forward movement of the polymer — a phenomenon known as bridging.

As can be seen in Figure 11.2, the extruder is also equipped with a gearbox, thrust-bearing mechanism, and motor. The screw derives its power for rotation from an electric motor. Power requirements are in the range of 1 hp per 5 to 10 lb per hour of material passing through the extruder. The electric motor has a speed of about 1700 rpm, which is too fast for a direct connection to the screw, which has speeds in the range 20 to 200 rpm. It is therefore necessary to employ a gear reducer, which provides varying screw speeds by matching the speed of the drive motor and the rotating screw. Variable speed units can usually provide a change of screw speed from 8 to 1. This enables control of extruder output since there is a direct relation between the extruder pumping rate and the screw speed.

The extruder screw is designed to develop the pressure required to pump the molten polymer through the die. This pressure also acts on the screw. Since the thrust bearing mechanism supports the drive mechanism into which fits the shank of the screw, the thrust-bearing mechanism resists the axial thrust exerted by the molten polymer on the screw. Pressures of up to 2000 psi can be developed in many extruder operations.

Some polymers contain unreacted volatile monomer, moisture, and entrapped gases, which when emitted during the extrusion process could potentially contribute to poor product quality. Some extruders are provided with vents to remove these volatiles from the melt before it reaches the die. The vent is a hole in the barrel about screw diameter size, and located about three-fourths of the distance from the feed throat to the front of the barrel. In some cases, a vacuum is attached to the vent to facilitate gas removal.

The screw for a vented extruder, referred to as a two-stage screw, has a special design. The screw section under the vent has a deep channel. The section of screw before the vent can be considered a normal screw with feed, transition, and metering sections. The section following the vent consists of a short transition zone and a metering section. The channels of the metering section after the vent are deeper than those of the metering section before the vent. This, coupled with the relatively higher channel depth under the vent, means that the channel is partially filled at and beyond the vent section and aids gas removal.

Example 11.1: The barrel of a 6-in. extruder must be able to withstand an internal pressure of 10,000 psi without elastic deformation of 0.15%, at which the alloy will crack in tension. If the barrel material is made of hard steel, estimate the thickness of the tube.

Solution: The stresses existing in the walls of a cylindrical vessel under pressure can be reduced to (1) the tangential or hoop stress σ_h, (2) the radial stress σ_r, and (3) the longitudinal stress σ_l. For a cylinder under internal pressure, σ_h and σ_l are tensile stresses while σ_r is a compressive stress. σ_l is generally smaller (about half the magnitude) than σ_h and is not considered in connection vessel failure. For a thick-walled vessel (i.e., where $r_o/r_i = R > 1.2$):

$$\sigma_n = \frac{p_i(R^2+1)}{R^2-1}$$

where P_i = internal pressure. The maximum allowable hoop stress $\sigma_h = E\varepsilon$.

$$\text{E for hard alloy} = 30 \times 10^6 \text{ psi}$$

$$\varepsilon = \text{strain}$$

$$\sigma_n = 0.0015 \times 30 \times 10^6$$

$$= 4.5 \times 10^4 \text{ psi}$$

$$4.5 \times 10^4 = \frac{10^4(R^2+1)}{R^2-1}$$

$$4.5(R^2-1) = R^2+1$$

$$R^2(4.5-1) = 5.5$$

$$R = 1.25$$

$$r_o = 1.25 r_i = 7.5 \text{ in.}$$

$$r_o - r_i = 1.5 \text{ in.}$$

B. EXTRUSION PROCESSES

The die shapes the polymer extrudate into the desired article. There are a large number of extrusion processes, the simplest of which is compounding. In one variation of extrusion, the die has a series of holes and as the polymer exudes from these holes it is cooled and cut into pellets. In pipe extrusion, the extrudate exiting the die is vacuum-sized and quenched in a water trough. Polymer melts, being viscoelastic, recover the stored elastic energy as they emerge from the die — a process known as die swell. In pipe extrusion, die swell must be controlled to ensure that the pipe dimensions meet standard codes. Profile extrusion is similar to pipe extrusion. However, unlike pipes whose shape is round, hollow, and uniform, the shapes of profiles depend on the end product, which usually is the raw material for downstream processing.

Fibers of various gauges and lengths are obtained in fiber extrusion. These can range from monofilament such as fishing lines to hundreds of continuous filaments extruded from the same die and drawn to hair-sized thickness. The continuous filaments may be crimped for added bulk or made into staples. Almost all electrical wire and cable insulation is currently done by covering the wires or cables with one or more layers of thermoplastic insulation. These take different forms: wire strands may be covered with several layers by successive insulation; multiple preinsulated wires may be covered to form a single cable; or several bore wires may be drawn through the die simultaneously and covered with insulation forming a ribbon cable. In coextrusion, two or more different materials or the same material with two different colors are extruded through the same die so that one material flows over and coats the other. Coextrusion is used to achieve different objectives: a solid cap may be extruded over a foamed core for overall weight reduction or to obtain insulation in addition to a serviceable outer surface; for materials with surfaces sensitive to color, a virgin cap may be extruded over reground, or several different materials may be coextruded so that each material contributes a derived property to the end product.

III. INJECTION MOLDING[2-4]

Injection molding is one of the processing techniques for converting thermoplastics, and recently, thermosetting materials, from the pellet or powder form into a variety of useful products. Forks, spoons, computer, television, and radio cabinets, to mention just a few, are some of these products. Simply, injection molding consists of heating the pellet or powder until it melts. The melt is then injected into and held in a cooled mold under pressure until the material solidifies. The mold opens and the product is ejected. The injection molding machine must, therefore, perform essentially three functions:

1. Melt the plastic so that it can flow under pressure.
2. Inject the molten material into the mold.
3. Hold the melt in the cold mold while it solidifies and then eject the solid plastic.

These functions must be performed automatically under conditions that ideally should result in a high quality and cost-effective part. Injection molding machines have two principal components to perform the cyclical steps in the injection molding process. These are the injection unit and the clamp unit (Figure 11.4). We now describe the operation of the various units of the injection molding machine that perform these functions.

A. THE INJECTION UNIT

The injection unit essentially has two functions: melt the pellet or powder and then inject the melt into the mold. It consists of the hopper, a device for feeding process material; a heated cylinder or chamber where the material is melted; and a device for injecting the molten material into the mold. In the early days when the amount of processed material was relatively small, the two functions of melting and injecting the polymer were accomplished by using a simple plunger machine (Figure 11.5). In this system, a measured volume of the plastic material is delivered into the heated cylinder from the hopper while the ram is retracted. At the beginning of the injection cycle, the plunger pushes forward and forces the material through the heated cylinder compacting it tightly behind and over the centrally located spreader or torpedo. The material is melted by heat convection and conduction. The sustained forward motion of the plunger forces the melt through the nozzle of the cylinder into the mold.

In the plunger-type machine, material flow in the cylinder is essentially laminar. Consequently there is hardly any mixing in this system and, as such, large temperature gradients exist in the melt, and color

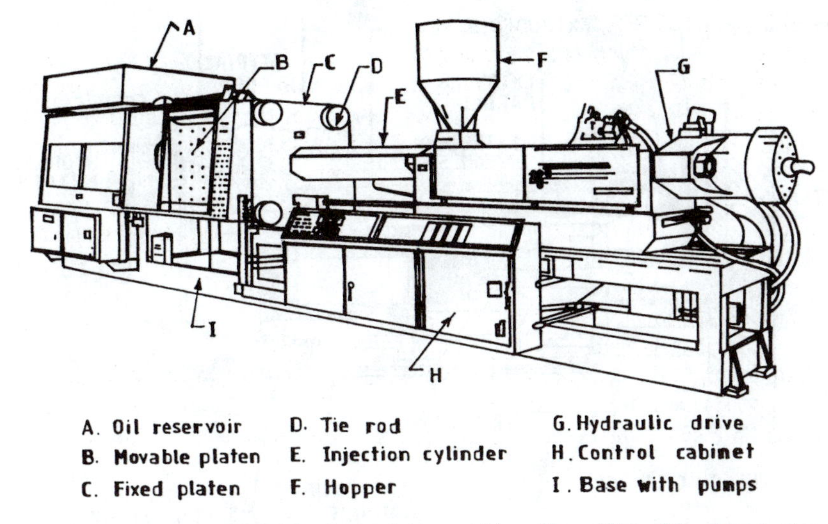

Figure 11.4 Major parts of a typical injection-molding machine. (From Weir, C.L., *Introduction to Molding,* Society of Plastics Engineers, CT, 1975. With permission.)

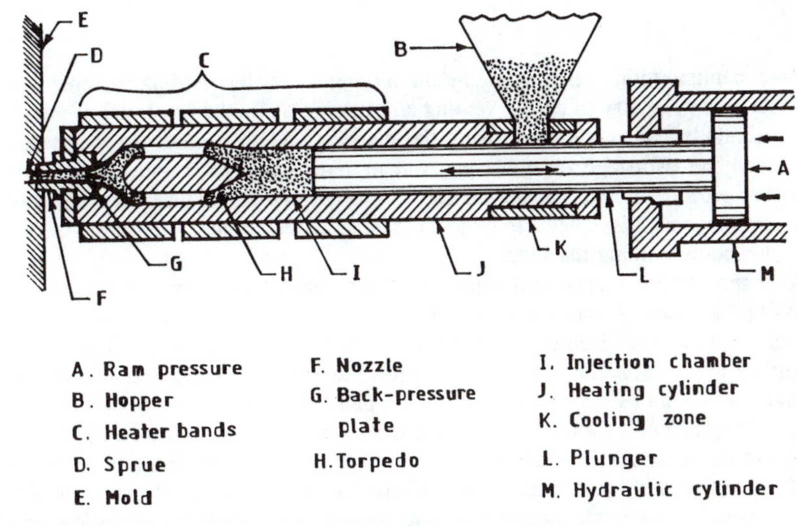

Figure 11.5 Schematic diagram of a plunger-type injection molding machine. (From Weir, C.L., *Introduction to Molding,* Society of Plastics Engineers, CT, 1975. With permission.)

blending is thus problematic. Also, as a result of the friction between the cold resin pellets in the neighborhood of the hopper and the barrel walls, a considerable loss of pressure, up to 80% of the total ram pressure, occurs. This necessitates long injection time. As indicated above, the resin is melted by heat conduction from the walls of the cylinder and the resin itself. Since plastics are poor heat conductors, high cylinder temperatures are required to achieve fast resin plasticization. This can result in the degradation of the material. To avoid such possible material deterioration, the heating of the cylinder is limited, and this also limits the plasticizing capacity of plunger-type injection machines.

Today, the plunger has been replaced almost totally by the plasticating screw. As described in Section II, the screw consists basically of three sections: the feed, the transition zone, and the metering section. Melting normally starts halfway down the length of the screw at which point the depth of the screw flights decreases initiating the compression of the melt. This marks the beginning of the transition zone, which terminates at that metering section — the point where the depth of the flights is minimum. A

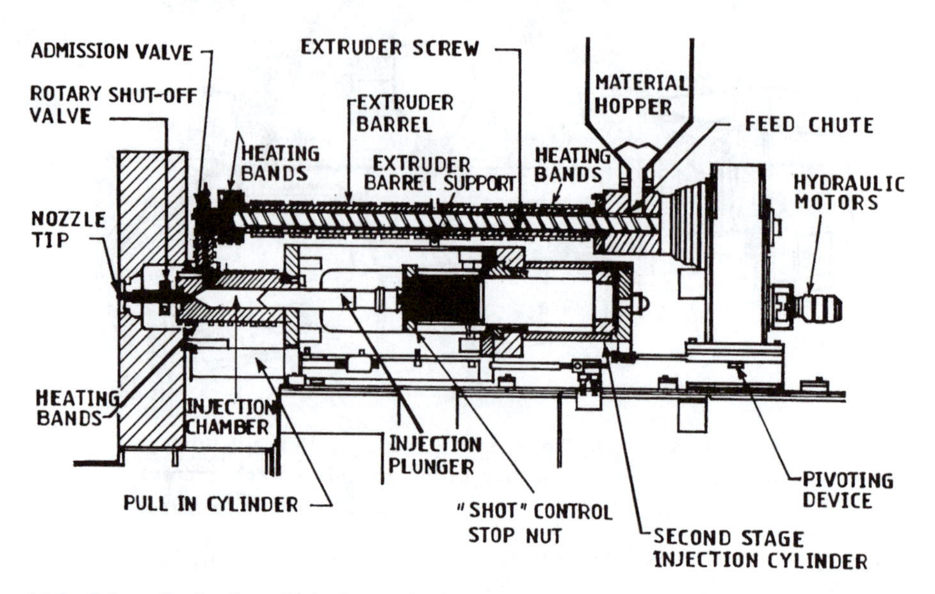

Figure 11.6 Schematic drawing of injection end of two-stage screw-plunger machine. (From Kaufman, H.S. and Falcetta, J.J., Eds., *Introduction to Polymer Science and Technology,* John Wiley & Sons, New York, 1977. With permission.)

number of screw configurations exist, including the mixing screw, the marblizing screw, and that with a decompression zone geometry to permit venting and ridding the melted material of volatiles.

There are essentially two methods of using the plasticating screw. The first is the screw-plunger system also called the two-stage or screw-pot system. The screw rotates in the heated barrel and consequently plasticizes the polymer material. The plasticized material is then transferred into a second heated cylinder from which it is injected into the mold by a plunger. Figure 11.6 shows the basic features of a screw-pot injection molding machine.

In the second approach developed in the mid-1950s, a reciprocating screw is employed. As the screw rotates, it picks up the material from the hopper. The material is melted primarily by the shearing action of the screw on the resin. The electrical heating bands attached to the barrel essentially provide start-up and compensate for heat losses. The rotation of the screw moves the melted material forward ahead of the screw. The pumping action of the screw generates back-pressure, which forces back the screw, the screw drive system, and the hydraulic motor when enough melted material required for a single shot has accumulated in front of the screw. As the screw moves backward, it continues to turn until it hits a limit switch, which stops the rotation and backward movement. The location of the limit switch is adjustable and determines the shot size. Meanwhile the clamp ram has moved forward with the movable platen and closed the mold. When the backward movement of the screw is stopped, the hydraulic cylinders bring the screw forward rapidly and inject the melted material through the nozzle into the mold cavity. A ball check or check ring on the end of the screw (valve) prevents the melt from leaking into the flights of the screw during injection. Consequently, most of the melt is forced into the mold (Figure 11.7). The mold is cooled and the part solidifies. At the end of the predetermined period, the mold opens as the movable platen returns to its initial position and the part is ejected. The screw starts rotating, beginning the next cycle.

The plasticating screw-type machines obviate most of the problems associated with the plunger system. The resin is heated mainly by viscous heat generation as opposed to thermal conduction from cylinder walls. In addition, in contrast to the block of material in the barrel in plunger-type machines, only a thin layer of material exists between the screw and the barrel walls and thus the resin is heated faster. Also in the screw-type machines, the melt is thoroughly mixed. Consequently, these machines produce a melt with a more uniform temperature and homogeneous color, high injection pressures, faster injection speeds and shorter cycles, and therefore higher production rates.

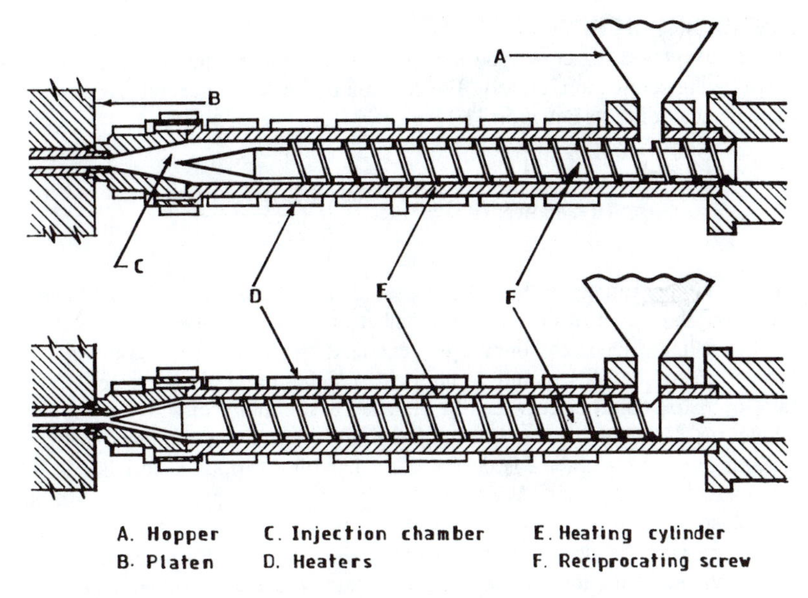

Figure 11.7 Schematic diagrams of a screw-type injection-molding machine: (top) screw retracted, before injection; (bottom) screw forward, at end of shot. (From Weir, C.L., *Introduction to Molding,* Society of Plastics Engineers, CT, 1975. With permission.)

Example 11.2: Explain the following observations:

a. Screw-machine-made end products generally have better physical properties, e.g., lower shrinkage, than the same products made from plunger-type machines.
b. Polymers like PVC and HDPE are easier to handle in screw machines than plunger machines.

Solution:

a. To a very large extent, high injection pressures create internal stresses in the molded part. The resulting part, consequently, has inferior physical properties and is vulnerable to rapid quality deterioration in use. Low injection pressures are therefore desirable. In plunger-type machines, there is usually excessive pressure loss. Therefore, injection pressures are necessarily high in these machines. In contrast, the same parts can be molded in reciprocating screw machines with much lower pressures. Indeed, pressures can be reduced to as much as one-half to two-thirds those of the plunger-type machines, particularly with high viscosity materials. Thus, parts produced in reciprocating screw machines have better physical properties than the same part made from a plunger-type machine.
b. Although screw machines are more complicated than plunger type machines, they are also more flexible and their operation and adjustment to varying molding conditions are simpler. Resin hang-up and the attendant decomposition and the loss of physical properties due to overheating occur very rarely in screw machines. Also, it is much easier to keep the melt temperature within limits by the regulation of screw speed and barrel heating. Thus heat-sensitive polymers like PVC are easier to handle in screw-type machines. In addition, some polymers require a greater heat input to effect plasticization. For example, the heat requirement for high-density polyethylene (HDPE) is about 310 Btu/lb. As a result of the greater shearing action and the attendant viscous heat generated in screw-type machines, polymers like HDPE are easier to handle in these machines.

B. THE PLASTICIZING SCREW

Given the relative advantages of screw type machines over plunger-type machines, it is instructive to examine the plasticizing screw more closely. The rotation of the screw generates a torque related to the power requirements according to Equation 11.1.

$$\text{Horsepower (hp)} = \frac{[\text{Torque (ft-lb)}][\text{Rpm}]}{5252} \tag{11.1}$$

where rpm is revolution per minute of the screw. Recall that torque is the product of the tangential force and the distance from the center of the rotating member. It is clear from Equation 11.1 that there is an inverse relation between the speed and the torque generated by the screw. Generally, a motor of a given horsepower has a fixed speed. Consequently, a unit with a higher torque (lower speed) has a larger frame than that with a lower torque. The screw can be attached to the motor directly or through a gear train. The latter provides an alternative mechanism of changing the speed and therefore the torque generated by the screw. A range of torque values is desirable for handling various materials due to the different processing characteristics of polymers. For example, much higher torque is required to plasticize polycarbonate than polystyrene.

For a given horsepower, the slower the screw speed, the higher the torque developed. However, the yield strength, and therefore the torque a screw can sustain, varies as the cube of the root diameter of the screw. It follows that if the torque generated is too high, the yield strength of the screw material may be exceeded, causing a shearing of the screw. On the other hand, if the screw speed (shear rate) is too high, the material may be degraded. As indicated earlier, plasticization of the polymer is a result of the viscous energy developed from the work done by the rotating screw. In other words, the screw output is related to the energy developed from the shearing forces. Therefore, if different screw designs have similar efficiency, the maximum screw output is determined largely by the power rating of the screw.

Example 11.3: A plasticizing screw driven by a 50-hp motor is used to raise the temperature of the polymer feed from 25°C to the processing temperature. Assuming a mechanical efficiency of 60% and neglecting the energy required to pump the polymer and that from the heating bands, estimate the screw output if:

a. The processing temperature is 440°F and the specific heat of the polymer material is 0.8 Btu/lb °F.
b. The processing temperature is 500°F and the specific heat of the polymer material is 0.8 Btu/lb °F.
c. The processing temperature is 500°F and the specific heat of the polymer material is 0.4 Btu/lb °F.

Solution: a. Since the energy from the heating bands and the energy required to pump the polymer are to be neglected, the power supplied by the motor is converted into work by the screw. The viscous energy dissipated goes into melting the polymer.

Power = $C_p Q (T_m - T_f)$
T_m = temp of polymer melt
T_f = temp of polymer feed
C_p = specific heat of polymer material (assumed constant)
Q = Screw output (in mass/time)

If Power = hp, Q = lb/h, then since 1 Btu/h = 0.0004 hp, hp = $0.0004\, C_p\, Q\, [T_m - T_f]$.

$$Q = \frac{\text{hp}}{0.0004\, C_p\, (T_m - T_f)}\ (\text{lb/h})$$

$$= \frac{50 \times 0.6}{0.0004 \times 0.8\,(440 - 72)}$$

$$= 255 \text{ lb/h}$$

$$Q = \frac{50 \times 0.6}{0.0004 \times 0.8\,(500 - 72)}$$

$$= 219 \text{ lb/h}$$

$$Q = \frac{50 \times 0.6}{0.0004 \times 0.4\,(500 - 72)}$$

$$= 438 \text{ lb/h}$$

Example 11.4: Explain the following observations:

a. It is generally necessary to mold plastics at the lowest possible melt temperature.
b. If two screws with different diameters but with the same L/D ratio are operated under identical conditions, the polymer material has a longer residence time in the larger diameter screw machine even though the machine output is the same. However, it is preferable sometimes to operate machines with the larger diameter screws.

Solution:

a. From Problem 11.3 it is obvious that raising the material processing temperature lowers output. Also, it increases the cycle time because the mold cooling time is longer. Therefore, to maximize output and increase overall productivity, it is best to mold at the lowest possible melt temperature.
b. Productivity or machine output is the primary concern of injection molders. However, high screw speeds and hence high productivity can result in material degradation and poor product quality. Consequently, it is generally prudent to operate machines at reasonably slower speeds. But because of the inverse relation between torque and screw speed (Equation 11.1) for the same power input drive, the magnitude of the torque developed by the smaller diameter screw will be higher than that of the larger diameter screw and may be high enough to shear the screw. Therefore, even when machine output is the same, larger diameter screws may be preferable to smaller ones to prevent possible screw damage.

C. THE HEATING CYLINDER

The heating or plasticizing cylinder of the injection molding machine is the primary element of the machine. It is here that the polymer material is softened or conditioned for injection into the mold cavity where it is shaped. The temperature and pressure of the melt as it leaves the cylinder nozzle into the mold cavity are two important variables that determine product quality. Cylinders are generally rated on the basis of their plasticizing capacity which is the rate at which the given cylinder can condition a given polymer material into a state suitable for injection. As may be expected, for a given cylinder, this varies with the particular plastic material being processed because the molding (softening or melting) temperature, specific heat, thermal conductivity, and specific gravity — all of which contribute to the complex heat generation and transfer processes which occur during processing — differ for various materials. Consequently, the plasticizing capacity of a machine is rated conventionally on the basis of one material — general purpose polystyrene, which is taken as the standard. The machine capacity with respect to other materials is then related to this standard material using the relative specific gravities of the two materials.

Table 11.2 gives the specific gravities of some plastic materials, while Table 11.3 shows the appropriate cylinder heater inputs and power requirements of typical machine sizes.

Table 11.2 Specific Gravities of Some Polymers

Resin	Specific Gravity
ABS	1.04
Acetal	1.42
Cellulose acetate	1.27
Cellulose acetate butyrate	1.19
Cellulose propionate	1.21
Ethyl cellose	1.10
Nylon	1.14
Polycarbonate	1.20
Polyethylene	
Low density	0.92
High density	0.95
Poly(methyl methacrylate)	1.19
Polypropylene	0.90
Polystyrene	
General purpose	1.05
High impact	1.04
Polyvinyl chloride	
Plasticized	1.20
Unplasticized	

From Weir, C.L., *Introduction to Molding,* Society of Plastics Engineers, CT, 1975. With permission.

Table 11.3 Cylinder-Heater Inputs and Power Requirements Typical Machine Sizes

Shot Size (oz)	Plasticizing Capacity (lb/h)	Approximate Heat Input (kW)	Screw Power Input (hp)
1	17.5	2.0–2.5	3.5
3	45	2.8–4.1	5
6	80	4.4–5.2	10
9	125	8.5–9.0	15
12	220	9.0–10.5	25
24	250	10.6–12.0	40
40	400	15–20	50
60	440	20–25	60
80	700	25–30	75
100	750	30–40	90
150	850	40–50	100
200	1000	60–70	125
225	1500	68–75	150
350	1800	75–100	200

From Weir, C.L., *Introduction to Molding,* Society of Plastics Engineers, CT, 1975. With permission.

Example 11.5: An injection molding machine is rated 100 oz per shot. What is the plasticizing capacity (lb/h) of this machine if the total cycle time is 30 s? What is the average residence time of the material in the heating cylinder for an inventory weight of 50 oz? Estimate the plasticizing capacity of the heating cylinder for low density polyethylene.

Solution: The machine output or plasticizing capacity, Q, is the ratio of the shot weight W to the cycle time t_c.

$$Q = \frac{W(oz)}{t_c(s)}$$

$$\frac{W(oz)}{t_c(s)}\ \frac{1\ lb}{16\ oz}\ \frac{3600\ s}{1\ h}$$

$$= 225\frac{W}{t_c}\frac{lb}{h}$$

$$Q = \frac{225 \times 100\ oz}{30} = 750\ lb/h$$

Residence or contact time in the heating cylinder, t_c, is given by the inventory weight, I_w, divided by the machine output Q.

$$t = \frac{I_w}{Q}$$

$$t(s) = 225\left(\frac{I_w\ oz}{Q\ lb/h}\right) = \frac{225 \times 50}{750} = 15\ s$$

$$\text{Plasticizing capacity for LDPE} = \frac{0.93}{1.05} \times 750$$

$$= 657\ lb/h$$

D. THE CLAMP UNIT

The clamp unit or press end of the injection molding machine performs three functions: opens and closes the mold at appropriate times during the molding cycle; ejects the molded part; and provides enough pressure to prevent the mold from opening due to the pressure developed in the mold cavity as it is filled with the melt by the injection unit. Injection pressures in the plasticating cylinder can range from 15,000 to 20,000 psi in a given system. As a result of possible pressure drops in the cylinder and the nozzle, the effective pressure within the mold cavity may be reduced to 25 to 50% of this value. Consequently, the force needed to resist the premature opening of the mold and obtain an acceptable part can be quite large. For example, assuming a 50% pressure drop and using the upper possible pressure in the cylinder stated above, the melt pressure in the mold cavity becomes 10,000 psi. This translates into a clamp force of 5 tons/in^2 of the projected area of the part. However, as a result of greater degree of homogenization achievable in screw-type machines, the clamp tonnage required in these machines is generally less than in plunger-type machines.

The halves of the mold are attached to the platens, one of which is stationary and one of which moves as the clamp mechanism is opened or closed. Molds are generally designed so that the ejection side of the mold (mold core) is on the movable platen and the injection side of the mold (mold cavity) is on the stationary platen, which must provide an entry for the nozzle of the plasticizing chambers. When the mold opens, the movable platen must be moved sufficiently for the part to be ejected. Shrinkage usually accompanies part solidification and this results in the part sticking to the core as the mold opens. Consequently, the movable platen is provided with an ejector or knockout system to eject the part. The ejector usually consists of a hydraulically actuated ejector plate mounted off the back face of the movable platen. The ejector system causes the knockout plate to change its location relative to the rest of the mold. The ejector pins attached to this plate push against the molded part and eject it from the mold. The force required to eject the part is usually less than 1% of the nominal clamp force; its magnitude depends on the part geometry, material, and packing pressure.

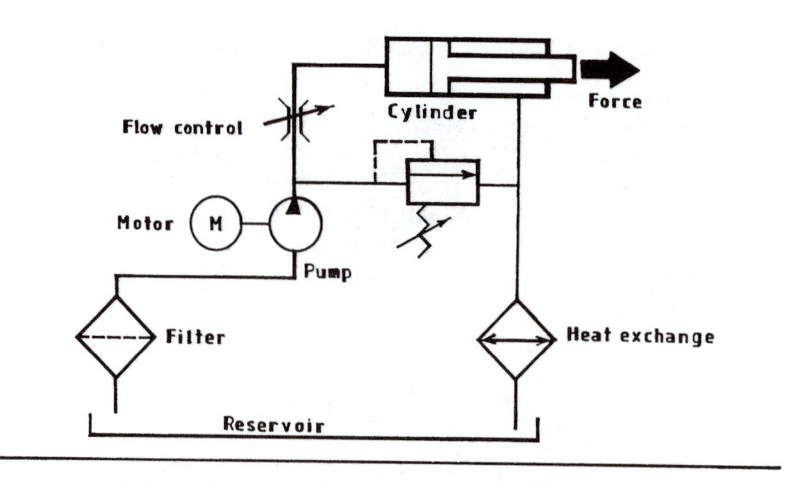

Figure 11.8 Elements of hydraulic system.

There are essentially two categories of clamp systems for moving and locking the movable platen. These are the mechanical (toggle) and hydraulic systems. The toggle system, which is used mainly in small tonnage (about 500 tons) equipment, is relatively inexpensive but requires good maintenance because wear reduces the clamping force. Hydraulic systems, used primarily in medium and large equipment (1500 to 2000 tons), are relatively slow acting but have the advantage of providing virtually limitless pressure selection, which can be followed continuously with a pressure gauge. Machine flexibility permits the necessary adjustment to match mold thickness and settings.

E. AUXILIARY SYSTEMS

The operation of the injection and clamp units and other components of the injection molding machine (opening and closing of the mold and melting and injection of the polymer material) requires power, which is supplied by an electric motor. The orderly delivery of this power depends on auxiliary systems: the hydraulic and control systems. The hydraulic system, the muscle for most machines, transmits and controls the power from the electric motor to the various parts of the machine. Machine functions are regulated by a careful control of the flow, direction, and pressure of the hydraulic fluid. The elements of the hydraulic system for most injection molding machines are essentially the same: fluid reservoir, pumps, valves, cylinders, hydraulic motors, and lines (Figure 11.8).

The pump driven by the electric motor draws oil from the reservoir and delivers it to the system through the suction filter. The restriction of oil flow by the control valve and/or the resistance to movement of the cylinder compress the oil and lead to pressure buildup. A pressure buildup of appropriate magnitude drives the hydraulic cylinder, and if this pressure reaches that set for the relief valve, the valve opens and excess fluid is bypassed to the reservoir. The opening of the relief valve decompresses the oil, converting the excess pressure energy into heat, which raises the oil temperature. The oil is cooled by passing it through a heat exchanger.

The injection molding operation involves a sequence of carefully ordered events, e.g., precise heat control and appropriate timing of injection pressures and cycles. The control system is the nerve center responsible for the orderly execution of the various machine functions. Over the years, the control system has developed from a collection of relays that perform logic, plug-in timers for timing functions, and plug-in temperature controllers for cylinder temperature regulation to solid-state circuitry for the control of these functions to the current microcomputer control, which has greatly enhanced process control.

F. THE INJECTION MOLD

The injection mold is a series of steel plates, which when assembled produces the cavity that defines the shape of the molded part. Conventional molds consist of the mold frame, components, runners, cooling channels, and ejector system. The mold frame is a collection of steel plates that contain mold components and runners, cooling and ejection systems. Components are parts inserted into either bored holes or cutout pockets in the mold frame. The polymer melt enters into the mold cavity or cavities through the runners, which are passages cut into the mold frame. The hot polymer material in the mold

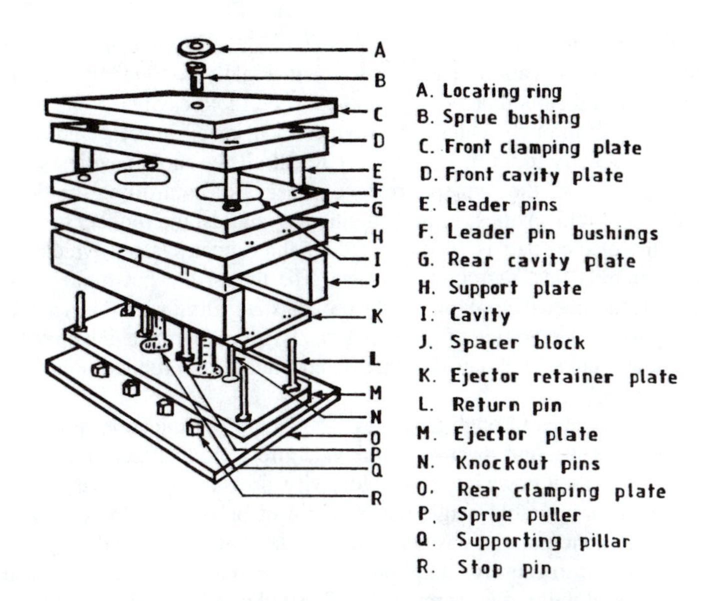

Figure 11.9 Exploded view of a six-plate mold base. (From Weir, C.L., *Introduction to Molding,* Society of Plastics Engineers, CT, 1975. With permission.)

is cooled by a coolant, usually water, which circulates through the cooling channels drilled at strategic locations into the mold frame and components for proper mold temperature control. When the material has cooled (hardened) sufficiently, the mold opens and the hardened part is removed from the mold by the ejector system.

Since the quality of the molded part depends to a large extent on the mold, it is essential to expand on the functions of the various parts of the mold. Details of mold components are shown in Figure 11.9. The register or retainer ring, which is fitted into the stationary platen, aligns the mold with the cylinder nozzle. Sometimes, the retainer ring, as the name suggests, also acts as the retainer of the sprue bushing within the mold. The sprue is the pathway through which the molten polymer material is introduced into the mold cavity area or the runner system from the cylinder nozzle. The sprue is generally made as small as practical since large diameter sprues have been known to require longer cooling time and consequently prolong cycle time beyond what is usually necessary for part thickness. In some newer mold designs, the sprue is completely eliminated.

The front clamp plate, which houses the retaining ring and sprue bushings, is used to support the stationary half of the mold, including the front cavity plate. The cavity (cavities) of the part to be molded is contained in the cavity plate in either of two ways. In one case the cavity is drilled directly into the steel plate, and in the other the cavity plate provides sockets for the insertion of cavities that have been constructed separately. The rear cavity plate, mounted on the rear support plate, contains the core section of the molded part or second half of the mold cavity. The leader (guide) pins and guide pin bushings are housed in the front cavity and rear cavity plates, respectively. The passage of the guide pins through the bushings ensures that the plates are properly aligned during closing of the mold. The knockout and reset pins are mounted on a series of plates that compose the ejector plate. At appropriate times during the molding cycle, these pins pass through holes in the cavity and cavity retainer plates and make contact with either the molded part to effect its ejection or the stationary cavity plate to initiate the movement of the pin plate back in readiness for the next injection shot. The movable half of the mold is anchored to the movable platen by the rear clamp plate.

As the mold cavity is filled with polymer melt, the pressure increase within the cavity can produce stresses of up to 10,000 to 180,000 psi in the mold cavity material. The resulting deformation is substantial but can be accommodated provided the elastic limit of the material is not exceeded. However, where the dimensional tolerance of the molded part is critical, it is imperative that the mold material modulus is sufficiently stiff to ensure part dimensional accuracy. Maintenance of proper temperature of the mold cavity and core is also necessary in this respect as well as for the production of a molded part with good physical and mechanical properties. Dimensional accuracy of the part also demands that the dimensions of the mold be slightly larger than those of the molded part so as to compensate for the shrinkage that normally accompanies the solidification of a polymer melt. Finally, a good surface appearance of the molded part such as luster and smoothness requires a polished mold cavity.

Runners are the channels through which the polymer melt is fed into the mold cavities from the cylinder nozzle. In a multicavity mold, it is necessary to fill all the mold cavities simultaneously and uniformly. Control of the size of the runners provides a means of controlling the flow resistance and balancing the flow into the mold cavities. In most multicavity molds, the runners form part of the mold frame. Consequently, the ejected part is accompanied by the runner system, which must be removed and, in the case of thermoplastics, reground for reuse. The use of the hot runner mold whereby the runner channels are heated to keep the polymer in the molten state, eliminates this need for plastic runner separation and avoids possible generation of scrap material. With proper machine operation, a hot runner mold requires a smaller amount of melt per shot than an equivalent cold runner mold, leading to reduced injection time and faster cycles.

Runners feed the mold cavities through the gates. The gate of an injection mold is one of its most vital components. The size, type, and location of the gate affect production rate and the quality of the molded part. As the polymer melt flows into the mold cavity, pressure in the cavity builds up and, if not controlled, can develop an undesirable magnitude. Control of pressure in the cavity requires that the gate freeze as soon as sufficient material has been fed into the mold cavity and only to the point where available pressure is sufficient to cause resumption and maintenance of adequate melt flow. If the gate freezes prematurely, the mold will not be completely filled and/or adequately packed to compensate for shrinkage. This produces a warped and unacceptable part. If, on the other hand, the mold cavity is filled and the gate takes too long to freeze, the resultant back pressure causes the still-hot polymer melt to flow back out of the mold into the nozzle and cylinder when the ram or screw is withdrawn.

Finally, to conclude this section, we briefly address mold cooling and the associated shrinkage and possible warpage of the molded part. As we discussed earlier, to complete the injection cycle the polymer melt injected into the mold cavity must be cooled to such a stage that the solidified part can be ejected. Ideally the temperature distribution of the mold surface should be uniform so as to ensure uniform temperature reduction, even shrinkage, and reduced warpage tendency of the molded part. This demands that the size and distribution of the cooling lines be such that cooling rate is highest near the gating and sprue where the melt temperature is highest and lowest in areas farthest from these points.

You will recall from our discussion in Chapter 5 that polymers undergo a reduction in specific volume (shrinkage) as they undergo a phase transformation from the molten to the solid state. The severity of this dimensional change varies with the nature of the polymer: crystalline polymers are more seriously affected than amorphous polymers. Therefore, a certain amount of mold shrinkage is inevitable in polymer molding operations. For most applications, this can be accommodated by proper mold design. However, for sophisticated parts and those parts that require close tolerance, mold shrinkage assumes an added importance. This must be addressed not only through mold design but by a more careful control of operating conditions.

Polymer molecules generally assume a random orientation in the melt. As the melt is forced through the gate during mold filling, polymer molecules tend to lose this random orientation and align themselves along the direction of flow. The polymer material that comes into contact with the much colder mold surface chills rapidly and becomes frozen in place. Meanwhile, the melt not yet in contact with the mold surface continues to move. The frictional forces (shearing stresses) between the moving and stationary polymer material, which are generally high, stretch the polymer molecules in the direction of flow. This orientation of the polymer molecules results in orientation strains or built-in stresses in the molded part. The relaxation of these stresses in the ejected part leads to a further or postmold shrinkage of the molded part. Uneven shrinkage causes the part to warp, particularly in large, thick, and flat molded articles and may lead to part rejection.

Example 11.6: Explain the following observations:

a. Nonferrous metals are used in injection molds for cavity and core components but must be properly supported on steel forms.
b. The dimensions of the mold cavity are generally made larger than those of the molded part to allow for the shrinkage that occurs when the molten polymer cools. Typical shrinkage allowances for some plastics are shown in Table E11.6.

Table E11.6 Shrinkage Allowances for Some Polymers

Polymer	Shrinkage Allowance (in./in.)
Poly(methyl methacrylate)	0.006
Nylon	0.023
Polyethylene	0.023
Polystyrene	0.006

Solution:

a. The surface finish of the mold cavity contributes to the surface appearance (luster and smoothness) of an injection-molded part. Consequently, mold cavities must be properly polished. Nonferrous metals are generally softer and easier to polish than hardened steel. To ensure the dimensional accuracy of the finished part, the surface temperature of injection mold cavities and cores must be maintained at appropriate temperatures during operation. Nonferrous metals with their high thermal conductivities ensure uniform temperature distribution. However, nonferrous metals are not sufficiently stiff and therefore are susceptible to permanent deformation that may affect not only the surface appearance but also the dimensional accuracy of the molded part. Consequently, it is often necessary to support nonferrous metals on the much stiffer hardened steel.

b. On cooling from the viscous state (melt), crystalline polymers shrink more than amorphous polymers because the process of crystallization involves a considerable contraction of volume. Poly(methyl methacrylate) and polystyrene are amorphous, while nylon and polyethylene are crystalline.

Example 11.7: Explain and comment on the following observation: prevention of excessive shrinkage of a molded part requires molding at low injection temperatures, running a cold mold, or operating at high melt pressures, whereas reduction of part warpage calls for maximum injection pressure.

Solution: Polymeric materials exhibit a marked decrease in specific volume (shrink) as they are cooled from the melt. Crystalline materials shrink more than amorphous materials as a result of the more ordered molecular arrangement. The amount of shrinkage is determined by the rate of quenching or temperature drop from the injection temperature to the mold temperature. Lowering the injection temperature reduces this temperature drop and consequently reduces shrinkage. High mold temperature allows the polymer melt to cool more slowly, promotes Brownian movement, and thus facilitates a more ordered molecular arrangement. On the other hand, running a cold mold essentially freezes in the amorphous arrangement of the melt. This in effect translates into a higher specific volume or a reduction in part shrinkage.

As the hot material comes into contact with the cold mold surface, it solidifies and shrinks. Operating at high injection pressures results in a compression of the material and therefore permits more material to flow into the mold cavity and compensate for the shrinkage. The increased mold packing that is associated with high injection pressures means that more material can be packed into a given volume, therefore reducing shrinkage.

Warpage is due largely to the internal stress developed in the molded part during the molding operation. Molding at high injection temperatures tends to diminish the elastic memory of the polymer melt and thus reduce the tendency to create stresses that might cause warpage. As indicated above, the faster the cooling rate of the polymer melt, the more disorder the molecular arrangement of the resulting solidified material and consequently the greater the built-in stresses in the molded part. Operating high mold temperatures allows the molecular rearrangement that is required for stress relief and thus a reduction of the internal stresses that cause part warpage. Low injection pressures lead to the generation of much lower orientation strain.

It is obvious from this discussion that procedures for reducing part shrinkage work in direct opposition to those required for preventing its warpage. Therefore, the injection molding of a trouble-free part requires a careful balancing of these operating conditions.

IV. BLOW MOLDING[4,5,8]

Blow molding is a process used extensively for the production of bottles and other hollow plastic items with thin walls. Blow-molded objects may range in size from less than 1 oz to a few hundred gallons.

A. PROCESS DESCRIPTION

The blow molding process consists of a sequence of steps leading to the production of a hollow tube or parison from a molten thermoplastic resin. This is then entrapped between the two halves of a mold of the desired shape. Air, usually at about 100 psi, is blown into the soft parison, expanding it against the contours of the cold mold cavity. The part is cooled and removed from the mold, and where necessary the excess plastic material or flash accompanying the molded part is trimmed and reclaimed for reuse.

The blow molding process therefore involves essentially two properly synchronized operations: parison formation from the plastic material and blowing the parison into the shape of the desired part. There are two techniques for plasticizing the resin for parison formation. These are extrusion blow molding (which is the most common method and which is characterized by scrap production) and injection blow molding. The latter process is versatile and scrap free and is beginning to be more understood and accepted by processors.

B. EXTRUSION BLOW MOLDING

In extrusion blow molding, an extruder, as described in Section II, is used to plasticize the resin and form the parison. The process may be continuous or intermittent. In the continuous process, a continuous parison is formed at a rate synchronized with the rates of part blowing, cooling, and removal. Two general mold clamp mechanisms are used for part formation from the extruded parison. In the first arrangement or shuttle system, the blowing station is situated on one or both sides of the extruder. As soon as an appropriate length of parison is extruded, the clamp mechanism moves from the blowing station to a position under the die head, captures and cuts the parison, and then returns to the blowing station for part blowing, cooling, and removal. This ensures that there is no interference with parison formation. In the second or rotary system, a number of clamping stations are mounted on a vertical or horizontal wheel. As the wheel rotates at a predetermined rate, blowing stations successively pass the parison head(s) where it is entrapped for subsequent part formation. In this case, parison entrapment and blowing, part cooling, and removal occur simultaneously in a number of adjacent blowing stations.

In the intermittent extrusion process, molding, cooling, and part removal take place under the extrusion head. An extruder system, which may be of the reciprocating screw, ram accumulator, or accumulator head type, extrudes the parison in a downward direction where it is captured at the proper time between the two halves of the mold. The part is then formed and ejected and a new cycle begins (Figure 11.10). As the name suggests, in the intermittent extrusion blow molding process, parison formation is not continuous. For example, with the reciprocating screw machine, after the parison is extruded, melt is accumulated in front of the screw causing a retraction of the screw. After the molded part has cooled and the mold opens and ejects the part, the screw is immediately pushed forward by hydraulic pressure, forcing the melt into the die to initiate the formation of the next parison.

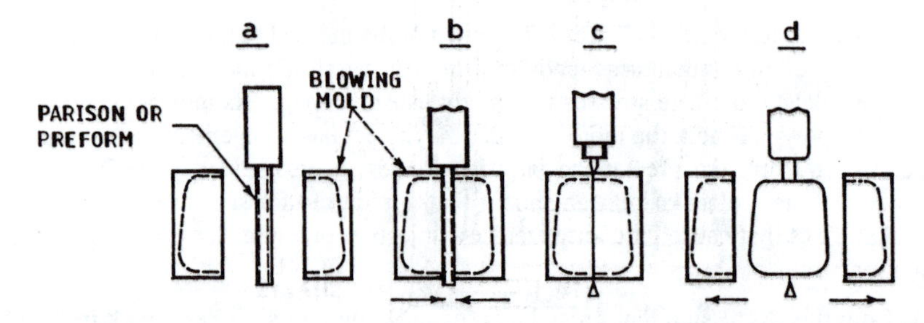

Figure 11.10 Schematic of the blowing stage. (a) The molten, hollow tube — the parison or preform — is placed between the halves of the mold; (b) the mold closes around the parison; (c) the parison, still molten, is pinched off and inflated by an air blast that forces its wall against the inside contours of the cooled mold; (d) when the piece has cooled enough to have become solid, the mold is opened and the finished piece is ejected. (From Kaufman, H.S. and Falcetta, J.J., Eds., *Introduction to Polymer Science and Technologies,* John Wiley & Sons, New York, 1977. With permission.)

C. INJECTION BLOW MOLDING

The injection blow molding process is a noncontinuous cyclic process consisting essentially of two phases. In the first phase, a preform is molded by injecting melted plastic into a steel mold cavity where it is kept hot and conditioned. In the second or subsequent phase, the preform is metered into the blow mold where the blowing operation takes place to form the final part. The major advantages of injection blow molding are the quality of the molded part and productivity. There is no flash production. Therefore, the molded part neither has a pinch-off scar from flash nor requires additional trimming or other finishing steps for waste retrieval. Also, the molded parts show hardly any variation in weight, wall thickness, and volume from the accurately molded preform. However, only blow-molded parts with limited size and shape and without handles are feasible with the injection blow molding process.

V. ROTATIONAL MOLDING[4,5]

A. PROCESS DESCRIPTION

Rotational molding is a process used for producing hollow, seamless products having heavy and/or complex shapes. In rotational molding a premeasured amount of powder or liquid polymer is placed in the bottom half of the mold, and the two halves of the mold are locked together mechanically. The mold is then rotated continuously about its vertical and horizontal axes to distribute the material uniformly over the inner surface of the mold (Figure 11.11). The rotating mold then passes through a heated oven. As the mold is heated, the powdered polymer particles fuse forming a porous skin that subsequently melts and forms a homogeneous layer of uniform thickness. In the case of a liquid polymer, it flows over and coats the mold surface and then gels at the appropriate temperature. While still rotating axially, the mold passes into a cooling chamber where it is cooled by forced air and/or water spray. The mold is then moved to the work station and opened, and the finished solid part whose outside surfaces and contour faithfully duplicate those of the inner mold surface is removed. The mold is recharged for the next cycle.

B. PROCESS VARIABLES

Different types of heating systems, including hot air, molten salts, or circulation of oil through a jacketed mold have been used. The essential requirement of any heating system is to ensure that the mold is heated uniformly and at a properly controlled rate so that the desired part thickness is obtained without causing resin degradation. Given the potential hazards and maintenance problems associated with the use of molten salts, the use of hot oil has gained wide commercial acceptance because of the relatively cleaner, cheaper, and safer operation involved.

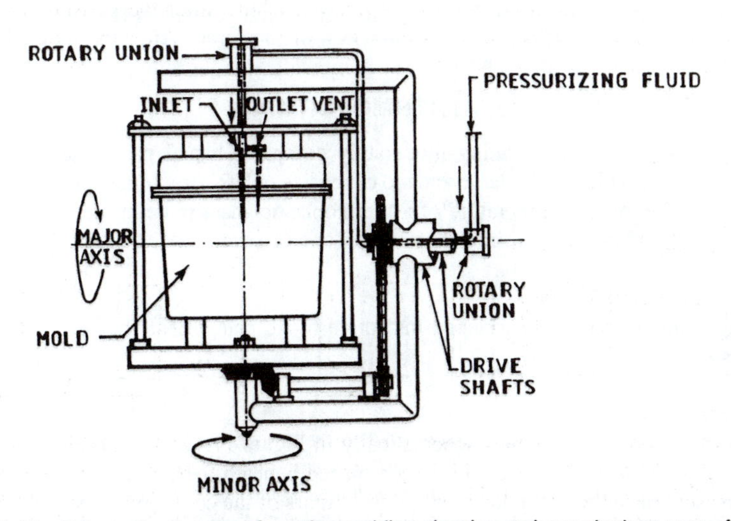

Figure 11.11 Schematic diagram of rotation molding showing major and minor axes of rotation.

While the heating cycle has virtually no effect on the properties of the finished part, the cooling rate determines part shrinkage, final density, brittleness, and other physical properties. Mold cooling is accomplished by the use of forced air and/or application of water spray.

Rotational molding machines vary from the simple one-arm rotocast system with capability for producing parts of up to 20 gal size to the industrially predominant three-arm machines that are capable of producing up to 5000 gal capacity tanks. In the latter case, each arm is always in one of the three stations: load-unload, oven, or cooling. More recently, shuttle-type machines with molds mounted on large self-driven shuttle carts have been developed. The carts move on tracks from the oven to the cooling chamber. These machines can make products of much larger capacities. Rotational molding machines with microprocessor controls and other solid-state devices for regulating operating variables such cycle time, oven temperature, rotational speed ratio (ratio of the speed of major to minor axis), and fan and water on–off times are now available.

The rotational molding operation involves neither high pressure nor shear rates. In addition, precise metering of materials is not crucial. Therefore, rotational molding machinery is relatively cheap and has a more extended lifetime. Other advantages of rotational molding include favorable cost–performance (productivity) ratio, absence of additional finishing operations even of complex parts, minimal scrap generation, and capability for simultaneous production of multiple parts and colors.

Example 11.8: Explain the variation in the physical properties of rotational molded parts from polypropylene and polystyrene shown in Table E11.8 with changes in the cooling cycle

Table E11.8 Physical Properties of Rotational Molded Parts

	Polypropylene Cooling Time		Polystyrene Cooling Time	
Property	**10 min**	**2 min**	**10 min**	**2 min**
Specific gravity	0.96	0.90	1.20	1.19
Shrinkage	0.040	0.015	0.003	0.004
Elongation at break (%)	300	100	1.5	1.3

Solution: A reduction in cooling time means faster cooling rates of the rotational molded parts. This is accompanied by marked changes in the physical properties of the part from polypropylene (crystalline polymer). On the other hand, the cooling rate has little effect on the part from polystyrene (amorphous polymer). A longer cooling time permits a greater ordered molecular arrangement in the crystalline polymer. The resultant enhanced crystallinity leads to higher specific gravity and shrinkage and reduced brittleness. Cooling rates do not seriously affect molecular arrangement in the amorphous polymer.

VI. THERMOFORMING[4,5]

Thermoforming is a process for forming moderately complex shaped parts that cannot be injection molded because the part is either very large and too expensive or has very thin walls. It consists essentially of two stages: elevation of the temperature of a thermoplastic sheet material until it is soft and pliable and forming the material into the desired shape using one of several techniques.

A. PROCESS DESCRIPTION

Thermoforming techniques may be grouped into three broad categories: vacuum, mechanical, and air blowing processes.

1. Vacuum Forming

The vacuum forming process is shown schematically in Figure 11.12. The plastic sheet is clamped in place mechanically and heated. A vacuum is then placed beneath the hot elastic sheet, and this makes atmospheric pressure push the sheet down onto the contours of the cold mold. The plastic material cools down, and after an appropriate time the cooled part is removed.

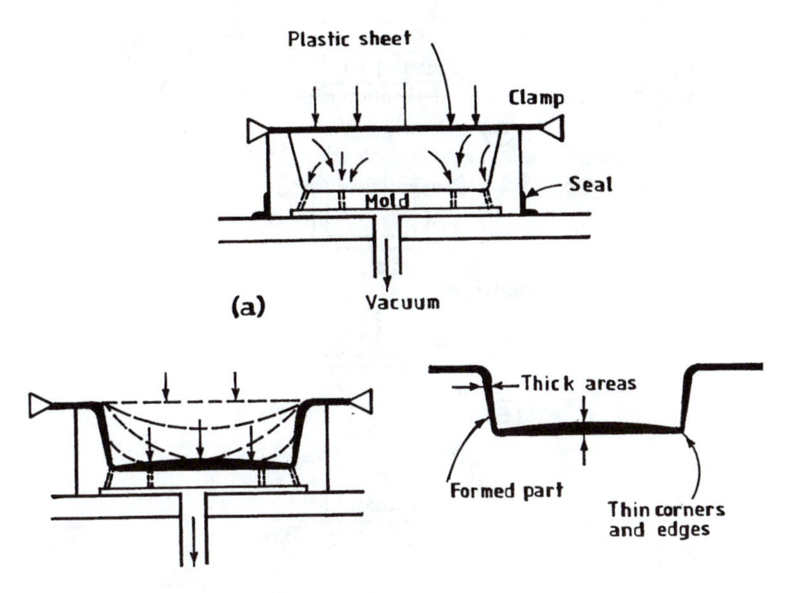

Figure 11.12 Steps in vacuum forming process.

2. Mechanical Forming

In this case, a hot sheet is stretched over a mold or matched molds without the use of air or pressure. For example, in matched mold forming, the heated sheet is clamped over a female mold or draped over the mold force (male mold) (Figure 11.13). The two molds are then closed. The resulting part has excellent dimensional accuracy and good reproduction of the mold detail, including any lettering and grained surfaces.

3. Air Blowing Process

Here compressed air is used to form the sheet. In one variation, a plastic sheet is heated and sealed across the female cavity (Figure 11.14). Air at controlled pressure is introduced into the mold cavity. This blows the sheet upward into an evenly stretched bubble. A plug which fits roughly into the mold cavity descends on the sheet. When the plug reaches its lowest possible position, a vacuum or, in some cases, air under pressure is used to complete part formation.

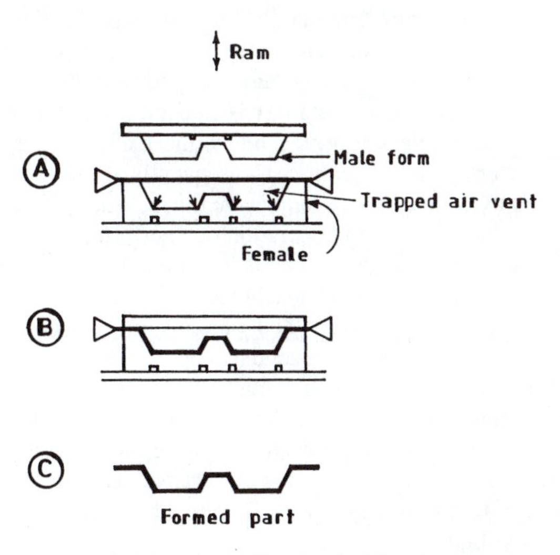

Figure 11.13 Steps in mechanical forming process.

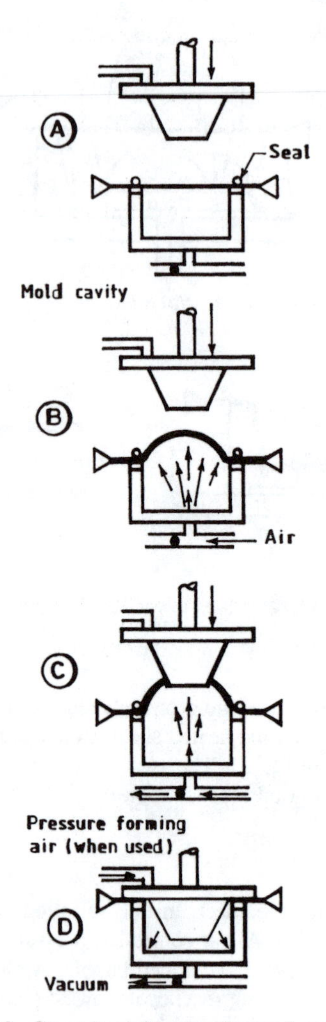

Figure 11.14 Steps in air-blowing thermoforming process.

From the foregoing discussion, it is evident that the basic steps in the thermoforming process involve a sequence of operations: loading the sheet or web through the system in increments and heating, forming, and cooling. Other secondary functions that may be integrated into the process include trimming and other finishing operations. Thermoforming machinery is categorized on the basis of the arrangement for performing these operations. In single station machines, the loading, heating, forming, cooling, and unloading operations are performed in succession. Consequently, even though these machines are versatile, they are characterized by comparatively long cycle times. With in-line machines, which are quite popular in the packaging industry, the different operations are performed simultaneously in a multistation in-line system. In this case, therefore, the cycle time is determined by the longest operation in the entire process. The third category of thermoforming machines is the rotary type, which is supported on a horizontal circular frame that has three work (and possibly other secondary stations for the loading–unloading, heating, and forming (possibly secondary) operations. By rotating the cut sheets sequentially from station to station, the rotary machine configuration is well suited for high production rates.

Molds for the thermoforming process may be made of wood, metal, or epoxy and are relatively cheap. They are provided with vents to allow trapped air to escape and release possible pressure buildup. Temperature control, as we shall see in the following discussion, determines part quality. It is, therefore, crucial that mold temperature is controlled properly. The mold is consequently provided with channels for the passage of the cooling liquid.

B. PROCESS VARIABLES

Thermoplastic materials, in general, can be stretched when hot. However, the degree of success in forming a part from a hot, stretched sheet material depends on the particular resin, its molecular weight, and processing conditions like forming temperature and speed. We recall that the thermal response of thermoplastic polymers depends first on whether the material is amorphous or crystalline and to some extent on their molecular weights. When heated, crystalline polymers undergo a rather abrupt phase transformation from the solid to the fluid state, while amorphous polymers undergo a more gradual transformation. For a given resin, the degree of fluidity depends on molecular weight and the processing temperature. The higher the molecular weight, the higher the melt strength and consequently the greater the capacity of the material to be deep-drawn, which is a necessary condition for the proper formation of intricate parts. On the other hand, materials that are too fluid at the forming temperature are susceptible to tearing, leading to the production of bad quality parts. Given the influence of forming temperature on part quality, it is necessary to ensure that sheets have a uniform temperature distribution. This also calls for a uniform sheet thickness. A variation in sheet thickness causes a nonuniform sheet temperature distribution resulting in uneven pulling and possible tearing of the sheet. It is also essential to recognize that variation of shrinkage with sheet orientation can generate forming problems. If there is excessive differential shrinkage due to sheet orientation, the pull from the clamping frames during forming becomes unbalanced and the sheet could be pulled out.

VII. COMPRESSION AND TRANSFER MOLDING[4,5]

The two most widely used methods for molding thermoset are compression and transfer molding. Thermosets, you will recall, undergo a permanent set, i.e., become essentially insoluble and infusible under the action of heat. Consequently, techniques for fabricating thermosets must take due cognizance of and make allowance for the fact that sprues, runners, and gates are not reusable and therefore constitute rejects.

A. COMPRESSION MOLDING

In compression molding, a preweighed amount of material is loaded into the lower half of a heated mold or cavity. The force plug (plunger) is lowered into the cavity, and pressure, which can range from 20 to 1000 tons, is applied to the powder (Figure 11.15). Under heat and pressure, the powder melts and flows into all parts of the mold cavity, the resin cross-links thus becoming irreversibly hardened. After an appropriate time, the mold is opened and the part is ejected while still hot (usually under gravity) and allowed to cool outside the mold.

The machinery for compression molding is relatively simple, consisting essentially of two platens, which when brought together, apply heat and pressure to the mold material to form a part of desired shape. The

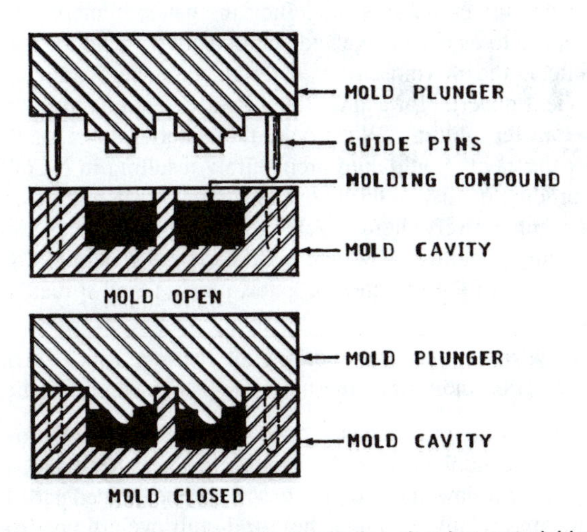

Figure 11.15 Schematic of a compression molding operation showing material before and after forming.

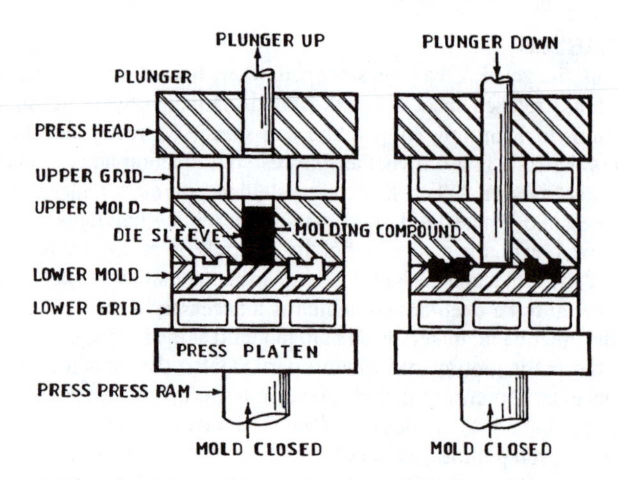

Figure 11.16 Schematic of a transfer molding operation.

platens move vertically, with the cavity usually mounted at the bottom so molding material can be loaded into it. Material may also be a preform, in which case it has been preheated. Material flow, cure time, and the ultimate properties of the molded part depend on the mold temperature, which therefore must be adequately controlled. A number of heating systems are employed for heating the molds. Steam heating provides uniform mold temperature but is limited to temperatures below 200°C. Electrical heating is both clean and easy and enjoys wide usage. Circulated hot oil also provides uniform heat. Most thermosets emit volatile gases or moisture during cure, and escape avenues must be provided for such volatiles to facilitate processing and obtain good quality parts. This is usually accomplished by opening the mold slightly. With modern compression molding machines various degrees of automation for feeding the material and ejecting the molded part are possible. This is achieved by the use of microprocessor controllers.

One of the major advantages of compression molding is that it is relatively inexpensive because of its simplicity. Also, since there are no sprues, runners, and gates, material waste is reduced considerably. The consistency of part size is good and the absence of gate and flow marks reduces finishing costs.

B. TRANSFER MOLDING[4,5]

In transfer molding, (Figure 11.16) a measured charge of preheated thermoset material is placed in a separate or auxiliary heated chamber called the pot. A plunger is then used to force the molten material out of the pot through the runner system into the closed mold cavity where curing occurs. As the material enters the mold, the air from the mold cavity escapes through vents located strategically on the mold. At the end of the cure cycle, the entire shot, including the gates, runners, sprues, and excess material remaining in the pot (referred to as cull) is ejected simultaneously with the molded part. If held for too long in the pot or cylinder, the thermosetting material could cure prematurely into a solid mass. Consequently, only sufficient material for a single shot is loaded at one time. In addition, preheating the material is necessary in transfer molding. When cold, the material flows relatively slowly. In this case, the first material to enter the cavity could cure prematurely resulting in overall improper material mix. This can result in poor product quality in terms of the surface finish and mechanical properties.

Transfer molding has comparatively shorter cycle and loading times than compression molding. Thick sections that cure evenly can be molded; however, mold costs are generally higher, and greater volumes of scrap are generated because of the presence of gates, runners, and sprues.

Examples 11.9: Explain why compression molding of thermoplastics is limited to small quantity production while screw injection molding is thermoset is also currently used on a limited scale.

Solution: In compression molding of thermoset, once the resin is cured the molded part can be ejected while still hot and allowed to cool outside the mold. On the other hand, compression molding of thermoplastics involves heating followed by cooling to solidify the molded part. Therefore, in compression molding of thermoplastics the relatively longer heat-and-chill cycle times involved are uneconomical for large volume production.

In screw injection molding of thermoplastics, the viscous heat generated assists in melting the resin material; during processing the resin is always held in the cylinder during and between shots. However, with thermoset, to avoid resin precure, not only must the barrel and material temperatures be carefully controlled, the resin cannot be held at high temperature for any length of time before molding.

VIII. CASTING[5]

In the casting process, a liquid material is poured into a mold and allowed to solidify by physical (e.g., cooling) or chemical (e.g., polymerization) means resulting in a rigid object that generally reproduces the mold cavity detail with great fidelity. A large number of resins are available and a variety of molds and casting methods are used in casting processes. Therefore, the choice of liquid material, mold type, and casting technique is determined by the particular application.

A. PROCESS DESCRIPTION

In casting processes, the resin material is added with an appropriate amount of hardener, catalyst, or accelerator, mixed manually or mechanically, and then poured into a mold, which is normally coated with a mold-release agent. Air is removed if necessary and the resin is allowed to solidify. The casting process is relatively slow and employs comparatively cheap equipment. Molds for casting processes are fabricated from a wide variety of materials, including wood, plaster and clay, glass, metal, rubber and latex (for flexible molds), and plastics. To facilitate the removal of the cast part from the mold, mold-releasing agents such as high melting waxes, silicone oils, greases, and some film-forming agents are used to coat the mold. Among other considerations, the choice of mold-release agents is based upon the absence of interaction between the resin system and the release agent.

The setting of the resin material is generally exothermic and is usually accompanied by a reduction in volume. For example, in the casting of acrylic, the amount of heat evolved is about 13.8 kcal/mol, while the volume reduction can be as large as 21%. The quantity of heat liberated depends on the size, but is independent of the shape of the casting. However, the rate of heat dissipation depends both on the size and shape of the casting.

Therefore, different methods have to be employed in casting thick or thin sections. Thin sections can be cast at room temperature with minimal possibility of cracking of the casting since heat can be dissipated rapidly. On the other hand, with thick sections, particularly those whose shapes limit the rate of heat removal, large temperature gradients exist between the interior and exterior sections of the casting. This generates internal stresses and enhances the possibility of cracking of the cast part. In this case, therefore, where the quantity of heat liberated cannot be changed, the rate of heat dissipation is controlled by carefully selecting the hardener and a mold material with appropriate thermal properties and by using a possible stepwise increase in casting temperature. To reduce shrinkage and the attendant built-in stresses in the cast part, flexibilizers, diluents, flexibilized resins or hardeners, or fillers are used.

B. CASTING PROCESSES

1. Casting of Acrylics

Acrylic castings are prepared with polymers derived from methacrylic ester monomers. They usually consist of poly(methyl methacrylate) or its copolymers as the major component modified with small amounts of other monomers. Acrylic castings are made by heating the monomers or partially polymerized syrups containing 0.02 to 0.1% radical initiator (e.g., peroxide or azo-bis-isobutyronitrile) in a mold. Syrups are prepared by dissolving finely divided polymer in the monomer. As indicated earlier, the polymerization reaction involved in the acrylic casting process is exothermic and is usually accompanied by about 15 to 21% volume shrinkage. Because of the exothermic nature of this reaction, measures are usually taken to deal with the problems of heat removal and residual stresses in the casting.

The outstanding optical properties of acrylic casting make acrylic sheets invaluable in applications where excellent transparency and resistance to UV are imperative. The fabrication of acrylic sheets therefore predominates in the acrylic casting process. Cast acrylic resins also account for most of the embodiments used for decorative or study purposes. In this case, hard polymers of ethyl or methyl methacrylate are used. Rods and tubes are also prepared from acrylic castings.

2. Casting of Nylon

The nylon casting process consists basically of four steps. These are the melting of the monomer, which is usually lactam flakes, the adding of the catalyst and activator, the mixing of the melts, and the casting process itself. Cocatalyzed anionic polymerization is currently the most widely used nylon casting method. The cocatalysts are strong bases and their salts with imides and lactams.

Since absorbed water can cause catalyst decomposition and hence incomplete polymerization and since lactam monomer flakes are highly hygroscopic, the melting of monomer is carried out under appropriately controlled temperature and humidity conditions. All additives are also completely dried and then mixed with the monomer in stainless steel vessels while flushing with inert gas under thermostatically controlled temperature. Molds can be of the single type fabricated from silicone rubber, epoxy, or sheet steel or the more expensive tool steel used in tight tolerance cast-to-size parts casting.

The nylon casting process is relatively more economical and is a more practical technique for the production of large and thick parts than comparable extrusion and injection molding processes. In addition, the crystallinity and molecular weight of cast nylon are higher than those of extruded or molded nylon. Consequently, cast nylon has a much higher modulus and heat deflection temperature, improved solvent resistance, and better hygroscopic characteristics and dimensional stability.

IX. PROBLEMS

11.1. A small business entrepreneur is considering buying an extruder for a 2-in. (ID) PVC pipe production. A two 8-h/d shift operation for a 300-day work year is planned. If the extruder is to be operated to produce 20 ft of pipe per hour and pipe thickness is 0.5 in., suggest the power rating for the electric motor appropriate for this extruder. PVC has a density of 1.44 g/cm^3.

11.2. Explain the following observations.

a. When complicated shapes are to be extruded, they are usually made from amorphous polymers rather than crystalline polymer.
b. Under comparable thermal conditions, extrusion rate is generally higher with amorphous polymers than crystalline polymers.
c. Many extruders are now equipped with dehumidified hopper driers.

11.3. The following table shows some thermal properties of three thermoplastic polymers. Compute the screw output for each polymer material if the screw drive power input is 60 hp and the mechanical efficiency is 75%. Assume that the heat requirement for molding the materials is satisfied by the viscous dissipation and feed temperature is 77°F.

Polymer	Avg Molding Temp. (°F)	Specific Heat (Btu/lb °F)	Heat of Fusion (Btu/lb)
Polystyrene	500	0.45	0
Polyethylene	440	0.91	104
Nylon 6,6	530	0.63	56

11.4. For a 2.5-in.-diameter screw at 200 rpm the maximum permissible drive input power is 40 hp. What is the maximum permissible drive input power for a 4.5-in.-diameter screw at (a) 150 rpm; (b) 200 rpm.

11.5. A syringe 6 in. long, 1 in. in diameter, and 0.1 in. thick is to be injection molded. The stress generated in the mold cavity and core material as the mold is filled with the plastic melt is 180,000 psi. Estimate the change in volume of the syringe if the mold cavity and core materials are made of:

a. Steel
b. Copper
c. Aluminum

Assume that the volume change is due essentially to the change in the core diameter.

11.6. For crystalline polymers, high mold temperatures result in enhanced tensile strength but reduced clarity of the molded part. Explain.

11.7. Decide which of the two processes (continuous extrusion blow molding or intermittent extrusion blow molding) is more suitable for the production of bottles from:

a. Polyethylene
b. Poly(vinyl chloride)

Explain the basis of your decision.

REFERENCES

1. Richardson, P.N., *Introduction to Extrusion*, Society of Plastics Engineers, CT, 1974.
2. Weir, C.L., *Introduction to Molding,* Society of Plastics Engineers, CT, 1975.
3. Bernhardt, Ed., *Processing of Thermoplastic Materials,* Robert E. Krieger, New York, 1974.
4. Kaufman, H.S. and Falcetta, J.J., Eds., *Introduction to Polymer Science and Technology,* John Wiley & Sons, New York, 1977.
5. *1989/90 Modern Plastics Encyclopedia,* McGraw-Hill, New York, 1990.
6. Frados, J., Ed., *The Story of the Plastics Industry,* 13th ed., Society of the Plastics Industry, New York, 1977.
7. Chastain, C.E., A plastics primer, in *1969/70 Modern Plastics Encyclopedia,* McGraw-Hill, New York, 1970.
8. Dunham, R.E., *Plast. Eng.,* 48(8), 21, 1992.
9. Mock, J., *Plast. Eng.,* 39(12), 17, 1983.

PART III: PROPERTIES AND APPLICATIONS

Chapter 12

Solution Properties of Polymers

I. INTRODUCTION

In order to gain insight into the polymer dissolution process, let us briefly review the dissolution of low-molecular-weight (simple) substances. We know, for example, that while oil will not mix with water, an oil stain in clothing can be removed rather easily by using hydrocarbon solvents like naphtha. On the other hand, ordinary table salt, or sodium chloride, dissolves readily in water but not in gasoline. As will become evident presently, the physical phenomena associated with the solubilities of various substances in different solvents are intimately tied with the nature of the solutes and solvents. For example, in molecular crystals, the attractive forces are of the dipole–dipole or London dispersion type, which is relatively weak and therefore fairly easy to break apart. Consequently, this type of solid dissolves to an appreciable degree in nonpolar solvents, where the molecules are held together by London-type attractive forces also. However, crystals will not dissolve to any great extent in polar solvents since the strong attraction between the polar solvent molecules cannot be overcome by the much weaker solute–solvent interaction forces. By similar arguments, polar solutes and ionic solids are soluble only in polar solvents. They are insoluble in nonpolar solvents because the weak solute–solvent interaction is not strong enough to overcome the strong attractive forces between the solute molecules and hold them apart. In essence, therefore, when a solute dissolves in a solvent, solute–solute molecular contacts are replaced by solute–solvent contacts. Consequently, for solute particles to enter into solution, the solute–solvent forces of attraction must be sufficient to overcome the forces that hold the solid together.

It follows from the above discussion that polymers, by virtue of their macromolecular nature, will be soluble only in selected solvents. The polymer solution process is certainly more complex than that of simple compounds. The dissolution of both simple compounds and polymers depends on the nature of the solute and solvent, but in addition the dissolution of polymers is affected by the viscosity of the medium, polymer texture, and molecular weight. Dissolution of a polymer is necessarily slow and is a two-staged process: first, the solvent molecules diffuse into the polymer producing a swollen gel; second, the gel breaks down slowly forming a true solution. In some cases and depending on the nature of the polymer, only the first step occurs. However, if the polymer–polymer interaction forces can be overcome by polymer–solvent attraction, then the second stage will follow, albeit slowly. For example, unvulcanized rubber will dissolve in solvents in which vulcanized rubber will only swell. In other cases, materials with strong polymer–polymer intermolecular forces due to, say, cross-linking (phenolics), crystallinity (Teflon), or strong hydrogen bonding (native cellulose) will not dissolve in any solvent at ordinary temperatures and will exhibit only a limited degree of swelling.

II. SOLUBILITY PARAMETER (COHESIVE ENERGY DENSITY)

From thermodynamic considerations, it is possible to predict whether or not a given solute will be soluble in a given solvent using the relation:

$$\Delta G_m = \Delta H_m - T\Delta S_m \tag{12.1}$$

where ΔG_m, ΔH_m, and ΔS_m are free energy, heat, and entropy of mixing, respectively. Solubility will occur if the free energy of mixing ΔG_m is negative. The entropy of mixing is believed to be always negative. Therefore, the sign and magnitude of ΔH_m determine the sign of ΔG_m. If we consider an ideal solution of two small spherical molecules with identical size and intermolecular forces, molecules of one type can replace neighbors with molecules of another type without changing the total energy of the system. This interchangeability of neighbors is the source of the configurational entropy term or the entropy of mixing ΔS_m. Since we are dealing with ideal conditions, the heat of mixing ΔH_m is zero because the two types of molecules have the same force fields, and consequently, $\Delta G_m = -T\Delta S_m$. However, departure from ideality normally occurs because the intermolecular forces operative between similar and dissimilar molecules give rise to a finite heat of mixing. In this case, the energy of mixing associated with the formation of contact between two dissimilar molecules can be shown to be positive. Therefore, mixing is generally endothermic for nonpolar molecules in the absence of strong intermolecular attraction such as hydrogen bonding.

Using similar arguments, Hildebrand and Scott[1] showed that

$$\Delta H_m = V\phi_1\phi_2\left[\left(\Delta E_1^v/V_1\right)^{1/2} - \left(\Delta E_2^v/V_2\right)^{1/2}\right]^2 \tag{12.2}$$

where V, V_1, V_2, are the volumes of the solution and components and the subscripts 1 and 2 denote the solvent and polymer, respectively. ΔE^v is the molar energy of vaporization and ϕ_1 and ϕ_2 are volume fractions. In terms of the heat of mixing per unit volume, Equation 12.2 becomes

$$\frac{\Delta H_m}{V} = \phi_1\phi_2\left[\left(\Delta E_1^v/V_1\right)^{1/2} - \left(\Delta E_2^v/V_2\right)^{1/2}\right]^2 \tag{12.3}$$

The quantity $\Delta E/V$ is referred to as the cohesive energy density (CED): its square root is the solubility parameter (δ). Thus

$$CED = \frac{\Delta E}{V} = \delta^2 \tag{12.4}$$

Equation 12.3 may be rewritten:

$$\frac{\Delta H_m}{V} = \phi_1\phi_2\left[\delta_1 - \delta_2\right]^2 \tag{12.5}$$

To a first approximation and in the absence of strong intermolecular forces like hydrogen bonding, a polymer is expected to be soluble in a solvent if $\delta_1 - \delta_2$ is less than 1.7–2.0. Equation 12.5 is valid only when ΔH_m is zero or greater. It is invalid for exothermic mixing, that is, when ΔH_m is negative. Typical values of δ for various types of solvents are shown in Table 12.1. Values for some polymers were listed in Table 3.7. The magnitude of the enthalpy of mixing can be conveniently estimated from these tables.

Table 12.1 Values for Different Solvents

Solvent	δ	Solvent	δ	Solvent	δ
Poorly hydrogen bonded		**Moderately hydrogen bonded**		**Strongly hydrogen bonded**	
n-Pentane	7.0	Diethyl ether	7.4	2-Ethylhexanol	9.5
n-Heptane	7.4	Diisobutyl ketone	7.8	Methyl isobutyl carbinol	10.0
Apco thinner	7.8	*n*-Butyl acetate	8.5	2-Ethylbutanol	10.5
Solvesso 150	8.5	Methyl propionate	8.9	*n*-Pentanol	10.9
Toluene	8.9	Dibutyl phthalate	9.3	*n*-Butanol	11.4
Tetrahydronaphthalene	9.5	Dioxane	9.9	*n*-Propanol	11.9
O-Dichlorobenzene	10.0	Dimethyl phthalate	10.7	Ethanol	12.7
1-Bromonaphthalene	10.6	2,3-Butylene carbonate	12.1	Methanol	14.5
Nitroethane	11.1	Propylene carbonate	13.3		
Acetonitrile	11.8	Ethylene carbonate	14.7		
Nitromethane	12.7				

From Burrel, H. and Immergut B., in *Polymer Handbook*, Brandrup, J. and Immergut E.M., Eds., John Wiley & Sons, New York, 1967. With permission.

Table 12.2 Molar Attraction Constants, E (cal cm^3)/mol

Group	E	Group	E
$-CH_3$	148	NH_2	226.5
$-CH_2-$	131.5	–NH–	180
>CH–	86	–N–	61
>C<	32	C=N	354.5
$CH_2=$	126.5	NCO	358.5
–CH=	121.5	–S–	209.5
>C=	84.5	Cl_2	342.5
–CH= aromatic	117	Cl primary	205
–C= aromatic	98	Cl secondary	208
–O– ether, acetal	115	Cl aromatic	161
–O– epoxide	176	F	41
–COO–	326.5	Conjugation	23
>C–O	263	*cis*	–7
–CHO	293	*trans*	–13.5
$(CO)_2O$	567	six-membered ring	–23.5
–OH–	226	ortho	9.5
OH aromatic	171	meta	6.5
–H acidic dimer	–50.5	para	40

From Hoy, K.L., *J. Paint Technol.,* 42, 76, 1970. With permission.

In addition, the solubility parameter can be estimated from the molar attraction constants, E, using the structural formula of the compound and its density (Table 12.2). For a polymer:

$$\delta_2 = \frac{\rho \Sigma E}{M} \tag{12.6}$$

where ρ and M are the density and molecular weight, respectively, of the polymer repeating unit.

Example 12.1: Estimate the solubility parameters of the following polymers:

a. Low-density polyethylene (LDPE)
b. High-density polyethylene (HDPE)
c. Polypropylene (PP)
d. Polystyrene (PS)

Solution: From Equation 12.6,

Polymer	Repeating Unit	M	ρ	ΣE	δ
a. LDPE	$-CH_2-CH_2-$	28	0.92	$\frac{0.92[131.5+131.5]}{28}$	8.6
b. HDPE	$-CH_2-CH_2-$	28	0.95	$\frac{0.93[131.5+131.3]}{28}$	8.9
c. PP	$-CH_2-CH(CH_3)-$	42	0.90	$\frac{0.90[131.5+86+148]}{42}$	7.8
d. PS	$-CH_2-CH(C_6H_5)-$	104	1.04	$\frac{1.04[131.5+86+(5\times 117)+98]}{104}$	9.0

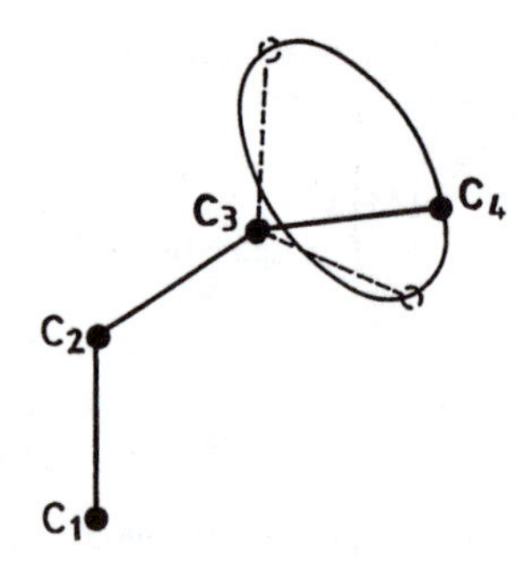

Figure 12.1 A segment of a polymer chain, showing four successive chain atoms. The first three of these define a plane, and the fourth can lie anywhere on the indicated circle perpendicular to and bisected by the plane.

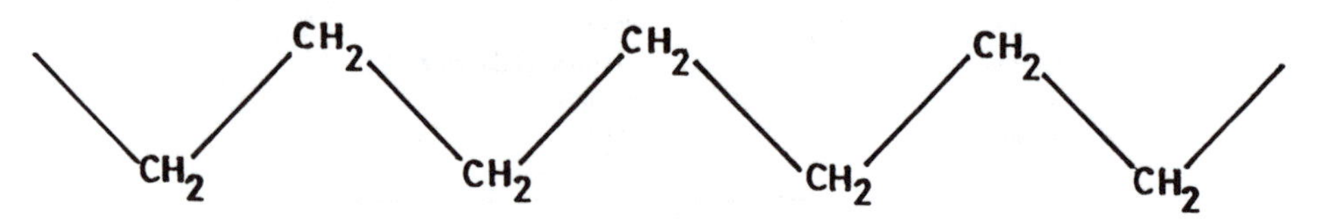

Figure 12.2 Fully extended chain with every carbon atom in *trans* location and in the same plane.

III. CONFORMATIONS OF POLYMER CHAINS IN SOLUTION

Molecules in the dissolved, molten, amorphous, and glassy states of macromolecules exist as random coils. This is a result of the relative freedom of rotation associated with the chain bonds of most polymers and the myriad number of conformations that a polymer molecule can adopt. As a consequence of the random coil conformation, the volume of a polymer molecule in solution is many times that of its segments alone. The size of the dissolved polymer molecule depends quite strongly on the degree of polymer–solvent contact. In a thermodynamically good solvent, a high degree of interaction exists between the polymer molecule and the solvent. Consequently, the molecular coils are relatively extended. On the other hand, in a poor solvent the coils are more contracted. Many properties of macromolecules are dictated by the random coil nature of the molecules. We now discuss briefly the conformational properties of polymer chains.

First, we need to develop a realistic physical picture of a polymer molecule. To do this, let us consider the properties of a single molecule in a dilute solution isolated from its neighbors by the molecules of the solvent. Let us consider initially a short segment of this molecule consisting only of four methylene groups, as shown in Figure 12.1. We define a plane by the first three carbon atoms C_1 to C_3. Since there is free rotation about the C–C bond, the fourth carbon atom, C_4, can be found in any position on the circle shown in the figure. Of course, some positions are more probable than others since absolutely free rotation about bonds is precluded by steric hindrance. Each successive atom on the chain can, in turn, occupy any random position on similar cycles based on the position of the preceding atom. It is easy to visualize, therefore, that for a molecule composed of thousands of atoms, the number of possible conformations is virtually limitless. One of these conformations is the fully extended chain, in which each successive carbon atom is coplanar and translocated with respect to the earlier atoms in the chain. The conventional formula for polymethylene expressed this configuration (Figure 12.2).

A. END-TO-END DIMENSIONS

Any physical property of a polymer molecule that depends on its conformation can ordinarily be expressed as a function of some sort of average dimension. The polymer dimension that is most often used to describe its spatial character is the displacement length, which is the distance from one end of the molecule to the other. For the fully extended chain, this quantity is referred to as the contour length. Given the extremely large number of possible conformations and number of chains, a statistical average, such as the root-mean-square end-to-end distance, $(\bar{r}_2)^{1/2}$, is required to appropriately express this quantity.

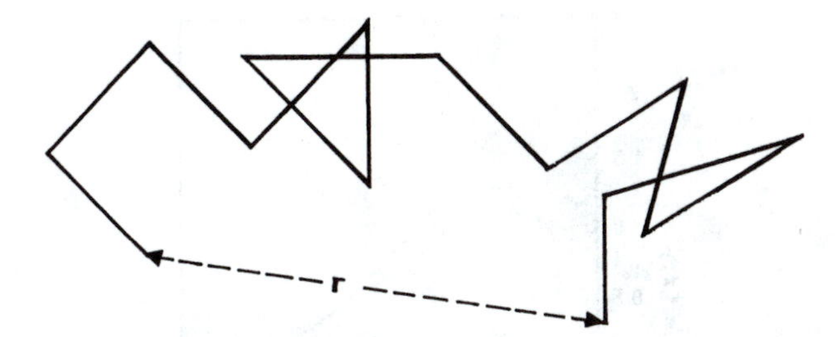

Figure 12.3 Freely jointed model of a polymer molecule with fixed and equal bond length and unrestricted value of bond angle.

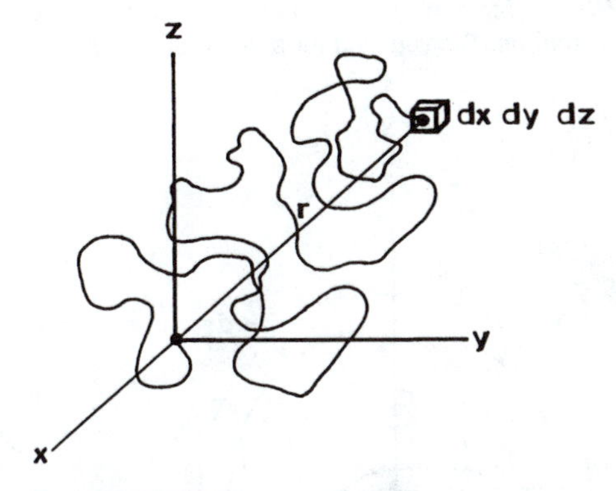

Figure 12.4 A highly schematic representation of a random-coil polymer chain, with one end at the origin of a coordinate system and the other in a volume element dx dy dz at a distance $(x^2 + y^2 + z^2)^{1/2}$ from the origin.

Another way of expressing the effective size of a molecule is the radius of gyration, $(\overline{S}^2)^{1/2}$. This is the root-mean-square distance of the elements of the chain from its center of gravity. For linear polymers, the root-mean-square end-to-end distance has a simple relation with its radius of gyration $(S^2)^{1/2}$, given by Equation 12.7:

$$\overline{r^2} = 6\,\overline{s^2} \tag{12.7}$$

B. THE FREELY JOINTED CHAIN

We begin by considering first a hypothetical freely jointed chain consisting of n bonds of fixed length l, jointed in a linear sequence without any restriction on the magnitude of the bond angles. Since the bond angles are free to assume all values with equal probability and rotations about bonds are similarly free, a given bond can assume all directions with equal probability regardless of the directions of its neighbors in the chain. Such a chain is illustrated in Figure 12.3. We are aware, however, that no real polymer approximates this model. The problem of determining the end-to-end distance, r, is reduced to that of random flight that occurs in diffusion theory. The question is determining the probability of finding one end of the chain in a volume element dx dy dz at a distance r from the other end (Figure 12.4). It can be shown that the solution of this random flight problem, for a very long chain unperturbed by self-interactions of long-range and external constraints, the probability per unit volume W(r), is a Gaussian distribution function shown in Figure 12.5. This shows that there is a much greater chance of finding the two ends close to each other and that as the two ends move farther apart, the probability decreases continuously. Another way of interpreting the curve is in terms of density distribution of chain ends. That is, if one end of the chain is located at the origin, the probability of finding the other end in a unit volume close to the origin is highest. On the other hand, granted that one end of the chain is at the origin, we want to find out the probability that the other end of the chain is in a spherical shell of thickness dr and at a distance r (Figure 12.6). This is given by Equation 12.8:

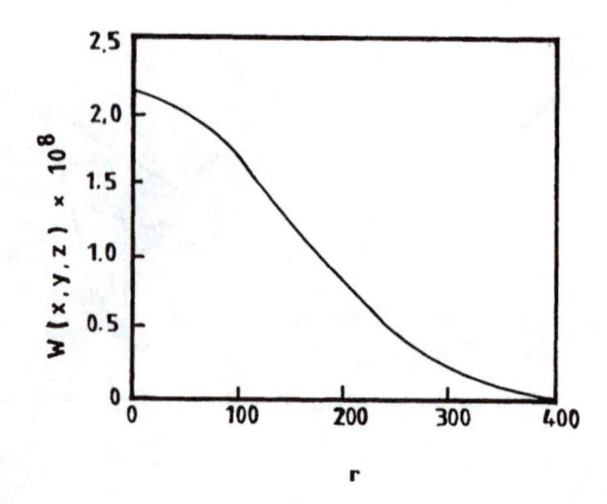

Figure 12.5 The probability W(x, y, z) of finding the end of the chain of Figure 12.4 in the volume element dx dy dz as a function of r (in angstrom units) calculated for a chain of 10^4 links, each 2.5 Å long.

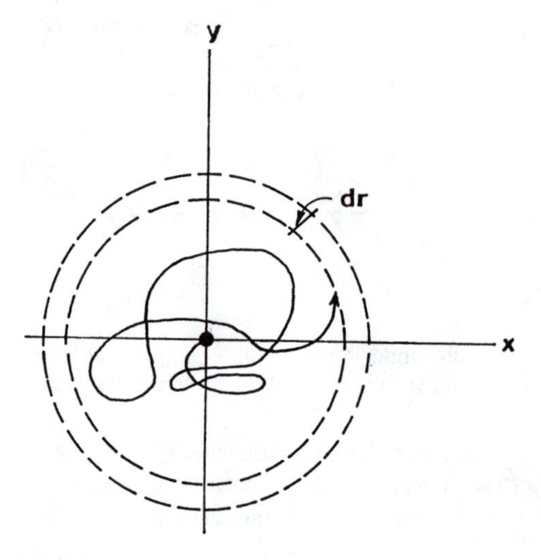

Figure 12.6 Spherical shell thickness dr at distance r from origin.

$$\int_0^\infty W(r) 4\pi r^2 \, dr = 1 \tag{12.8}$$

The integral is equal to one since there is a definite chance of finding the desired end in space. However, since the chain length has a finite value, the upper limit of the integral should appropriately be the contour length. That is,

$$\int_0^\infty W(r) 4\pi r^2 \, dr = \overline{r^2} \tag{12.9}$$

This is shown in Figure 12.7, which demonstrates that the maximum probability corresponds to the most probable dimension for the chain. Assuming the root-mean-square end-to-end distance represents the most probable chain dimension, then, according to random flight theory,

$$\left(\overline{r_f^2}\right)^{1/2} = ln^{1/2} \tag{12.10}$$

Here r^2 is the square of the magnitude of the end-to-end distance averaged over all conformations, and f denotes the result of random flight calculation.

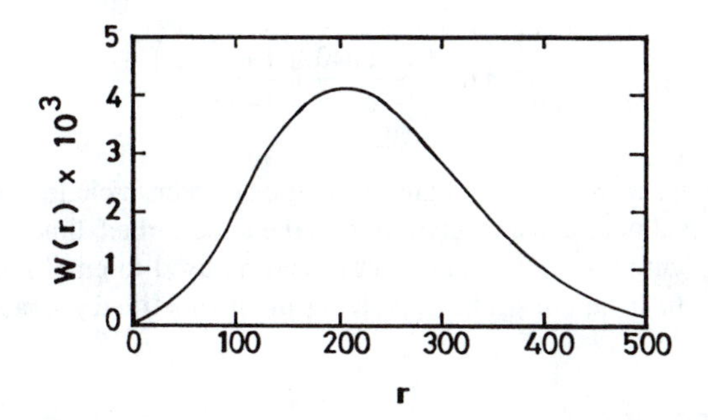

Figure 12.7 The probability of W(r) of finding the end of the chain of Figure 12.4 in a spherical shell of thickness dr at a distance r from the origin.

C. REAL POLYMER CHAINS

The freely jointed model discussed in the previous section grossly underestimates the true dimensions of real polymer molecules because it ignores restrictions to completely free rotation arising from fixed bond angles and steric effects (short-range interactions). Also, it fails to account for the effects of long-range interactions that result from the inability of two chain segments to physically occupy the same space at the same time.

1. Fixed Bond Angle (Freely Rotating)

In real polymer chains, the direction assumed by a given bond depends strongly on that of its immediate predecessor and to a smaller extent on the orientation of nearby bonds. While the structure of the chain unit determines the ultimate nature of the restrictions on given a bond, the overall effect of these short-range interactions is to expand the conformation of the real polymer chain relative to that obtained from the random flight model of the same contour length. The effect of fixed bond angles is a modification of the expression for the unrestricted bond angles from random flight (Equation 12.11):

$$\bar{r}^2 = nl^2\left(\frac{1-\cos\theta}{1+\cos\theta}\right) \tag{12.11}$$

where θ is the bond angle.

2. Fixed Bond Angles (Restricted Rotation)

Restriction to free rotation about bonds due to steric interferences between successive units of the chain leads to a further expansion of chain dimensions. Let us amplify this by considering Figure 12.8. Here θ_i is the valence (bond) angle between bonds i and i + 1; α_i is its supplement (i.e., $\alpha_i = 180 - \theta_i$) and ϕ_i is the angle of rotation about i. It is a measurement of the dihedral angle between the planes defined by the two planes i – 1, i; and i + 1. The direction of bond i + 1 is, as we have seen from the above discussion, a function of bond angle θ_i. In addition, it also depends on the direction of bond i – 1 through the angle of rotation ϕ_i. However, as a result of hindrance to free rotation, ϕ_i cannot assume all values from 0 to 2π with equal probability; it is limited to certain preferred values. The same argument holds for each bond in relation to its predecessor. When some conformations are preferred over others as a result of restriction to free rotation, Equation 12.11 becomes

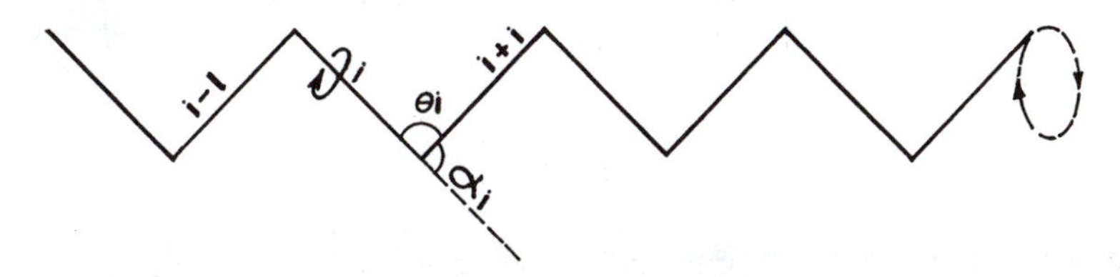

Figure 12.8 Rotation about bond i.

$$\overline{r^2} = nl^2\left(\frac{1-\cos\theta}{1+\cos\theta}\right)\left(\frac{1+\cos\phi}{1-\cos\phi}\right) \tag{12.12}$$

The dimension obtained from random flight calculation, which includes the effects of bond angles and hindrances to rotation about bonds, is referred to as the unperturbed dimension of the polymer chain. It is represented by the symbol $(\overline{r_0^2})^{1/2}$. The subscript zero is used to emphasize the condition that the molecule is subject only to local constraints involving the geometrical character of the bond structure and restricted rotation.

3. Long-Range Interactions

The freely jointed model assumes implicitly that two elements of the same molecule, possibly remote along the chain, can occupy the same position in space at the same time. In real polymer chains conformations in which this exists are impossible. Each segment of a real polymer chain exists within a volume that excludes all other segments. The number of such forbidden conformations that must be excluded is greater for the more compact arrangements with smaller values of $\overline{r^2}$. The net effect of such long-range interaction is to expand the actual chain dimension $(\overline{r^2})^{1/2}$ over its unperturbed dimensions, $(r_0^2)^{1/2}$ by an expansion coefficient defined by Equation 12.13:

$$\left(\overline{r^2}\right)^{1/2} = \alpha\left(\overline{r_0^2}\right)^{1/2} \tag{12.13}$$

The magnitude of α depends on the environment of the polymer molecule. In a thermodynamically good solvent, where there is strong polymer–solvent interaction, α is large. By the same token, α is small in a poor solvent. The value of α is therefore an indirect measure of the magnitude of the polymer–solvent interaction forces, or the solvent power. When α is 1, these forces become zero and by definition the polymer assumes its unperturbed conformation. However, polymer–solvent interaction forces and, consequently α, depend on temperature. For a given solvent, the temperature at which $\alpha = 1$ is referred to as Flory temperature θ. When a solvent is used at $T = \theta$, it is called theta (θ) solvent. Alpha (α) increases with n (or M) with solvent power and with temperature. Other parameters for characterizing the dimensions of polymer molecules are summarized in Table 12.3. The physical significance of the parameters is illustrated by a brief discussion of the Flory characteristic ratio and the Stockmayer–Kurato ratio.

a. Flory's Characteristic Ratio (C_∞)

The quantity α represents the effect of "long-range interactions." It describes the osmotic swelling of the chain due to polymer–solvent interaction. On the other hand, r_0^2 represents the effect of "short-range

Table 12.3 Parameters Characterizing Chain Dimensions

Parameter	Relation to Unperturbed Dimension
Flory's characteristic ratio (C_∞)	$C_\infty = \dfrac{\overline{r_0^2}}{nl^2}$
Stockmayer–Kurato ratio (σ)	$\sigma = \dfrac{\left(\overline{r_0^2}\right)^{1/2}}{\left(r_{0f}^2\right)^{1/2}}$
Characteristic length (a)	$a = C_\infty^{1/2}\, l j^{1/2}$ j = number of backbone bonds per monomer unit (usually 2)
Kratky–Porod persistence length (a_p)	$a_p = \dfrac{1}{2}\left(C_\infty + 1\right)$

Table 12.4 C_∞ and σ for Some Polymers

Polymer	Solvent	Temperature (°C)	C_∞	σ
Polyethylene	Decalin	140	6.8	1.84
Polypropylene				
Isotactic	Tetralin	140	5.2	1.61
Syndiotactic	Heptane	30	6.1	1.75
Atactic	Decalin	135	5.3	1.63
Poly(methylmethacrylate)	Benzene	21	9.0	2.12
Polystyrene				
Isotactic	Benzene	30	10.5	2.30
Atactic	Cyclohexane	34	10.4	2.28

From Kurata, M., Tsunashima, Y., Iwama, M., and Kamada, K., in *Polymer Handbook*, 2nd ed., Brandrup, J. and Immergut, E.H., Eds., John Wiley & Sons, New York, 1975. With permission.

interactions" induced by bond angle restrictions and steric hindrances to internal rotation. Flory's characteristic ratio, C_∞, is defined by

$$C_\infty = \frac{\overline{r_0^2}}{nl^2} \tag{12.14}$$

It is a measure of the effect of short-range interactions.

b. Stockmayer–Kurato Ratio (σ)

The freely rotating state is a hypothetical state of the chain in which the bond angle restrictions are operative but in which there are no steric hindrances to internal rotation. The Stockmayer–Kurato ratio, α, reflects such rotational isomerism preferences. That is, it is a measure of the effect of steric hindrance to the average chain dimension. It is given by

$$\sigma = \frac{\left(\overline{r_o^2}\right)^{1/2}}{\left(\overline{r_{0f}^2}\right)^{1/2}} \tag{12.15}$$

Here $(r_{0f}^2)^{1/2}$ is the root-mean-square end-to-end distance of the hypothetical chain with the same bond angles but with free rotation around valence cones. Table 12.4 lists the values of C_∞ and α for some polymers.

Example 12.2: A polyethylene molecule has a degree of polymerization of 2000. Calculate (a) the total length of the chain and (b) the contour length of the planar zigzag if the bond length and valence angle are 1.54 Å and 110°, respectively.

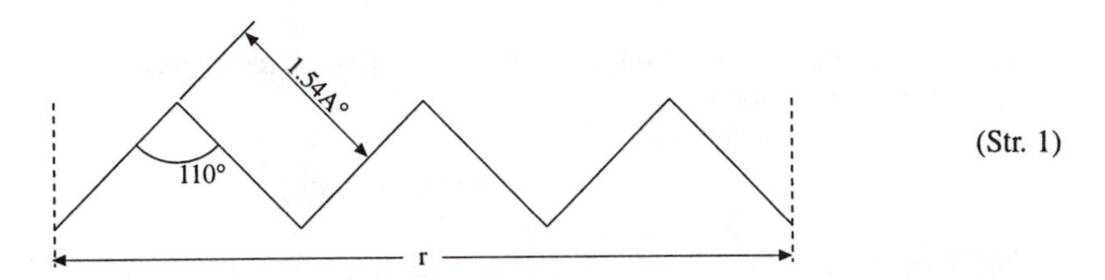

(Str. 1)

Solution:

a. The total length of the chain, L, is the sum of the length of each bond, l. It is the total distance traversed going from one end of the chain to the other following the bonds.

$$L = nl$$

where n = the number of bonds. Each monomer contributes the equivalent of 2 C–C bonds. Therefore,

$$n = 2 \times DP$$

$$= 2 \times 2000 = 4000$$

$$L = 1.54 \text{ Å} \times 4000 = 6.16 \times 10^3 \text{ Å}$$

b. The contour length is that of the fully extended chain conformation.

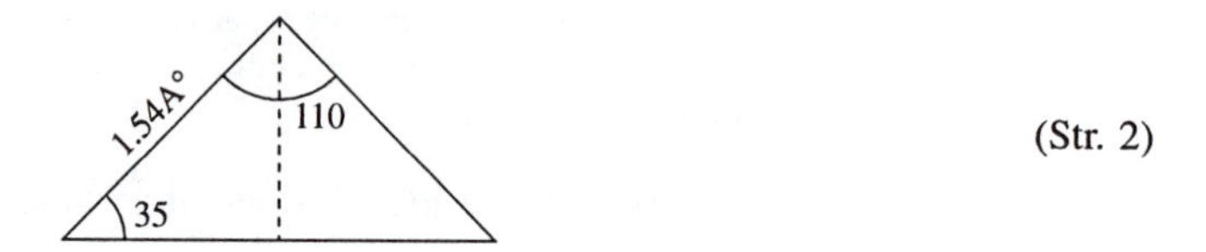

(Str. 2)

r = n (1.54 cas 35)
= 2 × DP (1.54 cas 35)
= 2 × 2000 × 1.54 cas 35
= 5.05 × 10^3 A^8

IV. THERMODYNAMICS OF POLYMER SOLUTIONS

As may be expected, polymers behave differently toward solvents than do low-molecular-weight compounds. Studies of the solution properties of polymers provide useful information about the size and shape of polymer molecules. In this section we discuss how some of the molecular parameters discussed in the previous sections are related to and can be calculated from thermodynamic quantities. We start with a discussion of the simplest case of an ideal solution. This is followed by a treatment of deviations from ideal behavior.

A. IDEAL SOLUTION

Consider a binary mixture of two types of molecules that are roughly identical in size, shape, and external force field. Such a mixture constitutes an ideal solution. Thus, one of the components of an ideal solution may replace another without seriously disturbing the circumstances of immediate neighbors in the solution. Raoult's law provides an appropriate basis for the treatment of an ideal solution. Raoult's law states that the activity, a, of a solvent in the solution is equal to its mole fraction n_1:

$$a_1 = \frac{N_1}{N_1 + N_2} = n_1 \tag{12.16}$$

where N_1, N_2 are the number of solvent and solute molecules, respectively. The free energy of mixing ΔG_{mix} is given by Equation 12.17:

$$\Delta G_{mix} = \Delta H_{mix} - T\Delta S_{mix} \tag{12.17}$$

For the ideal solution, since the intermolecular force fields around the two types of molecules are the same, $\Delta H_{mix} = 0$. Thus Equation 12.17 becomes

$$\Delta G_{mix} = -T\Delta S_{mix} \tag{12.18}$$

The entropy of an ideal solution is greater than that of the pure components because the number of possible arrangements of the molecules of the components of a solution is much greater than that for the molecules of a pure components. If N_0 is total number of molecules (i.e., $N_0 = N_1 + N_2$), then the total number of possible combinations of N_0 taken N_1 at a time is

$$W = \frac{N_0!}{N_1!\,N_2!} \tag{12.19}$$

From the Boltzmann relation,

$$\begin{aligned}\Delta S_{mix} &= k \ln W \\ &= k \ln \frac{N_0!}{N_1!\,N_2!}\end{aligned} \tag{12.20}$$

$$\Delta S_{mix} = k \ln (N_1 + N_2)! - k \ln N_1! - k \ln N_2! \tag{12.21}$$

Using Stirling's approximation, $\ln N! = N \ln N - N$, and rearranging in terms of mole fractions, Equation 12.20 becomes

$$\Delta S_{mix} = k[N_1 \ln n_1 + N_2 \ln n_2] \tag{12.22}$$

According to Raoult's law, the partial vapor pressure of each component in a mixture is proportional to its mole fraction. Thus for a binary solution consisting of solvent and a polymer, the partial pressure of the solvent in the solution, P_1, is related to that of the pure solvent P_1^0 by

$$P_1 = n_1 P_1^0 \tag{12.23}$$

Since the molecular weight of a polymer is usually at least three orders of magnitude greater than that of the solvent, for a small weight fraction of the solvent $N_1 \gg N_2$, consequently the mole fraction of the solvent approaches unity very rapidly. This, in effect, means that following Raoult's law, the partial pressure of the solvent in the solution should be virtually equal to that of the pure solvent over most of the composition range. Available experimental data do not confirm this expectation even if volume fraction is substituted for mole fraction. Polymer solutions exhibit large deviations from the ideal law except at extreme dilutions, where ideal behavior is approached as an asymptotic limit.

B. LIQUID LATTICE THEORY (FLORY–HUGGINS THEORY)

One of the reasons for the failure of the ideal solution law is the assumption that a large polymeric solute molecule is interchangeable with the smaller solvent molecule. The law also neglects intermolecular forces since the heat of mixing (ΔH_{mix}) is assumed to be zero. The Flory–Huggins theory attempted to remedy these shortcomings in the ideal solution law.[5,6–8]

1. Entropy of Mixing

In order to calculate the entropy of mixing of a polymer solution, the polymer chain is assumed to be composed of x chain segments, where x is the ratio of the molar volumes of the solute polymer and the solvent. Each chain segment represents the portion of the polymer molecule equal in size to a solvent molecule. This means that a polymer chain segment can replace a solvent molecule in the liquid lattice and vice versa. However, unlike a solution containing an equal proportion of a monomeric solute, a polymer solution requires a set of x contiguous or consecutive lattice cells to accommodate the polymer molecule (Figure 12.9). A further assumption is that the solution is sufficiently concentrated that the occupied lattice sites are randomly distributed instead of being sparse and widely separated, which would exist in a very dilute solution.

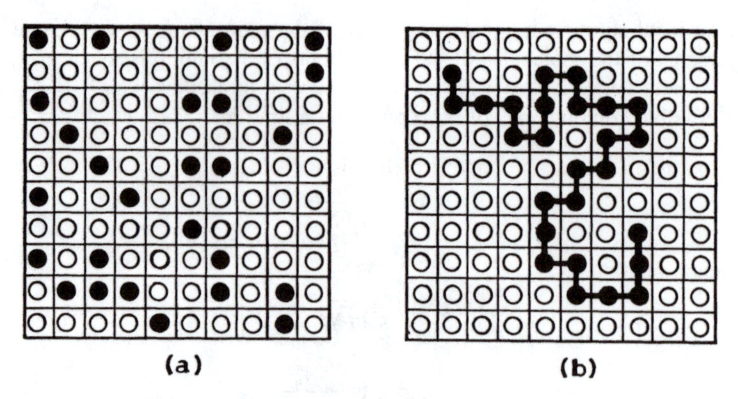

Figure 12.9 Binary solution of a monomer showing distribution of solute in lattice cells (a) compared with that of a polymer, which requires a set of contiguous cells to accommodate the solute (b).

The entropy of mixing of a polymer solution is relatively smaller than that of an equivalent proportion of a monomeric solute. This is because the macromolecular nature of the polymer molecule severely restricts the number of possible arrangements of polymer segments in the lattice sites. Once a given segment occupies a site, the number of sites available for adjacent segments becomes seriously limited. The dissolution of the polymer is conceived to occur in two consecutive steps: first, the polymer is disoriented, and then the disoriented polymer mixes with the solvent. The entropy of mixing of the disoriented polymer and solvent has been shown by Flory to be given by

$$\Delta S_{mix} = k\left[N_1 \ln v_1 + N_2 \ln v_2\right] \tag{12.24}$$

where the subscripts 1 and 2 denote the solvent and polymer, respectively, and v_1 and v_2 are their volume fractions defined as

$$v_1 = \frac{N_1}{N_1 + xN_2} \tag{12.25}$$

$$v_2 = \frac{xN_2}{N_1 + xN_2} \tag{12.26}$$

2. Heat and Free Energy of Mixing

To derive an expression for the heat of mixing of a polymer solution, the pure solvent and pure liquid polymer are taken as reference states. The heat of mixing is considered to be the difference between the total interaction energy in the solution relative to that of the pure components. It arises from the replacement of some of the solvent–solvent and polymer–polymer contacts in the pure components with solvent–polymer contacts in the solutions. As the distance between uncharged molecules increases, the forces between them decrease very rapidly. Consequently, interactions between elements that are not immediate neighbors can be safely neglected. If only the energies developed by first-neighbor elements are considered, then the heat of mixing of polymer solution, like that of ordinary solutions, is given by

$$\Delta H_{mix} = \chi_1 \, kTN_1 \, v_2 \tag{12.27}$$

where χ characterizes the interaction energy per solvent molecule divided by kT. Combining Equation 12.24 with that of the configurational entropy (Equation 12.27) gives the Flory–Huggins expression for the free energy of mixing of a polymer solution:

$$\Delta G_{mix} = kT(N_1 \ln v_1 + N_2 \ln v_2 + \chi_1 N_1 v_2) \tag{12.28}$$

Quantities that are determinable from experiments can be derived from Equation 12.28. For example, by differentiating the expression with respect to N_1, the number of solvent molecules, and multiplying the result by Avogadro's number, the relative partial molar free energy $\overline{\Delta G_1}$ is obtained:

$$\overline{\Delta G_1} = RT\left[\ln(1 - v_2) + (1 - 1/x)v_2 + \chi_1 v_2^2\right] \tag{12.29}$$

We note that $\overline{\Delta G_1}$ is expressed on a per mole basis. The activity of the solvent, a_1, and the osmotic pressure of the solution, π, are given by Equations 12.30 and 12.31, respectively:

$$\ln a_1 = \ln(1 - v_2) + (1 - 1/x)v_2 + \chi_1 v_2^2 \tag{12.30}$$

$$\pi = -\frac{RT}{V_1}\left[\ln(1 - v_2) + (1 - 1/x)v_2 + \chi_1 v_2^2\right] \tag{12.31}$$

where V_1 is the molar volume of the solvent. Expanding the logarithmic term and neglecting higher order terms, the expression for the osmotic pressure becomes

$$\pi = \frac{RT}{V_1}\left[\frac{v_2}{x} + \left(\frac{1}{2} - \chi_1\right)v_2^2 + \frac{1}{3}v_2^3 + \cdots\right] \tag{12.32}$$

The above thermodynamic expressions for a binary solution of a polymer in a solvent include the dimensionless parameter χ_1. Its value can be determined by measuring any of the experimentally obtainable quantities, like solvent activity or the osmotic pressure of the solution. The constancy of χ_1 over a wide composition range would be a confirmation of the validity of the Flory–Huggins theory. Figure 12.10 represents such a plot obtained by the measurement of solvent activities for various systems.

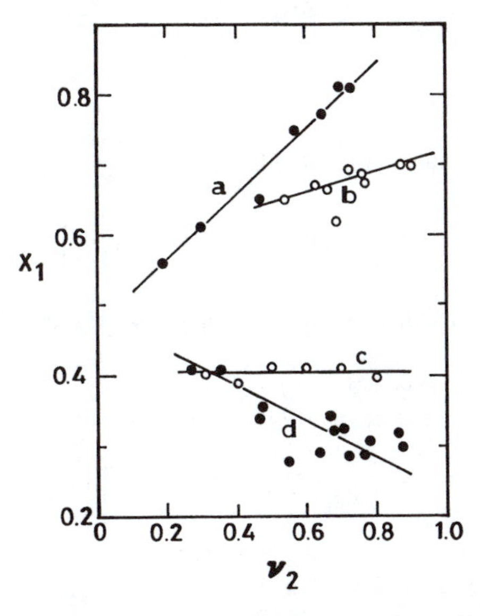

Figure 12.10 Experimentally observed variation of χ_1 with concentrations: (a) poly(dimethylsiloxane) in benzene; (b) polystyrene in benzene; (c) rubber in benzene; curve (d) polystyrene in toluene. [Curve (a) from Newing, M.J., *Trans. Faraday Soc.*, 46, 613, 1950; (b) and (d) from Bawn, C.E.H., Freeman, R.F.J., and Kamaliddin, A.R., *Trans. Faraday Soc.*, 46, 677, 1950; (c) from Gee, G. and Orr, W.J.C., *Trans. Faraday Soc.*, 42, 507, 1946; and Gee, G., *J. Chem. Soc.*, p. 280, 1947.

Only in the case of the nonpolar rubber–benzene system was the predicted constancy of χ_1 observed; other systems showed marked deviations from theory.

Equation 12.29 can be separated into contributions from the heat of dilution and configurational entropy, as shown in Equations 12.33 and 12.34, respectively:

$$\overline{\Delta H_1} = RT\chi_1 v_2^2 \tag{12.33}$$

$$\overline{\Delta S_1} = -R[\ln(1-v_2)+(1-1/x)v_2] \tag{12.34}$$

where $\overline{\Delta H_1}$, is the relative partial molar heat content while $\overline{\Delta S_1}$ is the relative partial molar configurational entropy of the solvent in solution. $\overline{\Delta H_1}$ was determined from the heat of mixing obtained from calorimetric methods. The predicted concentration dependence of $\overline{\Delta H_1}$ was also not observed.

C. DILUTE POLYMER SOLUTIONS (FLORY–KRIGBAUM THEORY)

The Flory–Huggins lattice model assumed a uniform density of lattice occupation. This assumption holds only for concentrated solutions; it is invalid for dilute polymer solutions. According to the Flory–Krigbaum model, a very dilute polymer solution consists of loose domains or clusters of polymer chain segments separated by intervening regions of pure solvent. Each such cloud is assumed to be approximately spherical with an average density that is maximum at the center and that decreases with increasing distance from the center in an approximately Gaussian function. Each molecule within a domain or occupied volume tends to exclude all other molecules. Long-range thermodynamic interactions occur between segments within such an excluded volume. The expression for the excess, relative, partial molar-free energy for these interaction is given by the relation:

$$\overline{\Delta G_1} = RT(\kappa_1 - \psi_1)v_2^2 \tag{12.35}$$

κ_1 and χ_1 are heat and entropy parameters also given by

$$\overline{\Delta H_1} = RT\kappa_1 v_2^2 \tag{12.36}$$

$$\overline{\Delta S_1} = R\psi_1 v_2^2 \tag{12.37}$$

If x is assumed to be infinite, Equation 12.29 reduces to

$$\overline{\Delta G_1} = RT[\ln(1-v_2)+v_2+\chi_1 v_2^2] \tag{12.38}$$

Expanding the logarithmic term and, as before, neglecting terms of order higher than 2 (i.e., $\ln(1-v_2) = -v_2 - \frac{v_2^2}{2} - \ldots$), Equation 12.38 becomes

$$\overline{\Delta G_1} = -RT\left[\left(\frac{1}{2}-\chi_1\right)v_2^2\right] \tag{12.39}$$

By comparing Equations 12.35 and 12.39, it is obvious that

$$\kappa_1 - \psi_1 = \frac{1}{2} - \chi_1 \tag{12.40}$$

Table 12.5 Behavior of Polymer Molecules with Change in Thermodynamic Parameters

Thermodynamic Parameter	Behavior of Polymer Molecules
$\kappa_1 = 0$ ($\theta = 0$)	A thermal solvent, $\delta\Delta G>0$; increase in segment concentration within volume element causes a decrease in entropy; solution wants to dilute itself; polymer coils move apart
$\kappa_1 > 0$ ($T < \theta$)	Polymer–polymer contacts preferred to polymer–solvent contacts; spontaneous concentration
$\kappa_1 = \psi_1$ ($T = \theta$)	$\delta\Delta G = 0$; excluded volume effects eliminated; polymer in an unperturbed state
$\kappa_1 < 0$ ($T > \theta$)	Exothermic solution; $\delta\Delta G = 0$; spontaneous dilution; expansion of polymer coil occurs as a result of interaction with solvent

The deviations from ideality in a polymer solution can be eliminated by selecting a temperature where $\overline{\Delta H}_1 = T\overline{\Delta S}_1$ or when $\kappa_1 = \psi_1$. The temperature at which these conditions prevail is called the Flory or theta temperature, θ, and is defined by

$$\theta = \frac{\kappa_1 T}{\psi_1} \tag{12.41}$$

It follows that

$$\psi_1 - \kappa_1 = \psi_1(1 - \theta/T) \tag{12.42}$$

and

$$\overline{\Delta G_1} = -RT\psi_1(1 - \theta/T)v_2^2 \tag{12.43}$$

We note for emphasis (also from Equation 12.43) that at the temperature $T = \theta$ the excess, relative, partial molar-free energy, ΔG_1, due to polymer–solvent interactions is zero and deviations from ideality vanish.

The change in free energy δ (ΔG) in a volume element when two molecules are brought together depends algebraically on the magnitude of $\psi_1 - \kappa_1$ or ψ_1 $(1 - \theta/T)$. Since the entropy of dilution is usually positive, the sign of the free energy change will depend on the relative magnitudes of ψ_1 and κ_1 or if θ/T is greater than unity. Table 12.5 summarizes the expected behavior of polymer molecules with thermodynamic parameters.

It follows from the above discussion that as the solvent is made poorer, i.e., as ψ_1 $(1 - \theta/T)$ decreases, the excluded volume shrinks and at $T = \theta$ it disappears entirely. In other words, as the solvent becomes poorer, polymer–polymer repulsion diminishes and at the θ-point the net interaction becomes zero. Where $T < \theta$, polymer molecules attract each other and the excluded volume is negative. When the temperature is much lower than the θ-point, precipitation occurs.

D. OSMOTIC PRESSURE OF POLYMER SOLUTIONS

Osmotic pressure, as indicated earlier, is one of the quantities that can be obtained experimentally from the Flory–Huggins and Flory–Krigbaum theories. Before we illustrate how thermodynamic parameters characteristic of polymers can be derived from osmotic pressure measurements, let us first explain very briefly the basis of these measurements.[13]

Consider the apparatus shown schematically in Figure 12.11. The semipermeable membrane, represented by the dashed line, allows the passage of solvent but not the solute. Suppose in the first instance that both sides of the tube contain only the pure solvent. At equilibrium the levels of the liquid in both arms would be at the same height and the external pressure would be P_A. In this case, the chemical potential of the solvent on both sides would be the same. Suppose a solute is now added to the right-hand side. Since it cannot pass through the semipermeable membrane, it must remain on right-hand side. The chemical potential of the solvent on the right-hand side (solution) is now less than that of the solvent on the left-hand side (pure solvent). If the external pressure on the right-hand side is maintained

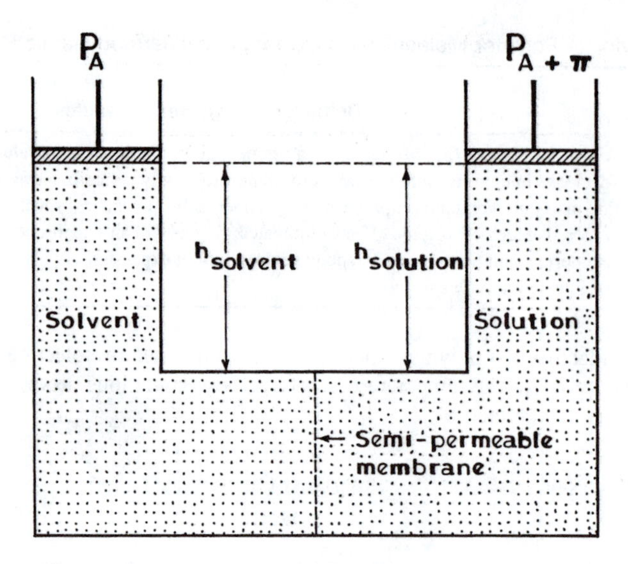

Figure 12.11 Schematic representation of an osmometer.

at P_A, the liquid level on the right-hand side will rise as the solvent passes from the left (higher chemical potential) to the right (lower chemical potential) to equalize the chemical potential on both sides. However, this flow of solvent can be prevented if the external pressure on the solution is increased so as to keep the liquid levels the same on both sides. The additional pressure is the osmotic pressure, π, of the solution. It arises as the driving force for solvent flow in response to the reduction of the chemical potential of the solvent due to the addition of a solute.

From Equation 12.32, the osmotic pressure is given by

$$\pi = \frac{RT}{V_1}\left[\frac{v_2}{x} + \left(\frac{1}{2} - \chi_1\right)v_2^2 + \frac{1}{3}v_2^3 + \cdots\right] \tag{12.32}$$

It is generally more convenient to replace volume fraction with the concentration, C, of the solution expressed in weight per unit volume:

$$v_2 = C\bar{v}_2 \tag{12.44}$$

where v_2 is the (partial) specific volume of the polymer. Since $x = V_2/V_1$, then

$$\frac{v_2}{xV_1} = \frac{c\bar{v}_2}{xV_1} = \frac{C}{M_2} \tag{12.45}$$

where M_2 is the molecular weight or the polymer. For heterogeneous polymers, M_2 is replaced by the number-average molecular weight $\overline{M}_n$. From Equation 12.45

$$v_2 = \frac{c\,x\,V_1}{M_2} \tag{12.46}$$

$$x = \frac{\bar{v}_2\,M_2}{V_1} \tag{12.47}$$

Substituting Equations 12.46 and 12.47 successively in Equation 12.32 and rearranging terms yields

$$\frac{\pi}{c}=\frac{RT}{M_2}\left[1+\left(\frac{1}{2}-\chi_1\right)\frac{\bar{v}_2^2\,M_2}{V_1}c+\frac{\bar{v}_2^3\,M_2}{3\,V_1}c^2+\cdots\right] \tag{12.48}$$

In applications to osmotic data, Equation 12.48 is most frequently preferred in the following forms:

$$\frac{\pi}{c}=RT\left[A_1+A_2c+A_3c^2+\cdots\right] \tag{12.49}$$

$$\frac{\pi}{c}=\frac{RT}{M_2}\left[1+\Gamma c+g\Gamma^2c^2+\cdots\right] \tag{12.50}$$

where $A_1 = 1/M_2$ and A_2 and Γ are the second virial coefficients given by

$$\Gamma=\frac{\bar{v}_2^2}{V_1}\left(\frac{1}{2}-\chi_1\right) \tag{12.51}$$

$$g\Gamma^2=\frac{\bar{v}_2^3\,M_2}{3\,V_1} \tag{12.52}$$

Osmotic pressure data can, therefore, be used for the determination of the number-average molecular weight, M_n, or the solvent–polymer interaction parameter χ_1. This depends, of course, on knowing the values of the densities and specific volumes of the polymer and the solvent. In good solvents, g is approximately 1/4 in which case Equation 12.50 becomes

$$\left(\frac{\pi}{c}\right)^{1/2}=\left(\frac{RT}{M_2}\right)^{1/2}\left[1+\frac{1}{2}\Gamma c\right] \tag{12.53}$$

In poor solvents, $g \cong 0$. In either case, an appropriate plot of π/c vs. c gives $M_2(\overline{M}_n)$ and χ_1.

Other thermodynamic parameters can be obtained from osmotic pressure. For example, the chemical potential of the solvent in the solution is given by $-\chi/\pi V_1$. From the foregoing discussion, it is evident that the thermodynamic behavior of the dilute polymer solution depends on the following factors:

1. Molecular weight
2. The interaction parameter ψ_1 and κ_1 or ψ_1 and θ, which characterize the segment–solvent interactions
3. The size or configuration of the molecules in solution

The first factor is usually determined from the coefficient A_1. A_2 (or Γ_2) depends on all three factors. Therefore, to evaluate the thermodynamic functions that depend on A_2, it is necessary to determine the size of the molecule in solution independently. The parameter α is related to the thermodynamic quantities according to Equation 12.54:

$$\alpha^5-\alpha^3=4C_M\,\psi_1\,(1-\theta/T)\,M^{1/2} \tag{12.54}$$

where C_M represents all the numerical and molecular constants. The expansion factor α^3 and $\bar{r}_0^2/M$ may be determined from suitable measurements of intrinsic viscosities. The quantity $\psi_1(1-\theta/T)$ may then be deduced from the relevant equation. If measurements are made over a limited range of temperature in the vicinity of θ, $\psi_1(1-\theta/T)$ may be resolved into its components θ and ψ_1 from a plot of A_2 against the reciprocal of the absolute temperature.

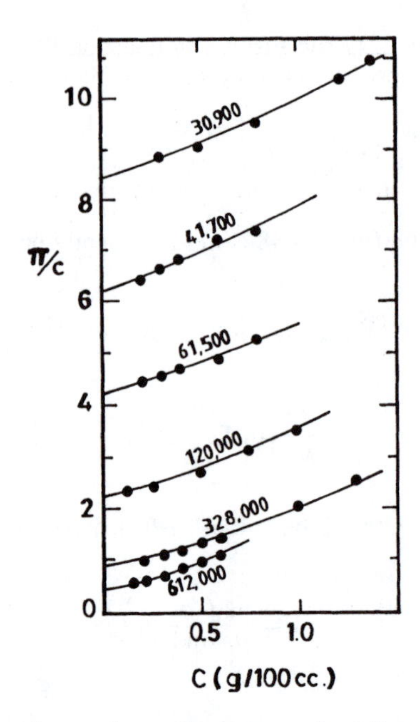

Figure 12.12 Concentration dependence of osmotic pressure of solutions of polystyrene in toluene at 30°C. Numbers on each curve denote polymer molecular weights. (From Flory, P.J., *Principles of Polymer Chemistry*, Cornell University Press, Ithaca, NY, 1953. With permission.)

Typical results of osmotic pressure measurements are shown in Figure 12.12. The experimental data are in consonance with the predicted positive curvature in good solvents. However, these data also show that the quantity $\psi_1(1 - \theta/T)$ decreases with increasing molecular weight. This is contrary to theory, which predicts that these thermodynamic parameters should characterize the inherent segment–solvent interaction independent of the entire molecular structure. The theory predicts that α increases without limit with molecular weight and that χ_1 is constant over a wide range of polymer concentrations. However, experimental data show that χ_1 increases with increasing volume fraction v_2. In spite of these shortcomings, the Flory–Huggins intermolecular interaction theory is in reasonably satisfactory agreement with experimental data within the approximations made in the theory. It provides a semiquantitative description of polymer solutions from which parameters characteristic of polymers may be derived. Table 12.6 gives values of χ_1 for some polymer–solvent systems.

Table 12.6 Polymer–solvent Interaction Parameter at 25°C

Polymer	Solvent	χ_1
Poly(dimethylsiloxane)	Cyclohexane	0.42
	Chlorobenzene	0.47
Polyisoprene (natural rubber)	Benzene	0.42
Poly(methyl methacrylate)	Chloroform	0.377
	Tetrahydrofuran	0.447
Polystyrene	Ethylbenzene	0.40
	Methyl ethyl ketone	0.47
	Cyclohexane	0.505

From Wolf, B.A., in *Polymer Handbook,* 2nd ed., Brandrup, J. and Immergut, E.M., Eds., John Wiley & Sons, New York, 1967. With permission.

Example 12.3: The data on the osmotic pressure of a sample of poly(vinyl acetate) in methyl ethyl ketone at 25°C are shown below:

Weight Fraction of Polymer	Pressure in cm of Solution
0.0021	0.40
0.0032	0.61
0.0057	1.23
0.0061	1.44
0.0082	2.10
0.0083	2.25
0.0093	2.52
0.0100	2.76
0.0114	3.54
0.0122	3.73

The densities of the polymer and solvent are 1.190 and 0.800, respectively; the solution densities can be calculated by assuming additivity of the volumes of the components. Using both linear and square-root plots, calculate the polymer molecular weight and the second virial coefficients, A_2 and Γ.

Solution: Basis: 1 g of solution:

$$\text{Vol of solution } V_s = \frac{\omega_2}{\rho_2} + \frac{1-\omega_2}{\rho_1} \left(\text{in cm}^3\right)$$

where ω_2 = weight fraction of polymer; ρ_1 and ρ_2 are the densities of the solvent and polymer, respectively.

$$\text{Concentration of Solution, } C = \frac{\omega_2}{V_s} \left(\text{g/cm}^3\right)$$

From these relations, data can be transformed into a Figure E12.3A,B. Now

$$\frac{\pi}{c} = RT\left[\frac{1}{M_n} + A_2 c + A_3 c^2 + \cdots\right]$$

$$\left(\frac{\pi}{c}\right)^{1/2} = \left(\frac{RT}{\overline{M}_n}\right)^{1/2} [1 + 0.5\,\Gamma c]$$

From Figure E12.3a,

$$\frac{RT}{\overline{M}_n} = \text{intercept on } \pi/C \text{ as } c \to 0$$

$$\text{Intercept on } \pi/c = 194 \, \frac{\text{cm solution}}{\text{g/cm}^3}$$

$$\overline{M}_n = \frac{RT}{\text{intercept}} = \left(82.06 \, \frac{\text{cm}^3\,\text{atm}}{\text{mol-K}}\right)(298\text{ K})\left(\frac{1}{194 \, \frac{\text{cm solution}}{\text{g/cm}^3}}\right)$$

$$= \left(82.06 \, \frac{\text{cm}^3 - 1033 \text{ cm } H_2O}{\text{mol-K}}\right)(298\text{ K})\left(\frac{1}{194 \times 0.8 \, \frac{\text{cm } H_2O}{\text{g/cm}^3}}\right)$$

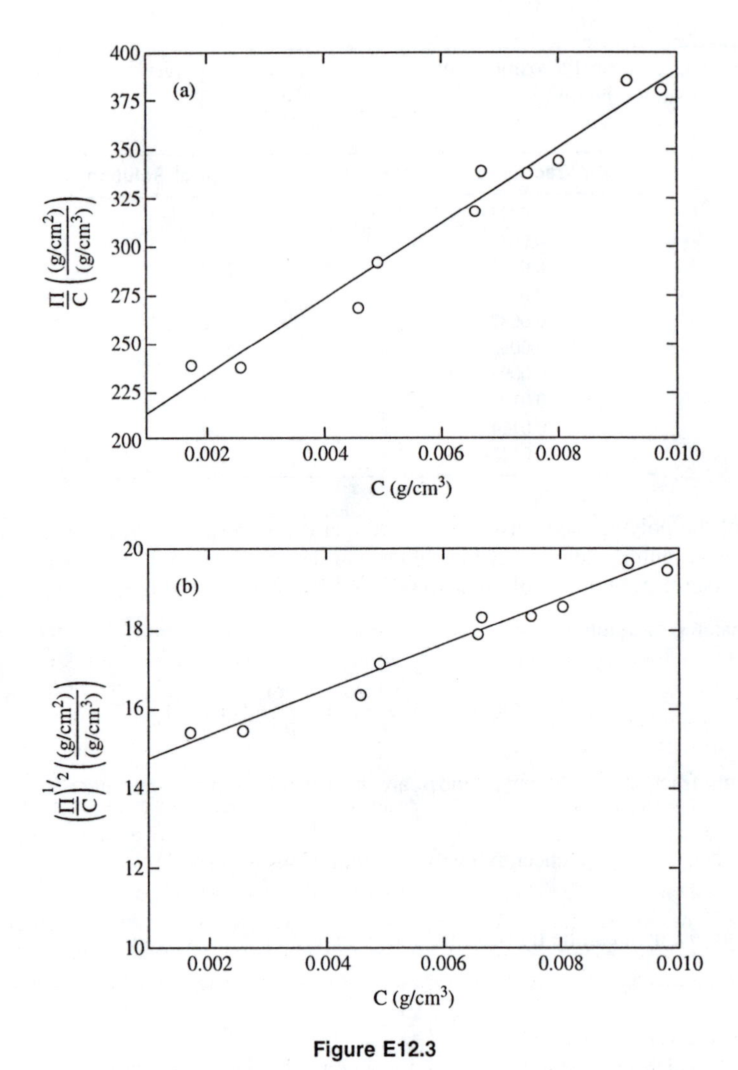

Figure E12.3

Note: 1 atm = 1033 cm H_2O; cm solution xρ_s = cm H_2O. Since the solution is dilute it is assumed that the density of solution is approximately that of the solvent.

$$\overline{M}_n = 1.63 \times 10^5 \text{ g/mol}$$

$$\text{Slope} = \text{RTA}_2$$

$$A_2 = \frac{\text{slope}}{\text{RT}} = \frac{19648 \text{ cm solution}/\left(\text{g/cm}^3\right)^2}{82.06\left(\text{cm}^3 - \text{atm/mol–K}\right)(298 \text{ K})}$$

$$= \frac{19648 \times 0.8 \text{ cm H}_2\text{O}/\left(\text{g/cm}^3\right)^2}{82.06\left(\text{cm}^3 - \text{cm H}_2\text{O/mol–K} \times 1033\right)(298 \text{ K})}$$

$$= 6.2 \times 10^{-4} \text{ mol} - \text{cm}^3/\text{g}^2$$

From Figure E12.3b,

$$\text{Slope} = 563 = 0.5\,\Gamma\left(\frac{RT}{\overline{M}_n}\right)^{1/2}$$

$$\Gamma = \frac{563}{0.5}\left(\frac{RT}{\overline{M}_n}\right)^{-1/2}$$

$$\left(\frac{RT}{M_n}\right)^{1/2} = \text{intercept at}\left(\frac{\pi}{c}\right)^{1/2} \text{ as } c \to 0$$

$$= 14.22$$

$$\Gamma = \frac{563}{0.5 \times 14.22} = 79.2\ \text{cm}^3/\text{g}$$

V. SOLUTION VISCOSITY

Rheology by definition is the science of deformation and flow of matter. Rheological measurements provide useful behavioral and predictive information for various products in addition to knowledge of the effects of processing, formulation changes, and aging phenomena. Material processability can also be determined through rheological studies. Rheology deals with those properties of materials that determine their response to mechanical force. For solids, as we shall see in subsequent chapters, this involves elasticity and plasticity. For fluids, on the other hand, rheological studies involve viscosity measurements. Viscosity is a measure of the internal friction of a fluid. For example, it has been observed that even at low concentrations of a dissolved polymer, the viscosity of a solution relative to that of the pure solvent is increased appreciably. This is due to the unusual size and shape of polymer molecules and the nature of their solutions. Thus measurements of the viscosity of polymer solutions can provide information about monomer molecular weight, molecular weight distribution, and other material characterization parameters. Before we deal with the relation between solution viscosity and polymer characterization parameters, we discuss briefly the various terms used to describe viscosity.

A. NEWTON'S LAW OF VISCOSITY[16,17]

Consider a fluid, which may be a gas or liquid, contained between two large parallel plates of area A and separated by the distance Y (Figure 12.13). The system is initially at rest. Now suppose the lower

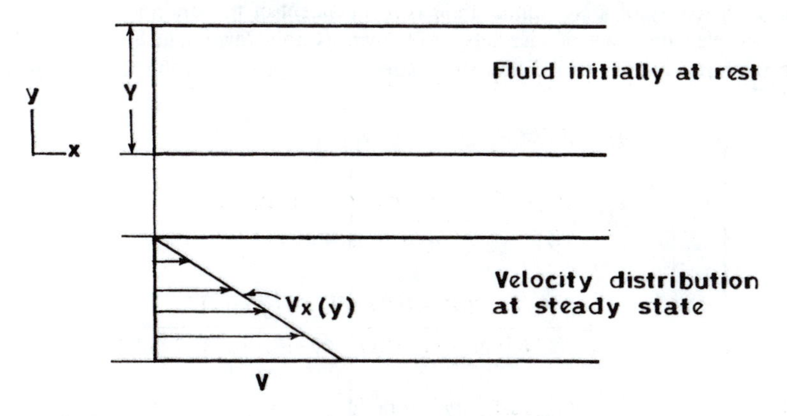

Figure 12.13 Laminar velocity profile for fluid between two plates.

plate is set in motion in the x-direction at a constant velocity. The fluid gains momentum with time, and at steady state a constant force, F, is required to maintain the motion of the lower plate. For laminar flow, this force is given by Equation 12.55:

$$\frac{F}{A} = \eta \frac{V}{Y} \tag{12.55}$$

According to this equation, the force per unit area is proportional to the velocity decrease in the distance Y. This constant of proportionality, η, is called the viscosity of the fluid. It is desirable and convenient to rewrite Equation 12.55 in a form that permits us to give a molecular interpretation to the meaning of viscosity (Equation 12.56):

$$\tau_{yx} = -\eta \frac{dV_x}{dy} \tag{12.56}$$

Equation 12.56 states that the shear stress is proportional to the negative of the local velocity gradient. This is Newton's law of viscosity, and fluids that exhibit this behavior are referred to as Newtonian fluids. According to this law, in the neighborhood of the surface of the moving plate (i.e., at y = 0), the fluid acquires a certain amount of x-momentum. This fluid, in turn, transmits some of its momentum to the adjacent layer of fluid causing it to remain in motion in the x-direction; in effect, the x-momentum is transmitted in the y-direction. The velocity gradient is a measure of the speed at which intermediate layers move with respect to each other. For a given stress, fluid viscosity determines the magnitude of the local velocity gradient. Fluid viscosity is due to molecular interaction; it is a measure of a fluid's tendency to resist flow, and hence it is usually referred to as the internal friction of a fluid.

Using the chain rule, the velocity gradient in Equation 12.56 can be interpreted differently:

$$\frac{dV}{dy} = \frac{d}{dy}\left(\frac{dx}{dt}\right) = \frac{d}{dt}\left(\frac{dx}{dy}\right) = \frac{d\gamma}{dt} = \dot{\gamma} \tag{12.57}$$

where γ is the strain rate. In a more general form, Equation 10.56 becomes

$$\tau = -\eta\dot{\gamma} \tag{12.58}$$

In this form, Newton's law simply states that for laminar flow, the shear stress needed to maintain the motion of a plane of fluid at a constant velocity is proportional to the strain rate. At a given temperature, the viscosity of a Newtonian fluid is independent of the strain rate (Figure 12.14). Fluids that do not obey Newton's law of viscosity are known as non-Newtonian fluids. For non-Newtonian fluids, when the strain rate is varied, the shear stress does not vary in the same proportion, i.e., the

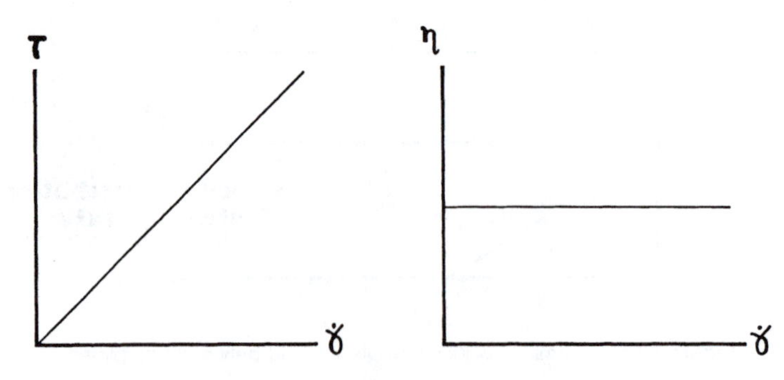

Figure 12.14 Behavior of Newtonian fluids.

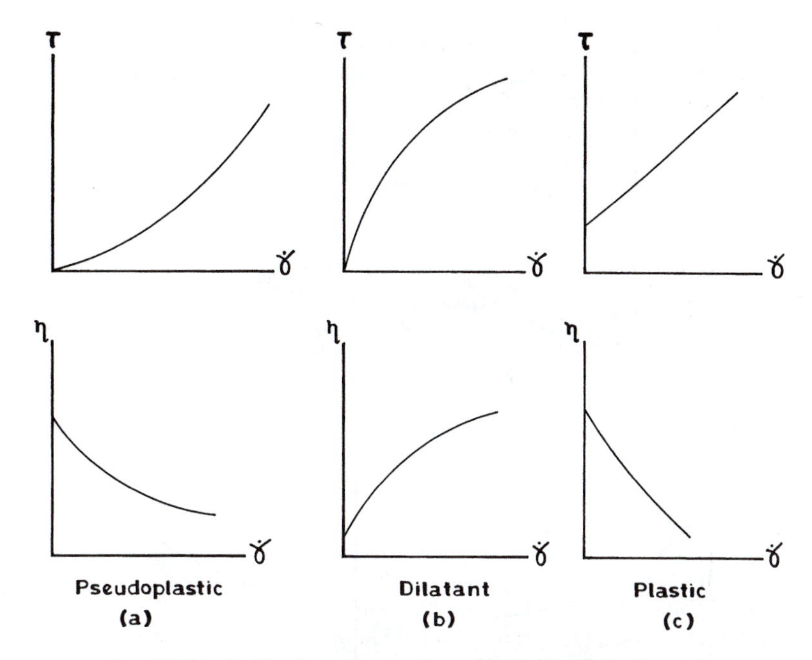

Figure 12.15 Some types of non-Newtonian behavior.

viscosity is not independent of the strain rate. In physical terms, non-Newtonian behavior means that as molecules pass each other, their size, shape, and cohesiveness determine how much force is required to maintain the movement. When the strain rate is changed, molecular alignment may change also as will the force required to maintain motion. There are several types of deviation from Newtonian behavior. Each is characterized by the way the fluid viscosity changes in response to variations in the strain rate (Figure 12.15). Pseudoplastic fluids display a decrease in viscosity with increasing strain rate, while a dilatant fluid is characterized by an increase in viscosity with increasing strain rate. For fluids that exhibit plastic behavior, a certain amount of stress is required to induce flow. The minimum stress necessary to induce flow is frequently referred to as the yield value. In addition, some fluids will show a change of viscosity with time at a constant strain rate and in the absence of a chemical reaction. Two categories of this behavior are encountered: thixotropy and rheopexy. A thixotropic fluid undergoes a decrease in viscosity, whereas a rheopectic fluid displays an increase in viscosity with time under constant strain rate (Figure 12.16). Table 12.7 gives examples of fluids that display Newtonian and non-Newtonian behavior.

B. PARAMETERS FOR CHARACTERIZING POLYMER SOLUTION VISCOSITY[12]

The viscous flow of a polymer solution involves a shearing action in which different layers of the solution move with differing velocities. As we observed earlier, there is a pronounced increase in the viscosity of a polymer solution relative to that of the pure solvent even at low concentrations of the polymer. In this respect, the polymer solute behaves as a colloidal dispersion, which is known to retard the flow of adjacent layers of a liquid under shearing force. For spherical colloidal particles, the viscosity of the solution, η, relative to that of the pure solvent, η_0, is referred to as the relative viscosity, given by

$$\eta_r = \eta/\eta_0 \tag{12.59}$$

The flow of fluids through a tube of uniform cross-section under an applied pressure is given by the Hagen–Poiseuille's law:

$$Q = \frac{\pi R^4 \Delta P}{8\eta L} \tag{12.60}$$

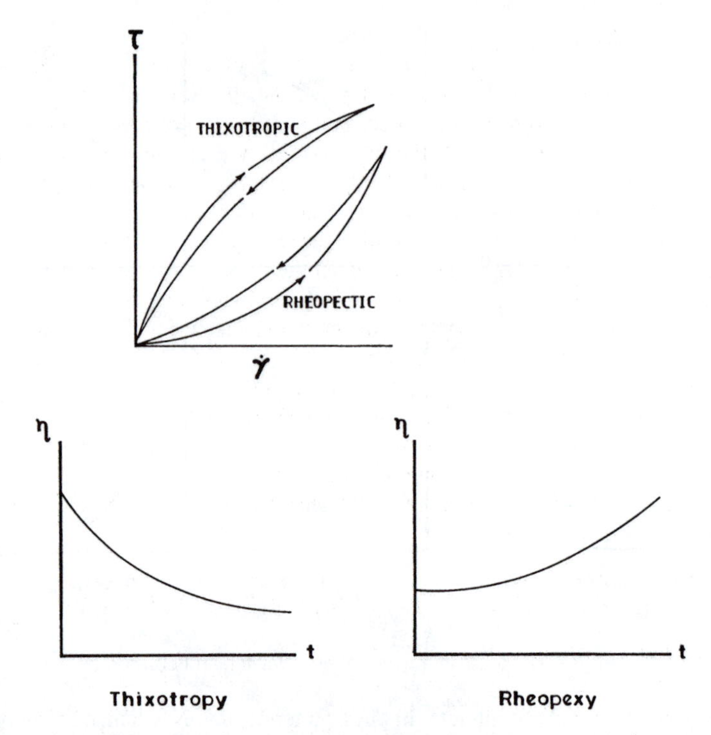

Figure 12.16 Change of viscosity with time under constant strain rate.

Table 12.7 Examples of Newtonian and Non-Newtonian Fluids

Viscosity Type	Example
Newtonian	All gases, water, thin motor oils
Non-Newtonian	
Pseudoplastic	Paints, emulsions, dispersions
Dilatant	In fluids containing high levels of deflocculated solids such as clay slurries, candy compounds, corn starch in water, sand/water mixtures
Plastic	Tomato catsup
Thixotropic	Greases, heavy printing inks, paints
Rheopectic	Rarely encountered

where Q is the volume flow rate, ΔP is the pressure drop across the tube of length L and radius R. Some of the assumptions made in the derivation of this law include

1. The flow is laminar, which means that the dimensionless quantity called Reynold's number, R_e, < 2100.
2. The fluid is incompressible, i.e., its density is constant.
3. The flow is independent of time, i.e., steady-state conditions prevail.
4. The fluid is Newtonian.

The viscosity of a liquid or solution can be measured by using a viscometer whose design is based on the Hagen–Poiseuille law. Essentially, this involves the measurement of the flow rate of the liquid through a capillary tube which is part of the viscometer. Consequently, by measuring the flow time of the solution, t, and that of the pure solvent, t_0, the relative viscosity can be determined:

$$\eta_r = \eta/\eta_0 = t/t_0 \tag{12.61}$$

As indicated above, the viscosity of the polymer solution is always greater than that of the pure solvent. This fractional increase in the viscosity resulting from the dissolved polymer in the solvent is referred to as the specific viscosity η_{sp}, given by

Table 12.8 Various Viscosity Terms

Viscosity Term	Expression
Relative viscosity	$\eta_r = \eta/\eta_0 = t/t_0$
Specific viscosity	$\eta_{sp} = (\eta - \eta_0)/\eta_0 = \eta_r - 1$
Reduced viscosity	$\eta_{sp}/C = (\eta_r - 1)/C$
Intrinsic viscosity	$[\eta] = \lim_{c\to 0} \eta_{sp}/C$

$$\eta_{sp} = \frac{\eta - \eta_0}{\eta_0} = \eta_r - 1 \tag{12.62}$$

According to Einstein's viscosity relation for rigid spherical particles in solution,

$$\frac{\eta - \eta_0}{\eta_0} = 2.5\left(\frac{n_2}{V}\right)V_e \tag{12.63}$$

where n_2/V is the number of polymer molecules per unit volume and $V_e = \frac{4\pi}{3}R_e^3$; R_e is the radius of an equivalent hydrodynamic sphere that would enhance the viscosity of the solvent medium to the same extent as would the actual polymer molecule. The quantity n_2/V may be written as

$$\frac{n_2}{V} = \frac{CN_A}{M} \tag{12.64}$$

where C and M are the concentration and molecular weight, respectively, of the polymer, and N_A is Avogadro's number. It is obvious from Equations 12.63 and 12.64 that both the relative viscosity and specific viscosity increase with increasing concentration of the polymer. The specific viscosity normalized with respect to the concentration, η_{sp}/C, is referred to as the reduced specific viscosity or, simply, reduced viscosity. It measures that capacity with which a given polymer enhances the specific viscosity. The intrinsic viscosity $[\eta]$ is the limiting value of the reduced viscosity at infinite dilution:

$$[\eta] = \lim_{c\to o} \eta_{sp}/c \tag{12.65}$$

When $\eta_r < 2$, it has been found that a linear relation exists between the reduced viscosity and polymer concentration. For a given polymer–solvent system, this linear dependence is described adequately by the equation

$$\frac{\eta_{sp}}{c} = [\eta] + k^1[\eta]^2 c \tag{12.66}$$

where k^1 is referred to as the Huggins constant, with a value usually in the range 0.35 to 0.40. Table 12.8 summarizes the various viscosity terms described above.

C. MOLECULAR SIZE AND INTRINSIC VISCOSITY

The Einstein viscosity relation given by Equation 12.62 may be written as

$$\frac{\eta - \eta_0}{\eta_0} = 2.5\left(\frac{N_A C}{M}\right)V_e \tag{12.67}$$

or

$$[\eta] = \frac{\eta - \eta_0}{\eta_0 C} = 2.5\frac{N_A}{M}V_e = 2.5\frac{N_A}{M}\frac{4\pi}{3}R_e^3 \tag{12.68}$$

Equation 12.67 predicts that the specific viscosity is proportional to the volume of the equivalent hydrodynamic sphere. The Einstein viscosity relation was derived for rigid spherical particles in solution. However, real polymer molecules are neither rigid nor spherical. Instead the spatial form of the polymer molecule in solution is regarded as a random coil. Theories based on this characteristic form of polymer molecules have resulted in the expression

$$[\eta] = \Phi \left(\bar{r}^2\right)^{3/2} M \tag{12.69}$$

where ϕ is considered to be a universal constant with a value of 2.1 (±0.2) × 10^{23}, $\bar{r}^2$ is the mean-square end-to-end distance of the polymer coil expressed in cm, and $[\eta]$ is in cm^3/g. According to Equation 12.68, the intrinsic viscosity is proportional to the ratio of the effective hydrodynamic volume of the molecule to its molecular weight. Specifically, it states that the effective volume is proportional to the linear dimensions of the randomly coiled polymer chain. Therefore, to understand the factors that influence the intrinsic viscosity, the quantity in Equation 12.68 may be separated into its component parts. Recall from our discussion in Section III.B that the net effect of long-range interaction is to expand the actual chain dimension $(r_0^2)^{1/2}$ by an expansion factor α given by the relation:

$$\left(\overline{r^2}\right)^{1/2} = \alpha\left(\overline{r_0^2}\right)^{1/2} \tag{12.13}$$

Equation 12.69 may therefore be rewritten as

$$[\eta] = \Phi \left(\bar{r}_0^2/M\right)^{3/2} M^{1/2} \alpha^3 \tag{12.70}$$

For a linear polymer of a given structural unit, the quantity $\bar{r}^2_0/M$ is independent of M. Consequently, Equation 12.70 becomes

$$[\eta] = K M^{1/2} \alpha^3 \tag{12.71}$$

where

$$K = \Phi \left(\overline{r_0^2}/M\right)^{3/2} \tag{12.72}$$

K is a constant independent of the polymer molecular weight and of the solvent.

From Equation 12.71, the intrinsic viscosity depends on the molecular weight as a result of the factor $M^{1/2}$ and also through the dependence of the expansion factor α^3 on molecular weight. By choosing a theta-solvent or θ temperature, the influence of the molecular expansion due to intramolecular interactions can be eliminated. Under these conditions, $\alpha = 1$, and the intrinsic viscosity depends only on the molecular weight. Thus Equation 12.71 is reduced to:

$$[\eta]_\theta = K M^{1/2} \tag{12.73}$$

This relation has been confirmed experimentally. It follows, therefore, that since Φ is regarded as a universal constant, the average dimensions of polymer molecules in solution can be estimated from knowledge of their intrinsic viscosities and molecular weight (Equation 12.71). Specifically, the unperturbed dimensions can be calculated from the value of K.

D. MOLECULAR WEIGHT FROM INTRINSIC VISCOSITY

Polymers possess the unique capacity to increase the viscosity of the solvent in which they are dissolved. Within a homologous series of linear polymers, the higher the molecular weight the greater the increase in viscosity for a given polymer concentration. In other words, this capacity to enhance viscosity or intrinsic viscosity is a reflection of the molecular weight of the dissolved polymer. Consequently, intrinsic

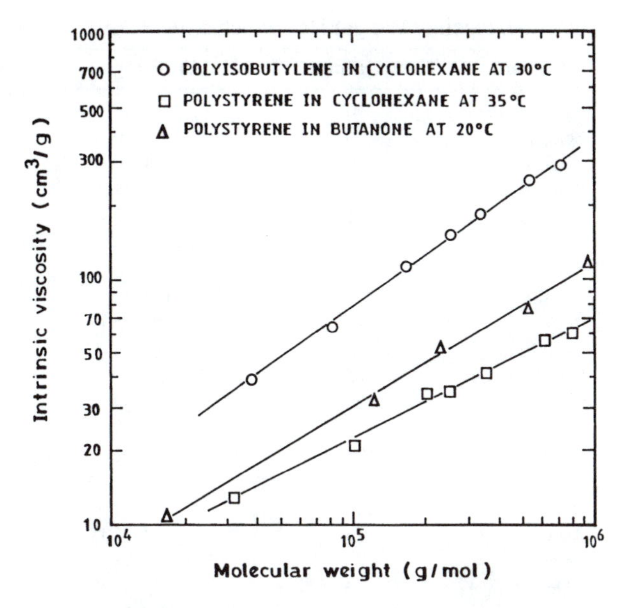

Figure 12.17 Intrinsic viscosity–molecular weight relationship. (From Allocock, H.R. and Lampe, F.W., *Contemporary Polymer Chemistry,* Prentice-Hall, Englewood Cliffs, NJ, 1981. With permission.)

viscosity measurements provide a tool for characterization of polymer molecular weight. However, since intrinsic viscosity does not provide absolute values of molecular weight, the relation between intrinsic viscosity and molecular weight has to be established empirically by comparison with molecular weights determined from absolute methods such as osmometry, light scattering, and ultracentrifugation. A linear relation has been found to exist between the logarithms of the intrinsic viscosities of different molecular weight fractions of a given polymer and the logarithms of the molecular weights of these fractions. This is illustrated for polyisobutylene in cyclohexane at 30°C and for polystyrene in cyclohexane at 35°C and in ethylethylketone at 25°C, as shown in Figure 12.17. The slopes of these curves for a given polymer depend on the solvent and, for a given polymer–solvent pair, on the temperature. It has also been established that the slopes of such plots for all polymer–solvent systems fall within the range of 0.5 to 1.0. The linear relation between $\log[\eta]$ and $\log M$ may then be written as

$$[\eta] = K\,M^{a} \tag{12.74}$$

where K and a are constants determined from the intercept and slope of plots such as in Figure 12.17.

The relation given in Equation 12.74 is referred to as the Mark–Houwink equation. Using the equation, it is possible to calculate the molecular weight from intrinsic viscosity measurements as long as K and a have been established for a particular temperature.

It must be reemphasized that the Mark–Houwink equation applies to fractionated samples of a given polymer. This means that, strictly speaking, it covers only a narrow molecular weight range. However, it is relatively easier in practice to use intrinsic viscosity measurements for the determination of the molecular weights even for unfractionated polymers. For such molecularly heterogeneous polymers, the appropriate relation becomes

$$[\eta] = K\,\overline{M}_v^{a} \tag{12.75}$$

where $\overline{M}_v$ is the viscosity average molecular weight given by

$$\overline{M}_v = \left[\sum \omega_i M_i^{a}\right]^{1/a} = \left[\frac{\sum N_i M_i^{1+a}}{\sum N_i M_i}\right]^{1/a} \tag{12.76}$$

Table 12.9 Intrinsic Viscosity–Molecular Weight Relationship, $[\eta] = KM^a$

Polymer	Solvent	Temperature (°C)	Mol-Wt Range × 10^{-4}	K × 10^3 (ml/g)	a
Polybutadiene	Cyclohexane	40	4–17	28.2	0.70
	Benzene	30	5–50	33.7	0.715
	Toluene	30	5–16	29.4	0.753
Natural rubber	Benzene	30	8–28	18.5	0.74
	Toluene	25	7–100	50.2	0.667
Polyethylene					
Low pressure	Decalin	135	3–100	67.7	0.67
High pressure	Decalin	70	0.2–3.5	38.73	0.738
Polyisobutylene	Benzene	25	0.05–126	83.0	0.53
	Cyclohexane	25	14–34	40.0	0.72
	Diisobutylene	25	0.4–2.5	130.0	0.50
	Toluene	25	14–34	87.0	0.56
Polypropylene (atactic)	Decalin	135	2–39	15.8	0.77
	Benzene	25	6–31	27.0	0.71
	Cyclohexane	25	6–31	16.0	0.80
Polypropylene (isotactic)	Decalin	135	2–62	11.0	0.80
Poly(methyl methacrylate)	Acetone	25	2–780	5.3	0.73
	Benzene	25	2–740	5.5	0.76
	Chloroform	25	40–330	3.4	0.83
Polystyrene (atactic)	Benzene	20	0.6–520	12.3	0.72
	Cyclohexane	34.5	14–200	84.6	0.50
Polystyrene (isotactic)	Benzene	30	4–37	10.6	0.735
	Toluene	30	15–71	9.3	0.72
	Chloroform	30	9–32	25.9	0.734

From Kurata, M., Tsunashima, Y., Iwama, M., and Kamada, K., in *Polymer Handbook,* 2nd ed., Brandrup, J. and Immergut, E.H., Eds., John Wiley & Sons, New York, 1975. With permission.

where $\omega_i = c_i/c$ = weight fraction of species i; N_i, M_i, and C_i are the number of molecules, molecular weight, and concentration, respectively, of the same species; and $C = \sum c_i$ is the total concentration of all species. Table 12.9 lists the values for K and a for some selected polymer–solvent systems.

Example 12.4: Explain the variation in the value of K with temperature for polystyrene measured in different solvents.

Solvent	θ(K)	K × 10^4 (at T = θ)
Cyclohexane	307	8.2
Methycyclohexane	343.5	7.6
Ethylcyclohexane	343	7.5

Solution: K is a constant that is essentially independent of the molecular weight of the polymer and the character of the solvent medium, as amply demonstrated by the above data. However, K shows a decrease with increasing temperature. From Equation 12.72, K is proportional to the factor $\bar{r}^2_0/M$. We recall that the unperturbed root-mean-square end-to-end distance $(\bar{r}^2_0)^{1/2}$ is expanded invariably to greater dimensions relative to completely free rotation as a result of the effects of hindrances to free rotation. As the temperature is increased, the tendency to completely free rotation is enhanced as the effects of these hindrances are diminished. Consequently, K also decreases.

Example 12.5: Given the following values of η_r for a polyisobutylene sample of molecular weight 1,500,000:

Solvent	Temp (°C)	C = 0.05	0.10	0.15	0.20g/dl
Cyclohexane	30	1.282	1.611	1.988	2.412
Diisobutylene	20	1.173	1.365	1.578	1.809
Benzene	25	1.066	1.136	1.209	1.287

a. Determine [η] and k′ in each solvent.
b. Calculate $(\bar{r}^2_0/M)^{1/2}$, C_∞ and the value α_n in each solvent. Note that 25°C is the θ-temperature in benzene and assume $\Phi = 2.6 \times 10^{21}$ in this case.

Solution:

$$\eta_{sp} = \eta_r - 1$$

From the plot of η_{sp}/C vs. C, i.e., Figure E12.5

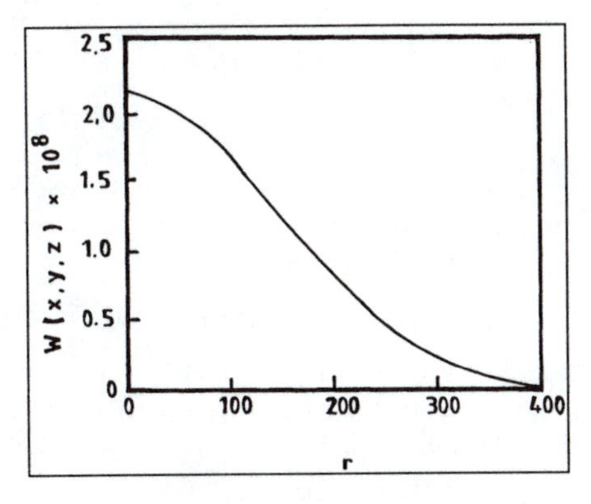

Figure E12.5

$$\text{Hexane: } [\eta] = 5\cdot 15 \text{ slope} = k'[\eta]^2 = 9.75$$

$$k' = 9.75/(5.15)^2 = 0.368$$

	[η]	Slope	k′
Hexane	5.15	9.75	0.368
Diisobutylene	3.20	4.25	0.415
Benzene	1.28	0.8	0.488

b. Since 25°C θ is the temperature in benzene,

$$[\eta]_\theta = KM^{1/2} = 1.28$$

$$K = \frac{[\eta]_\theta}{M^{1/2}} = \frac{1.28}{\left(1.5 \times 10^6\right)^{1/2}} = 1.045 \times 10^{-3}$$

$$K = \Phi\left(\frac{\overline{r_0^2}}{M}\right)^{3/2}$$

or

$$\left(\frac{\overline{r_0^2}}{M}\right)^{3/2} = \frac{K}{\Phi} = \frac{1.045 \times 10^{-3}}{2.6 \times 10^{21}}$$

$$= 0.402 \times 10^{-24}$$

$$\left(\frac{\overline{r_0^2}}{M}\right)^{1/2} = 0.738 \times 10^{-8}$$

$$C_\infty = \frac{\overline{r_0^2}}{nl^2}$$

$$\overline{r_0^2} = \left(0.738 \times 10^{-8}\right)^2 M$$

$$= \left(0.738 \times 10^{-8}\right)^2 \left(1.5 \times 10^6\right)$$

$$= 0.817 \times 10^{-10}$$

Polyisobutylene: $\left[\!-CH_2-C(CH_3)_2-\!\right]_n$ (Str. 3)

Molecular weight of monomer = 56 = Mo.

$$n = 2\,DP = \frac{2\,M}{M_o} = \frac{2 \times 1.5 \times 10^6}{56}$$

$$= 5.357 \times 10^4$$

$$l = 1.54 \times 10^{-8}\ \text{cm}$$

$$C_\infty = \frac{0.817 \times 10^{-10}}{5.357 \times 10^6 \times 2.372 \times 10^{-16}}$$

$$= 6.43$$

$$[\eta] = [\eta]_\theta \, \alpha_n^3$$

Cyclohexane:

$$\alpha_n = \left([\eta]/[\eta_0]_\theta\right)^{1/3} = \left(\frac{5.15}{1.28}\right)^{1/3} = 1.59$$

Dilsobutylene:

$$\alpha_n = \left(\frac{3.20}{1.28}\right)^{1/3} = 1.36$$

Benzene:

$$\alpha_n = \left(\frac{1.28}{1.28}\right)^{1/3} = 1.0$$

VI. PROBLEMS

12.1. Which of the following is the most suitable solvent for (a) natural rubber and (b) polyacrylonitrile: *n*-pentane, toluene, *o*-dichlorobenzene, nitroethane, and nitromethane? The densities of natural rubber and polyacrylonitrile are 1.1 and 1.15, respectively.

12.2. Explain the following observations.

a. An excellent example of a homogeneous polymerization is the bulk polymerization of methyl methacrylate, while heterogeneous polymerization is exemplified by bulk polymerization of vinyl chloride.
b. Polyacrylonitrile is prepared industrially either by precipitation polymerization of acrylonitrile in water or its solution polymerization in dimethyl formamide.

12.3. Show that the end-to-end distance of the freely jointed chain is expanded by a factor of $\sqrt{2}$ when there is restriction to bond angles. Assume that the bond angle is tetrahedral.

12.4. The root-mean-square end-to-end distance of poly(acrylic acid) with molecular weight 1,000,000 in a θ solvent at 30°C is 670 Å. Taking the C–C bond length as 1.53 Å and all backbone bond angles as tetrahedral, calculate:

a. Flory's characteristic ratio
b. The Stockmayer–Kurata ratio
c. The Kratky–Porod persistence length
d. The contour length of the planar zigzag

12.5. How does the root-mean-square end-to-end distance of a flexible polymer depend on:

a. Molecular weight at the Flory temperature θ
b. Temperature at the Flory point
c. Molecular weight, well above the Flory temperature for a good solvent

12.6. Calculate the vapor pressure over a 50% solution (by volume) of poly(vinyl acetate) in methylethyl ketone at 25°C using the lattice theory and assuming that the value of χ_1 derived from the dilute solution data is valid in the concentrated solution. The vapor pressure of pure methylethyl ketone at 25°C is 100 mm Hg.

12.7. The size of a polyisobutylene molecule in the pure material is given experimentally as

$$\left(\frac{\overline{r_0^2}}{M}\right)^{1/2} = 0.795 \text{ at } 24°C$$

$$= 0.757 \text{ at } 95°C$$

where r_0 is in angstroms and M is molecular weight. Compare these results to those obtained by assuming free rotation about the valence bonds. Explain why the experimental values:

a. Are higher than the calculated values
b. Decrease with increasing temperature

If the root-mean-square end-to-end distance of the polymer molecule is 1000 Å, what is its molecular weight at 24°C?

12.8. Compute the root-mean-square end-to-end length for a polystyrene molecule having a molecular weight of 10^6. Assume free rotation on the valence cone.

12.9. Polyisobutylene has a molecular weight of 885,000 and density 0.8 g/cm³. If

$$\left(\frac{\overline{r_0^2}}{M}\right)^{1/2} = 800 \times 10^{-11}$$

a. What is the approximate contour length of this polymer molecule (length of the planar zigzag structure)?
b. What is the approximate size of the molecule if a compact shape is assumed? Compact shape density is 0.8 g/cm³ (what is the radius of an equivalent sphere).
c. What is the approximate size of the molecule if a hindered random coil is assumed?
d. Discuss the relative magnitudes of parts b and c.

12.10. The ratios of the unperturbed end-to-end dimensions measured at 60°C to that calculated assuming completely free rotation $(\overline{r^2_0}/\overline{r^2_{0f}})^{1/2}$ for natural rubber and gutta-percha are 1.71 and 1.46, respectively. Explain.

12.11. The following data are for a narrow molecular weight fraction of poly(methyl methacrylate) in acetone at 30°C (density acetone = 0.780 g/cm³). Plot appropriately and estimate $[\eta]$, k′ (Huggins constant), $\overline{M}_n$, and the second virial coefficient. Knowing that $[\eta]_\theta = 4.8 \times 10^{-4} M^{0.5}$ for this polymer in a theta solvent at 30°C, calculate from these data $(\overline{r^2_0})^{1/2}$, $(\overline{r^2_2})^{1/2}$ and α, the expansion factor.

Concentration C, (g/100 ml)	Osmotic Pressure (cm solvent)	η_r, Relative Viscosity
0.275	0.457	1.170
0.338	0.592	—
0.344	0.609	1.215
0.486	0.867	—
0.96	1.756	1.629
1.006	2.098	—
1.199	2.710	1.892
1.536	3.728	—
1.604	3.978	2.330
2.108	5.919	2.995
2.878	9.713	—

12.12. The following data are available for polymers A and B in the same solvent at 27°C.

Conc, C_A (g/dl)	Osmotic pressure, (cm of solvent)	Conc, C_B (g/dl)	Osmotic Pressure (cm of solvent)
0.320	0.70	0.400	1.60
0.660	1.82	0.900	4.44
1.000	3.10	1.400	8.95
1.900	9.30	1.800	13.0

Solvent density = 0.85 g/cm^3; polymer density = 1.15 g/cm^3.

a. Estimate $\overline{M}_n$ and the second virial coefficient for each polymer.
b. Estimate M_n for a 25:75 mixture of A and B.
c. If $\overline{M}_w$/-M_n = 2.00 for A and for B, what is $\overline{M}_w$/-M_n for the mixture in part b?

REFERENCES

1. Hildebrand, J.H. and Scott, R.L., *The Solubility of Nonelectrolytes,* Van Nostrand-Reinhold, New York, 1950.
2. Burrel, H. and Immergut, B., in *Polymer Handbook,* Brandrup, J. and Immergut, E.M., Eds., John Wiley & Sons, New York, 1967.
3. Hoy, K.L., *J. Paint Technol.,* 42, 76, 1970.
4. Kurata, M., Tsunashima, Y., Iwama, M., and Kamada, K., Viscosity molecular-weight relationships and unperturbed dimensions of linear chain molecules, in *Polymer Handbook,* 2nd ed., Brandrup, J. and Immergut, E.H., Eds., John Wiley & Sons, New York, 1975.
5. Flory, P.J., *J. Chem. Phys.,* 10, 51, 1942.
6. Huggins, M.L., *Ann. N.Y. Acad. Sci.,* 43, 1, 1942.
7. Huggins, M.L., *J. Phys. Chem.,* 45, 151, 1942.
8. Huggins, M.L., *J. Am. Chem. Soc.,* 64, 1712, 1942.
9. Newing, M.J., *Trans. Faraday Soc.,* 46, 613, 1950.
10. Bawn, C.E.H., Freeman, R.F.J., and Kamaliddin, A.R., *Trans. Faraday Soc.,* 46, 677, 1950.
11. Gee, G. and Orr, W.J.C., *Trans. Faraday Soc.,* 42, 507, 1946.
12. Gee, G., *J. Chem. Soc.,* p. 280, 1947.
13. Klotz, M., *Chemical Thermodynamics,* Prentice-Hall, Englewood Cliffs, NJ, 1950.
14. Flory, P.J., *Principles of Polymer Chemistry,* Cornell University Press, Ithaca, NY, 1953.
15. Wolf, B.A. in *Polymer Handbook,* 2nd ed., Brandrup, J. and Immergut, E.M., Eds., John Wiley & Sons, New York, 1967.
16. Bird, R.B., Stewart, W.E., and Lightfoot, E.N., *Transport Phenomena,* John Wiley & Sons, New York, 1960.
17. Anon., *More Solutions to Sticky Problems,* Brookfield Engineering Laboratories, Stoughton, MA, 1985.
18. Allocock, H.R. and Lampe, F.W., *Contemporary Polymer Chemistry,* Prentice-Hall, Englewood Cliffs, NJ, 1981.

Chapter 13

Mechanical Properties of Polymers

I. INTRODUCTION

In Chapter 1, we observed that plastics have experienced a phenomenal growth since World War II, when they assumed enhanced commercial importance. This explosive growth in polymer applications derives from their competitive costs and versatile properties. Polymers vary from liquids and soft rubbers to hard and rigid solids. The unique properties of polymers coupled with their light weight make them preferable alternatives to metallic and ceramic materials in many applications. In the selection of a polymer for a specific end use, careful consideration must be given to its mechanical properties. This consideration is important not only in those applications where the mechanical properties play a primary role, but also in other applications where other characteristics of the polymer such as electrical, optical, or thermal properties are of crucial importance. In the latter cases, mechanical stability and durability of the polymer may be required for the part to perform its function satisfactorily.

The mechanical behavior of a polymer is a function of its microstructure or morphology. Polymer morphology itself depends on many structural and environmental factors. Compared with those of metals and ceramics, polymer properties show a much stronger dependence on temperature and time. This strong time and temperature sensitivity of polymer properties is a consequence of the viscoelastic nature of polymers. This implies that polymers exhibit combined viscous and elastic behavior. For example, depending on the temperature and stress levels, a polymer may show linear elastic behavior, yield phenomena, plastic deformation, or cold drawing. An amorphous polymer with T_g below ambient temperature may display nonlinear but recoverable deformation or even exhibit viscous flow. Given the complexity of polymer response to applied stresses or strains, it is imperative that, for their judicious use, those who work with polymers have an elementary knowledge of how polymer behavior is influenced by structural and environmental factors. We have already dealt with the structures of polymers in earlier chapters. We devote succeeding sections to discussing the mechanical properties of polymers in the solid state.

II. MECHANICAL TESTS

Polymer components, like other materials, may fail to perform their intended functions in specific applications as a result of

1. Excessive elastic deformation
2. Yielding or excessive plastic deformation
3. Fracture

Polymers show excessive elastic deformation, particularly in structural, load-bearing applications, due to inadequate rigidity or stiffness. For such failure, the controlling material mechanical property is the elastic modulus. As we shall see in subsequent discussions, the elastic moduli of some polymers are subject to some measure of control through appropriate structural modification.

Failure of polymers in certain applications to carry design loads or occasional accidental overloads may be due to excessive plastic deformation resulting from the inadequate strength properties of the polymer. For the quantification of such failures, the mechanical property of primary interest is the yield strength and the corresponding strain. The ultimate strength, along with the associated strain, also provides useful information.

Cracks constitute regions of material discontinuity and frequently precipitate failure through fracture. Fracture may occur in a sudden, brittle manner or through fatigue (progressive fracture). Brittle fracture occurs in materials where the absence of local yielding results in a build-up of localized stresses, whereas fatigue failure occurs when parts are subjected to alternating or repeated loads. Fatigue fractures occur without visible signs of yielding since they occur at strengths well below the tensile strength of the material.

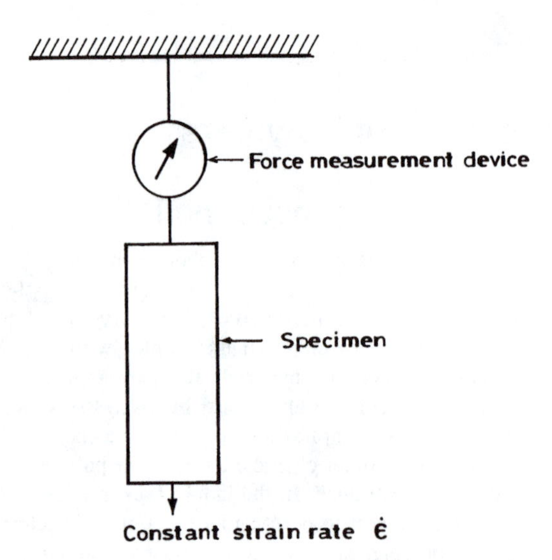

Figure 13.1 Schematic of stress–strain test.

As we said earlier, polymers will continue to be used in a variety of end-use situations. Therefore, to ensure their successful performance in these applications, it is necessary to clearly understand their mechanical behavior under a variety of stress conditions. Particular cognizance must be taken of the relatively high sensitivity of polymer failure modes to temperature, time, and loading history. For good design, it is important to be able to relate design load and component dimensions to some appropriate material property that defines the limits of the load-bearing capability of the polymer material. A variety of test methods exist for predicting mechanical performance limits under a variety of loading conditions. These range from simple tension, compression, and shear tests to those designed to test complex stress states and polymer time–temperature response. Elaborate treatment of polymer deformation behavior under complex stress states would require complex mathematical analysis, which is beyond the scope of this volume. Our discussion emphasizes problems of a one-dimensional nature, and cases of nonlinear deformation will be treated in an elementary fashion.

A. STRESS–STRAIN EXPERIMENTS

Stress–strain experiments have traditionally been the most widely used mechanical test but probably the least understood in terms of interpretation. In stress–strain tests the specimen is deformed (pulled) at a constant rate, and the stress required for this deformation is measured simultaneously (Figure 13.1). As we shall see in subsequent discussions, polymers exhibit a wide variation of behavior in stress–strain tests, ranging from hard and brittle to ductile, including yield and cold drawing. The utility of stress–strain tests for design with polymeric materials can be greatly enhanced if tests are carried out over a wide range of temperatures and strain rates.

B. CREEP EXPERIMENTS

In creep tests, a specimen is subjected to a constant load, and the strain is measured as a function of time. The test specimen in a laboratory setup can be a plastic film or bar clamped at one end to a rigid support while the load is applied suddenly at the other end (Figure 13.2). The elongation may be measured at time intervals using a cathetometer or a traveling microscope. Measurements may be conducted in an environmental chamber.

Creep tests are made mostly in tension, but creep experiments can also be done in shear, torsion, flexure, or compression. Creep data provide important information for selecting a polymer that must sustain dead loads for long periods. The parameter of interest to the engineer is compliance (J), which is a time-dependent reciprocal of modulus. It is the ratio of the time-dependent strain to the applied constant stress [$J(t) = \varepsilon(t)/\sigma_0$]. Figure 13.3 shows creep curves for a typical polymeric material.

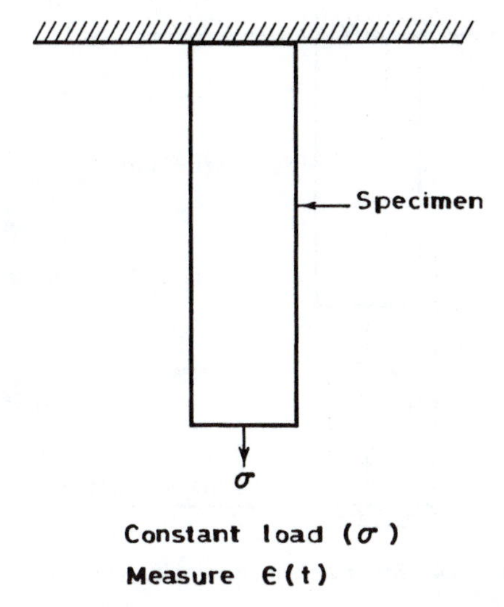

Figure 13.2 Schematic representation of creep experiment.

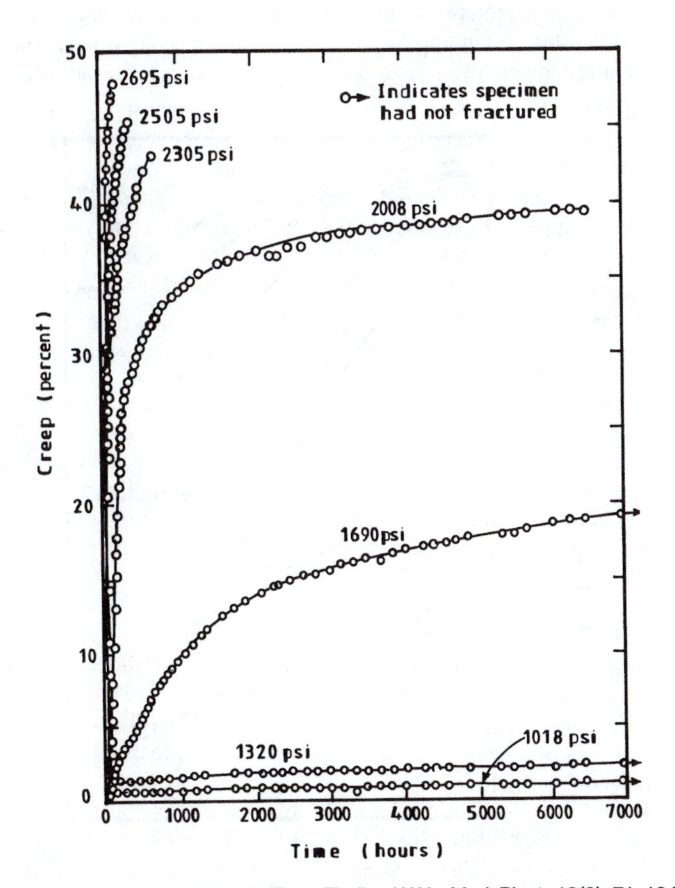

Figure 13.3 Creep of cellulose acetate at 45°C. (From Findley, W.N., *Mod. Plast.*, 19(8), 71, 1942. With permission.)

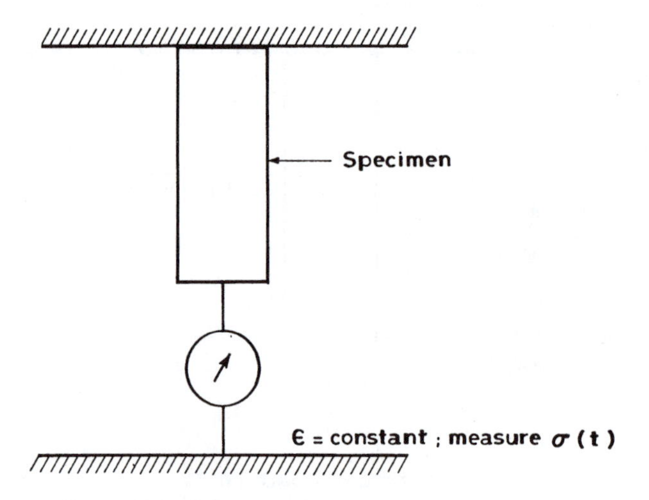

Figure 13.4 Schematic of stress relaxation experiment.

C. STRESS RELAXATION EXPERIMENTS

In stress relaxation experiments, the specimen is rapidly (ideally, instantaneously) extended a given amount, and the stress required to maintain this constant strain is measured as a function of time (Figure 13.4). The stress that is required to maintain the strain constant decays with time. When this stress is divided by the constant strain, the resultant ratio is the relaxation modulus ($E_r(t,T)$, which is a function of both time and temperature. Figure 13.5 shows the stress relaxation curves for PMMA at

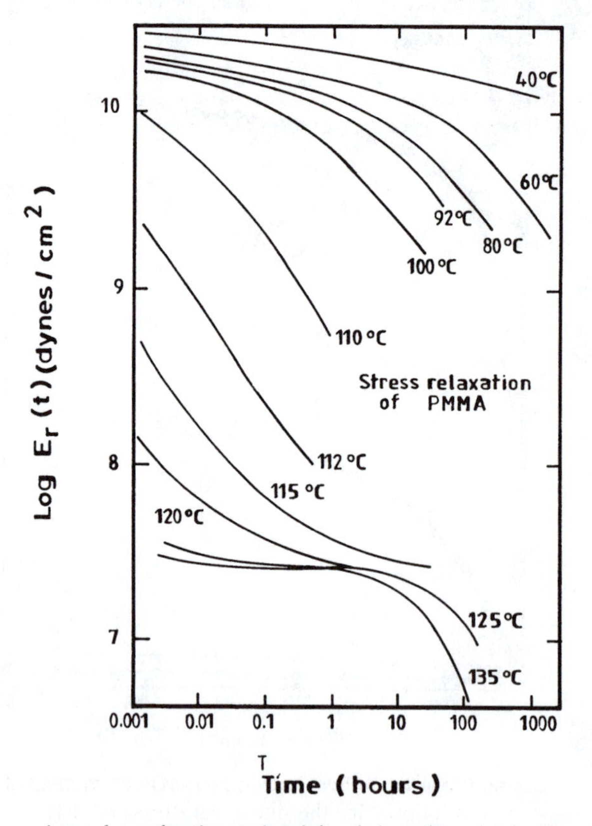

Figure 13.5 Log $E_r(t)$ vs. log t for unfractionated poly(methyl methacrylate). (From McLoughlin, J.R. and Tobolsky, A.V., *J. Colloid Sci.*, 7, 555, 1952. With permission.)

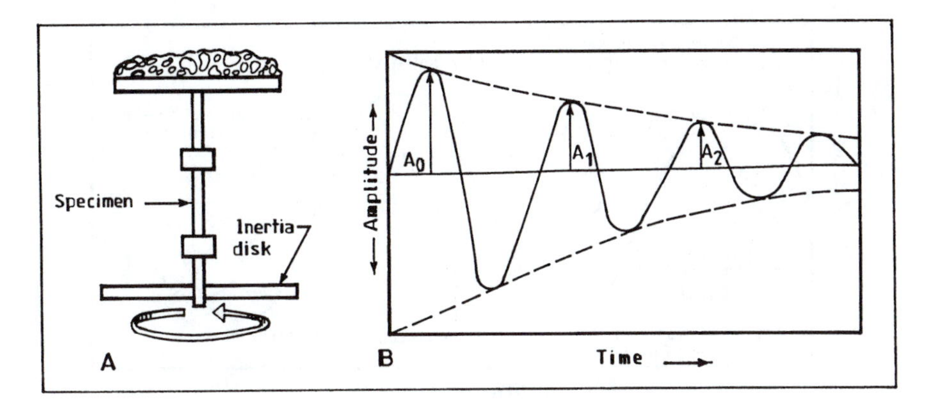

Figure 13.6 Torsion pendulum (A) is used to get data for typical response curve (B), indicating decreasing amplitude of oscillation. (From Fried, J.R., *Plast. Eng.*, 38(7), 27, 1982. With permission.)

different temperatures. Stress relaxation data provide useful information about the viscoelastic nature of polymers.

D. DYNAMIC MECHANICAL EXPERIMENTS

In dynamic mechanical tests, the response of a material to periodic stress is measured. There are many types of dynamic mechanical test instruments. Each has a limited frequency range, but it is generally possible to cover frequencies from 10^{-5} to 10^{6} cycles per second. A popular instrument for dynamic mechanical measurements is the torsion pendulum (Figure 13.6A). A polymer sample is clamped at one end, and the other end is attached to a disk that is free to oscillate. As a result of the damping characteristics of the test sample, the amplitude of oscillation decays with time (Figure 13.6B).

Dynamic mechanical tests provide useful information about the viscoelastic nature of a polymer. It is a versatile tool for studying the effects of molecular structure on polymer properties. It is a sensitive test for studying glass transitions and secondary transitions in polymer and the morphology of crystalline polymers.

Data from dynamic mechanical measurements can provide direct information about the elastic modulus and the viscous response of a polymer. This can be illustrated by considering the response of elastic and viscous materials to imposed sinusoidal strain, ε:

$$\varepsilon = \varepsilon_0 \sin(\omega t) \tag{13.1}$$

where ε_0 is the amplitude and ω is the frequency (in radians per second, $\omega = 2\pi f$; f is in cycles per second). For a purely elastic body, Hooke's law is obeyed. Consequently,

$$\sigma = G\varepsilon_0 \sin(\omega t) \tag{13.2}$$

where G is the shear modulus. It is evident from Equations 13.1 and 13.2 that for elastic bodies, stress and strain are in phase.

Now consider a purely viscous fluid. Newton's law dictates that the shear stress is given by $\sigma = \eta\dot{\varepsilon}$, that is,

$$\sigma = \eta\varepsilon_0\ \omega \cos(\omega t) \tag{13.3}$$

In this case, the shear stress and the strain are 90° out of phase. The response of viscoelastic materials falls between these two extremes. It follows that the sinusoidal stress and strain for viscoelastic materials are out of phase by an angle, say δ. The behavior of these classes of materials is illustrated in Figure 13.7.

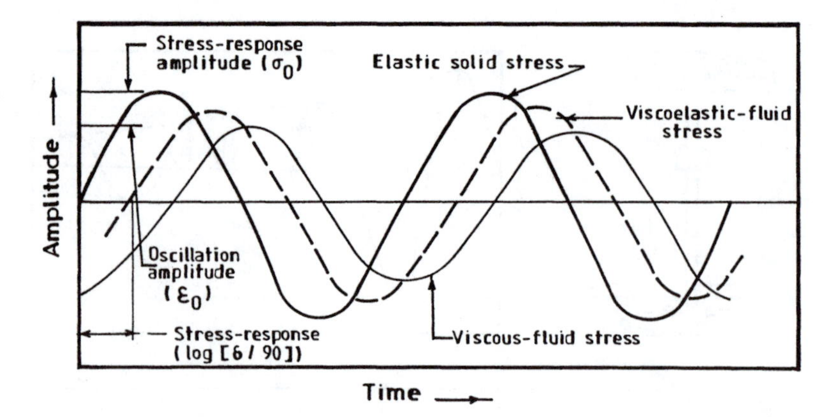

Figure 13.7 The phase relation is shown between dynamic strain and stress for viscous, elastic, and viscoelastic materials.

The lag angle between stress and strain is defined by the dissipation factor or tan δ given by

$$\tan \delta = \frac{G''}{G'} \tag{13.4}$$

where G′ is the real part of the complex modulus (G″ = G′ + iG″), and G″ is the imaginary part of the modulus. In physical terms, tan δ denotes material damping characteristics. It is a measure of the ratio of the energy dissipated as heat to the maximum energy stored in the material during one cycle of oscillation.

A convenient measure of damping is in terms of quantities determined from experiment. Thus,

$$\tan \delta = \frac{1}{\pi} \ln \frac{A_1}{A_2} \tag{13.5}$$

where A_1, and A_2 are the amplitudes of two consecutive peaks (Figure 13.7). Alternatively, this may be expressed in terms of log decrement (Δ) for free vibration instruments like the torsional pendulum.

$$\Delta = \ln \frac{A_1}{A_2} = \ln \frac{A_2}{A_3} = \frac{1}{n} \ln \frac{A_i}{A_{i+n}} \tag{13.6}$$

It follows from Equations 13.5 and 13.6 that

$$\Delta = \pi \tan \delta \tag{13.7}$$

E. IMPACT EXPERIMENTS

Polymers may also fail in service due to the effects of rapid stress loading (impact loads). Various test methods have been proposed for assessing the ability of a polymeric material to withstand impact loads. These include measurement of the area under the stress–strain curve in the high-speed (rapid) tensile test; the measurement of the energy required to break a specimen by a ball of known weight released from a predetermined height, the so-called falling ball or dart test; and the Izod and Charpy tests. The most popular of these tests methods are the Izod and Charpy impact strength tests. Essentially, the Izod test involves the measurement of the energy required to break a ½ × ½ in. notched cantilever specimen that is clamped rigidly at one end and then struck at the other end by a pendulum weight. In the case of the Charpy test, a hammerlike weight strikes a notched specimen that is rigidly held at both ends. The energy required to break the specimen is obtained from the loss in kinetic energy of the hammer

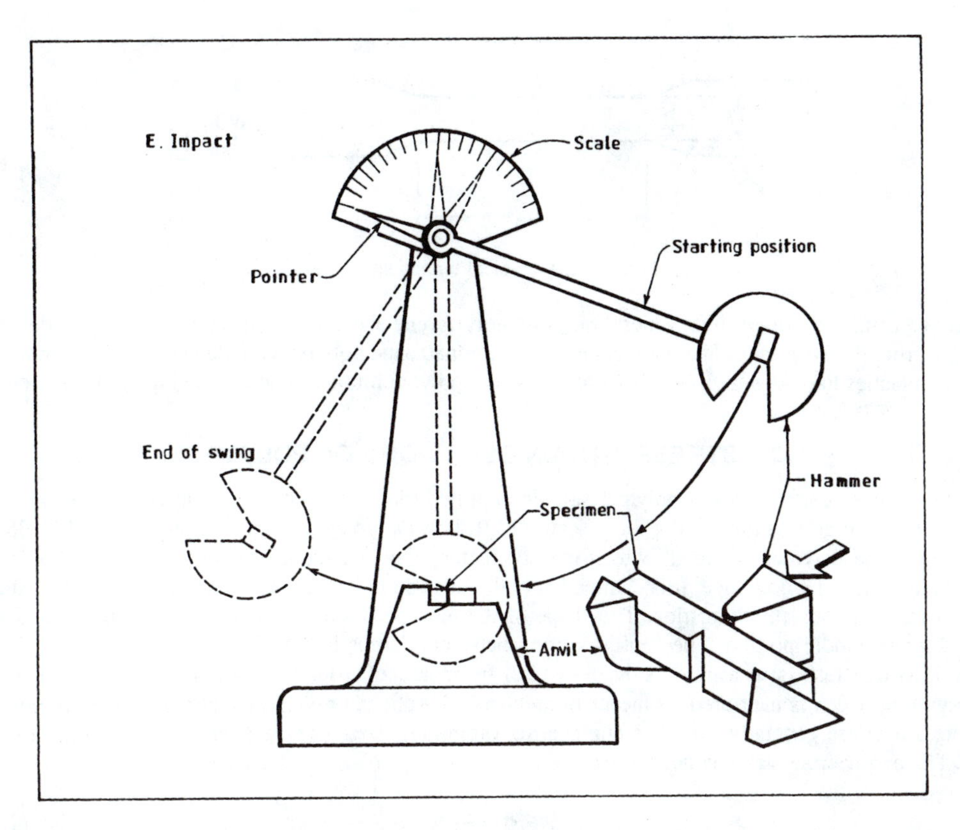

Figure 13.8 Schematic representation of impact test. (From Fried, J.R., *Plast. Eng.*, 38(7), 27, 1982. With permission.)

(Figure 13.8). Another test that is emerging as a substitute for the impact test is the measurement of fracture toughness, which in essence measures the resistance to failure of a material with a predetermined crack.

Impact tests provide useful information in the selection of a polymer for a specific application, such as determining the suitability of a given plastic as a substitute for glass bottles or a replacement for window glass. Table 13.1 gives values of impact energies for some polymers. It can be seen that, in

Table 13.1 Impact Energies of Some Polymers

Polymer	Grade	Impact Energy (J)
Polystyrene	General purpose	0.34–0.54
Polystyrene	Impact	0.68–10.80
Poly(vinyl chloride)	Rapid	0.54–4.07
Poly(vinyl chloride)	Plasticized	1.36–20.33
Polypropylene	Unmodified	0.68–2.71
Poly(methyl methacrylate)	Molding	0.41–0.68
Poly(methylmethacrylate)	High impact	1.90
Polyoxymethylene		1.90–3.12
Nylon 6,6		1.36–3.39
Nylon 6		1.36–4.07
Poly(propyleneoxide)		6.78
Polycarbonate		16.26–24.39
Polyethylene	Low density	21.70
Polyethylene	High density	0.68–27.10
Polytetrafluoroethylene		4.07
Polypropylene		0.68–2.71

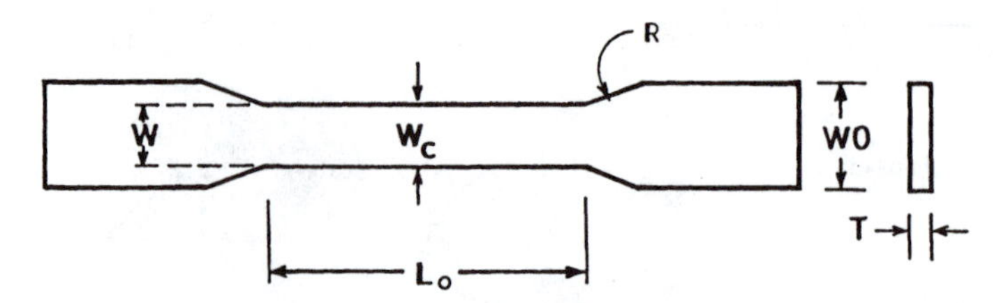

Figure 13.9 Typical tensile specimen.

general, brittle polymers like general-purpose polystyrene have low impact resistance, while most engineering thermoplastics like polyamides, polycarbonate, and polyoxymethylene and some commodity thermoplastics like polyethylene, polypropylene, and polytetrafluoroethylene have high impact strength.

III. STRESS–STRAIN BEHAVIORS OF POLYMERS

In stress–strain experiments, a polymer sample is pulled (deformed) at a constant elongation rate, and stress is measured as a function of time. Figure 13.9 shows a typical tensile specimen (ASTM D638M). Generally the polymer specimen, which may be rectangular or circular in cross-section, is molded or cut in the form of a dog bone. It is clamped at both ends and pulled at one of the clamped ends (usually downward) at constant elongation. The shape of the test specimen is designed to encourage failure at the thinner middle portion. The central section between clamps is called the initial gauge length, L_0. The load or stress is measured at the fixed end by means of a load transducer as a function of the elongation, which is measured by means of mechanical, optical, or electronic strain gauges. The experimental data are generally stated as engineering (nominal) stress (σ) vs. engineering (nominal) strain (ε). The engineering stress is defined as

$$\sigma = \frac{F}{A_0} \tag{13.8}$$

where F = the applied load
A_0 = the original cross-sectional area over the specimen

The engineering strain is given by

$$\varepsilon = \frac{L - L_0}{L_0} = \frac{\Delta L}{L_0} \tag{13.9}$$

where L_0 = original gauge length
ΔL = elongation or change in the gauge length
L = instantaneous gauge length

Engineering stress and strain are easy to calculate and are used widely in engineering practice. However, engineering stress–strain curves generally depend on the shape of the specimen. A more accurate measure of intrinsic material performance is plots of true stress vs. true strain. True stress σ_t is defined as the ratio of the measured force (F) to the instantaneous cross-sectional area (A) at a given elongation, that is,

$$\sigma_t = F/A \tag{13.10}$$

True strain is the sum of all the instantaneous length changes, dL, divided by the instantaneous length L.

$$\varepsilon_t = \int_{L_0}^{L} \frac{dL}{L} = \ln \frac{L}{L_0} \tag{13.11}$$

Some relation exists between true strain and engineering strain and between true stress and engineering stress. From Equation 13.11,

$$\varepsilon_t = \ln\frac{L}{L_0} = \ln\frac{L_0 + \Delta L}{L_0} = \ln(1+\varepsilon) \tag{13.12}$$

Up to the onset of necking, plastic deformation is essentially a constant volume process such that any extension of the original gauge length is accompanied by a corresponding contraction of the gauge diameter. Thus,

$$AL = A_0 L_0 \tag{13.13}$$

That is,

$$\frac{L}{L_0} = \frac{A_0}{A} \tag{13.14}$$

But from Equation 13.12 it follows that

$$\varepsilon_t = \ln\frac{A_0}{A} \text{ and } \frac{A_0}{A} = 1+\varepsilon \tag{13.15}$$

Now

$$\sigma_t = \frac{F}{A} = \frac{F}{A_0}\cdot\frac{A_0}{A}$$

That is,

$$\sigma_t = \sigma(1+\varepsilon) \tag{13.16}$$

For small deformations, true stress and engineering stress are essentially equal. However, for large deformations the use of true strain is preferred because they are generally additive while engineering strain is not.

A. ELASTIC STRESS–STRAIN RELATIONS

When a material is subjected to small stresses, it responds elastically. This means that

1. The strain produced is reversible with stress.
2. The magnitude of the strain is directly or linearly proportional to the magnitude of the stress for material that exhibits Hookean behavior. This relation between stress and strain is known as Hooke's law and may be written as

$$\frac{\text{Stress}}{\text{Strain}} = \text{Constant} \tag{13.17}$$

Since stress may act on a plane in different ways, this constant is defined in different ways depending on the applied force and the resultant strain. Two of the most important types of stress are shear stress, which acts in a plane, and tensile stress, which acts normally or perpendicular to the plane. Normal stresses may be tensile or compressive.

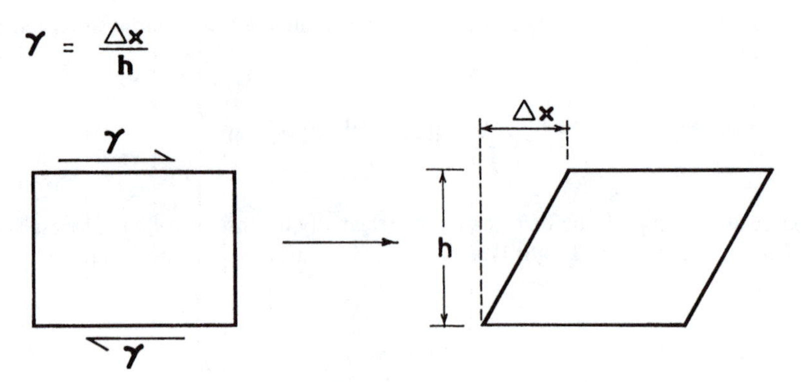

Figure 13.10 Generation of shear strain from simple shear.

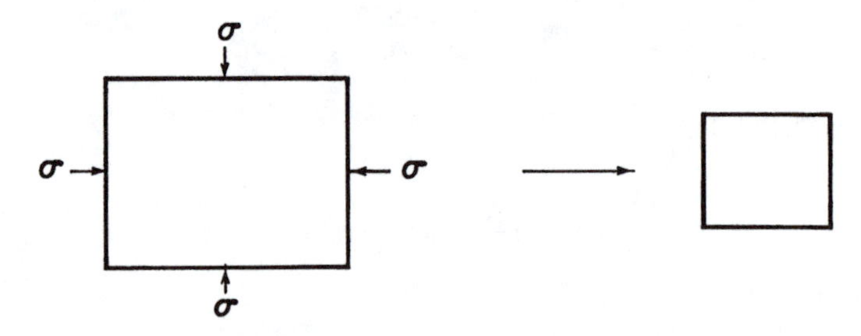

Figure 13.11 Pure dilatation.

Consider a parallelepiped that has been deformed by application of a force τ (Figure 13.10). The elastic shear strain γ is the amount of their ΔX divided by the distance, h, over which the shear has occurred:

$$\gamma = \frac{\Delta x}{h} \tag{13.18}$$

For small strains, this is simply the tangent of the angle of deformation. In pure shear, Hooke's law is expressed as

$$\tau/\gamma = G \tag{13.19}$$

where τ is the shear stress and G is the shear modulus. Deformation due to pure shear does not result in a change in volume, but produces a change in shape.

Now suppose the parallelepiped is subjected to equal normal pressure (compressive stress, $-\sigma$) in such a way that its shape remains unchanged but the volume, V, is changed by the amount ΔV (Figure 13.11). Deformation of this type is called pure dilatation, and Hooke's law for elastic dilatation is written as

$$\sigma = KD \tag{13.20}$$

where K is the bulk modulus and D is the dilatation strain given by ΔV/V. We reemphasize that pure dilatation does not produce a change in shape but produces a change in volume.

In a majority of cases, a body under stress experiences neither pure shear nor pure dilatation. Generally, a mixture of both occurs. Such a situation is exemplified by uniaxial loading which, of course, may be tensile or compressive. Here a test specimen is loaded axially resulting in a change in length, ΔL. The axial strain, ε, is related to the applied stress in an elastic deformation by Hooke's law:

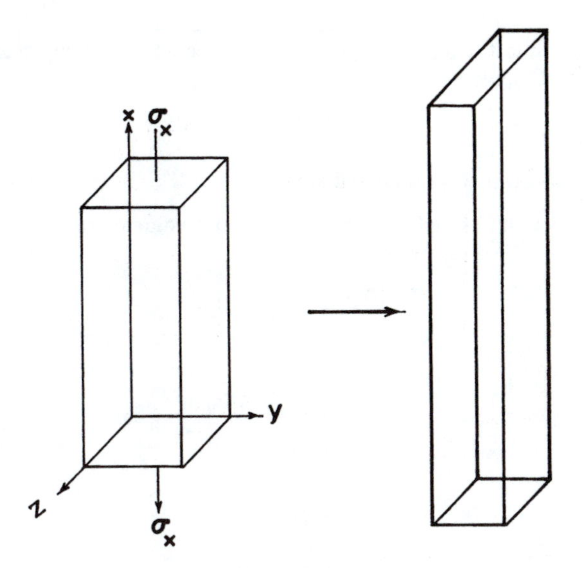

Figure 13.12 Axial elongation accompanied by transverse contractions.

$$\sigma = E\varepsilon \tag{13.21}$$

where E is Young's modulus (modulus of elasticity).

During extension, specimen elongation in the axial or longitudinal direction is accompanied by a contraction in the perpendicular transverse directions, given by the compressive strains $\varepsilon_y = \varepsilon_z$ as shown in Figure 13.12. Poisson's ratio, denoted by the parameter ν, is the ratio of the induced transverse strains to the axial strain.

$$\nu = -\frac{\varepsilon_y}{\varepsilon_x} = -\frac{\varepsilon_z}{\varepsilon_x} \tag{13.22}$$

The negative sign indicates that the strains ε_z and ε_y are due to contractions.

For incompressible materials, the volume of the specimen remains constant during deformation, and ν is 0.5. This is generally not true, although it is approached by natural rubber with $\nu = 0.49$. For most polymeric materials, there is a change in volume ΔV, which is related to Poisson's ratio by

$$\Delta V = (1 - 2\nu)\,\varepsilon V_0 \tag{13.23}$$

or, in general,

$$\nu = \frac{1}{2}\left[1 - \frac{1}{V}\frac{\partial V}{\partial \varepsilon}\right]$$

where V_0 is the initial (unstrained) volume and ΔV is the difference between the volume V at a given strain ε and the initial volume.

For materials that are isotropic and under deformations where Hooke's law is valid, the elastic constants and ν are related according to the following equations:

$$E = 2G(1 + \nu) \tag{13.24}$$

$$E = 3K(1 - 2\nu) \tag{13.25}$$

Example 13.1: In a tension test, a brittle polymer experienced an elastic engineering strain of 2% at a stress level of 35 MN/m^2. Calculate

a. The true stress
b. The true strain
c. The fractional change in cross-sectional area

Solution: Under elastic conditions, deformation is a constant-volume process, hence:

$$\sigma_t = \sigma(1+\varepsilon)$$

$$\varepsilon = 0.02$$

$$\sigma_t = 35\,(1+0.02)\ \text{MN/m}^2$$

$$= 35.7\ \text{MN/m}^2$$

$$\varepsilon_t = \ln\,(1+\varepsilon)$$

$$= \ln\,(1+0.02)$$

$$= 0.0198$$

$$\frac{A_0 - A}{A_0} = 1 - \frac{A}{A_0} = 1 - \frac{1}{1+\varepsilon} \text{ since } \frac{A_0}{A} = 1+\varepsilon$$

$$= 1 - \frac{1}{1.02}$$

$$= 0.0196$$

Example 13.2: Polypropylene has an elastic modulus of 2×10^5 psi and Poisson's ratio of 0.32. For a strain of 0.05, calculate the shear stress and the percentage change in volume.

Solution:

$$E = 2G\,(1+\nu)$$

$$G = \frac{E}{2(1+\nu)}$$

$$\tau = G\gamma = \frac{E}{2(1+\nu)}\gamma$$

$$= \frac{2 \times 10^5}{2(1+0.32)} \times 0.05$$

$$= 3.79 \times 10^3 \text{ psi}$$

$$\Delta V = (1-2\nu)\,\gamma V_0$$

$$\frac{\Delta V}{V_0} = (1-2\nu)\gamma$$

$$= [1 - 2(0.32)]\,0.05$$

$$= 1.8\%$$

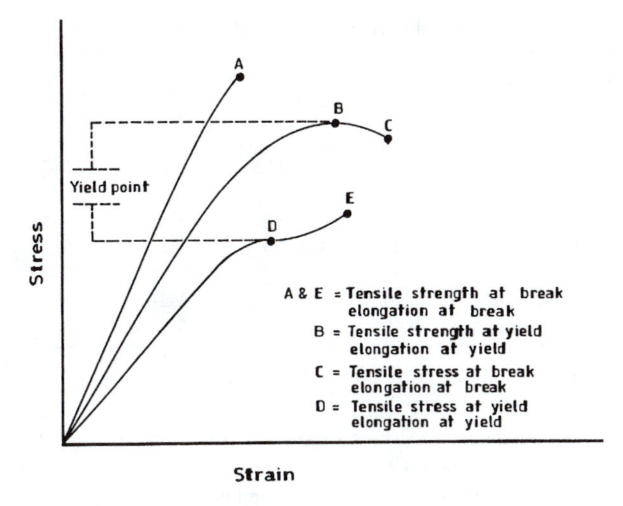

Figure 13.13 Engineering data from stress–strain tests.

IV. DEFORMATION OF SOLID POLYMERS

To relieve stress, all materials under the influence of external load undergo some deformation. In some cases, this deformation may be quite large and perceptible, like the boughs of a mango tree under the weight of its fruit. In some other cases, deformation may be imperceptibly small, for example, a fly perching uninvitedly on the dining table. Irrespective of our ability to observe it, some deformation always occurs when stresses are imposed on a material.

It is known that up to a certain limiting load, a solid will recover its original dimensions on the removal of the applied loads. This ability of deformed bodies to recover their original dimensions is known as elastic behavior. Beyond the limit of elastic behavior (elastic limit), a material will experience a permanent set or deformation even when the load is removed. Such a material is said to have undergone plastic deformation. For most materials, Hooke's law is obeyed within the elastic limit, that is, stress proportional to strain. However, proportionality between stress and strain does not always hold when a material exhibits elastic behavior. A typical example is rubber, which is elastic but does not show Hookean behavior over the entire elastic region.

Figure 13.13 illustrates the basic data on mechanical properties that are obtainable from stress–strain experiments. The gradient of the initial linear portion of the curve, within which Hooke's law is obeyed, gives the elastic, or Young's, modulus. The determination of the elastic limit is tedious and very frequently depends on the sensitivity of the strain-measuring devices employed. Consequently, it is common practice to replace it with the proportional limit, which defines the point where the nonlinear response is observed on the stress–strain curve. The maximum on the curve denotes the yield strength. For engineering purposes, this marks the limit of usable elastic behavior or the onset of plastic deformation. The stress at which fracture occurs (material breaks apart) is referred to as the ultimate tensile strength or, simply, tensile strength σ_B. The strains associated with the yield point or the fracture point are referred to as the elongation at yield and elongation at break, respectively. Typical values of these mechanical properties for selected polymers are shown in Table 13.2.

To emphasize the usefulness of stress-strain measurements, it is necessary to highlight the physical significance of the parameters defined above or the mechanical quantities derivable from these parameters:[4]

- Stiffness — Defines the ability to carry stress without changing dimension. The magnitude of the modulus of elasticity is a measure of this ability or property.
- Elasticity — Stipulates the ability to undergo reversible deformation or carry stress without suffering a permanent deformation. It is indicated by the elastic limit or, from a practical point of view, the proportional limit or yield point.
- Resilience — Defines the ability to absorb energy without suffering permanent deformation. The area under the elastic portion of the stress-strain curve gives the resilient energy.

Table 13.2 Typical Mechanical Properties of Selected Polymers

Polymer	Poisson Ratio	Elastic Modulus (10^3 psi)	Yield Strength (10^3 psi)	Ultimate Strength (10^3 psi)	Elongation to Fracture (%)
Polypropylene	0.32	1.5–2.25	3.4	3.5–5.5	200–600
Polystyrene	0.33	4–5	—	5.5–8	1–2.5
Poly(methylmethacrylate)	0.33	3.5–5	7–9	7–10	2–10
Polyethylene (LDPE)	0.38	0.2–0.4	1–2	1.5–2.5	400–700
Polycarbonate	0.37	3.5	8–10	8–10	60–120
Poly(vinyl chloride (PVC), rigid	0.40	3–6	8–10	6–11	5–60
Polytetrafluoroethylene	0.45	0.6	1.5–2	2–4	100–350

From Fried, J.R., *Plast. Eng.*, 38(7), 27, 1982. With permission.

- Strength — Indicates the ability to sustain dead load. It is represented by the tensile strength or the stress at which the specimen ruptures σ_B.
- Toughness — Indicates the ability to absorb energy and undergo extensive plastic deformation without rupturing. It is measured by the area under the stress-strain curve.

The response of material to applied stress may be described as ductile or brittle depending on the extent to which the material undergoes plastic deformation before fracture. Ductile materials possess the ability to undergo plastic deformation. For engineering purposes, an appropriate measure of ductility is important, because this property assists in the redistribution of localized stresses. On the other hand, brittle materials fail with little or no plastic deformation. Brittle materials have no ability for local yielding; hence local stresses build up around inherent flaws, reaching a critical level at which abrupt failure occurs. Figure 13.14 shows the broad spectrum of stress-strain behavior of polymeric materials, while Table 13.3 lists the general trends in the magnitude of various mechanical parameters typical of each behavior.

From the preceding discussion (Figure 13.14 and Table 13.3), it is obvious that polymers have a broad range of tensile properties. It is therefore instructive to examine these properties more closely and present current molecular interpretation at the observed properties. Figure 13.15 is a schematic representation of the macroscopic changes that occur in polymers that exhibit cold drawing during a tensile test.

At small strains, polymers (both amorphous and crystalline) show essentially linear elastic behavior. The strain observed in this phase arises from bond angle deformation and bond stretching; it is recoverable on removing the applied stress. The slope of this initial portion of the stress–strain curve is the elastic modulus. With further increase in strain, strain-induced softening occurs, resulting in a reduction of the instantaneous modulus (i.e., slope decreases). Strain-softening phenomenon is attributed to uncoiling

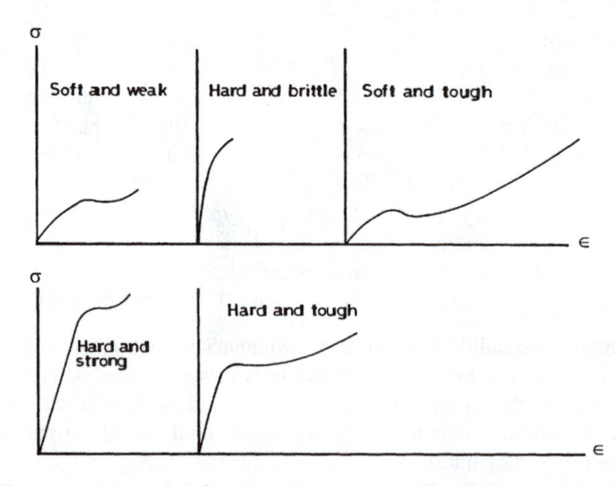

Figure 13.14 Typical stress–strain curves for polymeric materials. (From Winding, C.C. and Hiatt, G.D., *Polymer Materials,* McGraw-Hill, New York, 1961. With permission.)

Table 13.3 Characteristic Features of Polymer Stress–Strain Behavior

Material Stress-Strain Behavior	Elastic Modulus	Yield Point	Tensile Strength	Elongation at Break
Soft and weak (polymer gels)	Low	Low	Low	Moderate
Hard and brittle (PS)	High	Practically nonexistent	High	Low
Hard and strong (PVC)	High	High	High	Moderate
Soft and tough (rubbers and plasticized PVC)	Low	Low	Moderate	High
Hard and tough (cellulose acetate, nylon)	High	High	High	High

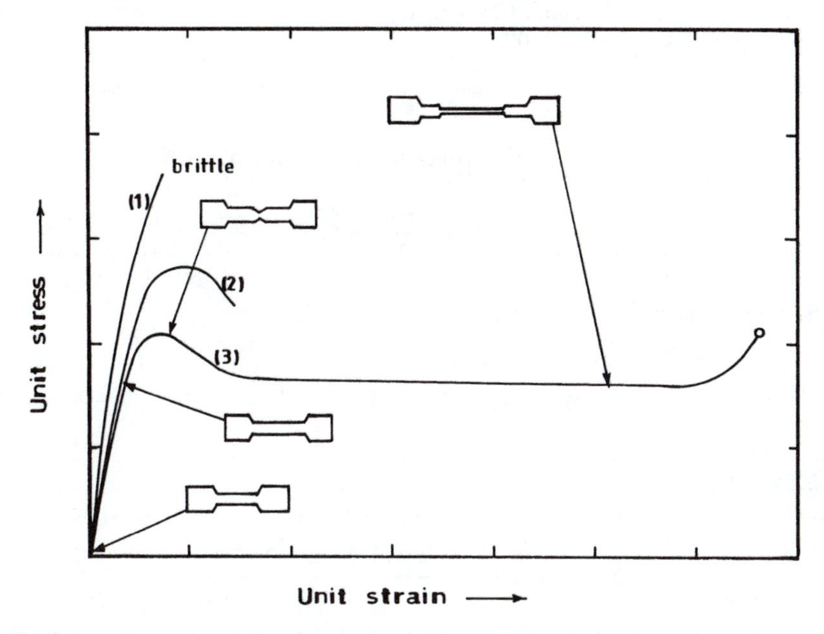

Figure 13.15 Schematic representation of macroscopic changes in tensile specimen shape during cold drawing. (From Fried, J.R., *Plast. Eng.*, 38(7), 27, 1982. With permission.)

and straightening of polymer chains, and the associated strain is recoverable. For hard, rigid polymers like polystyrene that show little or no yielding, a further increase in stress results in brittle failure (curve 1). In the case of ductile polymers, including engineering thermoplastics (polyamides, high-impact polystyrene), the stress–strain curve exhibits a maximum: the stress reaches a maximum value called the yield stress (more precisely, upper yield stress) and then decreases to a minimum value (drawing stress or lower yield stress). At this point, the sample may either rupture or experience strain hardening before failure. At the drawing stress, polymers that strain-harden require no further increase in applied stress to induce further elongation. It is believed that at the yield stress some slippage of polymer chains past each other occurs. The attendant deformation is recovered partially and slowly on the removal of the applied stress. What happens after the upper yield stress depends on the ability of the polymer material to strain-harden. The onset of necking is associated with an increase in the local stress at the necked region due to the reduction in the load-bearing cross-sectional area. This results in extensive deformation of the polymer material in the vicinity of the necked region, and the polymer chains in the amorphous regions undergo conformational changes and become oriented (stretched) in the direction of the applied tensile stress. The extended chains resist further deformation. If this orientation-induced hardening or resistance is sufficiently high to sustain or overcome the increased stress due to the reduction in the cross-sectional area (following the onset of necking), then further deformation (extension) of the specimen will occur only through the propagation of the neck along the sample. On the other hand, if the increased stress at the neck region increases faster than orientation hardening, then the necked region deepens continuously, leading ultimately to local failure at that region.

The molecular orientation of polymer chains is reflected in the observed changes of shape of the specimen (curve 3, Figure 13.15). Up to the yield stress, specimen deformation is essentially homogeneous. This means that deformation occurs uniformly over the entire gauge length of the specimen. At the yield stress, local instability ensues and the specimen begins to neck at some point along its gauge length. For specimens that exhibit orientation hardening before failure, the neck stabilizes; that is, the specimen shows no further reduction in cross-sectional area, but the neck propagates along the length of the gauge section until the specimen finally ruptures. The process of neck propagation is referred to as cold drawing.

Example 13.3: The mechanical properties of nylon 6,6 vary with its moisture content. A nylon specimen with a moisture content (MC) of 2.5% has an elastic modulus of 1.2 GN/m^2, while that for a sample of moisture content of 0.2% is 2.8 GN/m^2. Calculate the elastic energy or work per unit volume in each sample subjected to a tensile strain of 10%.

Solution:

$$\text{Work} = [\text{Force, (F)}][\text{increment of extension, (dL)}]$$

$$\text{Work per unit volume} = \frac{\text{FdL}}{\text{AL}} = \int_0^{\varepsilon} \sigma d\varepsilon$$

In the elastic region, Hooke's law holds, i.e., $\sigma = E\varepsilon$.

$$\text{Work per volume} = \int_0^{\varepsilon} E\varepsilon d\varepsilon = \frac{E\varepsilon^2}{2}$$

Sample 1 (2.5% MC):

$$\text{Elastic energy} = \frac{(1.2 \times 10^9\,\text{N/m}^2)(0.1)^2}{2}$$

$$= 0.6 \times 10^9 \times 10^{-2}\ \text{N-m/m}^3$$

$$= 6\ \text{MJ/m}^3$$

Sample 2 (0.2% MC):

$$\text{Elastic energy} = \frac{(2.8 \times 10^9\,\text{N/m}^2)(0.1)^2}{2}$$

$$= 14\ \text{MJ/m}^3$$

Example 13.4: Two nylon samples of moisture contents 2.5% and 0.2% have ε_B of 300% and 60%, respectively. Calculate the toughness of each sample if the stress-strain curve of nylon for plastic deformation is given by

$$\sigma = 8500\ \varepsilon^{0.1}\ \text{psi}$$

Comment on your results.

Solution:

$$\text{Work per unit volum (toughness)} = \int_0^{\varepsilon_B} \sigma d\varepsilon$$

$$= \int_0^{\varepsilon_B} 8500\varepsilon^{0.1} d\varepsilon$$

$$= 7727\varepsilon^{1.1} \text{ in.-lb/in.}^3$$

$$\text{Nylon 2.5\% MC, toughness} = 7727(3)^{1.1}$$

$$= 25{,}873 \text{ in.-lb/in.}^3$$

$$\text{Nylon 0.2\% MC, toughness} = 7727(0.6)^{1.1}$$

$$= 4405 \text{ in.-lb/in.}^3$$

As MC increased from 0.2 to 2.5%, toughness increased by about 500%.

V. COMPRESSION VS. TENSILE TESTS[6]

Recall that normal stresses can be either tensile or compressive. However, the main focus of our discussion so far has been on tensile tests. Let us now examine the behavior of polymers in compression.

Figures 13.16 and 13.17 are plots of the compressive stress–strain data for two amorphous and two crystalline polymers, respectively, while Figure 13.18 shows tensile and compressive stress–strain behavior of a normally brittle polymer (polystyrene). The stress–strain curves for the amorphous polymers are characteristic of the yield behavior of polymers. On the other hand, there are no clearly defined yield points for the crystalline polymers. In tension, polystyrene exhibited brittle failure, whereas in compression it behaved as a ductile polymer. The behavior of polystyrene typifies the general behavior of polymers. Tensile and compressive tests do not, as would normally be expected, give the same results. Strength and yield stress are generally higher in compression than in tension.

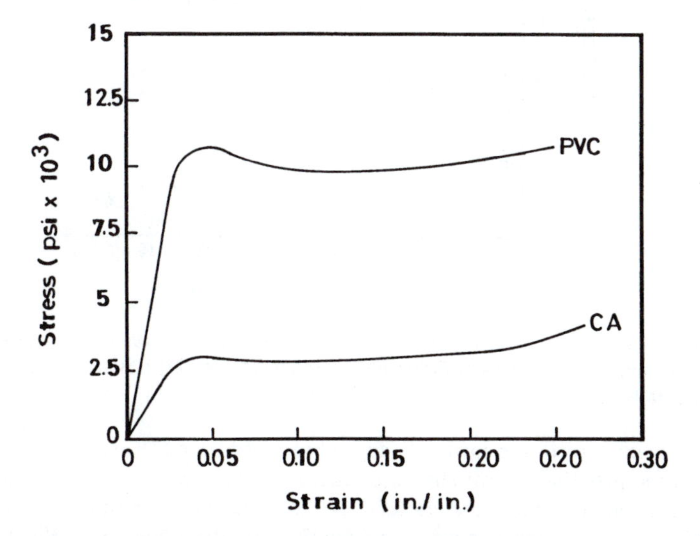

Figure 13.16 Compressive stress–strain data for two amorphous polymers: polyvinyl chloride (PVC) and cellulose acetate (CA). (From Kaufman, H.S. and Falcetta, J.J., Eds., *Introduction to Polymer Science and Technology,* John Wiley & Sons, New York, 1977. With permission.)

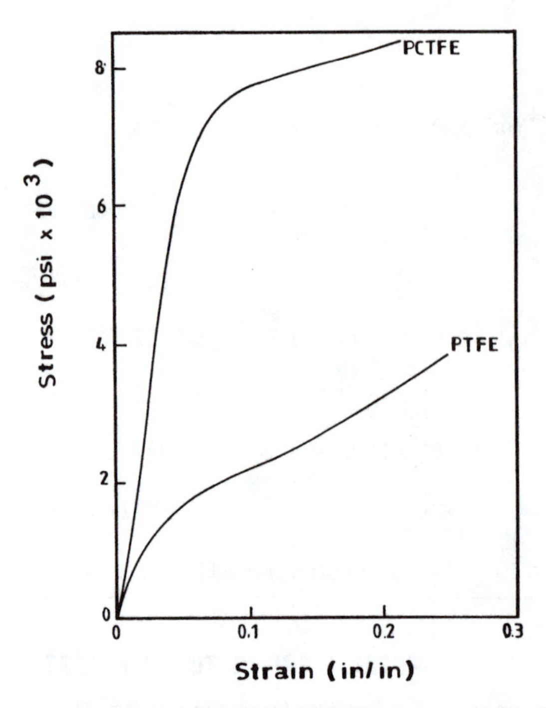

Figure 13.17 Compressive stress–strain data for two crystalline polymers: polytetrafluoroethylene (PTFE) and polychlorotrifluoroethylene (PCTFE). (From Kaufman, H.S. and Falcetta, J.J., Eds., *Introduction to Polymer Science and Technology,* John Wiley & Sons, New York, 1977. With permission.)

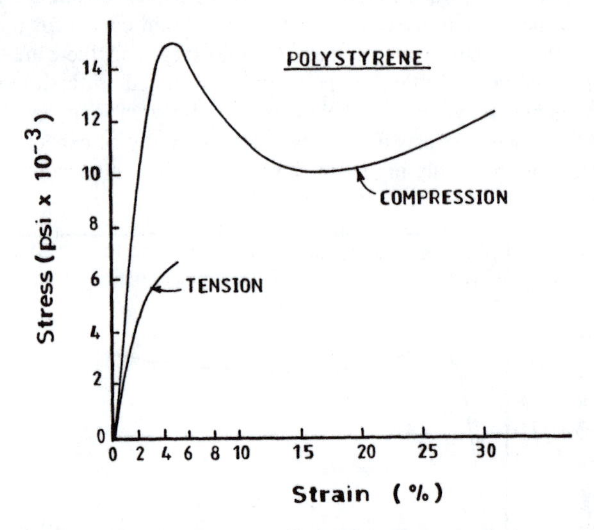

Figure 13.18 The stress–strain behavior of a normally brittle polymer, polystyrene, under tension and compression. (From Nielsen, L.E., *Mechanical Properties of Polymers and Composites,* Vol. 2, Marcel Dekker, New York, 1974. With permission.)

The tensile properties of brittle materials depend to a considerable extent on the cracks and other flaws inherent in the material. As we shall see later, for a material in tension, brittle fracture occurs through the propagation of one of these cracks. Since load cannot be transmitted through free surfaces, the presence of cracks essentially creates concentrations of stress intensity. When this tensile stress exceeds the fracture strength of the material, fracture occurs. It is apparent, therefore, that in contrast to tensile stresses, which open cracks, compressive stresses tend to close them. This could conceivably enhance the tensile strength.

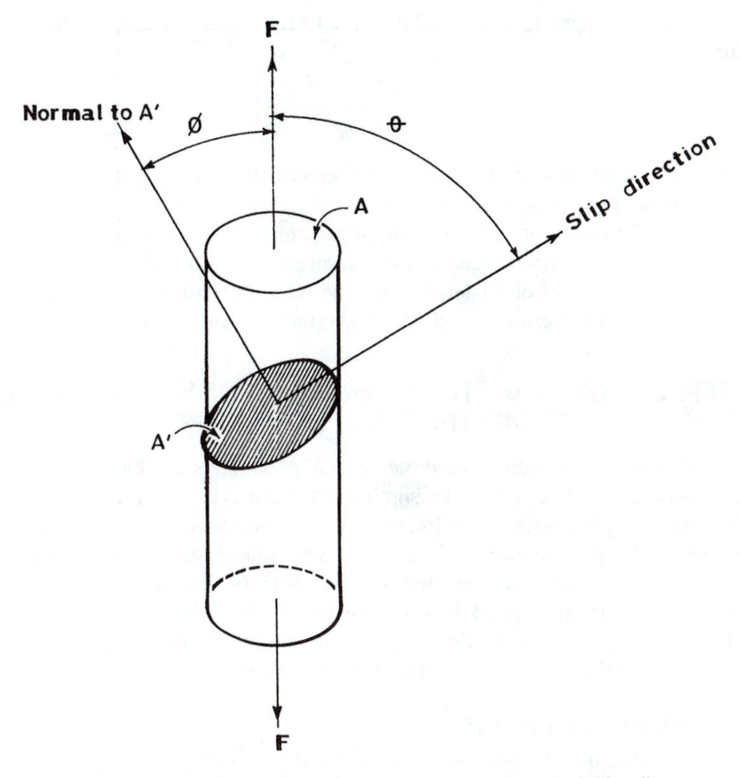

Figure 13.19 Generation of shear stress due to uniaxial loading.

The conditions for fracture are dictated by the magnitude of the imposed tensile stresses. On the other hand, the extent of permanent (plastic) deformation produced by a given load depends primarily on the magnitude of the shear stresses induced by the load. For example, the application of a uniaxial load, F, on a material generates shear stresses, τ, in certain geometric planes in particular directions. The magnitude of the generated shear stresses depends on the orientations of the planes and directions to the tensile axis. This can be illustrated by considering Figure 13.19.

Let us compute the shear stresses generated on the sectional plane A′ in a direction of angle θ to the tensile axis.[8] The normal to plane A′ makes an angle ϕ with the tensile axis. The load on A′ in the direction of deformation (slip) is F cos θ. Therefore, the stress generated on A′ is given by

$$\tau = \frac{F \cos \theta}{A/\cos \phi} \tag{13.26}$$

But by definition, F/A = σ, hence

$$\tau = \sigma \cos \theta \cos \phi \tag{13.27}$$

For a fixed value of ϕ, the minimum value of θ is $\pi/2 - \phi$. Thus Equation 13.27 becomes

$$\tau = \sigma \cos \phi \cos \left(\frac{\pi}{2} - \phi\right) = \sigma \cos \phi \sin \phi \tag{13.28}$$

The maximum value of τ occurs when $\phi = \pi/4$, that is,

$$(\tau)_{max} = \sigma (0.707)^2 = \frac{\sigma}{2} \tag{13.29}$$

Plastic deformation occurs when τ_{max} is at least equal to the yield strength of the material, i.e., for yielding to occur.

$$\tau_{max} \geq \sigma_y \tag{13.30}$$

The yield criterion for simple uniaxial stress dictates theoretically that for plastic deformation to occur, the imposed tensile stress must be at least twice the magnitude of the shear stresses generated. In other words, the tensile strength must be at least twice the shear strength. This is often not the case in practice; tensile strength is generally less than twice the shear strength. Also, the fact that contrary to theoretical prediction, the yield value for a given material is not the same in tension and in compression suggests that for polymers, plastic deformation may not be due entirely to shear stresses alone.

VI. EFFECTS OF STRUCTURAL AND ENVIRONMENTAL FACTORS ON MECHANICAL PROPERTIES

As we saw in the preceding discussion, several mechanical parameters can be derived from stress-strain tests. Two of these parameters are of particular significance from a design viewpoint. These are strength and stiffness. For some applications, the ultimate tensile strength is the useful parameter, but most polymer products are loaded well below their breaking points. Indeed, some polymers deform excessively before rupture and this makes them unsuitable for use. Therefore, for most polymer applications, stiffness (resistance to deformation under applied load) is the parameter of prime importance. Modulus is a measure of stiffness. We will now consider how various structural and environmental factors affect modulus in particular and other mechanical properties in general.

A. EFFECT OF MOLECULAR WEIGHT

In contrast to materials like metals and ceramics, the modulus of polymers shows strong dependence on temperature. Figure 13.20 is a schematic modulus–temperature curve for a linear amorphous polymer like atactic polystyrene. Five regions of viscoelastic behavior are evident: a hard and glassy region followed by a transition from the glassy to rubbery region. The rubbery plateau, in turn, is followed by a transition to the melt flow region. The glassy-to-rubbery transition is denoted by T_g, while the rubber-to-melt-flow transition is indicated by T_{fl}. There is a drop in modulus of about three orders of magnitude near the T_g. There is a further modulus drop at T_{fl}. If the T_g is above room temperature, the material will be a rigid polymer at room temperature. If, however, the T_g occurs below room temperature, the material will be rubbery and might even be a viscous liquid at room temperature.

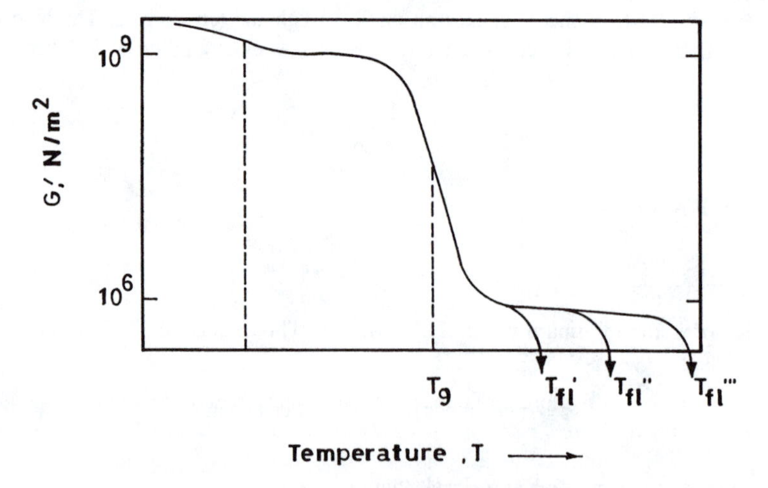

Figure 13.20 Schematic representation of the effect of molecular weight on shear modulus–temperature curve. T_g is the glass transition temperature while T_{fl} is the flow temperature. T_{fl}', T_{fl}'', T_{fl}''' represent low-, medium-, and high-molecular-weight materials, respectively.

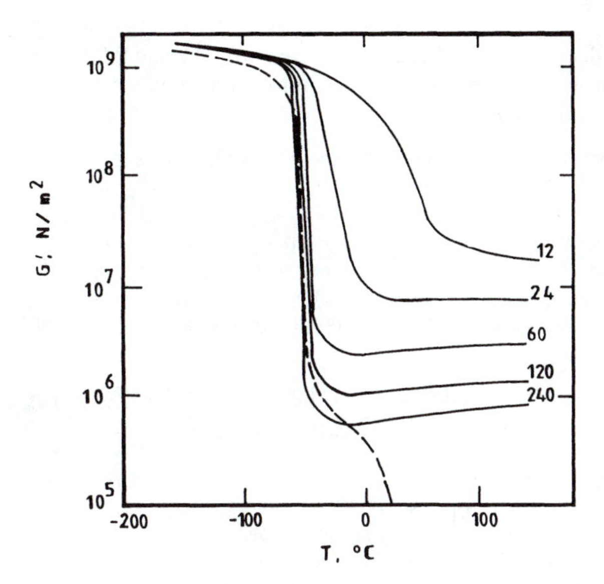

Figure 13.21 Effect of cross-linking on shear modulus of natural rubber: (———) cross-linked, the approximate mean number of chain atoms between successive cross-links is indicated; (----) noncross-linked. (From Heijboer, J., *Br. Polym. J.*, 1, 3, 1969. With permission.)

The molecular weight has practically no effect on the modulus in the glassy region (below T_g); the drop in modulus and the location of the T_g are also almost independent of the molecular weight. We recall that T_g is the onset of cooperative motion of chain segments. It is therefore to be expected that the number of chain entanglements (which increases with increasing molecular weight) will hardly affect such segmental mobility, and consequently T_g is independent of molecular weight. In contrast to the T_g, T_{fl} is strongly dependent on the molecular weight. Movement of entire molecules is associated with viscous flow. This obviously will depend on the number of entanglements. The higher the molecular weight (the higher the number of entanglements), the higher must be the temperature at which viscous flow becomes predominant over rubbery behavior. Consequently, for high molecular weight polymer, T_{fl} is high and the rubbery plateau long, whereas for low-molecular-weight polymers the rubbery plateau is absent or very short.

B. EFFECT OF CROSS-LINKING

Figure 13.21 shows the effect of cross-linking on the modulus of natural rubber cross-linked using electron irradiation. M_c is the average molecular weight between cross-links. It is a measure of the cross-link density; the smaller the value of M_c, the higher the cross-link density. In the glassy region, the increase in modulus due to cross-linking is relatively small. Evidently the principal effect of cross-linking is the increase in modulus in the rubbery region and the disappearance of the flow regions. The cross-linked elastomer exhibits rubberlike elasticity even at high temperature. Cross-linking also raises the glass transition temperature at high values of cross-link density. The glass-to-rubber transition is also considerably broadened.

Cross-linking involves chemically connecting polymer molecules by primary valence bonds. This imposes obvious restrictions on molecular mobility. Consequently, relative to the uncross-linked polymer, cross-linking increases polymer ability to resist deformation under load, i.e., increases its modulus. As would be expected, this effect is more pronounced in the rubbery region. In addition, the flow region is eliminated in a cross-linked polymer because chains are unable to slip past each other. Since T_g represents the onset of cooperative segmental motion, widely spaced cross-links will produce only a slight restriction on this motion. However, as the cross-link density is increased, the restriction on molecular mobility becomes substantial and much higher energy will be required to induce segmental motion (T_g increases).

Example 13.5: The densities of hard and soft rubbers are 1.19 and 0.90 g/cm^3, respectively. Estimate the average molecular weight between cross-links for both materials if their respective moduli at room temperature are 10^6 and 10^8 dynes/cm^2.

Solution: The average molecular weight between cross-links, M_c is given by

$$M_c = \rho \frac{RT}{G}$$

where ρ = density, R = gas constant, T = absolute temperature, and G = shear modulus.

$$M_c = \frac{\left(1.19\ \text{g/cm}^3\right)\left(8.31 \times 10^7\ \text{erg/mol K}\right)\left(300\ \text{K}\right)}{\left(10^8\ \text{dyne/cm}^2\right)}$$

$$= 300\left(\frac{\text{g}}{\text{mol}}\right)\left(\frac{\text{erg}}{\text{cm}^2}\right)\left(\frac{\text{cm}^3}{\text{dyne}}\right)$$

$$= 300\left(\frac{\text{g}}{\text{mol}}\right)\left(\frac{\text{erg}}{\text{cm}^3}\right)\left(\frac{\text{cm}^3}{\text{dyne-cm}}\right)$$

$$= 300\ \text{g/mol}$$

Soft rubber:

$$M_c = \frac{\left(0.90\ \text{g/cm}^3\right)\left(8.31 \times 10^7\ \text{erg/mol K}\right)\left(300\ \text{K}\right)}{10^7\ \text{dyne/cm}^2}$$

$$= 2250\ \text{g/mol}$$

C. EFFECT OF CRYSTALLINITY

Crystallinity in a polymer is the result of ordered molecular aggregation, with molecules held together by secondary valence bonds. Therefore, crystallinity may be viewed as a form of physical cross-linking which is thermoreversible. Since crystallites have much higher moduli than the amorphous segments, crystallites can also be regarded as rigid fillers in an amorphous matrix. The effect of crystallinity on modulus becomes readily understandable on the basis of these concepts. Figure 13.22 is an idealized modulus–temperature curve for various degrees of crystallinity. We observe that crystallinity has only a small effect on modulus below the T_g but has a pronounced effect above the T_g. There is a drop in modulus at the T_g, the intensity of which decreases with increasing degree of crystallinity. This is followed by a much sharper drop at the melting point. Crystallinity has no significant effect on the location of the T_g, but the melting temperature generally increases with increasing degree of crystallinity. These features are evident for two polyethylenes of different crystallinities (Figure 13.23). Alkathene is branched with a density of 0.92 g/cm^3, while the Ziegler polyethylene is linear and has a density of 0.95 g/cm^3. The greater crystallinity of the Ziegler polyethylene results in a higher modulus, especially above 0°C.

D. EFFECT OF COPOLYMERIZATION

In discussing the effect of copolymerization on modulus, it is necessary to make a distinction between random and alternating copolymers and block and graft comonomers. Random and alternating copolymers are necessarily homogeneous, while block and graft copolymers with sufficiently long sequences exhibit phase separation. Random and alternating copolymers show a single transition that is intermediate between those of the two homopolymers of A and B (Figure 13.24). Copolymerization essentially shifts

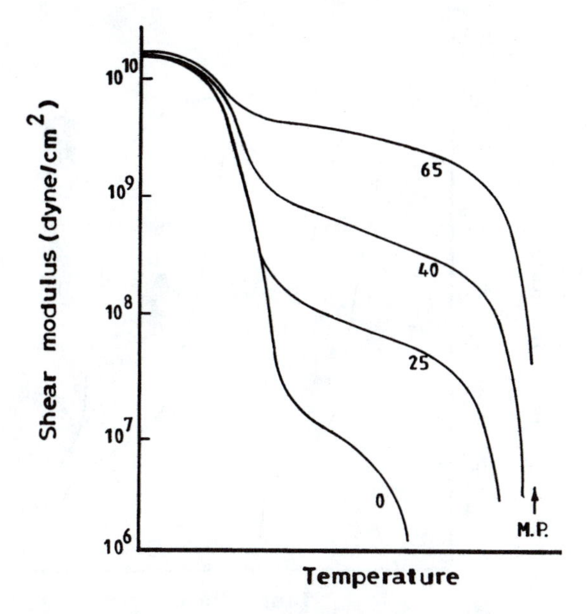

Figure 13.22 Effect of crystallinity on the modulus–temperature curve. The numbers of the curves are rough approximations of the percentage of crystallinity. Modulus units = dynes/cm². (From Nielsen, L.E., *Mechanical Properties of Polymers and Composites*, Vol. 2, Marcel Dekker, New York, 1974. With permission.)

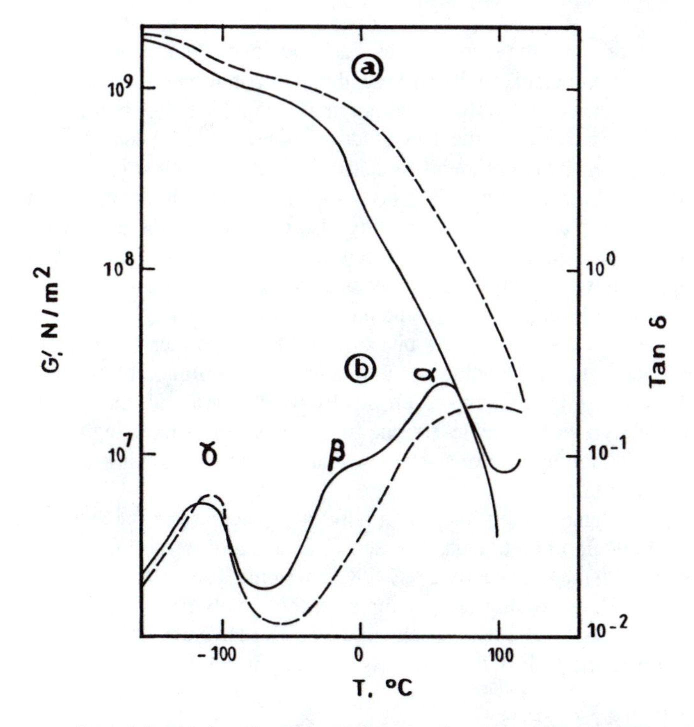

Figure 13.23 Shear modulus (a) and damping (b) at I Hz as a function of temperature for a branched and a linear polyethylene: (– – – –) Ziegler polyethylene; (———) Alkathene. (From Heijboer, J., *Brit. Polymer J.*, 1, 3, 1969. With permission.)

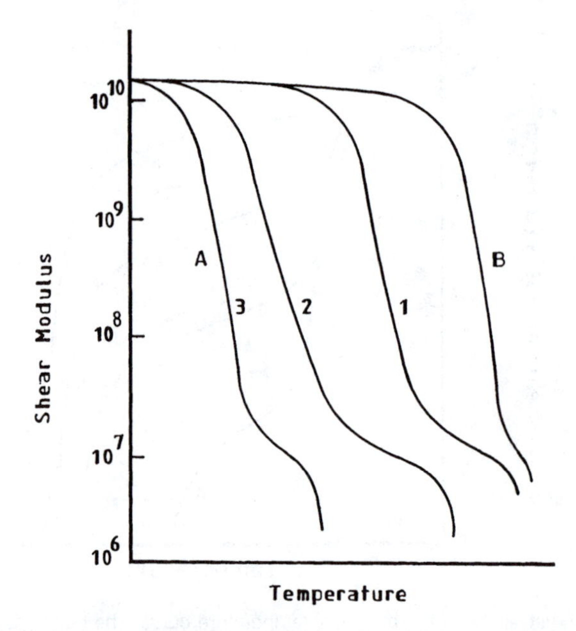

Figure 13.24 Effect of plasticization or copolymerization on the modulus–temperature curve. The curves correspond to different copolymer compositions. (B) Unplasticized homopolymer; (A) either a second homopolymer or plasticized B. (From Nielsen, L.E., *Mechanical Properties of Polymers and Composites,* Vol. 2, Marcel Dekker, New York, 1974. With permission.)

the modulus–temperature curve the same way as T_g. In addition, there is a broadening of the transition due to polymer composition heterogeneity. We recall that in copolymer the reactivities of the monomers are generally different. Consequently, the initial polymer formed is richer in the more reactive monomer while that formed later is richer in the less reactive monomer. This leads overall to a polymer of heterogeneous composition and consequently a distribution of glass transitions (broad transition region).

Block and graft copolymers, which exist as a two-phase system, have two distinct glass transitions, one for each of the homopolymers. Consequently, the modulus–temperature curve shows two steep drops. The value of modulus in the plateau between the two glass transitions depends upon the ratio of the components and upon which of the two phases is dispersed in the other. These features are illustrated in Figure 13.25 for a styrene–butadiene–styrene block copolymer. The glass transition of the butadiene phase near –80°C and that for the styrene phase near 110°C are clearly evident. Between the T_g of butadiene and the T_g of styrene, the value of the modulus is determined by the amount of polystyrene; the rubbery butadiene phase is cross-linked physically by the hard and glassy polystyrene phase. It is noteworthy that while styrene–butadiene–styrene block copolymers have high tensile strength, butadiene–styrene–butadiene copolymers have a very low tensile strength, showing that strength properties are dictated by the dispersed phase.

In both cases of copolymerization, there is a noticeable decrease in the slope of the modulus curve in the region of the inflection point. This, in essence, means a decrease in the modulus in the rubbery region. This contrasts with the chemically cross-linked systems where the modulus in the rubbery region shows some increase with increasing temperatures. In the copolymer system, the molecules are interconnected by physical cross-links due to secondary forces. These cross-links can be disrupted reversibly by heating, and this forms the basis of the new class of copolymers referred to as thermoplastic elastomers.

E. EFFECT OF PLASTICIZERS

Plasticizers are low-molecular-weight, usually high boiling liquids that are capable of enhancing the flow characteristics of polymers by lowering their glass transition temperatures. Modulus, yield, and tensile strengths generally decrease with the addition of plasticizers to a polymer. In general, on

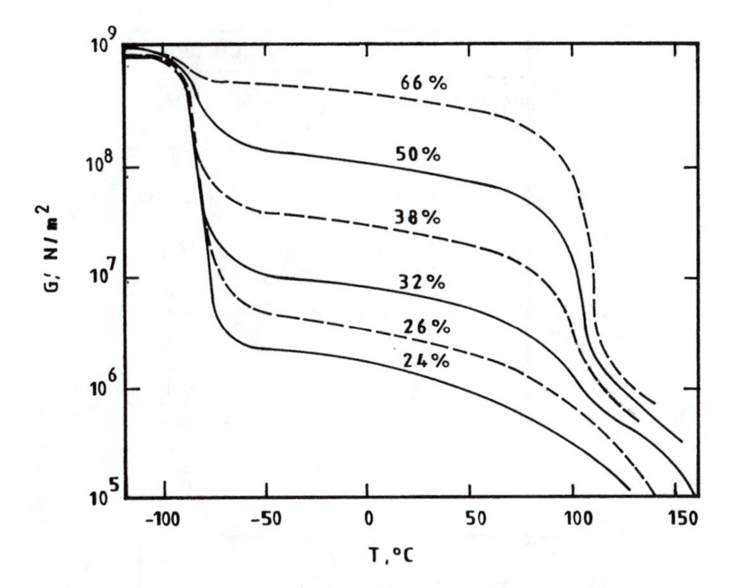

Figure 13.25 Shear modulus as a function of temperature for styrene–butadiene–styrene block copolymers. Wt.% styrene is indicated on each curve. (From Heijboer, J., *Br. Polymer J.*, 1, 3, 1969. With permission.)

plasticization a polymer solid undergoes a change from hard and brittle to hard and tough to soft and tough. This is exemplified by the use of dioctylphthalate to convert poly(vinyl chloride) from a rigid material, such as PVC pipes, to a soft one, as in car seat covers or a raincoat. Plasticization and alternating or random copolymerization have similar effects on modulus (Figure 13.24). In this case B is the unplasticized homopolymer, while curves 1, 2 and 3 represent increasing plasticization of B.

F. EFFECT OF POLARITY

The effect of polarity is shown in Figure 13.26, which compares poly(vinyl chloride) with polypropylene. The T_g of the polar poly(vinyl chloride) is about 90°C higher than that of the nonpolar polar polypropylene. As the methyl group and chlorine atom occupy about the same volume, the differences in mechanical behavior can only be due to the relative polarities of the two polymers. However, the effect of the substitution of the chlorine atom for the methyl group depends on the molecular environment of the chlorine atom. The further the chlorine atom is from the main chain, the smaller its effect on the T_g. This is illustrated in Figure 13.27, which compares the mechanical behavior of the following polymers:

Polymer	Structure
Poly(2-chloroethyl methacrylate)	$-CH_2-C(CH_3)(C(=O)-O-CH_2-CH_2-Cl)-$

Polymer	Structure
Poly(*n*-propyl methacrylate)	$-CH_2-C(CH_3)(C(=O)-O-CH_2-CH_2-CH_3)-$

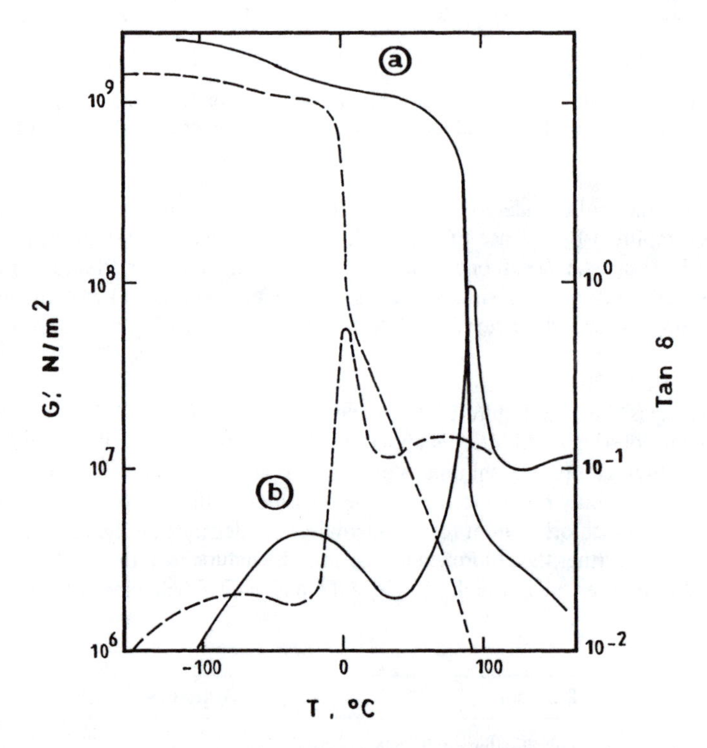

Figure 13.26 Shear modulus (a) and damping (b) as a function of temperature: (———) poly(vinyl chloride); (– – – –) polypropylene. (From Heijboer, J., *Br. Polym. J.*, 1, 3, 1969. With permission.)

In this case, the T_g of poly(2-chloroethyl methacrylate) is only 20°C higher than that of poly(*n*-propyl methacrylate). However, with the secondary transitions, which represent the movement of the side chains (chloroethyl and *n*-propyl groups, respectively), the effect on the location of the glass temperature is much larger. Also, the modulus of the polar polymer is much higher. In the glassy region, the magnitude of the modulus is determined primarily by secondary bonding forces. It is therefore to be expected that the modulus will be increased by the introduction of polar forces. Indeed, polarity is generally the most effective means for increasing modulus in the glassy region.

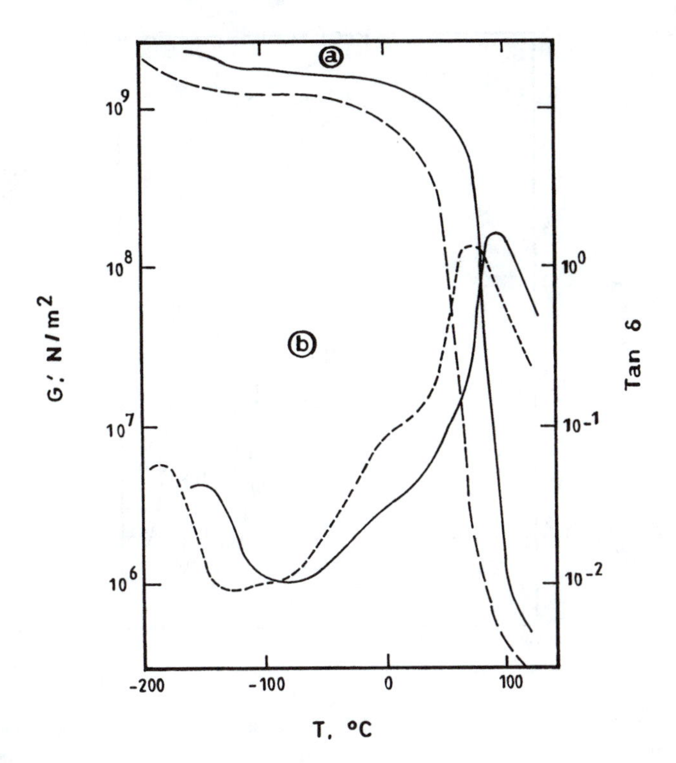

Figure 13.27 Shear modulus (a) and damping (b) at 1 Hz as a function of temperature: (———) poly(2-chloroethyl methacrylate); (- - - -) poly(*n*-propyl methacrylate). (From Heijboer, J., *Br. Polym. J.*, 1, 3, 1969. With permission.)

Example 13.6: Both polyoxymethylene (DuPont Delrin 550) and polyethylene (Calanese Fortiflex A70) show similar mechanical behavior but the glass transition temperature of polyoxymethylene is about 50°C higher than that of polyethylene. Explain.

Solution:

Polymer	Structure
Polyoxymethylene	$-CH_2-O-$
Polyethylene	$-CH_2-CH_2-$

The presence of the oxygen atom in the main chain of polyoxymethylene might have been expected to enhance its flexibility compared with polyethylene and hence reduce its T_g relative to that of polyethylene. However, we note that the dipole character of the C–O–C group produces polar forces between adjacent chains, which act over a longer range and are stronger than van der Waals forces. Thus, for polyoxymethylene the induced flexibility is more than offset by the increased bonding forces resulting from polarity.

G. STERIC FACTORS

In discussing the influence of steric features on mechanical properties, it is convenient to consider the side chains and the main chain separately. The effects of flexible side chains differ completely from those of stiff side chains. Long, flexible side chains reduce the glass transition temperature, while stiff side chains increase it. Long, flexible side chains increase the free volume and ease the steric hindrance from neighboring chains and as such facilitate the movement of the main chain. Figure 13.28 illustrates

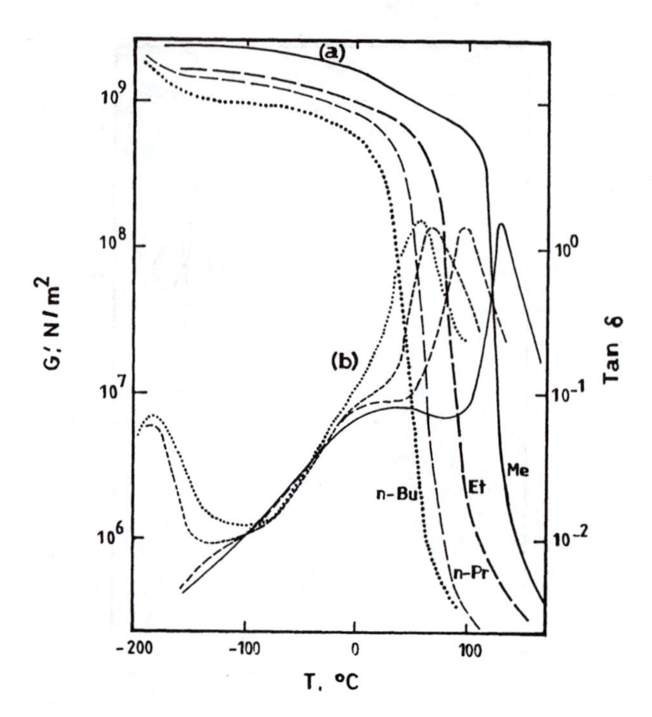

Figure 13.28 Shear modulus (a) and damping (b) at 1 Hz as a function of temperature for poly(*n*-alkyl methacrylate). (——) Polymethyl methacrylate; (– – –) polyethyl methacrylate; (— — —) poly(*n*-propyl methacrylate); (····) poly(*n*-butyl methacrylate). (From Heijboer, J., *Proc. Int. Conf. Physics Non-crystalline Solids,* North-Holland, Amsterdam, 1965, 231. With permission.)

the reduction in T_g and the increase in modulus in the glassy region with increase in length of the alkyl group for poly(*n*-alkyl methacrylate).

Branched side chains, particularly if the branched point is located close to the main chain, increase the glass transition temperature. This is illustrated in Figures 13.29 and 13.30 for a series of polyacrylates and polyolefins with branched side chains, respectively.

Structural changes within or near the main chain, even if minor, can produce a drastic effect on the mechanical properties of a polymer. Table 13.4 shows the increase in T_g and T_m (for the crystallizable polymers) by the introduction of rings into the main chain.

In spite of the possible increase in chain flexibility due to the elongation of the diacid by two methylene groups, polycyclamide has higher T_g and T_m than nylon 6,6 as a result of the substitution of stiff cyclohexylene 1,4 for four methylene groups in nylon 6,6. Trogamid T does not crystallize. However, the additional stiffening of the main chain due to the presence of methyl side groups leads to a further increase in the T_g. In general, the introduction of rings into the main chain provides a better structural mechanism for toughening polymers than chain stiffening by bulky side groups. For example, polycarbonate, polysulfone, and polyphenylene oxide all have high impact strength whereas poly(vinyl carbozole) with comparable T_g is brittle (Table 13.5).

H. EFFECT OF TEMPERATURE

The mechanical properties of polymers are generally more susceptible to temperature changes than those of ceramics and metals. As discussed in Section 12.5.1, the modulus of a polymer decreases with increasing temperature. However, the rate of decrease is not uniform; the drop in modulus is more pronounced at temperatures associated with molecular transitions. As the temperature is increased to a level that can induce some form of molecular motion, a relaxation process ensues and there is a drop in modulus. At temperatures below the T_g, the polymer is rigid and glassy. In the glassy region, there are usually one or more secondary transitions that are related either to movements of side chains or to restricted movements of small parts of the main chain. For the secondary transitions, the drop in modulus is about 10 to 50%, while primary transitions like T_g and T_m, which involve large-scale main chain

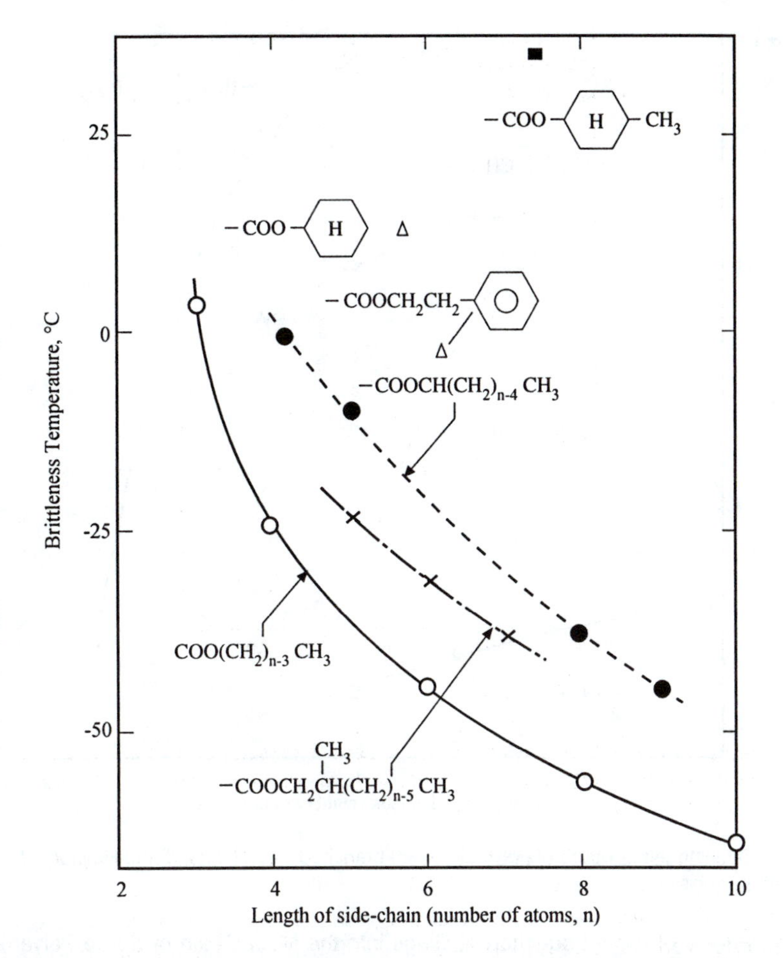

Figure 13.29 Brittleness temperatures for polyacrylates as a function of the total length of the side chain. (From Heijboer, J., *Br. Polym. J.*, 1, 3, 1969. With permission.)

motions, the drop in modulus varies from one to three orders of magnitude depending on the type of polymer. If no molecular relaxation process occurs over a particular temperature range, the modulus decreases rather slowly with increasing temperature. In this case, the decrease in modulus is due to the normal reduction in intermolecular forces that occurs with an increase in temperature.

The general effect of temperature on the stress-strain properties of a polymer is illustrated in Figure 13.31. Below the T_g, the modulus is high, as discussed above; there is essentially no yield point and consequently the polymer is brittle (i.e., has low elongation at break). The yield point appears at temperatures close to the T_g. As we shall see in the next section, a high speed of testing requires higher temperatures for the appearance of the yield point. In some polymers with pronounced secondary transitions, the yield point appears in the neighborhood of this transition temperature rather than at T_g. For example, in spite of a high T_g of 150°C, polycarbonate is remarkably tough at room temperature, and this is associated with a secondary transition at about –100°C. In general, therefore, as the temperature is increased, the modulus and yield strength decrease and the polymer becomes more ductile.

I. EFFECT OF STRAIN RATE

Polymers are very sensitive to the rate of testing. As the strain rate increases, polymers in general show a decrease in ductility while the modulus and the yield or tensile strength increase. Figure 13.32 illustrates this schematically. The sensitivity of polymers to strain rate depends on the type of polymer: for brittle polymers the effect is relatively small, whereas for rigid, ductile polymers and elastomers, the effects can be quite substantial if the strain rate covers several decades.

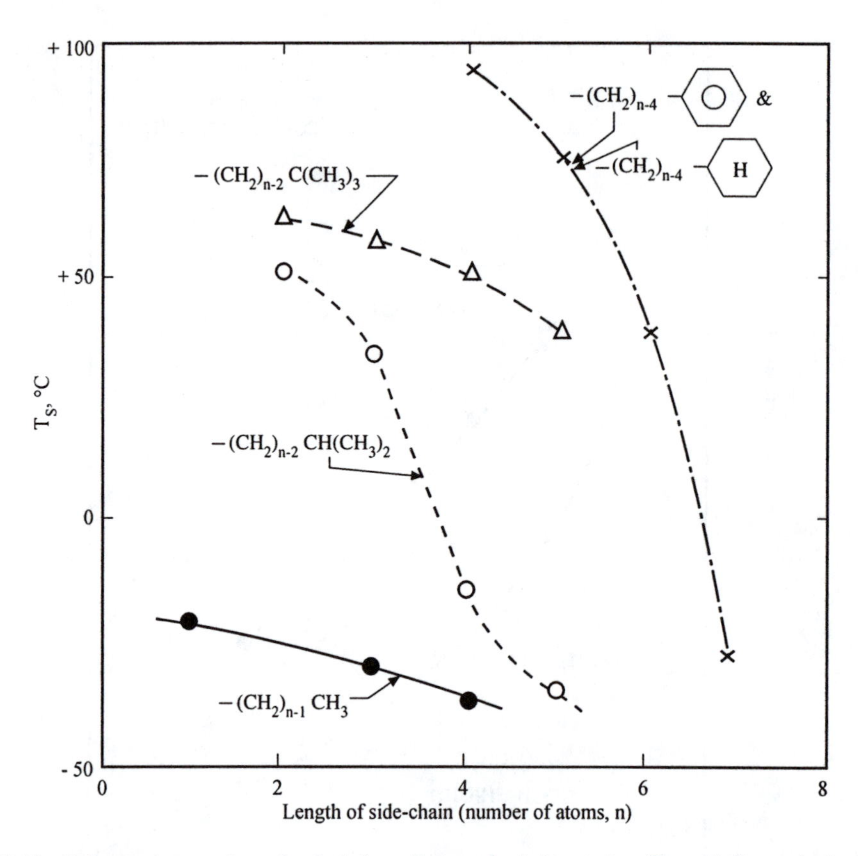

Figure 13.30 Softening temperature of polyolefins with branched side chains. (From Heijboer, J., *Br. Polym. J.*, 1, 3, 1969. With permission.)

Table 13.4 Effects of the Introduction of Rings into the Main Chain of Some Polyamides

Polymer	Structure	T_g °C	T_m °C
Nylon 6.6 (Zytel, Du Pont de Nemours)	$\left[-C(=O)-(CH_2)_4-C(=O)-N(H)-(CH_2)_6-N(H)-\right]_n$	75	260
Polycyclamide (Q.2 Tennessee Eastman Co.)	$\left[-C(=O)-(CH_2)_6-C(=O)-N(H)-CH_2-C_6H_{10}-CH_2-N(H)-\right]_n$	125	300
Trogamid T (Dynamit Nobel A.G.)	$\left[-N(H)-CH_2-C(CH_3)_2-CH(CH_3)-(CH_2)_2-N(H)-C(=O)-C_6H_4-C(=O)-\right]_n$	145	none

Table 13.5 Polymer Stiffening Due to the Introduction of Rings into the Main Chain

Polymer	Structure	T_g °C
Polycarbonate	$[-O-C_6H_4-C(CH_3)_2-C_6H_4-O-C(=O)-]$	150
Polysulfone	$[-O-C_6H_4-C(CH_3)_2-C_6H_4-O-C_6H_4-SO_2-C_6H_4-]$	190
Poly(phenylene oxide)	$[-C_6H_2(CH_3)_2-O-C_6H_2(CH_3)_2-O-C_6H_2(CH_3)_2-O-C_6H_2(CH_3)_2-O-]$	220
Poly(vinyl carbazole)	$-CH_2-CH(N\text{-carbazole})-$	208

Polymers show a similar response to temperature and strain rate (time), as might be expected from the time–temperature superposition principle (compare Figures 13.31 and 13.32). Specifically, the effect of decreasing temperature is equivalent to that of increasing the strain rate. As has become evident from our previous discussions, low temperature restricts molecular movement of polymers, and consequently they become rigid and brittle. Materials deform to relieve imposed stress. High strain rates preclude such deformation and therefore result in brittle failure.

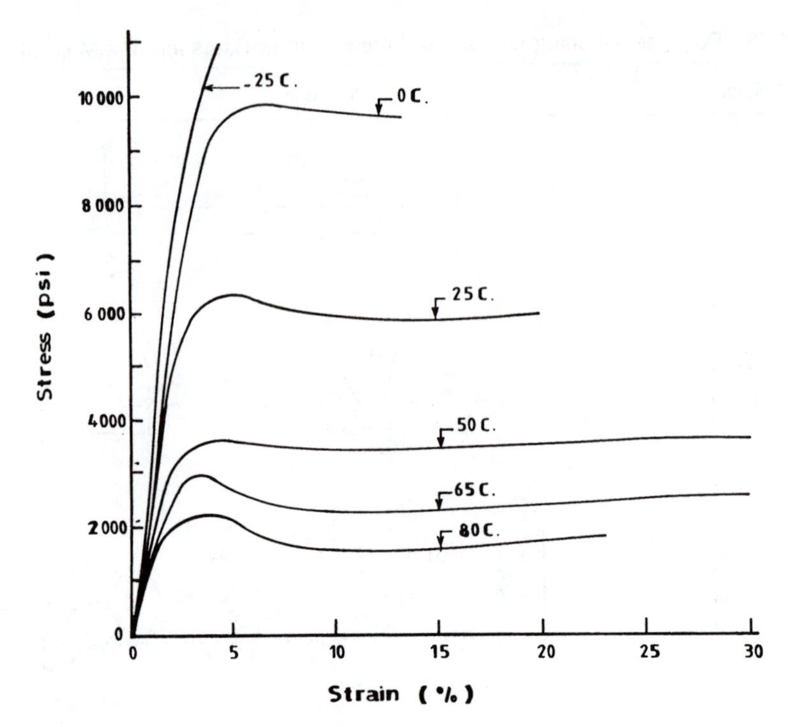

Figure 13.31 The stress–strain behavior of cellulose acetate at different temperatures. (From Carswell, T.S. and Nasor, H.K., *Mod. Plast.*, 21(6), 121, 1944. With permission.)

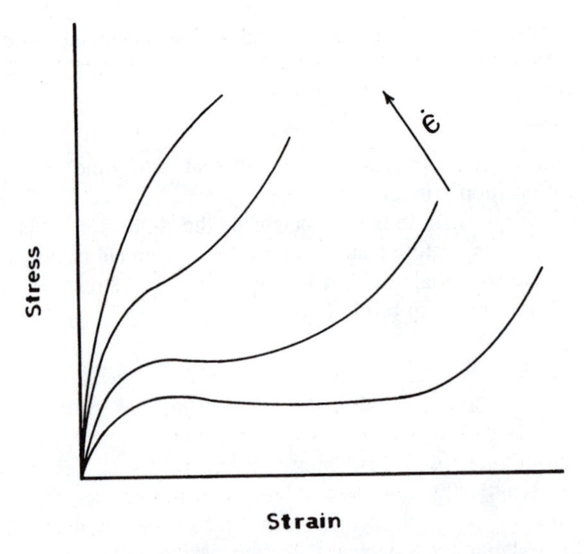

Figure 13.32 Schematic illustration of the effect of strain rate on polymers.

Example 13.7: Explain the observed differences in the tan δ of polystyrene tested at the two different frequencies shown in Figure E13.7.

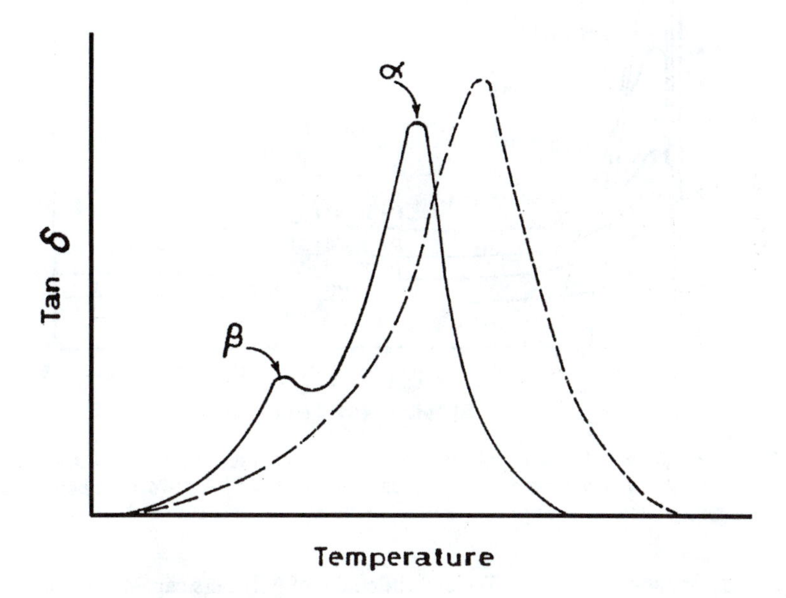

Figure E.13.7 Schematic representation of the variation of tan δ with temperature for atactic polystyrene tested at 0.1 Hz (——) and 50 Hz (---).

Solution: The mechanical properties of polymers depend on time and temperature. The time dependence is usually expressed as a frequency dependence, which to a first approximation is related to time by $2\pi \upsilon = 1/t$ where υ is the frequency. The combined dependence of molecular processes of viscoelastic materials on frequency and temperature can be described by an activation energy E_a. E_a is about 100 kcal/mol and 10 kcal/mol for primary and secondary transitions, respectively. This implies that the relaxation processes associated with the molecular motions shift to higher temperatures at higher frequencies; however, the secondary transition shifts more than the primary transition. Therefore, if tests are conducted at high frequencies, the resolution between the energy absorption peaks for primary and secondary transitions that are close to each other is poor. Thus, in this case, the β and α peaks, which are relatively distinct at 0.1 Hz, merge at 50 Hz, and there is a shift in the peak to higher temperatures.

J. EFFECT OF PRESSURE

The imposition of hydrostatic pressure on a polymer has a tremendous effect on its mechanical properties, as demonstrated by the stress–strain behavior of polypropylene in Figure 13.33. The modulus and yield stress increase with increasing pressure. This behavior is general for all polymers. However, the effect of pressure on the tensile strength and elongation at break depends on the polymer. The tensile strength tends to increase for ductile polymers, but decrease for some brittle polymers. The elongation at break increases for some ductile polymers, but decreases for most brittle polymers and polymers such as PE, PTFE, and PP, which exhibit cold drawing at normal pressures. In some brittle polymers like PS, brittle-ductile transition is induced beyond a certain critical pressure.

The increase in modulus and yield stress with increase in pressure is to be expected on the basis of the available free volume. An increase in pressure decreases the free volume (i.e., increases the packing density) and as such enhances the resistance to deformation (modulus) or delays the onset of chain

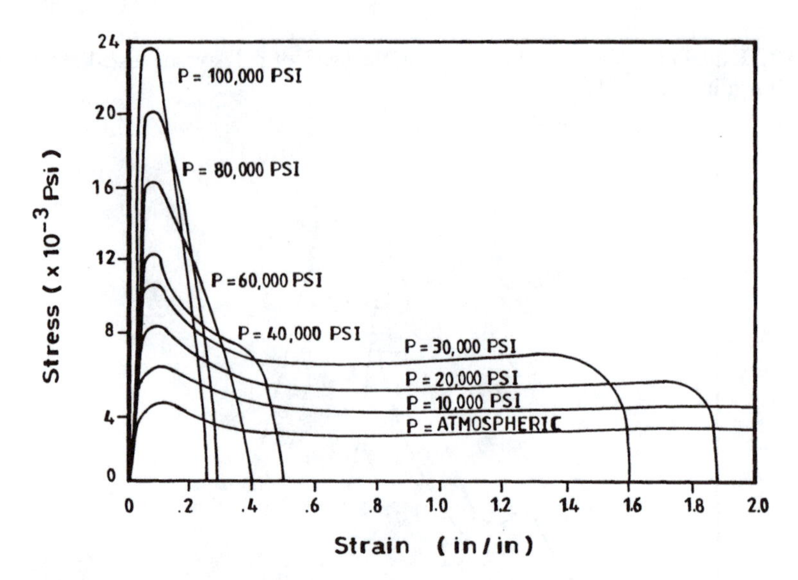

Figure 13.33 The stress–strain behavior of polypropylene at different pressures. (From Nielsen, L.E., *Mechanical Properties of Polymers and Composites,* Vol. 2, Marcel Dekker, New York, 1974. With permission.)

slippage or plastic deformation (yielding). The embrittlement of polymers capable of cold drawing with increase in pressure can be rationalized on the basis of the reduction in chain mobility associated with the decrease in free volume. Also, with an increase in pressure, the occurrence of relaxation processes responsible for the cold drawing phenomenon is delayed to higher temperatures. In other words, subjecting a polymer to increasing pressure at a given temperature has an effect on its mechanical behavior equivalent to reducing its temperature at a given pressure.

VII. POLYMER FRACTURE BEHAVIOR

Structural members can fail to perform their intended functions in three general ways: (1) excessive elastic deformation, (2) excessive plastic deformation, and (3) fracture. Modulus is a measure of the resistance to elastic deformation. We discussed the effects of various factors on modulus in the previous section. The stress levels encountered in most end-use situations preclude excessive plastic deformation. Although polymers are noted generally for the ductility, they are susceptible to brittle failure under appropriate conditions. Brittle fracture is usually catastrophic and involves very low strains. The major conditions responsible for brittle failure include low temperature, high rates of loading such as during shock or impact loading, and alternating loads. A polymer can rupture at loads much lower than would be dictated by its yield or tensile stress if subjected to alternating loads or to static loads for long duration. Therefore, to utilize a polymer properly, careful consideration must be given to possible brittle failure that can arise from environmental conditions and/or stress levels imposed on the polymer in use. We now discuss very briefly the fracture behavior in polymers.

A. BRITTLE FRACTURE

The theoretical fracture strength of a material can be deduced from the cohesive forces between the component atoms in the plane under consideration from a simple energy balance between the work to fracture and the energy require to create two new surfaces. It can be shown that the theoretical cohesive strength is given by

$$\sigma_c = \sqrt{\frac{E\gamma_s}{a_0}} \tag{13.31}$$

where σ_C = theoretical cohesive strength
E = Young's modulus
γ_S = surface energy per unit area
a_0 = equilibrium interatomic spacing of atoms in the unstrained state

Engineering materials, including polymers, generally have low fracture strengths relative to their theoretical capacity. The lower-than-ideal fracture strengths of engineering materials are generally attributed to the presence of flaws such as cracks, scratches, and notches inherent in these materials.

The first explanation of the discrepancy between observed fracture strength of crystals and the theoretical cohesive strength was proposed by Griffith.[12] He utilized an earlier analysis by Inglis,[13] who showed that the applied stress, σ_o, was magnified at the ends of the major axis of an elliptical hole in an infinitely wide plate (Figure 13.34) according to the following relation:

$$\sigma_t = \sigma_0 \left(1 - 2\sqrt{a/\rho}\right) \tag{13.32}$$

where ρ is the radius of curvature of the tip of the hole, 2a is the length of the major axis of the hole, and σ_t is the stress and the end of the major axis.

Griffith proposed that a brittle material contains a population of cracks and that the amplification of the local stresses at the crack tip was such that the theoretical cohesive strength was reached at nominal stress levels much below the theoretical value. The high stress concentrations lead to an extension of one of the cracks. The creation of two new surfaces results in a concomitant increase in surface energy, which is supplied by a decrease in the elastic strain energy. Griffith established the following criterion for the propagation of a crack: "A crack will propagate when the decrease in the elastic strain energy

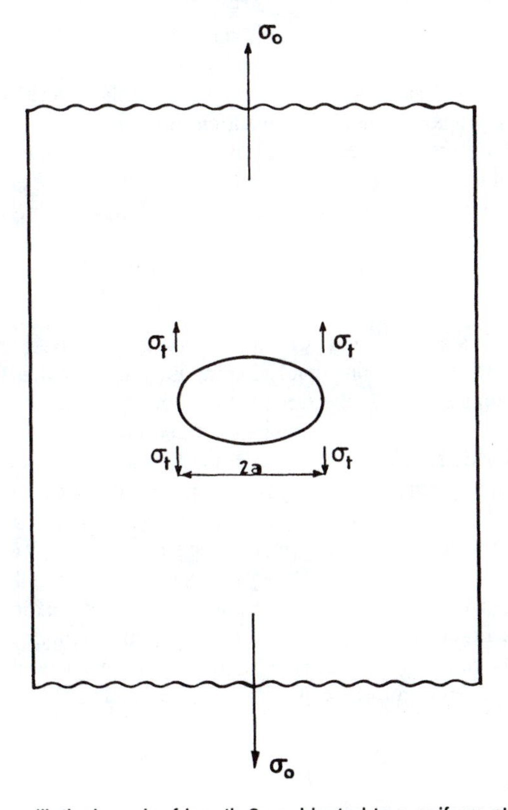

Figure 13.34 Model of an elliptical crack of length 2a subjected to a uniform stress σ_0 in an infinite plate.

is at least equal to the energy required to create the new surface."[12] He showed that the stress required for crack propagation is given by the relation:

$$\sigma_f = \sqrt{\frac{2E\gamma_s}{\pi a}} \quad \text{for plane stress} \tag{13.33}$$

$$\sigma_f = \sqrt{\frac{2E\gamma_s}{(1-\nu^2)(\pi a)}} \quad \text{for plane strain} \tag{13.34}$$

where E is Young's modulus, γ_s is the surface energy per unit area, ν is Poisson's ratio and a is half crack length.

The Griffith relation was derived for an ideally brittle material and the theory satisfactorily predicts the fracture strength of a material like glass, where the work to fracture is essentially equal to the increase in surface energy. On the other hand, the theory is inadequate for describing the fracture behavior of materials like polymers and metals, which are capable of extensive plastic deformation (i.e., materials whose fracture energies are several orders of magnitude greater than surface energy). A substantial amount of local plastic deformation invariably occurs at the crack tip.

To obviate this inadequacy, Griffith's equation had to be modified to include the energy expended in plastic deformation in the fracture process. Accordingly, Irwin[14] defined a parameter, G, the strain-energy-release rate or crack extension force, which he showed to be related to the applied stress and crack length by the equation:

$$\sigma = \sqrt{\frac{EG}{\pi a}} \tag{13.35}$$

At the point of instability, the elastic energy release rate G reaches a critical value, G_C, whereupon a previously stationary crack propagates abruptly, resulting in fracture.

A different approach from the strain-energy-release rate in the study of the fraction process is the analysis of the stress distribution around the crack tip. This gives the stress intensity factor or fracture toughness K, which for a sharp crack in an infinitely wide and elastic plate is given by

$$K = \sigma\sqrt{\pi a} \tag{13.36}$$

The stress intensity factor K is a useful and convenient parameter for describing the stress distribution around a crack. The difference between one cracked component and another lies in the magnitude of K. If two flaws of different geometries have the same value of K, then the stress fields around each of the flaws are identical. Since K is a function of applied load and crack size, it increases with load. When the intensity of the local tensile stresses at the crack tip attains a critical value, K_C, failure occurs. This critical value, K_C, defines the fracture toughness and is constant for a particular material, since cracking always occurs at a given value of local stress intensity regardless of the structure in which the material is used. K_C is a function of temperature, strain rate, and the state of stress, varying between the extremes of plane stress and plane strain. In addition, K_C also depends on the failure mode. The stress field surrounding a crack tip can be divided into three major modes of loading depending on the crack displacement, as shown in Figure 13.35. Mode I is the predominant mode of failure in real situations involving cracked components. This is the mode of failure that has received the most extensive attention. For mode I, G and K are related according to Equations 13.37 and 13.38.

$$G_{Ic} = \frac{K_{IC}^2}{E} \quad \text{(plane stress)} \tag{13.37}$$

$$G_{IC} = \frac{K_{IC}^2}{E}(1-\nu^2) \quad \text{(plane strain)} \tag{13.38}$$

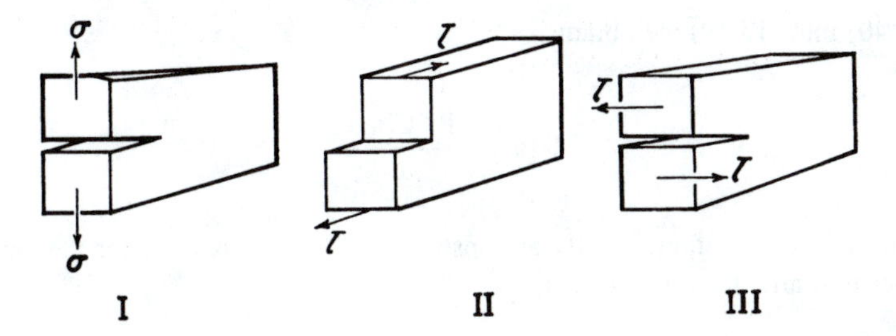

Figure 13.35 Basic modes of failure of structural materials: (I) opening or tensile mode; (II) sliding or in-plane shear mode; (III) tearing or antiplane shear mode.

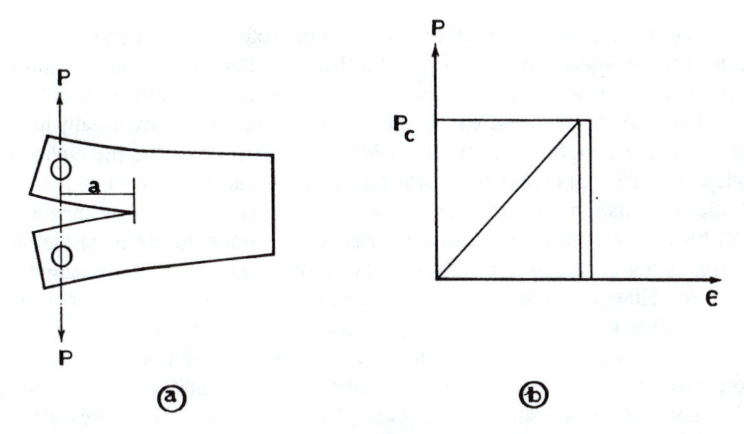

Figure 13.36 (a) Schematic of a specimen loading in mode I in Figure 13.35; (b) schematic of a corresponding load-displacement diagram.

B. LINEAR ELASTIC FRACTURE MECHANICS (LEFM)

Our discussion thus far has focused in a rather superficial way on the general evolution of the important area of fracture mechanics. The basic objective of fracture mechanics is to provide a useful parameter that is characteristic of the given material and independent of test specimen geometry. We will now consider how such a parameter, such as G_{Ic}, is derived for polymers. In doing so we confine our discussion to the concepts of linear elastic fracture mechanics (LEFM). As the name suggests, LEFM applies to materials that exhibit Hookean behavior.

To determine G_{Ic}, consider a precracked specimen (Figure 13.36a). In order to obtain the energy lost to a growing crack, we should examine the energy stored in the system before and after crack extension. This can be done by using the load (P)–displacement (ε) diagram (Figure 13.36b) and calculating the total energy before and after a finite amount of crack motion. This difference becomes, in the limit, the value of G_{Ic}.

The total energy $U(a_1)$ available to the specimen for an initial crack length, a, and a critical load, P_c, is given by

$$U(a_1) = \frac{P_c \varepsilon(a_1)}{2} \tag{13.39}$$

where ε is the displacement. The reciprocal of the slope of P–ε line is defined as the compliance of the specimen

$$C(a_1) = (a_1)/P_c \tag{13.40}$$

Substituting (13.40) into (13.39) we obtain

$$U(a_1) = \frac{P_c^2\, C(a_1)}{2} \tag{13.41}$$

Assuming the crack extends infinitesimally at constant load, the expression for the energy lost to the growing crack per unit area for a width, b, is

$$G_{IC} = \frac{dU}{dA} = \frac{1}{b}\frac{dU}{da} = \frac{P_c^2}{2b}\left(\frac{\partial C}{\partial a}\right) \tag{13.42}$$

Equation 13.42 is general and can be used to determine the cohesive fracture behavior of polymers. To do this, a large number of specimen geometries have been devised. However, we illustrate the use of Equation 13.42 in the assessment of the strength of adhesive joints using tapered double-cantilever beam (TDCB) specimen geometry. Most adhesive materials are thermosets, which are usually highly cross-linked and brittle. Thermosets are therefore ideally suited for the application of the principles of linear elastic fracture mechanics. As indicated earlier, structural members can fail to perform their intended functions by excessive elastic deformation, excessive plastic deformation, or fracture. Adhesive joints constitute a minute fraction of the total volume of a structure. Consequently, even large elastic or inelastic deformation within the bond line is not hazardous since their contribution to the overall deformation of the structure would be insignificant. However, rigid structural adhesives are particularly susceptible to brittle failures; hence, prevention of brittle fracture is the critical problem in adhesive-bonded structures.

Fracturing in brittle materials, we have seen from theories discussed above, is associated with the occurrence of preexisting flaws. These initial flaws, introduced in manufacture or service, may be dust particles, bubbles, and nonbonded areas in the case of adhesive joints. Fractures usually occur by a progressive extension of the largest of these flaws. Furthermore, structural adhesives are inherently brittle when tested uniaxially, a characteristic that is accentuated when they are used in thin layers where deformation is further restricted because of the multiaxial stress state imposed by the proximity of high modulus adherends. The combination of fracturing with very little permanent deformation and crack growth from preexisting flaws suggests that fracturing of adhesive joints can be described by the techniques of fracture mechanics. Using the concepts of linear elastic fracture mechanics and with a knowledge of the largest flaw contained in these materials, one can establish minimum toughness standards for structural adhesive joints. Mostovoy et al.[15,16] and Ripling et al.[17–19] have developed and applied the tapered double-cantilever beam (TDCB) to study strength and durability characteristics of adhesives and adhesive joints with metal adherends. Using beam theory from strength of materials, they established the following relation between specimen compliance and crack length.

$$\frac{\partial C}{\partial a} = \frac{8}{Eb}\left[\frac{3a^2}{h^3} + \frac{1}{h}\right] \tag{13.43}$$

where E = Young's modulus
ν = Poisson's ratio (assumed to be 1/3)
b = specimen width
h = beam height at the distance a from the point of loading

From Equations 13.42 and 13.43,

$$G_{IC} = \frac{4P_c^2}{EB^2}\left[\frac{3a^2}{h^3} + \frac{1}{h}\right] \tag{13.44}$$

From Equation 13.44, it can be seen that as the crack gets longer, i.e., as *a* increases, P_c decreases to maintain a constant value of G_{Ic}. Obviously, the calculation of G_{Ic} requires monitoring both P_c and a for each calculation of G_{Ic}. Testing can be simplified, however, if the specimen is contoured so that the

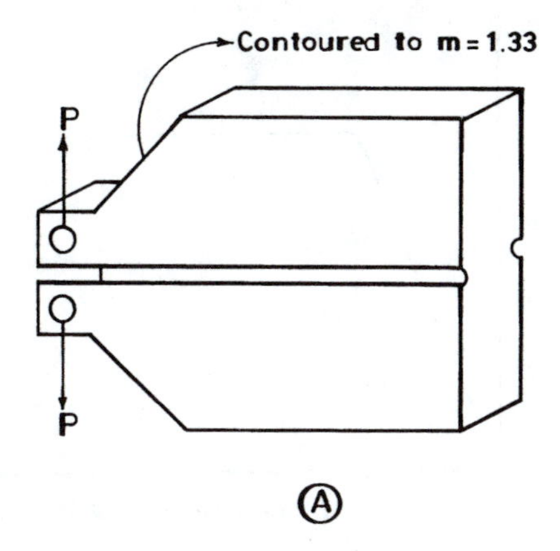

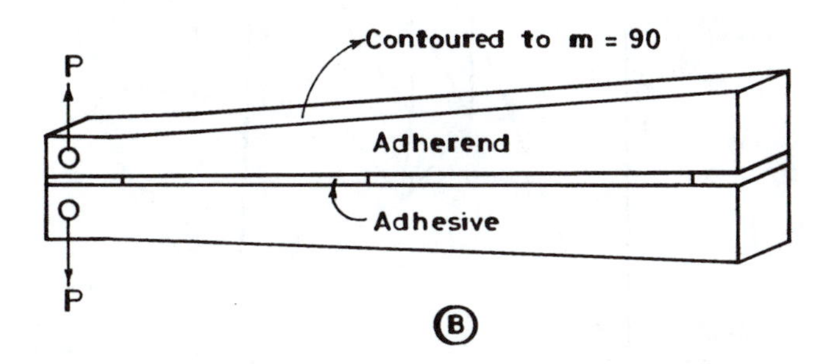

Figure 13.37 Schematic representation of TDCB specimens for testing of A) bulk materials and B) adhesives.

compliance changes linearly with crack length. If dC/da is a constant, the relation between G_{Ic} and P_c is independent of a and, hence, only P_c needs to be followed for the evaluation of G_{Ic}. To develop a linear compliance specimen, its height is varied so that the quantity $(3a^2/h^3 + 1/h)$ in Equation 13.44 is constant. Hence

$$\frac{3a^2}{h^3} + \frac{1}{h} = m \text{ (a constant)} \tag{13.45}$$

so that

$$G_{IC} = \frac{4P_c^2}{Eb^2} \cdot m \tag{13.46}$$

This type of contoured specimen is known as the tapered-double-cantilever beam (TDCB) specimen.

There are, of course, an infinite number of m values that can be chosen to satisfy Equation 13.46. The determination of m is governed by a basic assumption that the entire energy supplied to the specimen is concentrated on the crack line for crack extension. Thus the choice of m is such that the bending stresses on the adherend are minimized. This, in turn, depends on the modulus of the adherends relative to the adhesive. This necessitates an empirical determination of the appropriate m value through a procedure known as compliance calibration. Figure 13.37 is a schematic of the specimen geometries of TDCB for the determination of G_{Ic} for bulk adhesives and adhesive joints.

The shape of the load–displacement curves using this specimen geometry can reveal interesting information about the intrinsic nature of the adhesive. Generally, in evaluating G_{Ic}, a continuously increasing load is applied to the specimen and a P–ε diagram is obtained with the extensometer mounted directly on the sample. Two types of P–ε diagrams are usually obtained (Figure 13.38). Curves of type A are typical of rate-insensitive materials, while type B curves exemplify rate-sensitive ones. For type

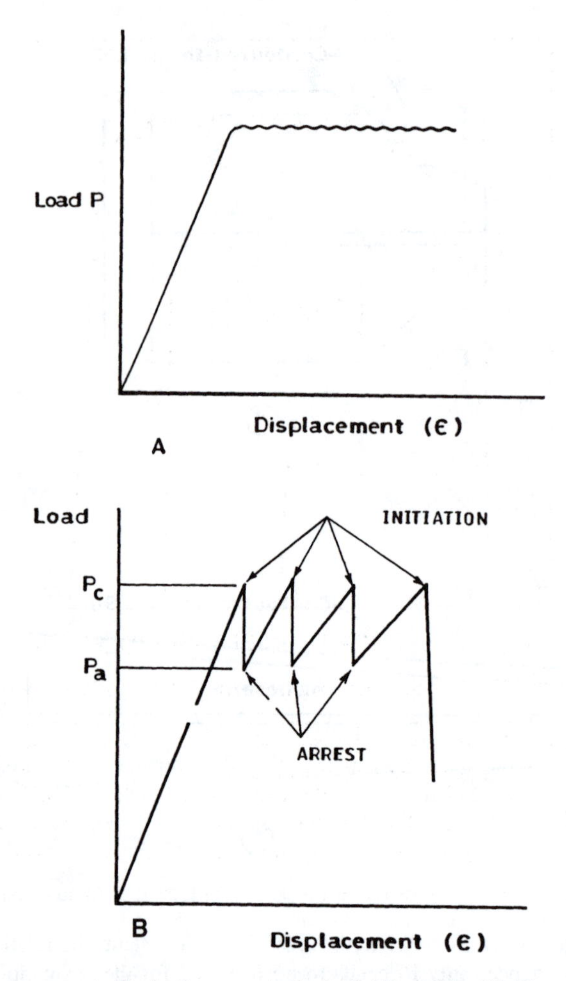

Figure 13.38 Load–displacement (P–ε) diagrams typical of (A) stable ("flat") propagation; (B) unstable ("peaked") propagation.

B curves, two load values are exhibited: P_c, the crack initiation load at the point of instability used in the calculation of G_{Ic}; and P_a, the crack arrest load used for estimating G_{Ia}. For rate-insensitive materials, G_{Ic} and G_{Ia} are identical, and the crack growth a is dictated by the cross-head speed, $\dot{\varepsilon}$.

On a given test machine and at a given cross-head speed, the difference between initiation and arrest fracture energies, $G_{Ic} - G_{Ia}$ indicates the energy released during crack propagation; it is a measure of the brittleness or resistance to catastrophic failure of the adhesive system. Figure 13.39 shows different adhesive joint failure modes for TDCB specimens. We notice that the systems A and B have about the same crack initiation loads. However, the two systems are different in their resistance to crack propagation. A is a hard and very brittle adhesive. Unlike A, B has an internal mechanism for arresting crack propagation; at the onset of instability the material undergoes plastic deformation, blunts the crack, and consequently prevents catastrophic failure. B is therefore not only hard but also tough. Using these arguments, the adhesive system C is hard and strong while D is soft and weak. D is characteristic of systems with a spongy adhesive layer.

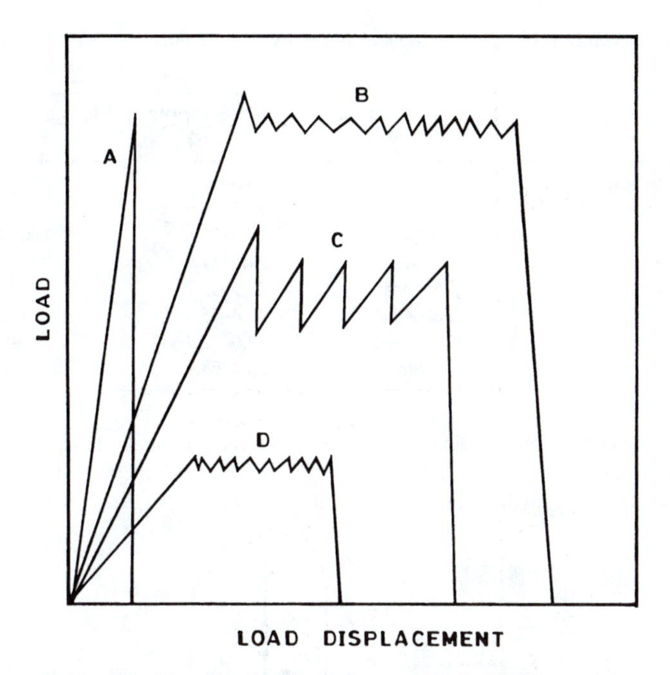

Figure 13.39 Possible adhesive joint failure modes for TDCB specimens tests.

Example 13.8: Urea–formaldehyde (UF) adhesive used in bonded wood products was modified by copolymerizing 10 mol% urea derivative of dodecanediamine (DDDU). The fracture energies of wood joints made with the unmodified and DDDU-modified adhesives were found to be 130 and 281 J/m^3, respectively. Explain the enhanced fracture energy of the wood joint bonded with the modified adhesive.

Solution:

$$H_2N-CO-NH-(CH_2)_{12}-NH-CO-NH_2$$

urea derivative of dodecanediamine (DDDU)

Cured urea–formaldehyde adhesive is characterized by the presence of methylene bridges between strongly hydrogen-bonded urea linkages. Consequently, cured UF adhesives are inherently stiff and brittle. Incorporation of DDDU with its 12 methylene groups into the resin structure results in cured UF adhesive with a more flexible network. The increased flexibility decreases internal stress and the associated flaws, and hence the fracture energy increases.

VIII. PROBLEMS

13.1. Calculate the resilience of polycarbonate with yield strength 62.05 × 10^6 N/m^2 and elastic modulus 24.13 × 10^8 N/m^2.

13.2. A polymer, T, when tested in uniaxial tension has its cross-sectional area reduced from 2.00 cm^2 to 1.0 cm^2. The cross-sectional area of a second polymer, S, under the same test changed from 20 cm^2 to 18 cm^2. Which of these two polymers would you choose for making children's toys, assuming other qualities are comparable for both polymers?

13.3. Explain the observed trend in the mechanical properties of the polymers shown in the following table.

Polymer	Elastic modulus (MN/m^2)	Tensile Strength (MN/m^2)	Elongation of Break %
Low-density polyethylene	138–276	10.3–17.2	400–700
High-density polyethylene	414–1034	17.2–39.9	100–600
Polytetrafluoroethylene	414	13.8–27.6	100–350
Polypropylene	1034–1551	24.1–37.9	200–600
Polystyrene	2758–3447	37.9–55.2	1–2.5
Poly(vinyl chloride)	2068–4436	41.4–75.8	5–60
Nylon 6,6	1241–2760	62.0–82.7	60–300
Polycarbonate	1243	55.2–68.9	60–120

13.4. Figure Q13.4 shows the stress–strain curve for a polymer material under uniaxial loading. The material deformed uniformly until N, where necking ensued. If the material obeys the relation $\sigma = K\varepsilon^n$ up to the point N, calculate the strength, σ_N, of the material at the onset of necking. $K = 10^6$ N/m^2, $\varepsilon = 0.5$.

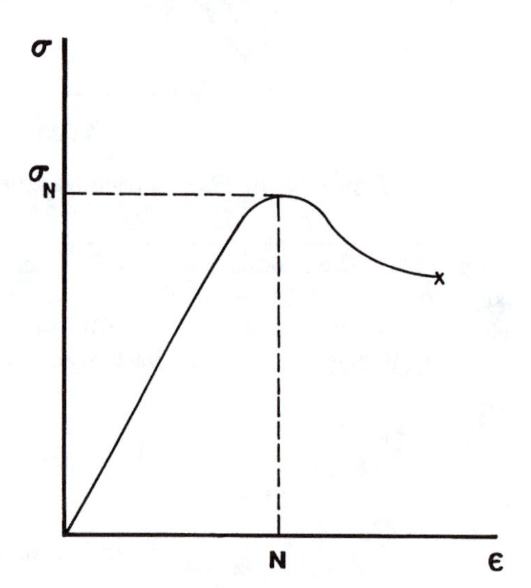

Figure Q13.4 Stress–strain curve for a polymer.

13.5. Calculate the maximum load a polymer sample in uniaxial tension can sustain before yielding when the maximum resolved shear stress (τ_{max}) is 10^6 N/m^2. The cross-sectional area of the sample is 10^{-4} m^2.

13.6. Suppose that yielding occurs in a polyethylene crystal when the critical resolved shear stress, τ_{max}, ≈ 27.58×10^6 N/m^2 is produced on the {110} type slip plane along a <111> type direction. If the tensile axis coincides with the <110> direction, what maximum axial stress must be applied to cause yielding on a {110} plane in a <111> direction?

13.7. Explain the following observations:

a. Although the refractive indices of polystyrene and polybutadiene differ considerably, materials derived from styrene–butadiene–styrene copolymers are transparent and have high tensile strength.
b. When natural rubber with chains of poly(methylmethacrylate) grafted to it is isolated from ethylacetate, it is hard, stiff, and nontacky. On the other hand, when the same polymer is isolated from hexane it is limp, flabby, and tacky.
c. Block copolymers of styrene and vinyl alcohol are soluble in benzene, water, and acetone.
d. Both poly(hexamethylene sebacamide) (nylon 6,10) and poly(hexamethylene adipamide) (nylon 6,6) are fiber-forming polymers. A copolymer consisting of hexamethylene terephthalamide and hexamethylene adipamide retains the qualities of nylon 6,6 fiber, whereas a copolymer consisting of hexamethylene sebacamide and 10 mol% hexamethylene terephthalamide is elastic and rubbery with drastic reductions in stiffness and hardness.

13.8. Two polypropylene materials have densities 0.905 and 0.87 g/cm^3. Sketch the modulus–temperature curves for these materials on the same graph.

13.9. Explain the difference in the glass transition temperatures between the following pairs of polymers.

a. Poly(vinyl methyl ether) and poly(vinyl formal)

$T_g = 13°C$ $T_g = 105°C$

b. Poly(*t*-butyl methacrylate) and poly(ethyl methacrylate)

$T_g = 135°C$ $T_g = 100°C$

13.10. How would you expect the stiffness of the series of poly(*p*-alkylstyrene) shown below to vary with n? Explain your answer very briefly.

$$—CH_2—CH— \\ | \\ C_6H_4 \\ | \\ (CH_2)_n \\ | \\ CH_3$$

(Str. 11)

13.11. Two materials formed by reaction injection molding have the following tensile and fracture properties:[26]

Material	E(MPa)	σ_u(MPa)	ε_u(%)	G_c(kJ/m^2)
S	266	16.8	222	7.12
H	532	15.3	17	0.06

Estimate the intrinsic flaw size in these materials assuming they were tested under (a) plane stress and (b) plane strain. Assume $\nu = 0.3$. Comment on your result. Which of these two materials will be more suitable for use in an application where the ability to withstand extensive abuse is a requirement?

13.12. Figure Q13.12 shows the mechanical properties of two urea–formaldehyde (UF) resins cured with NH_4Cl: (bottom) variation of the shear strength of bonded wood joints with cyclic wet–dry treatments of joints; (middle) development of internal stress with duration of resin cure at room temperature; (top) dynamic mechanical properties of resins. Discuss the interrelationships between the observed mechanical properties.

13.13. Figure Q13.13 shows the variation of fracture initiation (G_{Ic}) energies with cure of phenol–formaldehyde adhesive cured at 85°C. It is known that during cure of phenolic resins, dimethylene ether linkages are formed initially from the condensation of methylol groups. These ether linkages decompose subsequently according to the scheme shown. How do these reactions provide a possible explanation for the trend in G_{Ic}? A and B represent two different states of cure of the phenolic resin with the same fracture energy. What is the major difference in the fracture characteristics of materials in these states?

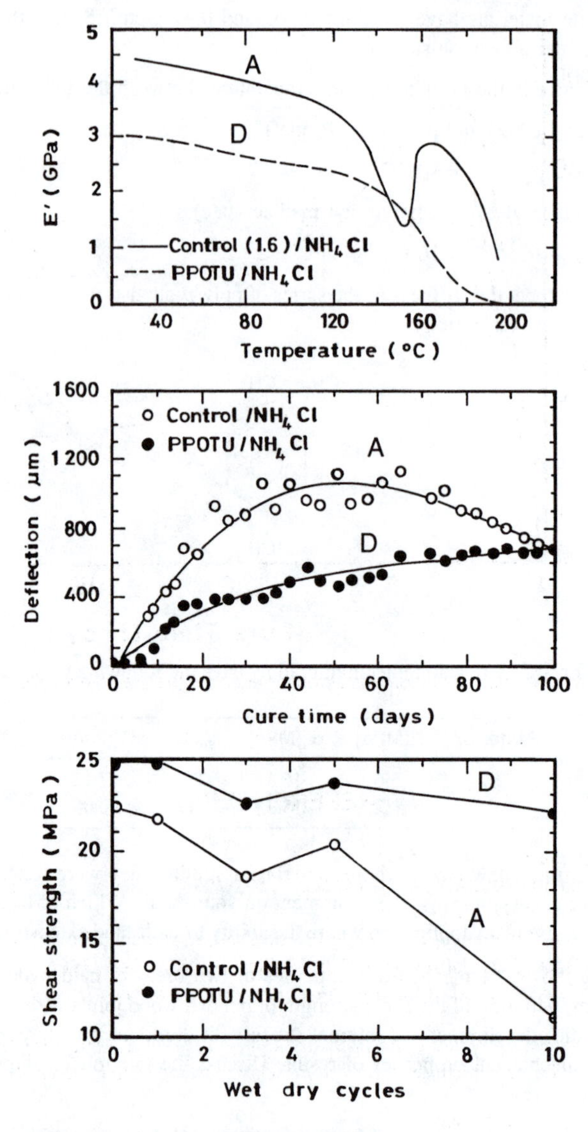

Figure Q13.12

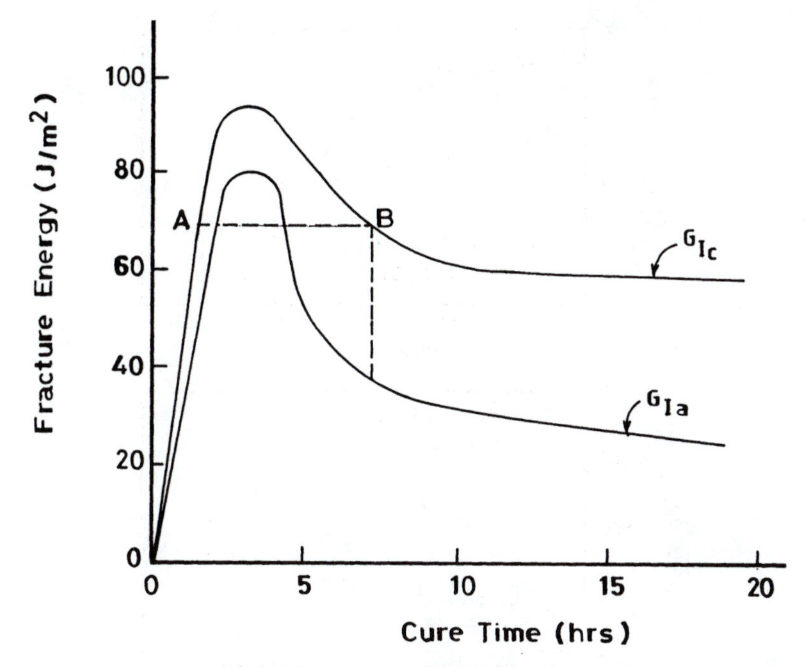

Figure Q13.13 Variation of fracture initiation and arrest energies with cure time for phenolic resin cured at 85° ( ).

REFERENCES

1. Findley, W.N., *Mod. Plast.*, 19(8), 71, 1942.
2. McLoughlin, J.R. and Tobolsky, A.V., *J. Colloid Sci.*, 7, 555, 1952.
3. Fried, J.R., *Plast. Eng.*, 38(7), 27, 1982.
4. Williams, D.J., *Polymer Science and Technology,* Prentice-Hall, Englewood Cliffs, NJ, 1971.
5. Winding, C.C. and Hiatt, G.D., *Polymer Materials,* McGraw-Hill, New York, 1961.
6. Kaufman, H.S. and Falcetta, J.J., (Eds.), *Introduction to Polymer Science and Technology,* John Wiley & Sons, New York, 1977.
7. Nielsen, L.E., *Mechanical Properties of Polymers and Composites,* Vol. 2, Marcel Dekker, New York, 1974.
8. Barrett, C.R., Nix, W.D., and Tetelman, A.S., *The Principles of Engineering Materials,* Prentice-Hall, Englewood Cliffs, NJ, 1973.
9. Heijboer, J., *Br. Polym. J.,* 1, 3, 1969.
10. Heiboer, J., *Proc. Int. Conf. Physics Non-crystalline Solids,* North-Holland, Amsterdam, 1965, 231.
11. Carswell, T.S. and Nasor, H.K., *Mod. Plast.,* 21(6), 121, 1944.
12. Griffith, A.A., *Philos. Trans. R. Soc. London,* A221, 163, 1920.
13. Inglis, C.E., *Proc. Institute of Naval Architects,* 55, 219, 1913.
14. Irwin, G.R. and Wells, A.A., *Metall. Rev.,* 10(58), 223, 1965.
15. Mostovoy, S., Crosley, P.B., and Ripling, E.J., *J. Mater.,* 2(3), 661, 1967.
16. Mostovoy, S. and Ripling, E.J., *J. Appl. Polym. Sci.,* 15, 641, 1971.
17. Ripling, E.J., Mostovoy, S., and Patrick, R. L., Adhesion ASTM STP 360, 5, 1963.
18. Ripling, E.J., Mostovoy, S., and Patrick, R.L., *Mater. Res. Stand.,* 64(3), 129, 1964.
19. Ripling, E.J., Corten, H.T., and Mostovoy, S., *J. Adhes.,* 3, 107, 1971.
20. Ryan, A.J., Bergstrom, T.D., Willkom, W.R., and Macosko, C.W., *J. Appl. Polym. Sci.,* 42, 1023, 1991.

Chapter 14

Polymer Viscoelasticity

I. INTRODUCTION

Traditional engineering practice deals with the elastic solid and the viscous liquid as separate classes of materials. Engineers have been largely successful in the use of materials like motor oil, reinforced concrete, or steel in various applications based on design equations arising from this type of material classification. However, it has become increasingly obvious that elastic and viscous material responses to imposed stresses represent the two extremes of a broad spectrum of material behavior. The behavior of polymeric materials falls between these two extremes. As we said in Chapter 13, polymers exhibit viscoelastic behavior. The mechanical properties of solid polymers show marked sensitivity to time compared with traditional materials like metals and ceramics. Several examples illustrate this point. (1) The stress–strain properties of polymers are extremely rate dependent. For traditional materials, the stress–strain behavior is essentially independent of strain rate. (2) Under a constant load, the deformation of polymeric material increases with time (creep). (3) When a polymer is subjected to a constant deformation, the stress required to maintain this deformation decreases with increasing time (stress relaxation). (4) The strain resulting from a polymer subjected to a sinusoidal stress has an in-phase component and an out-of-phase component. The phase lag (angle) between the stress and strain is a measure of the internal friction, which in principle is the mechanical strain energy that is convertible to heat. Traditional materials, for example, metals close to their melting points, exhibit similar behavior. However, at normal temperatures, creep and stress relaxation phenomena in metals are insignificant and are usually neglected in design calculations. In choosing a polymer for a particular end-use situation, particularly structural applications, its time-dependent behavior must be taken into consideration if the polymer is to perform successfully.

Our discussion of the viscoelastic properties of polymers is restricted to the linear viscoelastic behavior of solid polymers. The term *linear* refers to the mechanical response in which the ratio of the overall stress to strain is a function of time only and is independent of the magnitudes of the stress or strain (i.e., independent of stress or strain history). At the onset we concede that linear viscoelastic behavior is observed with polymers only under limited conditions involving homogeneous, isotropic, amorphous samples under small strains and at temperatures close to or above the T_g. In addition, test conditions must preclude those that can result in specimen rupture. Nevertheless, the theory of linear viscoelasticity, in spite of its limited use in predicting service performance of polymeric articles, provides a useful reference point for many applications.

To aid our visualization of viscoelastic response we introduce models that represent extremes of the material response spectrum. This is followed by the treatment of mechanical models that simulate viscoelastic response. These concepts are developed further by discussion of the superposition principles.

II. SIMPLE RHEOLOGICAL RESPONSES

A. THE IDEAL ELASTIC RESPONSE

The ideally elastic material exhibits no time effects and negligible inertial effects. The material responds instantaneously to applied stress. When this stress is removed, the sample recovers its original dimensions completely and instantaneously. In addition, the induced strain, ε, is always proportional to the applied stress and is independent of the rate at which the body is deformed (Hookean behavior). Figure 14.1 shows the response of an ideally elastic material.

The ideal elastic response is typified by the stress–strain behavior of a spring. A spring has a constant modulus that is independent of the strain rate or the speed of testing: stress is a function of strain only. For the pure Hookean spring the inertial effects are neglected. For the ideal elastic material, the mechanical response is described by Hooke's law:

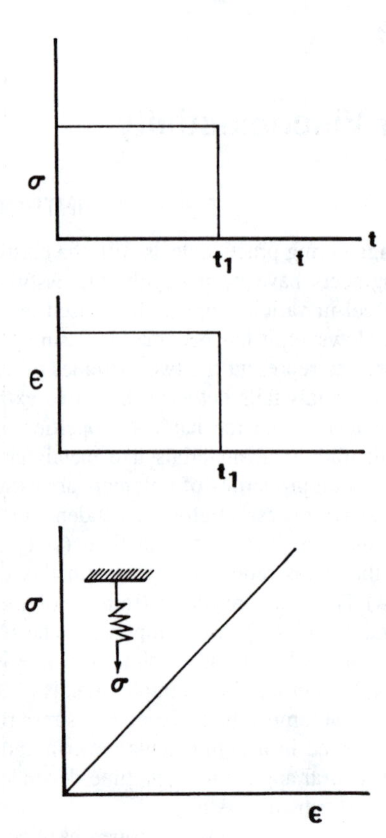

Figure 14.1 Ideal elastic response.

$$\sigma = E\varepsilon \tag{14.1}$$

where σ is the applied stress, ε is the strain, and E is Young's modulus.

B. PURE VISCOUS FLOW

Fluids have no elastic character; they cannot support a strain. The dominant characteristic of fluids is their viscosity, which is equivalent to elasticity in solids. According to Newton's law, the response of a fluid to a shearing stress τ is viscous flow, given by

$$\tau = \eta \frac{d\gamma}{dt} \tag{14.2}$$

where η is viscosity and dγ/dt is strain rate. Thus in contrast to the ideal elastic response, strain is a linear function of time at an applied external stress. On the release of the applied stress, a permanent set results. Pure viscous flow is exemplified by the behavior of a dashpot, which is essentially a piston moving in a cylinder of Newtonian fluid (Figure 14.2). A dashpot has no modulus, but the resistance to motion is proportional to the speed of testing (strain rate).

However, no real material shows either ideal elastic behavior or pure viscous flow. Some materials, for example, steel, obey Hooke's law over a wide range of stress and strain, but no material responds without inertial effects. Similarly, the behavior of some fluids, like water, approximate Newtonian response. Typical deviations from linear elastic response are shown by rubber elasticity and viscoelasticity.

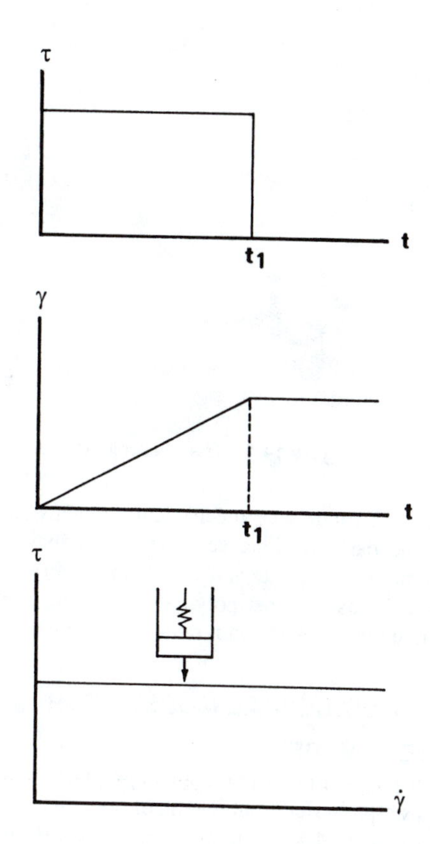

Figure 14.2 Pure viscous behavior.

C. RUBBERLIKE ELASTICITY

The response of rubbery materials to mechanical stress is a slight deviation from ideal elastic behavior. They show non-Hookean elastic behavior. This means that although rubbers are elastic, their elasticity is such that stress and strain are not necessarily proportional (Figure 14.3).

III. VISCOELASTICITY

Viscoelastic material such as polymers combine the characteristics of both elastic and viscous materials. They often exhibit elements of both Hookean elastic solid and pure viscous flow depending on the experimental time scale. Application of stresses of relatively long duration may cause some flow and irrecoverable (permanent) deformation, while a rapid shearing will induce elastic response in some polymeric fluids. Other examples of viscoelastic response include creep and stress relaxation, as described previously.

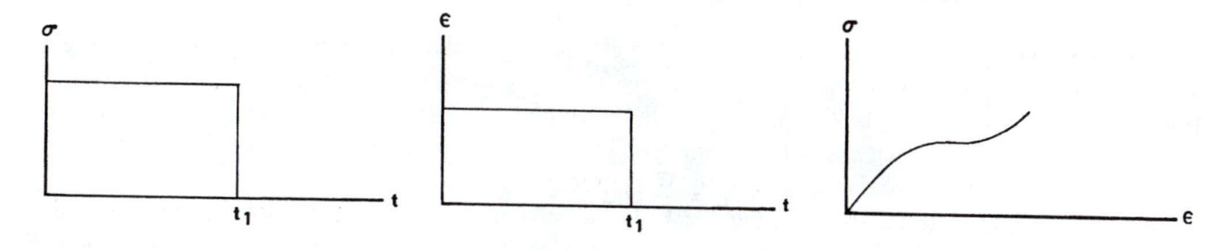

Figure 14.3 Rubber elasticity.

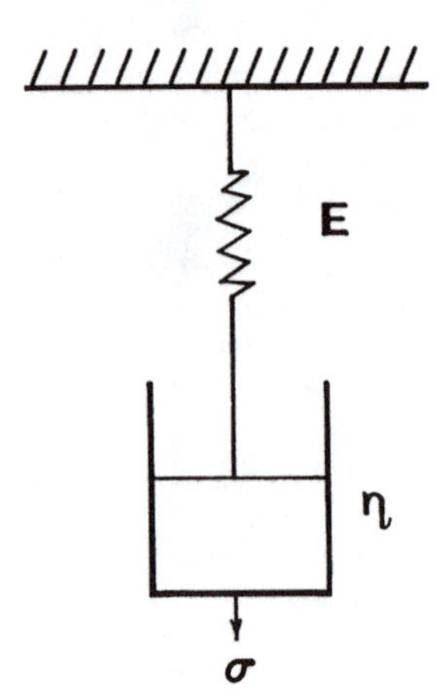

Figure 14.4 The Maxwell element.

It is helpful to introduce mechanical elements as models of viscoelastic response, but neither the spring nor the dashpot alone accurately describes viscoelastic behavior. Some combination of both elements is more appropriate and even then validity is restricted to qualitative descriptions; they provide valuable visual aids. In most polymers, mechanical elements do not provide responses beyond strains greater than about 1% and strain rates greater than 0.1 s^{-1}.

IV. MECHANICAL MODELS FOR LINEAR VISCOELASTIC RESPONSE

A. MAXWELL MODEL

To overcome the poor description of real polymeric materials by either the spring or the dashpot, Maxwell suggested a simple series combination of both elements. This model, referred to as the Maxwell element, is shown in Figure 14.4. In the Maxwell model, E, the instantaneous tensile modulus, characterizes the response of the spring while the viscosity, η, defines viscous response. In the following description we make no distinction between the types of stress. Thus, we use the symbol E even in cases where we are actually referring to shearing stress for which we have previously used the symbol G. This, of course, does not detract from the validity of the arguments.

In the Maxwell element, both the spring and the dashpot support the same stress. Therefore,

$$\sigma = \sigma_a = \sigma_d \tag{14.3}$$

where σ_s and σ_d are stresses on the spring and dashpot, respectively. However, the overall strain and strain rates are the sum of the elemental strain and strain rates, respectively. That is,

$$\varepsilon_T = \varepsilon_s + \varepsilon_d \tag{14.4}$$

or

$$\dot{\varepsilon}_T = \dot{\varepsilon}_6 + \dot{\varepsilon}_d \tag{14.5}$$

But

$$\dot{\varepsilon}_s = \frac{\dot{\sigma}}{E} \text{ and } \dot{\varepsilon}_d = \sigma/\eta \tag{14.6}$$

where $\dot{\varepsilon}_T$ is the total strain rate, while $\dot{\varepsilon}_s$ and $\dot{\varepsilon}_d$ are the strain rates of the spring and dashpot, respectively. The rheological equation of the Maxwell element on substitution of Equation 14.6 in 14.5 becomes

$$\dot{\varepsilon}_T = \frac{1}{E}\dot{\sigma} + \frac{1}{\eta}\sigma \tag{14.7}$$

As Equation 14.7 shows, the Maxwell element is merely a linear combination of the behavior of an ideally elastic material and pure viscous flow. Now let us examine the response of the Maxwell element to two typical experiments used to monitor the viscoelastic behavior of polymer.

1. Creep Experiment

In creep, the sample is subjected to an instantaneous constant stress, σ_0, and the strain is monitored as a function of time. Since the stress is constant, $d\sigma/dt$ is zero and therefore, Equation 14.7 becomes

$$\dot{\varepsilon}_T = \frac{1}{\eta}\sigma_0 \tag{14.8}$$

Solving the equation and noting that the initial strain is σ_0/E, the equation for the Maxwell element for creep can be written as

$$\varepsilon(t) = \frac{\sigma_0}{E} + \frac{\sigma_0}{\eta}t \tag{14.9}$$

$$\varepsilon(t) = \sigma_0\left[\frac{1}{E} + \frac{t}{\eta}\right] \tag{14.10}$$

On removal of the applied stress, the material experiences creep recovery. Figure 14.5 shows the creep and the creep recovery curves of the Maxwell element. It shows that the instantaneous application of a constant stress, σ_0, is initially followed by an instantaneous deformation due to the response of the spring by an amount σ_0/E. With the sustained application of this stress, the dashpot flows to relieve the stress. The dashpot deforms linearly with time as long as the stress is maintained. On the removal of the applied stress, the spring contracts instantaneously by an amount equal to its extension. However, the deformation due to the viscous flow of the dashpot is retained as permanent set. Thus the Maxwell element predicts that in a creep/creep recovery experiment, the response includes elastic strain and strain recovery, creep and permanent set. While the predicted response is indeed observed in real materials, the demarcations are nevertheless not as sharp.

2. Stress Relaxation Experiment

In a stress relaxation experiment, an instantaneous strain is applied to the sample. The stress required to maintain this strain is measured as a function of time. When the Maxwell element is subjected to an

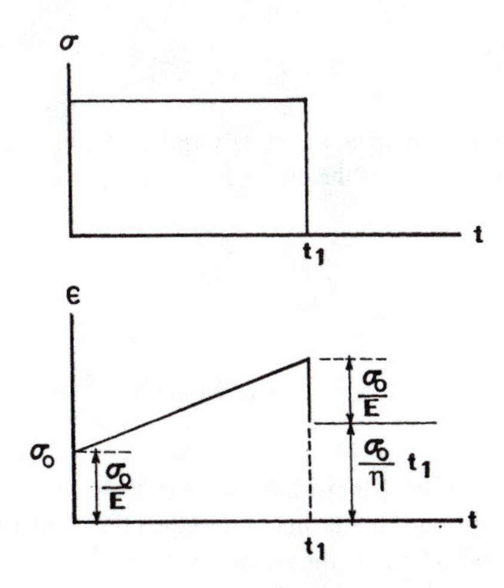

Figure 14.5 Creep and creep recovery behavior of the Maxwell element.

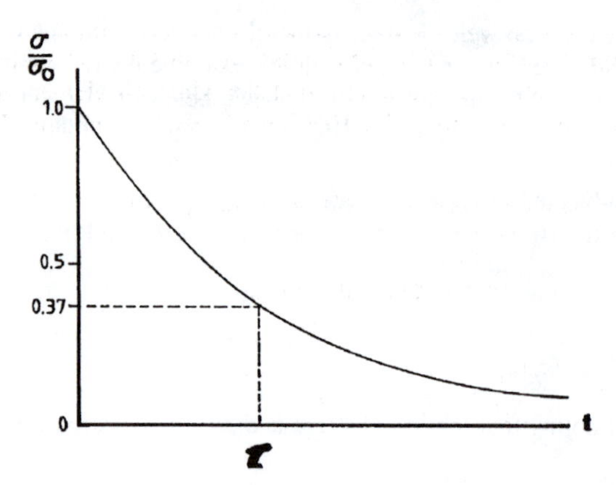

Figure 14.6 Relaxation time for the Maxwell element.

instantaneous strain, only the spring can respond initially. The dashpot will relax gradually, and consequently, the stress decreases with increasing time.

The rheological equation for the Maxwell element from Equation 14.7 is

$$\dot{\varepsilon}_T = \frac{1}{E}\dot{\sigma} + \frac{1}{\eta}\sigma \tag{14.7}$$

Since the strain is constant, ε_T is zero, thus Equation 14.7 is reduced to

$$\frac{1}{E}\dot{\sigma} + \frac{1}{\eta}\sigma = 0 \tag{14.11}$$

The solution to this first-order differential equation with the boundary condition that $\sigma = E\varepsilon_0$ at $t = 0$ is

$$\sigma = \sigma_0 \exp\left(-\frac{E}{\eta}t\right) \tag{14.12}$$

We define the quantity τ as the relaxation or response time and it is given as the ratio η/E. Equation 14.12 thus becomes

$$\sigma = \sigma_0 \exp(-t/\tau) \tag{14.13}$$

The relaxation time from Equation 14.13 is the time required for the stress to decay to 1/e or 37% of its initial value (Figure 14.6). If we divide the stress by the constant strain, ε_0, Equation 14.13 becomes

$$\frac{\sigma(t)}{\varepsilon_0} = \frac{\sigma_0}{\varepsilon_0} e^{-t/\tau}$$

or

$$E_r(t) = Ee^{-t/\tau} \tag{14.14}$$

E_r is the relaxation modulus. For the Maxwell element in a stress relaxation experiment, all the initial deformation takes place in the spring. The dashpot subsequently starts to relax and allows the spring to contract. For times considerably shorter than the relaxation time, the Maxwell element behaves essentially like a spring; while for times much longer than the relaxation time, the element behaves like a dashpot. For times comparable to the relaxation time, the response involves the combined effect of the spring and the dashpot.

Example 14.1: A polystyrene sample of 0.02/m^2 cross-sectional area is subjected to a creep load of 10^5 N. The load is removed after 30 s. Assuming that the Maxwell element accurately describes the behavior of polystyrene and that viscosity is 5×10^{10} P, while Young's modulus is 5×10^5 psi, calculate:

a. The compliance
b. The deformation recovered on the removal of the dead load
c. The permanent

Solution: a. The creep equation for the Maxwell element is

$$\varepsilon(t) = \sigma_0 \left[\frac{1}{E} + \frac{t}{\eta}\right] \qquad \text{or}$$

$$\frac{\varepsilon(t)}{\sigma_0} = \left[\frac{1}{E} + \frac{t}{\eta}\right] \text{ or compliance.}$$

$$E = 5 \times 10^5 \text{ psi} = 5 \times 10^5 \text{ psi} \times 6.894 \times 10^3 \left(N/m^2/psi\right)$$

$$= 3.45 \times 10^9 \ N/m^2$$

$$\eta = 5 \times 10^{10} \, P = 5 \times 10^9 \, \frac{N \cdot s}{m^2}$$

$$J = \frac{1}{3.45 \times 10^9} + \frac{30}{5 \times 10^9} \ m^2/N. = 6.29 \times 10^{-9} \ m^2/N.$$

b. The deformation recovered on the removal of load is due to the spring ε_s.

$$\varepsilon_s = \frac{\sigma_0}{E}$$

$$\sigma_0 = \frac{P_0}{A} = \frac{10^5}{0.02} \ N/m^2 = 5 \times 10^6 \ N/m^2$$

$$\varepsilon_s = \frac{5 \times 10^6}{3.45 \times 10^9} = 1.45 \times 10^{-3}$$

c. The permanent set is due to the viscous flow ε_d.

$$\varepsilon_d = \frac{\sigma_0}{\eta} t$$

$$= \frac{5 \times 10^6 \left(N/m^2\right)}{5 \times 10^9 \left(Ns/m^2\right)} \times 30 \ (s)$$

$$= 0.03$$

3. Dynamic Experiment

Let us consider the response of a Maxwell element subjected to a sinusoidal stress. The corresponding strain will be sinusoidal but out of phase with the stress by an angle δ, as discussed in Chapter 13. Thus,

$$\sigma = \sigma_0 \sin \cdot \omega t \tag{14.15}$$

Now the rheological equation for the Maxwell element is

$$\dot{\varepsilon}(t)=\frac{1}{E}\dot{\sigma}+\frac{1}{\eta}\sigma$$
$$\frac{d\varepsilon}{dt}=\frac{\sigma_0\omega}{E}\cos\omega t+\frac{\sigma_0}{\eta}\sin\omega t \tag{14.16}$$

Integration of Equation 14.16 between two time limits and noting that ε(o) is not necessarily zero yields

$$\tan\delta=\frac{1}{\tau\omega}$$
$$E^1=\frac{E\tau^2\omega^2}{1+\omega^2\tau^2} \tag{14.17}$$

$$E^{11}=\frac{E\tau\omega}{1+\omega^2\tau^2} \tag{14.18}$$

where $\tau = \eta/E$.

B. THE VOIGT ELEMENT

Since, as we saw above, the Maxwell element is not perfect, it seems logical to consider a parallel arrangement of the spring and the dashpot. This is the so-called Voigt or Voigt–Kelvin element (Figure 14.7).

The Voigt element has the following characteristics:

- The spring and the dashpot always remain parallel. This means that the strain in each element is the same.
- The total stress supported by the Voigt element is the sum of the stresses in the spring and the dashpot.

$$\sigma_T=\sigma_s+\sigma_d \tag{14.19}$$

Thus, the rheological equation for the Voigt element is given by

$$\sigma_T=E\varepsilon+\eta\frac{d\varepsilon}{dt} \tag{14.20}$$

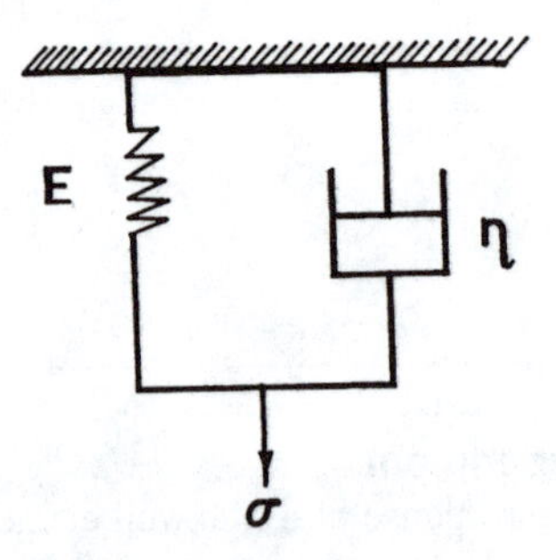

Figure 14.7 The Voigt element.

1. Creep Experiment

In a creep experiment, the applied stress is constant; consequently, Equation 14.20 becomes

$$\sigma_0 = E\varepsilon + \eta\frac{d\varepsilon}{dt} \tag{14.21}$$

This is a linear differential equation with solution [integrate between the limits $\varepsilon(o) = 0$ and $\varepsilon(\tau) = \varepsilon(t)$]:

$$\begin{aligned} \varepsilon(t) &= \frac{\sigma_0}{E}\left[1 - e^{-Et/\eta}\right] \\ \varepsilon(t) &= \frac{\sigma_0}{E}\left[1 - e^{-t/\tau}\right] \end{aligned} \tag{14.22}$$

or

$$J(t) = \frac{\varepsilon(t)}{\sigma_0} = J\left[1 - e^{-t/\tau}\right]$$

where

$$J = \frac{1}{E} = \text{reciprocal of modulus}$$

The creep and creep recovery curves for the Voigt elements are shown in Figure 14.8.

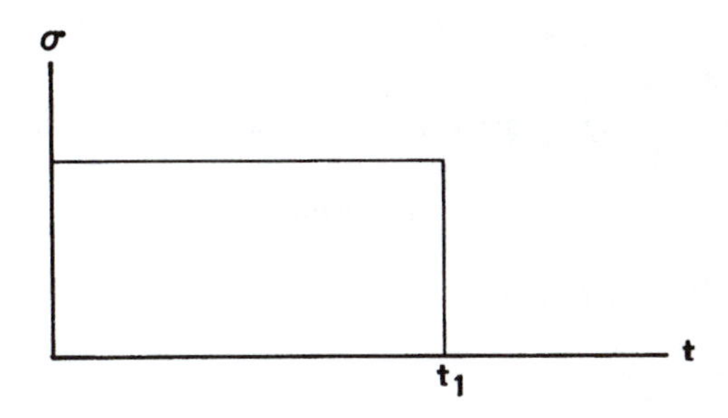

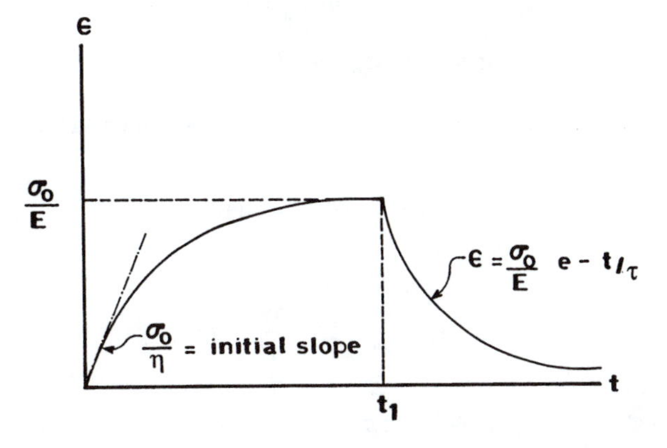

Figure 14.8 Creep and creep recovery curves for the Voigt element.

On the application of a sudden constant stress in a creep experiment, only the spring offers the initial resistance to deformation; the spring would elongate instantaneously if possible, but its deformation is constrained by that of the dashpot. Recall that for the Voigt element both the spring and the dashpot have equal strains. Therefore, the initial total stress is borne by the dashpot. Under the influence of the constant force, the dashpot begins to flow thus transferring part of the load to the spring. The transfer of load to the spring results in a concomitant decrease in the stress on the dashpot and hence a decrease in the strain rate, which is proportional to the magnitude of the stress experienced by the dashpot. Eventually, the element comes to its equilibrium strain. At this point the strain rate is zero; the resistance of the dashpot is therefore also zero which means that the entire stress is now supported by the spring. The equilibrium strain is simply the strain due to the spring (σ_0/E). If the load is removed after equilibrium, the strain decays exponentially.

We note that the Voigt model predicts that strain is not a continuous function of stress; that is, the element does not deform continuously with the sustained application of a constant stress. The strain approaches an asymptomatic value given by (σ_0/E). The strain of the element at equilibrium is simply that of an ideal elastic solid. The only difference is that the element does not assume this strain instantaneously, but approaches it gradually. The element is shown to exhibit retarded elasticity. In creep recovery, the Maxwell element retracts instantaneously but not completely, whereas the Voigt element exhibits retarded elastic recovery, but there is no permanent set.

2. Stress Relaxation Experiment

In a stress relaxation experiment (ε = constant), the rheological equation for the Voigt element reduces to

$$\sigma(t) = E\varepsilon \tag{14.23}$$

This is essentially Hooke's law. The Voigt model is not suited for simulating a stress relaxation experiment. The application of an instantaneous strain induces an infinite resistance in the dashpot. It would require an infinite stress to overcome the resistance and get the dashpot to strain instantaneously. This is obviously unrealistic.

3. Dynamic Experiment

Now consider the response of a Voigt element subjected to a sinusoidal strain:

$$\varepsilon = \varepsilon_0 \sin \omega t$$

The stress response of the Voigt element is

$$\sigma = \sigma_s + \sigma_d$$

$$\sigma_s = E\varepsilon_0 = E\varepsilon_0 \sin \omega t, \text{ and } \sigma_d = \eta\dot{\varepsilon} = \eta\omega\varepsilon_0 \cos \omega t \tag{14.24}$$

$$\sigma = E\varepsilon_0 \sin wt + \eta\omega\varepsilon_0 \cos wt$$

Since the sine term is the component in phase with the strain and the cosine term denotes the 90° out-of-phase term, then

$$\sigma' = E\varepsilon' \text{ and } \sigma'' = \eta\omega\varepsilon'' \tag{14.25}$$

Consequently,

$$E' = \frac{\sigma'}{\varepsilon'} = E'' = \frac{\sigma''}{\varepsilon''} = \eta\omega$$

$$\tan \delta = \frac{E''}{E'} \tag{14.26}$$

$$\tan \delta = \frac{\eta\omega}{E} = \tau\omega$$

Example 14.2: Comment on the physical significance of the quantities measured in dynamic mechanical (oscillating) experiments.

Solution: To appreciate the physical significance of the quantities measured in oscillatory experiments, we consider the energy changes in a sample undergoing cyclic deformation. We start by noting that in viscoelastic and purely viscous materials, the stress and strain are out of phase. Figure E.14.2 shows stress–strain representation of a viscoelastic material. The oscillatory strain is $\varepsilon = \varepsilon_0 \sin \omega t$.

a. *Purely elastic body:* the work per unit volume is

$$W = \sigma d\varepsilon$$

$$d\varepsilon = \varepsilon_0 \cos \omega t \, d(\omega t)$$

For purely elastic body,

$$\sigma = E\varepsilon = E\varepsilon_0 \sin \omega t$$

Therefore, the work done over the first quarter cycle of applied strain is given by

$$W = \int_0^{\pi/2} E\varepsilon \sin \omega t \, \varepsilon_0 \cos \omega t \, d(\omega t)$$

$$= E\varepsilon_0^2 \int_0^{\pi/2} \sin \omega t \cos \omega t \, d(\omega t)$$

$$W = \frac{E\varepsilon_0^2}{2}$$

For the second quarter (i.e., integrating from $\pi/2$ to π, the result is exactly the same except that the sign is negative. Thus for a full cycle, in the case of an elastic body, the energy stored in the first and third quarter cycles is recovered completely in the second and fourth cycles.

b. *Completely viscous flow:* in this case,

$$\sigma = \eta\varepsilon$$

$$W = \int_0^{2\pi} \eta\varepsilon_0\omega \cos \omega t \, \varepsilon_0 \cos \omega t \, d(\omega t)$$

$$= \pi\eta\omega\varepsilon_0^2$$

For a viscous body, the energy imparted is completely dissipated over the full cycle.

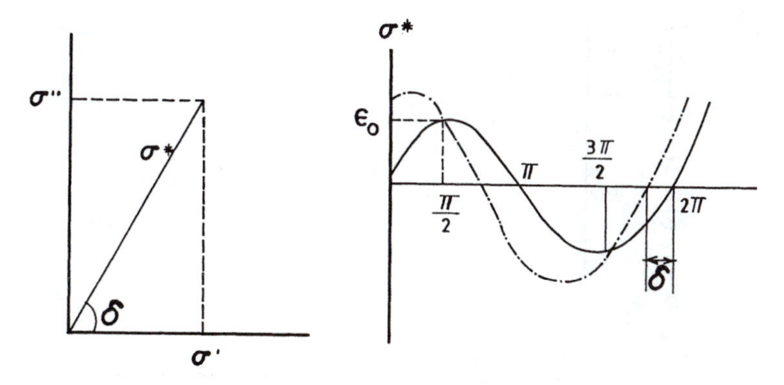

Figure E.14.2 Quantities in oscillatory experiments.

Therefore,

$$W = \int_0^{2\pi} \sigma d\varepsilon = \sigma * \varepsilon_0 \int_0^{2\pi} \sin(\omega t + \delta) \text{ dos } \omega t \, d(\omega t)$$

Integration of this equation shows that for the elastic component of the work per unit volume, there is no net energy lost or gained. The viscous component becomes

$$W = \pi \varepsilon_0^2 E'' = \pi \varepsilon_0^2 E' \tan \delta$$

This is the net energy loss through viscous heat generation in material.

Example 14.3: A Voigt element has parameters $E = 10^8$ N/m² and $\eta = 5 \times 10^{10}$ N · s/m². Sketch the creep curve for this element if the imposed constant stress is 10^8 N/m².

Solution:

$$J(t) = \frac{\varepsilon(t)}{\sigma_0} = J\left(1 - e^{-t/\tau}\right)$$

$$J = \frac{1}{E} = \text{compliance} = 10^{-8} \text{ m}^2/\text{N}$$

$$\tau = \eta/E = \text{retardation time} = \frac{5 \times 10^{10} \text{ Ns/m}^2}{10^8 \text{ N/m}^2}$$

$$= 500 \text{ s}$$

$$J(t) = \frac{\varepsilon(t)}{\sigma_o} = 10^{-8}\left(1 - e^{-t/500}\right)$$

$$E(t) = \sigma_o 10^{-8}\left(1 - e^{-t/500}\right)$$

$$= 1 - e^{-t/500} \text{ since } \sigma_o = 10^8$$

Using this equation a creep curve for several decades of time is shown in Figure E14.3.

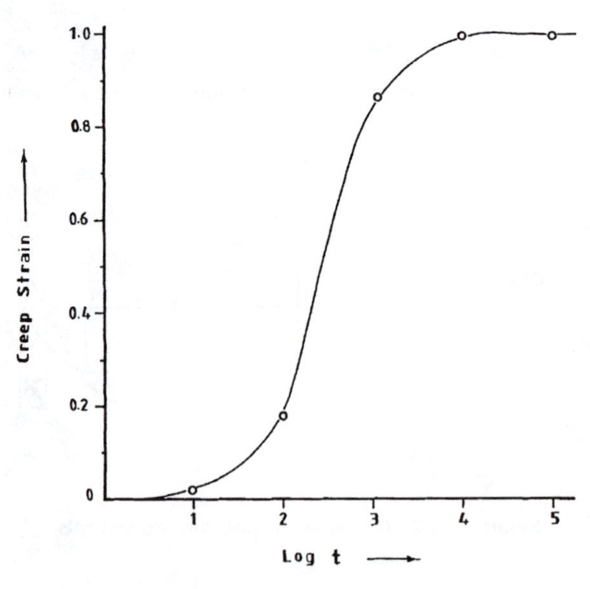

Figure E.14.3 Creep curve for Voigt element.

Example 14.4: Figure E.14.4 shows the loss modulus–temperature curves for the two materials A and B. Select either A or B for use as

a. Car tire
b. Engine mount

Explain the basis of your selection.

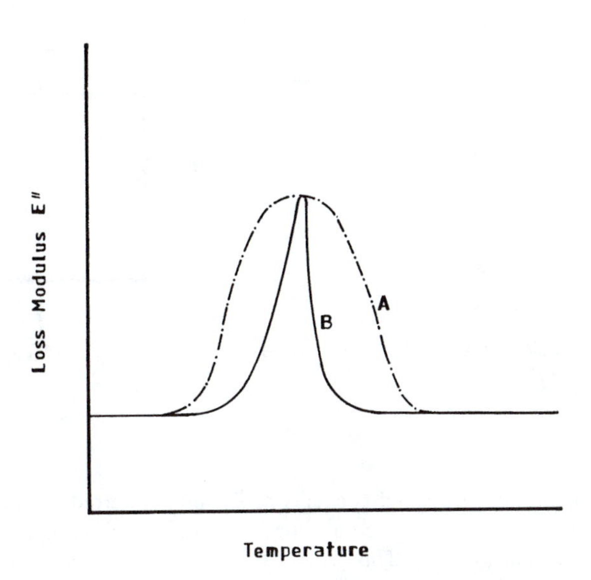

Figure E.14.4 Energy absorption profiles for two materials.

Solution: The area under the loss modulus–temperature curve is a measure of the damping capacity or ability to absorb energy of the material. Obviously, A has a higher damping capacity than B.

a. In car tires, temperature buildup contributes to rapid deterioration and wear of tire and loss of traction. Consequently, material B will be more suitable for use as a car tire since it will absorb less energy and hence result in less temperature buildup.
b. A critical requirement for an engine mount is the ability to absorb the vibrational loads from the engine. In this case, a material with the ability to dissipate the vibrational energy as heat would be preferable; that is, material A.

C. THE FOUR-PARAMETER MODEL

Neither the simple Maxwell nor Voigt model accurately predicts the behavior of real polymeric materials. Various combinations of these two models may more appropriately simulate real material behavior. We start with a discussion of the four-parameter model, which is a series combination of the Maxwell and Voigt models (Figure 14.9). We consider the creep response of this model.

Under creep, the total strain will be due to the instantaneous elastic deformation of the spring of modulus E, and irrecoverable viscous flow due to the dashpot of viscosity η_2, and the recoverable retarded elastic deformation due to the Voigt element with a spring of modulus E_3 and dashpot of viscosity η_3. Thus, the total strain is the sum of these three elements. That is,

$$\varepsilon(t) = \varepsilon_1 + \varepsilon_2 + \varepsilon_3 \tag{14.27}$$

$$\varepsilon(t) = \frac{\sigma_0}{E_1} + \frac{\sigma_0 t}{\eta_2} + \frac{\sigma_0}{E_3}\left[1 - \exp^{-t/\tau_3}\right] \tag{14.28}$$

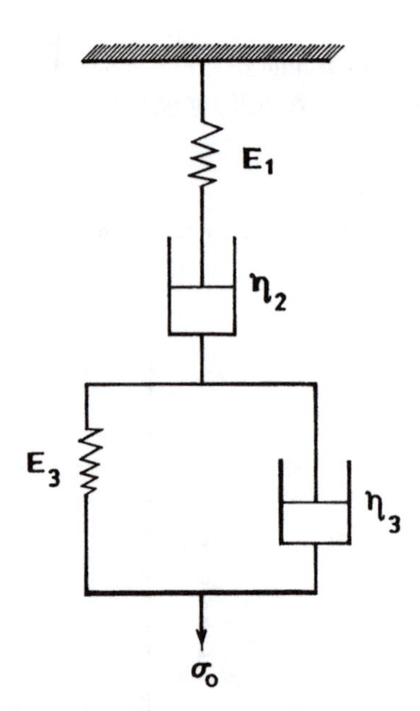

Figure 14.9 Schematic of the four-parameter model.

where σ_0 is the imposed constant stress and τ_3 equals η_3/E_3 and is referred to as the retardation time.

In creep recovery, say, the load is removed at time t_1; the deformation, σ_0/E_1, due to the spring of modulus E_1 is recovered instantaneously. This will be followed by the retarded elastic creep recovery due to the Voigt element given by ε_3 or

$$\varepsilon_3 = \frac{\sigma_0}{E_3}\left[1 - \exp - t_i/\tau_3\right] \tag{14.29}$$

Only the deformation due to the dashpot of viscosity η_2 is retained as a permanent set. The creep and creep recovery curve of this model is shown in Figure 14.10.

The four-parameter model provides a crude qualitative representation of the phenomena generally observed with viscoelastic materials: instantaneous elastic strain, retarded elastic strain, viscous flow, instantaneous elastic recovery, retarded elastic recovery, and plastic deformation (permanent set). Also, the model parameters can be associated with various molecular mechanisms responsible for the viscoelastic behavior of linear amorphous polymers under creep conditions. The analogies to the molecular mechanism can be made as follows.

1. The instantaneous elastic deformation is due to the Maxwell element spring, E_1. The primary valence bonds in polymer chains have equilibrium bond angles and lengths. Deformation from these equilibrium values is resisted, and this resistance is accompanied by an instantaneous elastic deformation.
2. Recoverable retarded elastic deformation is associated with the Voigt element. This arises from the resistance of polymer chains to coiling and uncoiling caused by the transformation of a given equilibrium conformation into a biased conformation with elongated and oriented structures. The process of coiling and uncoiling requires the cooperative motion of many chain segments, and this can only occur in a retarded manner.
3. Irrecoverable viscous flow is due to the Maxwell element dashpot η_3. This is associated with slippage of polymer chains or chain segments past one another.

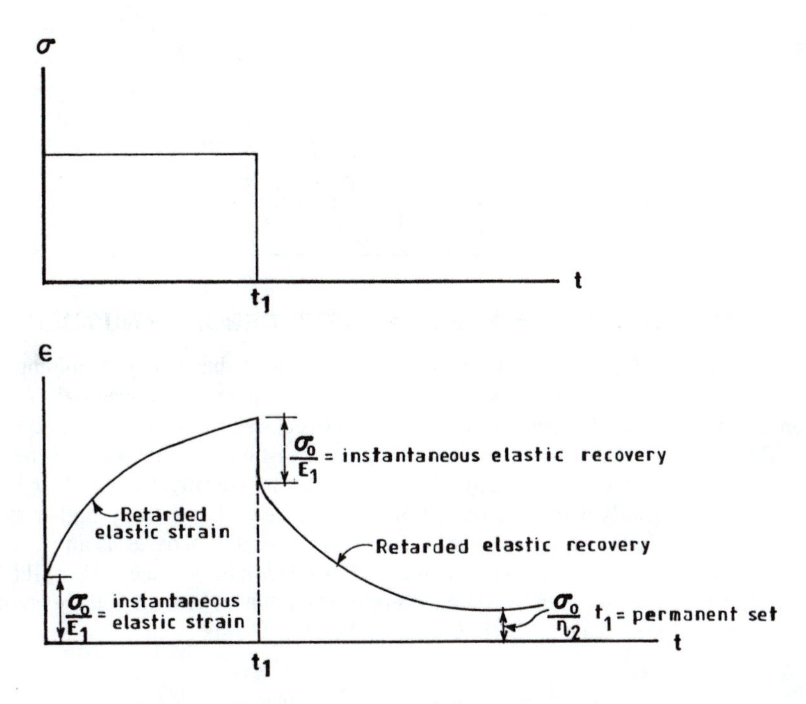

Figure 14.10 Creep response of the four-parameter model.

Example 14.5: The constants for a four-parameter model are

$$E_1 = 5 \times 10^8 \text{ N/m}^2,\ \eta_2 = 5 \times 10^{10} \text{ N.s/m}^2,\ E_3 = 10^8 \text{ N/m}^2, \text{ and } \eta_3 = 5 \times 10^8 \text{ N.s/m}^2$$

For creep and creep recovery experiments calculate:

a. The instantaneous elastic strain
b. The recoverable retarded elastic strain
c. The permanent set

Assume that the creep experiment lasted for 200 s and that the imposed stress is 10^8 N/m^2.

Solution:

a. Instantaneous elastic strain

$$\varepsilon_1 = \frac{\sigma_0}{E_1}$$

$$= \frac{10^8 \text{ N/m}^2}{5 \times 10^8 \text{ N/m}^2} = 0.2$$

b. Recoverable retarded elastic strain given by

$$\varepsilon_3 = \frac{\sigma_0}{E_3}\left[1 - e^{-t/\tau_3}\right];\ \tau_3 = \frac{\eta_3}{E_3} = \frac{5 \times 10^8 \text{ N.s/m}^2}{10^8 \text{ N/m}^2} = 5 \text{ s}$$

$$\varepsilon_3 = \frac{10^8}{10^8}\left[1 - e^{-200/5}\right]$$

$$= 1.0 - 4.25 \times 10^{-18} = 1.0$$

c. Permanent set

$$\varepsilon_2 = \frac{\sigma_0 t}{\eta_2}$$

$$= \frac{(10^8\ \text{N/m}^2)(200\ \text{s})}{5 \times 10^{10}\ \text{N.s/m}^2} = 0.4$$

V. MATERIAL RESPONSE TIME — THE DEBORAH NUMBER

A physical insight into the viscoelastic character of a material can be obtained by examining the material response time. This can be illustrated by defining a characteristic time for the material — for example, the relaxation time for a Maxwell element, which is the time required for the stress in a stress relaxation experiment to decay to e^{-1} (0.368) of its initial value. Materials that have low relaxation times flow easily and as such show relatively rapid stress decay. This, of course, is indicative of liquidlike behavior. On the other hand, those materials with long relaxation times can sustain relatively higher stress values. This indicates solidlike behavior. Thus, whether a viscoelastic material behaves as an elastic solid or a viscous liquid depends on the material response time and its relation to the time scale of the experiment or observation. This was first proposed by Marcus Reiner, who defined the ratio of the material response time to the experimental time scale as the Deborah number, D_n. That is,

$$D_n = \frac{\text{material response time}}{\text{experimental time scale (observation time)}} \quad (14.30)$$

A high Deborah number that is a long response time relative to the observation time implies viscoelastic solid behavior, whereas a low value of Deborah number (short response time relative to the time scale of experiment) is indicative of viscoelastic fluid behavior. From a conceptual standpoint, the Deborah number is related to the time one must wait to observe the onset of flow or creep. For example, the Deborah number of a wooden beam at 30% moisture is much smaller than that at 10% moisture content. For these materials the onset of creep occurs within a reasonably finite time. At the other extreme, the Deborah number of a mountain is unimaginably high. Millions of years must elapse before geologists find evidence of flow. This apparently is the genesis of Marcus Reiner's analogy ("The mountains flowed before the Lord" from the Song of Deborah, Book of Judges V).

It must be emphasized, however, that while the concept of the Deborah number provides a reasonable qualitative description of material behavior consistent with observation, no real material is characterized by a simple response time. Therefore, a more realistic description of materials involves the use of a distribution or continuous spectrum of relaxation or retardation times. We address this point in the following section.

VI. RELAXATION AND RETARDATION SPECTRA

Real polymers are not characterized by a simple response time. Instead, a distribution or continuous spectrum of relaxation or retardation times is required for a more accurate description of real polymers. Many complex models have been proposed to simulate the viscoelastic behavior of polymeric materials. We discuss two of these models.

A. MAXWELL–WEICHERT MODEL (RELAXATION)

The generalized model consists of an arbitrary number of Maxwell elements in a parallel arrangement (Figure 14.11).

Consider the generalized Maxwell model in a stress relaxation experiment. The strain in all the individual elements is the same, and the total stress is the sum of the stress experienced by each element. Thus,

$$\sigma = \sigma_1 + \sigma_2 + \sigma_3 + \ldots + \sigma_{n-1} + \sigma_n \quad (14.31)$$

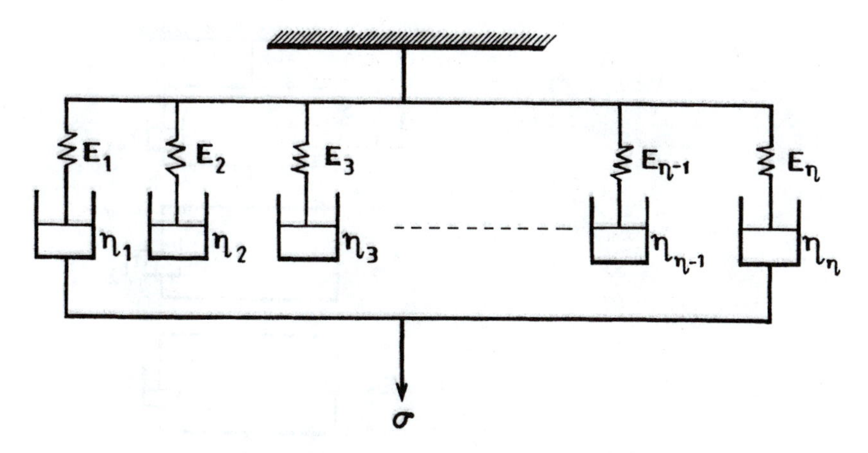

Figure 14.11 The Maxwell–Weichert model.

The individual stress in each element is given by

$$\sigma_i = \sigma_0 \, e^{-t/\tau_i} \tag{14.32}$$

This gives the stress relaxation of an individual element under a constant strain ε_0 as

$$\sigma_i(t) = \varepsilon_0 \, E_i \, e^{-t/\tau_i} \tag{14.33}$$

where $\tau_i = \eta_i/E_i$. For the Maxwell-Reichert model under a constant strain, ε_0,

$$\sigma(t) = \varepsilon_0 \sum_{i=1}^{n} E_i e^{-t/\tau_i} \tag{14.34}$$

or

$$E(t) = \sum_{i=1}^{n} E_i e^{-t/\tau_i} \tag{14.35}$$

If n is large, the summation in the equation may be approximated by the integral of a continuous distribution of relaxation times E(r).

$$E(t) = \int_0^{\infty} E(\tau) e^{-t/\tau} \, d\tau \tag{14.36}$$

If one of the Maxwell elements in the Maxwell–Weichert model is replaced with a spring or a dashpot of infinite viscosity, then the stress in such a model would decay to a finite value rather than zero. This would approximate the behavior of a cross-linked polymer.

B. VOIGT–KELVIN (CREEP) MODEL

The generalized Voigt element or the Voigt–Kelvin model is a series arrangement of an arbitrary number of Voigt elements (Figure 14.12). Under creep, the creep response of each individual element is given by

$$\varepsilon_i(t) = \sigma_0 J_i \left(1 - e^{-t/\tau_i}\right) \tag{14.37}$$

or

$$J_i(t) = \frac{\varepsilon_i(t)}{\sigma_0} = J_i \left(1 - e^{-t/\tau_i}\right) \tag{14.38}$$

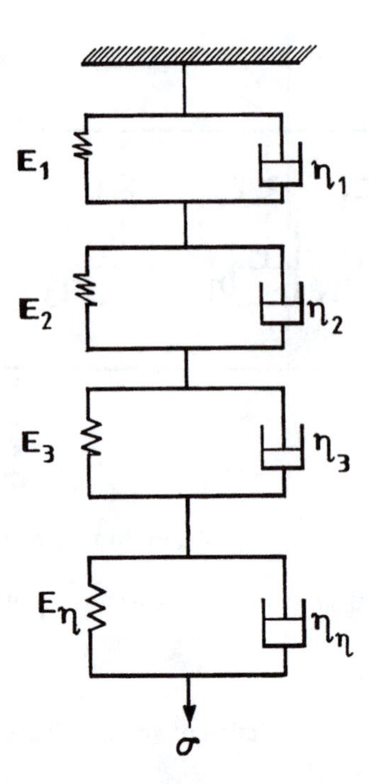

Figure 14.12 Voigt-Kelvin model.

where $J_i = \frac{1}{E_i}$ and is creep compliance. The response of a series of elements subjected to the same constant stress σ_0 becomes

$$J(t) = \frac{\varepsilon(t)}{\sigma_0} = \sum_{i=1}^{n} J_i\left(1 - e^{-t/\tau_i}\right) \tag{14.39}$$

For a large value of n (i.e., n $\rightarrow \infty$), the discrete summation in Equation 14.39 may be replaced by an integration over all the retardation times:

$$J(t) = \int_0^{\infty} J(\tau)\left(1 - e^{-t/\tau}\right)_{d\tau} \tag{14.40}$$

where J(t) is the continuous distribution of retardation times. If the generalized Voigt model is to represent a linear polymer (viscoelastic liquid), then the modulus of one of the springs must be zero. This element has infinite compliance and represents a simple dashpot in series with all the other Voigt elements.

Example 14.6: A polymer is represented by a series arrangement of two Maxwell elements with parameters $E_1 = 3 \times 10^9$ N/m^2, $t_1 = 1$ s, $E_2 = 5 \times 10^5$ N/m^2, and $t_2 = 10^3$ s. Sketch the stress relaxation behavior of this polymer over several decades (at least seven) of time.

Solution:

$$E_r(t) = \sum_{i=1}^{2} E_i e^{-t/\tau_i}$$

Determining log $E_r(t)$ when t varies from 0.01 to 10^4, a plot of log $E_r(t)$ vs. log t is shown in Figure E14.6.

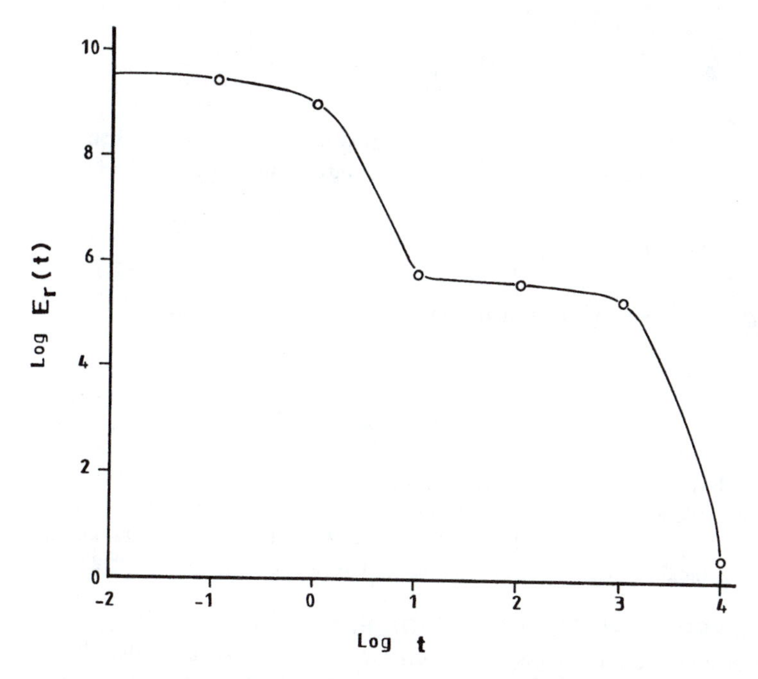

Figure E.14.6 Behavior of a 2-component Maxwell-Reichert model.

VII. SUPERPOSITION PRINCIPLES

In the following sections we discuss the two superposition principles that are important in the theory of viscoelasticity. The first is the Boltzmann superposition principle, which is concerned with linear viscoelasticity, and the second is time–temperature superposition, which deals with the time–temperature equivalence.

A. BOLTZMANN SUPERPOSITION PRINCIPLE

As discussed earlier for a Hookean solid, stress is a linear function of strain, while for a Newtonian fluid, stress is a linear function of strain rate. The constants of proportionality in these cases are modulus and viscosity, respectively. However, for a viscoelastic material the modulus is not constant; it varies with time and strain history at a given temperature. But for a linear viscoelastic material, modulus is a function of time only. This concept is embodied in the Boltzmann principle, which states that the effects of mechanical history of a sample are additive. In other words, the response of a linear viscoelastic material to a given load is independent of the response of the material to any load previously on the material. Thus the Boltzmann principle has essentially two implications — stress is a linear function of strain, and the effects of different stresses are additive.

Let us illustrate the Boltzmann principle by considering creep. Suppose the initial creep stress, σ_0, on a linear, viscoelastic body is increased sequentially to $\sigma_1, \sigma_2 \ldots \sigma_n$ at times $t_1, t_2 \ldots t_n$, then according to the Boltzmann principle, the creep at time t due to such a loading history is given by

$$\varepsilon(t) = J(t)\sigma_0 + J(t - t_1)[\sigma_1 - \sigma_0] + \ldots + J(t - t_n)[\sigma_n - \sigma_{n-1}] \tag{14.41}$$

Here J is compliance, whose functional dependence on time is denoted by the parentheses. The square brackets denote multiplication. For a continuous loading history, then, the creep is expressed by the integral:

$$\varepsilon(t) = \int_0^t J(t-\theta)[\dot{\sigma}(\theta)]\, d\theta \tag{14.42}$$

where $\dot{\sigma}(\theta)$ describes the stress history. A similar expression can be derived for stress relaxation. In this case, the initial strain is changed sequentially to ε_1, ε_2, and ε_n at times $t_1, t_2 \ldots t_n$; then the resultant stress is

$$\sigma(t) = E_r(t)\varepsilon_0 + E_r(t-t_1)[\varepsilon_1 - \varepsilon_0] + \ldots + E_r(t-t_n)[t_n - t_{n-1}] \tag{14.43}$$

For a continuous strain history, the Boltzmann expression becomes

$$\sigma(t) = \int_0^t E_r(t-\theta)\,[\varepsilon(\theta)]\, d\theta \tag{14.44}$$

Linear viscoelasticity is valid only under conditions where structural changes in the material do not induce strain-dependent modulus. This condition is fulfilled by amorphous polymers. On the other hand, the structural changes associated with the orientation of crystalline polymers and elastomers produce anisotropic mechanical properties. Such polymers, therefore, exhibit nonlinear viscoelastic behavior.

B. TIME–TEMPERATURE SUPERPOSITION PRINCIPLE

Structural engineering design with engineering materials usually requires that the structures maintain their integrity for long periods of time. In such designs the elastic modulus of structural components is an important parameter for relating design stresses to component dimensions. We know, of course, that the modulus of polymeric materials decreases with increasing time. Therefore, to ensure a safe and proper design, it is necessary to know the lower limit of modulus. Ideally, the most reliable information on modulus changes of a polymeric material should be developed over a long period in which test samples are subjected to conditions comparable to those that will be experienced by the material in real-time service. Accumulation of such long-term data is obviously inconvenient, expensive, and indeed hardly practical. Consequently, in engineering practice, reliance has to be placed necessarily on short-term data for the design for long-term applications.

Fortunately for linear amorphous polymers, modulus is a function of time and temperature only (not of load history). Modulus–time and modulus–temperature curves for these polymers have identical shapes; they show the same regions of viscoelastic behavior, and in each region the modulus values vary only within an order of magnitude. Thus, it is reasonable to assume from such similarity in behavior that time and temperature have an equivalent effect on modulus. Such indeed has been found to be the case. Viscoelastic properties of linear amorphous polymers show time–temperature equivalence. This constitutes the basis for the time–temperature superposition principle. The equivalence of time and temperature permits the extrapolation of short-term test data to several decades of time by carrying out experiments at different temperatures.

Time–temperature superposition is applicable to a wide variety of viscoelastic response tests, as are creep and stress relaxation. We illustrate the principle by considering stress relaxation test data. As a result of time–temperature correspondence, relaxation curves obtained at different temperatures can be superimposed on data at a reference temperature by horizontal shifts along the time scale. This generates a simple relaxation curve outside a time range easily accessible in laboratory experiments. This is illustrated in Figure 14.13 for polyisobutylene. Here, the reference temperature has been chosen arbitrarily to be 25°C. Data obtained at temperature above 25°C are shifted to the right, while those obtained below 25°C are shifted to the left.

The procedure for such data extrapolation is not arbitrary. The time–temperature superposition principle may be expressed mathematically for a stress relaxation experiment as

$$E_r(T_1, t) = E_r(T_2, t/a_T) \tag{14.45}$$

This means that the effect on the modulus of changing the temperature from T_1 to T_2 is equivalent to multiplying the time scale by a shift factor a_r which is given by

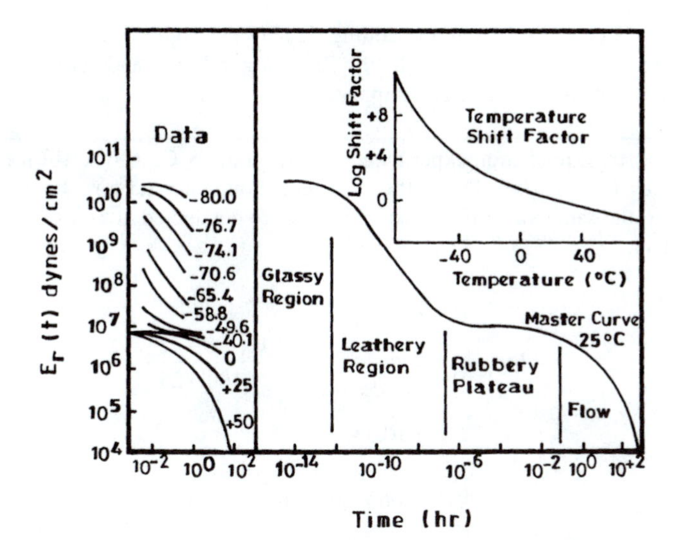

Figure 14.13 Time–temperature superposition for polyisobutylene. (From Tobolsky, A.V. and Catsiff, E., *J. Polym. Sci.*, 19, 111, 1956. With permission.)

$$a_T = t_T/t_{To} = \tau_T/\tau_{To} = \eta_T/\eta_{To} \tag{14.46}$$

where t_T is the time required to reach a particular mechanical response (modulus in this case) at temperature T, and t_{T0} is the time required to produce the same response at the reference temperature T_0. An important empirical relation correlating the shift factor with temperature changes has been developed by Williams, Landel, and Ferry, the so-called WLF equation.[3] The WLF equation, which is valid between T_g and T_g + 100°C, is given by the general expression

$$\log_{10} a_T = \frac{-C_1(T - T_0)}{C_2 + T - T_0} \tag{14.47}$$

where T_0 is the reference temperature and C_1 and C_2 are constants to be determined experimentally. The temperatures are in degrees Kelvin. It is common practice to choose the glass transition temperature, T_g, as the reference temperature. In this case, the WLF equation is given by

$$\log_{10} a_T = \frac{-17.44(T - T_g)}{51.6 + T - T_g} \tag{14.48}$$

Before shifting the curves to generate the master curve, it is necessary to correct the relaxation modulus at each temperature for temperature and density with respect to the reference temperature. That is,

$$E^{(t)}_{reduced} = \left(\frac{T_o}{T}\right)\left(\frac{\rho_o}{\rho}\right) E_r(t) \tag{14.49}$$

ρ_o = density at T_o
ρ = density at T

This correction is based on the theory of rubber elasticity, which postulates that the elastic modulus of rubber is proportional to the absolute temperature T and to the density of the material. It can be argued, of course, that this correction may only be necessary in the rubbery region, where the predominant response mechanism is chain coiling and uncoiling. It may not apply in the glassy region, where the mechanical response is governed essentially by the stretching of bonds and deformation of bond angles, or the viscous region, which involves chain slippage.

Example 14.7: In a stress relaxation experiment conducted at 25°C, it took 10^7 h for the modulus of the polymer to decay to 10^5 N/m^2. Using the WLF equation, estimate how long will it take for the modulus to decay to the same value if the experiment were conducted at 100°C. Assume that 25°C is the T_g of the polymer.

Solution:

$$\log_{10} a_T = \log \frac{t_{100}}{t_{25}} = \frac{-17.44(100-25)}{51.6+100-25}$$

$$= -10.33$$

$$\frac{t_{100}}{t_{25}} = 4.66 \times 10^{-11} \text{ h}$$

$$t_{100} = 4.66 \times 10^{-11} \times 10^{7} \text{ h}$$

$$= 4.66 \times 10^{-4} \text{ h}$$

VIII. PROBLEMS

14.1. The following three-parameter model is assumed to simulate the behavior of a certain polymer:

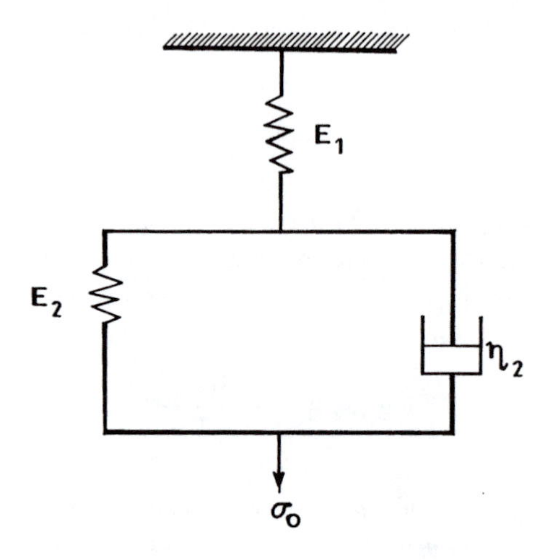

In a creep and creep recovery experiment:

a. What is the rheological equation that describes this model?

b. Sketch the creep and creep recovery curve.

14.2. Calculate the relaxation modulus after 10 s of the application of stress for a polymer represented by three Maxwell elements in parallel where:

$E_1 = 10^5$ N/m$_2$, $\tau_1 = 10$ s
$E_2 = 10^6$ N/m$_2$, $\tau_2 = 20$ s
$E_3 = 10^7$ N/m^2, $\tau_3 = 30$ s

14.3. A Voigt–Kelvin model consists of four elements with the following parameters:

Element No.	E(N/m²)	η (N.s/m²)
1	5×10^8	5×10^{10}
2	10^{10}	5×10^{10}
3	5×10^8	5×10^9
4	10^8	5×10^{10}

In a creep experiment, if the imposed stress is 10^8 N/m_2, determine the strain after 100 s.

14.4. The initial stress on a polymer in a creep experiment is 10^8 N/m². This load is increased by 10^7 N/m² and 10^6 N/m² after 10^2 and 10^3 s, respectively. Assuming that the Boltzmann superposition principle holds for this material, find the strain after 10^4 s. The creep compliance for the material is given by 10^{-8} $(1 - e^{-10^{-4}t})$.

14.5. In a stress relaxation experiment the modulus of polyisobutylene relaxed to 10^6 N/m² in 10^4 h at 0°C. If it is desired to cut the experimental time to 10 h, use the WLF equation to estimate the temperature at which the experiment must be conducted. The T_g for PIB is –70°C.

14.6. The relaxation time for a material that obeys the WLF equation at 0°C is 10^4 s. Its relaxation time at T_g is 10^{13} s. What is the relaxation time of this material at 25°C?

14.7. In a forced vibration experiment, the damping peak for polycarbonate occurred at 150°C at a frequency of 1 Hz. What would be the location of this peak if the frequency were 1000 Hz? Polycarbonate has a T_g of 150°C.

14.8. A tactic polystyrene has a T_g of 100°C. What are the relative rates of stress relaxation of this polymer at 150°C and 125°C?

REFERENCES

1. Reiner, M., *Phys. Today,* p. 62, January 1964.
2. Tobolsky, A.V. and Catsiff, E., *J. Polym. Sci.,* 19, 111, 1956.
3. Williams, M.L., Landel, R.F., and Ferry, J.D, *Am. Chem. Soc.,* 77, 3701, 1955.
4. Ferry, J.D., *Viscoelastic Properties of Polymers,* John Wiley & Sons, New York, 1980.
5. Tobolsky, A.V., *Properties and Structures of Polymers,* John Wiley & Sons, New York, 1960.
6. Aklonis, J.J. and MacKnight, W.J., *Introduction to Polymer Viscoelasticity,* John Wiley & Sons, New York, 1983.
7. Nielson, L.E., *Mechanical Properties of Polymers and Composites,* Vol. 1, Marcel Dekker, New York, 1974.

Chapter 15

Polymer Properties and Applications

I. INTRODUCTION

Many of the products we enjoy every day are derived from polymers even though we may not readily appreciate this. Polymers influence the way we live, work, and play. The evidence is all around us. There is no room in our house, no vehicle we ride in, no sports activity we participate in that does not use products made from polymers; the list of polymer products is virtually endless. In fact, the products are so diverse, so prevalent, and so ingrained in our lifestyles that we tend to take them for granted.

There are currently more than 40 families of polymers, many of whose uses still go unrecognized because of their diversity. Most consumers many be familiar with polymers in the form of housewares, toys, appliance parts, knobs, handles, electrical fixtures, toothbrushes, cups, lids, and packages. But few are aware that

- the lifelines of communications — television, radio telephone, radar — are based on polymers for insulation and other vital components;
- foams made from polymers have changed modern concepts of cushioning and insulation;
- polymers are vital to the electrical/electronics industry; and
- the practice of replacing heart valves, sockets and joints, and other defective parts of the body owes it success to the availability of suitable polymer materials.[1]

Polymers are true man-made materials that are the ultimate tribute to man's creativity and ingenuity. It is now possible to create different polymers with almost any quality desired in the end product — some similar to existing conventional materials but with greater economic value, some representing significant property improvements over existing materials, and some that can only be described as unique prime materials with characteristics unlike any previously known.[1]

The use of polymers has permeated every facet of our lives. Polymers contribute to meeting our basic human needs, including food, shelter, clothing, health, and transportation. It is hard to imagine what the world was like before the advent of synthetic polymers. It is difficult to contemplate what the world, with all its luxury and comfort, would be like without polymers. In this chapter we discuss the properties and applications of a number of them. First, however, the structure of and the raw materials for the polymer industry are briefly highlighted.

II. THE STRUCTURE OF THE POLYMER INDUSTRY[1]

The polymer industry is generally made up of a large number of companies. Some of these companies produce basic raw materials, others process these materials into end-use products, and still others finish these products in some way that may sometimes make them unrecognizable as polymers. The activities or functions of these companies are not mutually exclusive; indeed, they frequently overlap. For example, automotive and packaging, which depend on polymer products for their operations, may themselves be large processors of polymers. In addition, within the industry itself, polymer materials manufacturers may also be engaged in processing and finishing operations. In any case, the structure of the polymer industry can be resolved essentially into the following categories.

A. POLYMER MATERIALS MANUFACTURERS

As has been discussed in Chapter 1, polymers are high-molecular-weight compounds (macromolecules) built up by the repetition of small chemical units. These units, or repeating units as they are called, are derived from monomers. Monomers, as we shall see in the Section III, are obtained largely from fractions of petroleum or gas recovered during the petroleum refining process. The reaction of monomers, referred to as polymerization, leads to the polymer formation. The polymer materials manufacturer or supplier is involved in the transformation of monomers or basic feedstocks into polymer materials, which are then sold in the form of granules, powder, pellets, flakes, or liquids for subsequent processing into finished products.

A number of polymerization techniques are used in the transformation of monomers into plastics (Chapter 10). These include bulk, solution, suspension, and emulsion polymerization processes. Each of these polymerization techniques has its advantages and disadvantages and may be more appropriate for the production of certain types of polymer materials. For example, bulk polymerization is ideally suited for making pure polymer products, as in the manufacture of optical-grade poly(methyl methacrylate) or impact-resistant polystyrene, because of minimal contamination of the product. On the other hand, solution polymerization finds ready application when the end use of the polymer requires a solution, as in certain adhesives and coating processes.

B. MANUFACTURERS OF CHEMICALS, ADDITIVES, AND MODIFIERS

Very frequently, the polymer material from the materials manufacturer does not go directly to the processor. There is often an intermediate step that involves the addition of other materials (chemicals, additives, modifiers) that serve to impart special properties or enhance the qualities of the polymers or resin. For example, polymers can be integrally colored (with polymers or dyes), made more flexible (with plasticizers), more heat and light resistant (with stabilizers), or stronger and more impact resistant (with fiber reinforcements). These modifiers may be supplied by the companies that manufacture the plastics themselves or by companies that specialize in the production of one or more modifiers.

C. COMPOUNDING/FORMULATING

As we said above, various types of modifiers, chemicals, and additives are compounded with the base resin or polymer material before processing. The compounding is usually done by the materials manufacturer; however, this is also performed by a group within the polymer industry. Such companies buy the base resin from the materials manufacturer and then compound it specially for resale to the processor. The processor may also buy the base polymer, modifiers, chemicals, and additives for in-house compounding.

D. THE PROCESSOR

The heart of the polymer industry is the processor, who is responsible for turning polymer materials into secondary products like film, sheet, rod and tube, component parts, or finished end products. Processors may be classified into three general categories depending on the target market for their products:

- Custom processors, who do processing on a custom basis for end users
- Proprietary processors, who manufacture polymer products such as housewares and toys for direct sale to consumers
- Captive processors, who possess in-house production facilities; they are usually manufacturers that consume large volumes of polymer parts

A number of processing techniques are employed in the plastics industry to shape the plastic material into the desired end product. In fact, processors are more often categorized on the basis of the type of processing operations they perform. Polymer processing operations include molding (injection, blow, rotational, compression, and transfer); extrusion; calendering; thermoforming; and casting (Chapter 11).

E. THE FABRICATOR

Some of the products from the processor are not directly usable by the consumer. The fabricator in the polymers industry is engaged in turning secondary products such as film, sheet, rods, and tubes into end products. Using conventional machine tools and simple bending techniques, the fabricator creates products like jewelry, signs, and furniture from rigid polymers. Similarly, products like shower curtains, rainwear, inflatables, upholstery, and packaging overwrap are obtained from flexible film and sheeting by employing various die cutting and sealing methods.

F. THE FINISHER

A number of uses of polymers are not easily recognizable as such because the appearance of polymers is similar to that of conventional materials. Examples of such polymer products include plastic furniture with integral wood grain patterns and plastic automotive grilles that can be electroplated with a metallic surface. Production of such polymer products, including the large-volume printing of plastic film and sheets, is done by the finisher. The finishing of polymers includes the different methods of adding either decorative or functional surface effects to the polymer product. Color and decorative effects can be added to polymers prior to and during manufacturing.

Also, polymer parts, whether film and sheeting or rigid products, can be postfinished in a number of ways. Some companies within the polymers industry specialize in various finishing techniques. On the other hand, finishing, decorating, and assembly of the polymer end product can be done in-house by the processor or fabricator.

III. RAW MATERIALS FOR THE POLYMER INDUSTRY[2,3]

A finished polymer article may be made entirely from the neat or pure resin. More often, however, it is necessary to compound the resin with additives to improve its processing behavior and/or enhance product quality and service performance. In this section we examine, in a global way, the various feedstocks for the polymers industry. The raw materials can be divided essentially into the base polymers and the various additives. Polymer additives and reinforcements have been treated in Chapter 9. Therefore, only base polymers are discussed here.

The first plastic made was cellulose nitrate, which is a derivative of cellulose, obtained from wood pulp. The first truly synthetic polymer material was phenolic resin, which was synthesized from phenol and formaldehyde derived from coal. Today, the source of organic chemicals for the production of polymers has shifted from these traditional sources to petroleum and natural gas. Petroleum as a raw material for organic chemicals (petrochemicals) is relatively cheap, readily available in large tonnages, and more easily processed than the other main source of organic chemicals — coal.

There are two stages involved in the production of petrochemicals from petroleum. The first is the separation of petroleum, which is a mixture of hydrocarbons, into various fractions (mainly liquid fuels) by a process called fractional distillation. The second is the further refinement of certain fractions from the distillation process to form petrochemicals. Most petrochemicals are derived from three sources:

1. Various mixtures of carbon monoxide and hydrogen, known as synthesis gas, obtained from steam reforming of natural gas (methane) or, in a few cases, steam reforming of naphtha.
2. Olefins obtained by steam cracking (pyrolysis) of various feedstocks, including ethane, propane–butane (LPG), distillates (naphtha, gas oil), and even crude oil.
3. Aromatics — benzene, toluene, and xylene (BTX) — obtained from catalytic reforming.

These routes are the sources of the eight building blocks — ammonia, methanol, ethylene, propylene, butadiene, benzene, toluene, and xylene — from which virtually all large tonnage petrochemicals are derived. Figure 15.1 represents a simplified version of the production of these chemicals from petroleum and natural gas. Simplified flow diagrams for the production of some polymers from these basic petrochemicals are shown in Figures 15.2–15.9.

IV. POLYMER PROPERTIES AND APPLICATIONS

The rapid growth and widespread use of polymers are due largely to the versatility of their properties. As we saw in Chapter 3, polymer properties are attributable to their macromolecule nature and the gross configuration of their component chains as well as the nature and magnitude of interactions between the constituent chain atoms or groups. The major uses of polymers as elastomers, fibers, and plastics are a consequence of a combination of properties unique to polymers. For example, the elasticity of elastomers, the strength and toughness of fibers, and the flexibility and clarity of plastic films reflect their different molecular organizations. Throughout the discussion in the previous chapters, we have consistently emphasized the structural basis of the behavior of polymeric materials. We have discussed the variables necessary to define the mechanical, physical, and other properties of polymers. We now focus attention on the properties and applications of polymers. We start the discussion with the large-volume polymers — so-called general-purpose or commodity thermoplastics. The selection of polymers for discussion is based on convenience rather than on a consistent classification scheme.

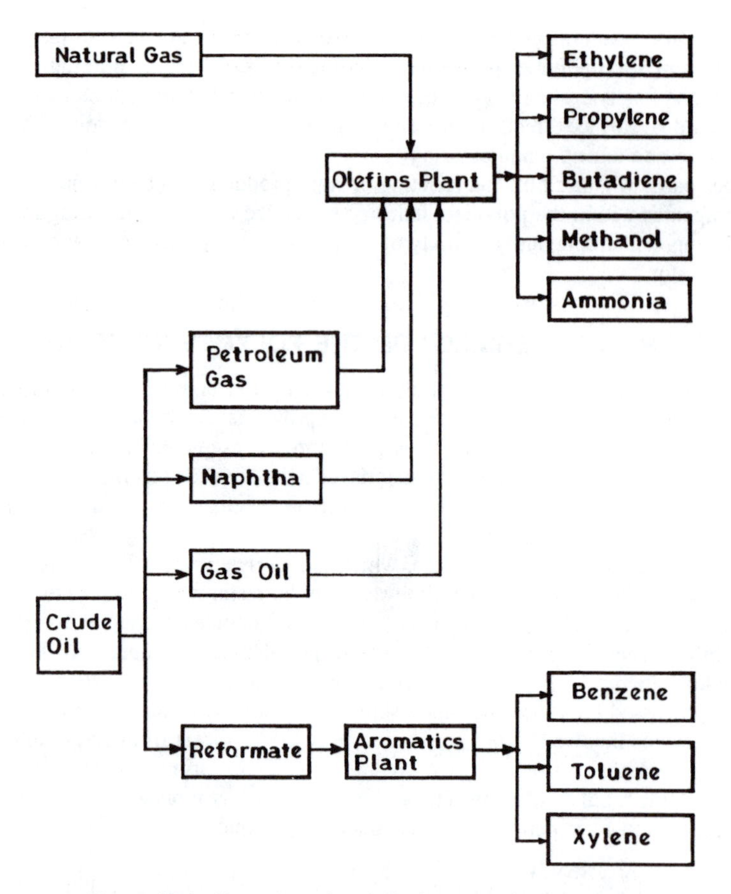

Figure 15.1 Base petrochemicals from petroleum and gas.

A. POLYETHYLENE

$$\left[CH_2 - CH_2 \right]_n \qquad \text{(Str. 1)}$$

Ethylene may be polymerized by a number of processes to produce different varieties of polyethylene. The most commercially important of these polymers are low-density polyethylene (LDPE); high-density polyethylene (HDPE); and, more recently, linear low-density polyethylene (LLDPE) and ultra-high-density polyethylene (UHDPE).

The first commercial ethylene polymer (1939) was low-density, low-crystalline (branched) polyethylene (LDPE), which is the largest of the thermoplastics produced in the U.S. LDPE is produced by free-radical bulk polymerization using traces of oxygen or peroxide (benzoyl or diethyl) and sometimes hydroperoxide and azo compounds as the initiator. To obtain a high-molecular-weight product, impurities such as hydrogen and acetylene, which act as chain transfer agents, must be scrupulously removed from the monomer. Polymerization is carried out either in high-pressure, tower-type reactors (autoclaves) or continuous tubular reactors operating at temperatures as high as 250°C and at pressures between 1000 and 3000 atm (15,000 to 45,000 psi). The exothermic heat of polymerization (about 25 kcal/mol) is controlled by conducting the polymerization in stages of 10 to 15% conversion. Solution polymerization of ethylene with benzene and chlorobenzene as solvents is also possible at the temperatures and pressures employed.

Polyethylene with limited branching, that is, linear or high-density polyethylene (HDPE), can be produced by the polymerization of ethylene with supported metal–oxide catalysts or in the presence of

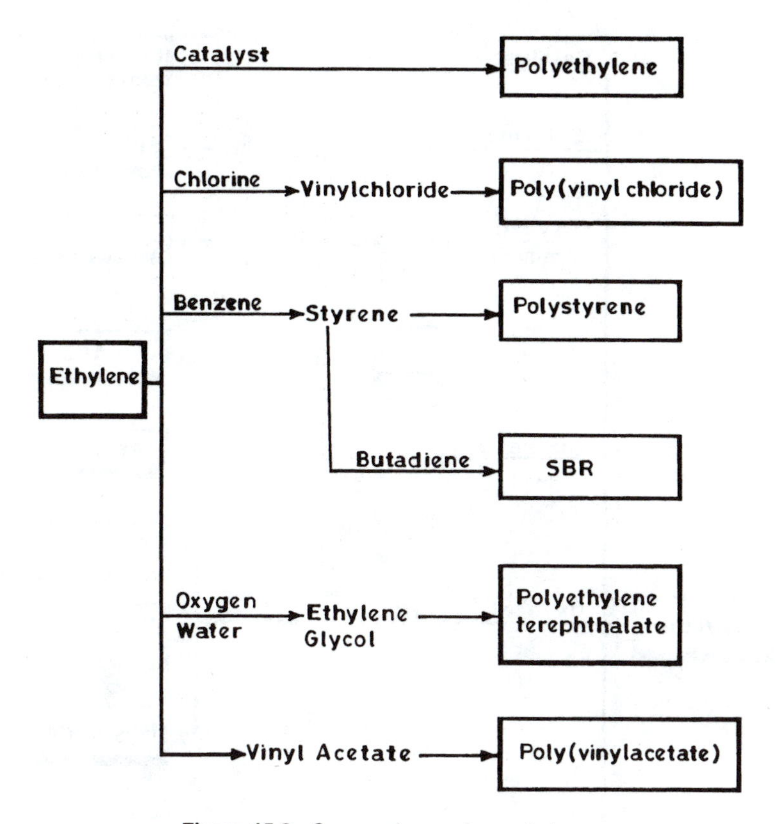

Figure 15.2 Some polymers from ethylene.

coordination catalysts. The first class of metal–oxide catalyst (Phillips type) consisted of chromium oxide (CrO_3) supported on alumina (Al_2O_3) or silica–alumina base. Polymerization is carried out at 100 atm and 60 to 200°C in hydrocarbon solvents in which the catalysts are insoluble using either fixed-bed, moving bed, fluidized-bed, or slurry processes. The coordination polymerization of ethylene utilizes Ziegler-type catalysts. These are complexes of aluminum trialkyls and titanium or other transition-metal halides (e.g., $TiCl_4$). Coordination polymerization of ethylene generally requires lower temperatures and pressures than those that involve the use of supported metal–oxide catalysts, typically 60 to 70°C and 1 to 10 atm. Linear, low-to-medium-density polyethylene (LLDPE) with shorter chain branches than LDPE is made also by a low-pressure process (Dow Chemical).

Low-density polyethylene is a partially crystalline solid with a degree of crystallinity in the 50 to 70% range, melting temperature of 100 to 120°C, and specific gravity of about 0.91 to 0.94 (Table 15.1). Free-radical polymerization of ethylene produces branched polymer molecules. Branches act as defects, and as such the level of side chain branching determines the degree of crystallinity, which in turn affects a number of polymer properties. The number of branches in LDPE may be as high as 10 to 20 per 1000 carbon atoms. Branching is of two different types. The first and predominant type of branching, which arises from intermolecular chain transfer, consists of short-chain alkyl groups such as ethyl and butyl. The second type of chain branching is produced by intermolecular chain transfer. This leads to long-chain branches that, on the average, may be as long as the main chain. High-density polyethylene, on the other hand, has few side chains, typically 1 per 200 carbon main chain atoms. Linear polyethylenes are highly crystalline, with a melting point over 127°C — usually about 135°C — and specific gravity in the 0.94 to 0.97 range.

The physical properties of LDPE depend on three structural factors. These are the degree of crystallinity (density), molecular weight (MW), and molecular weight distribution (MWD). The degree of crystallinity and, therefore, density of polyethylene is dictated primarily by the amount of short-chain branching. Properties such as opacity, rigidity (stiffness), tensile strength, tear strength, and chemical

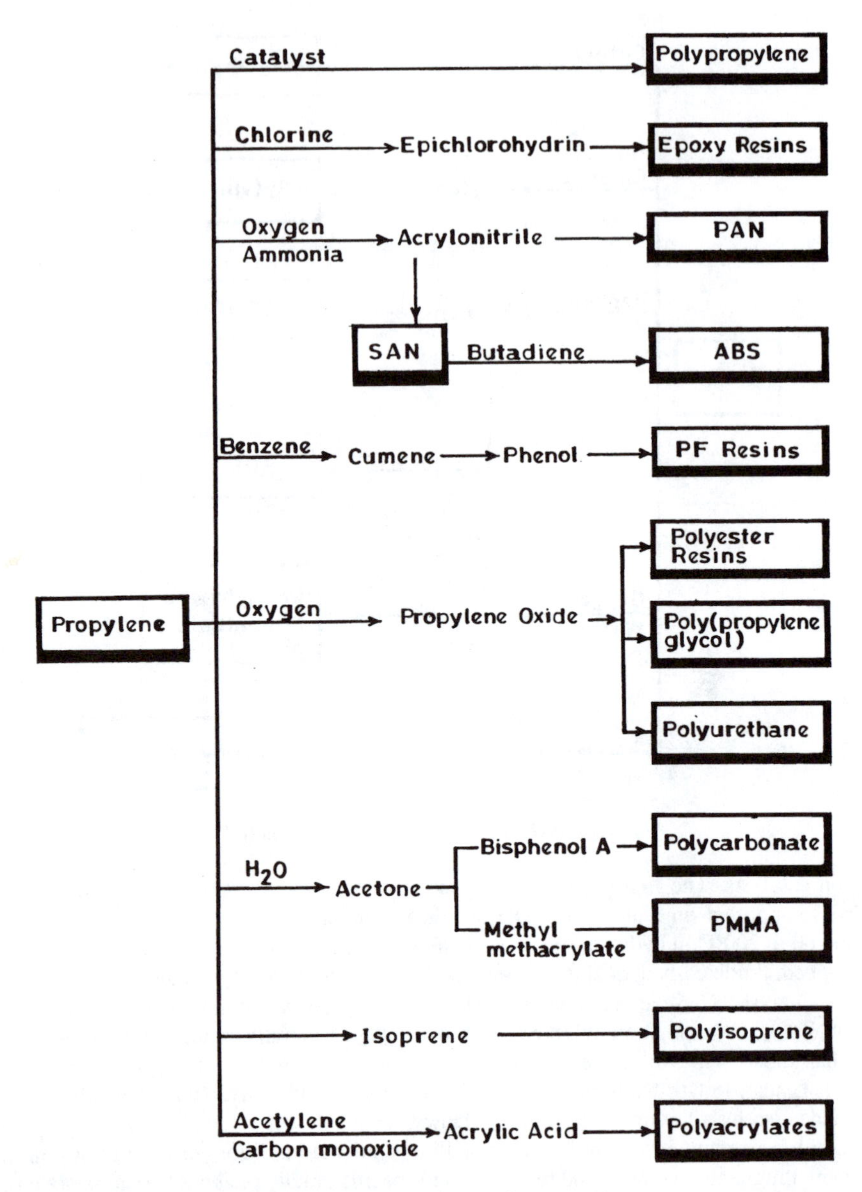

Figure 15.3 Some polymers from propylene.

resistance, which depend on crystallinity, increase as density increases (i.e., the amount of short-chain branching decreases). On the other hand, permeability to liquids and gases decreases and toughness decreases with increasing crystallinity.

Copolymerization with polar monomers such as vinyl esters (e.g., vinyl acetate, acrylate esters, carboxylic acids, and vinyl ethers) can be used to adjust crystallinity and modify product properties. Ester comonomers provide short-chain branches that reduce crystallinity. For example, LDPE films with increased toughness, clarity, and gloss have been obtained by the incorporation of less than 7% vinyl acetate. Films made from ethylene and ethyl acrylate (EEA) copolymers have outstanding tensile strength, elongation at break, clarity, stress cracking resistance, and flexibility at low temperatures. As we shall see later, ionomers display extreme toughness and abrasion resistance and improved tensile properties. With HDPE, control of branching is usually achieved by adding comonomers such as propylene, butene, and hexene during polymerization.

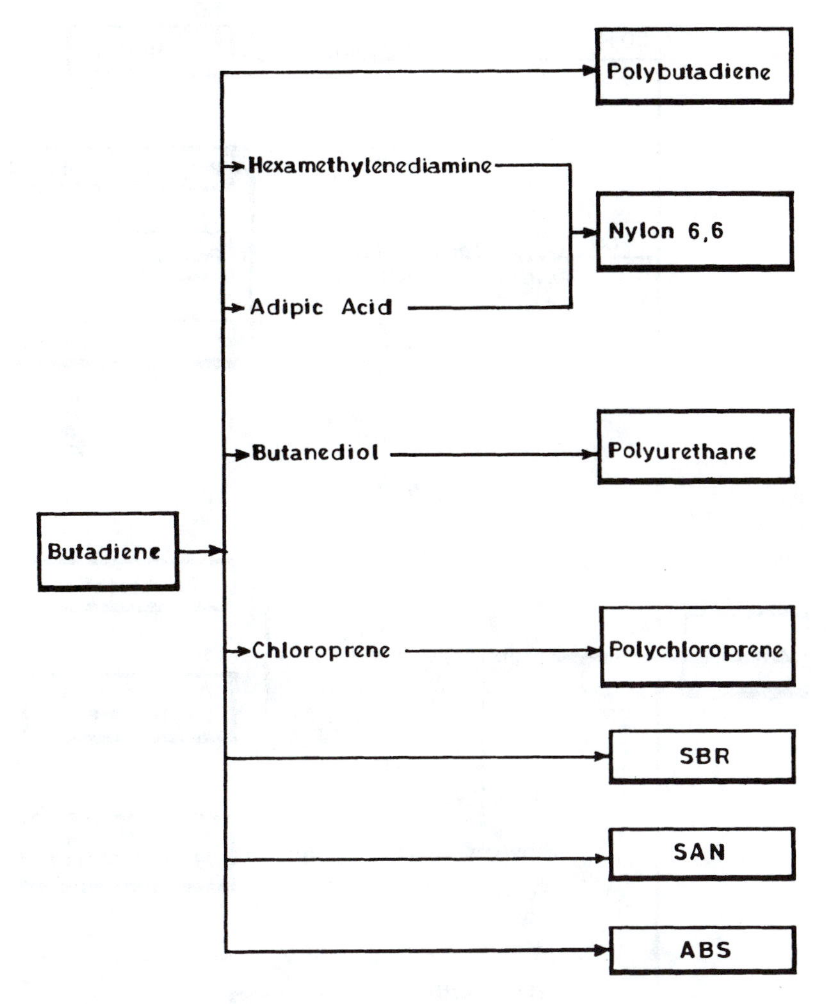

Figure 15.4 Some polymers from butadiene.

The molecular weight of LDPE is typically in the range of 6000 to 40,000. Melt index (MI) is used as a convenient measure of average molecular weight. Melt index designates the weight (in grams) of polymer extruded through a standard capillary at 190°C in 10 min (ASTM D 1238). Consequently, melt index is inversely related to molecular weight. Typical melt index values for LDPE are in the range of 0.1 to 109. As molecular weight increases, tensile and tear strength, softening temperatures, and stress cracking and chemical resistance increase, while processibility becomes more difficult and coefficient of friction (film) decreases.

The ratio $\overline{M}_w/\overline{M}_n$ is a measure of molecular weight distribution (MWD). The breadth of MWD is used to evaluate the influence of long-chain branching on the properties of polyethylene. Polyethylene with a narrow MWD has high impact strength, reduced shrinkage and warpage, enhanced toughness and environmental stress cracking resistance, but shows a decrease in the ease of processing.

Polyethylene and its copolymers find applications in major industries such as packaging, housewares, appliances, transportation, communications, electric power, agriculture, and construction. The majority of LDPE is used as thin film for packaging. Other uses include wire and cable insulation, coatings, and injection-molded products (Table 15.2).

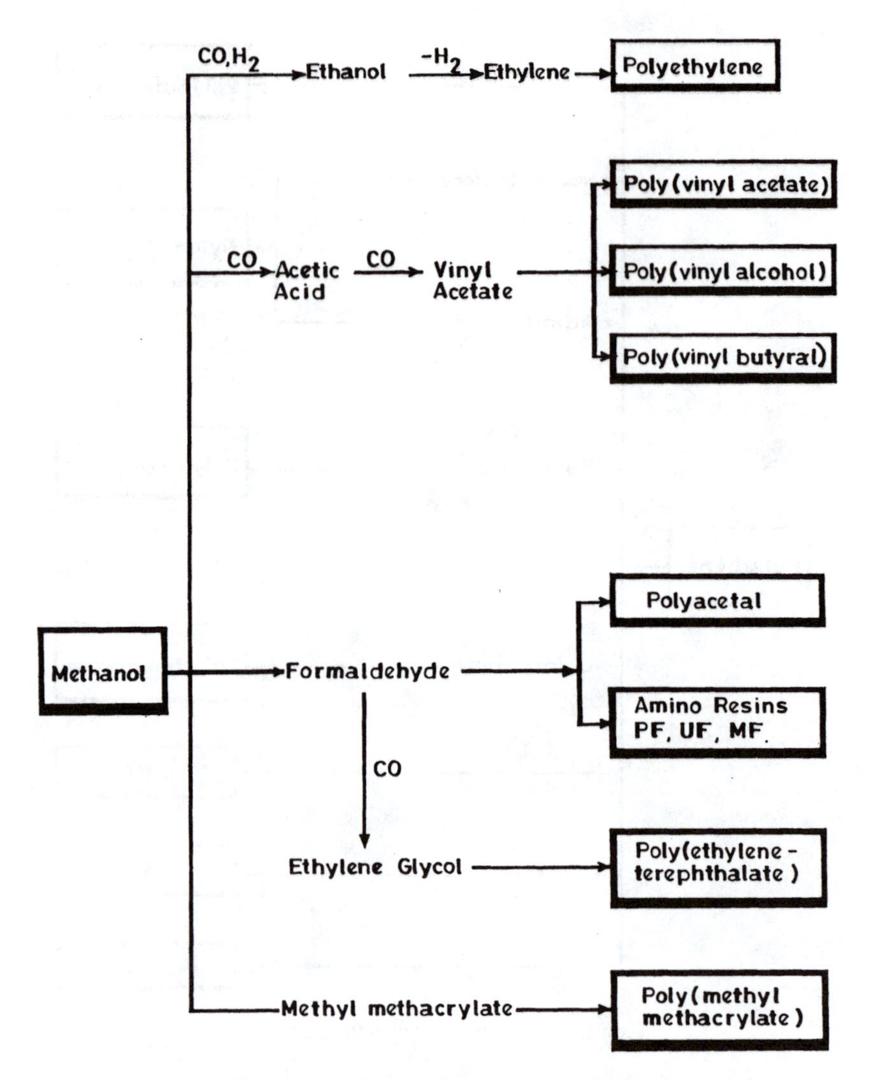

Figure 15.5 Some polymers from methanol.

Example 15.1: Explain the following observations.

a. LDPE is used mainly as thin film for packaging and sheets while HDPE is used predominantly in injection molding of crates, pails, tubs, and automobile gas tanks. EVA copolymer containing a high percentage of the ester comonomer is used in film applications such as disposable protective gloves.
b. EVA and EEA copolymers are sometimes preferable to conventional LDPE for wire and cable insulation, particularly for outdoor applications.

Solution:

a. LDPE is used when clarity, flexibility, and toughness are desired. LDPE possesses the desired combination of low density, flexibility, resilience, high tear strength, and moisture and chemical resistance, which are characteristics of a good film material. HDPE is used where hardness, rigidity, high strength, and high chemical resistance are required. EVA copolymers are used in articles requiring extreme flexibility and toughness and rubbery properties.

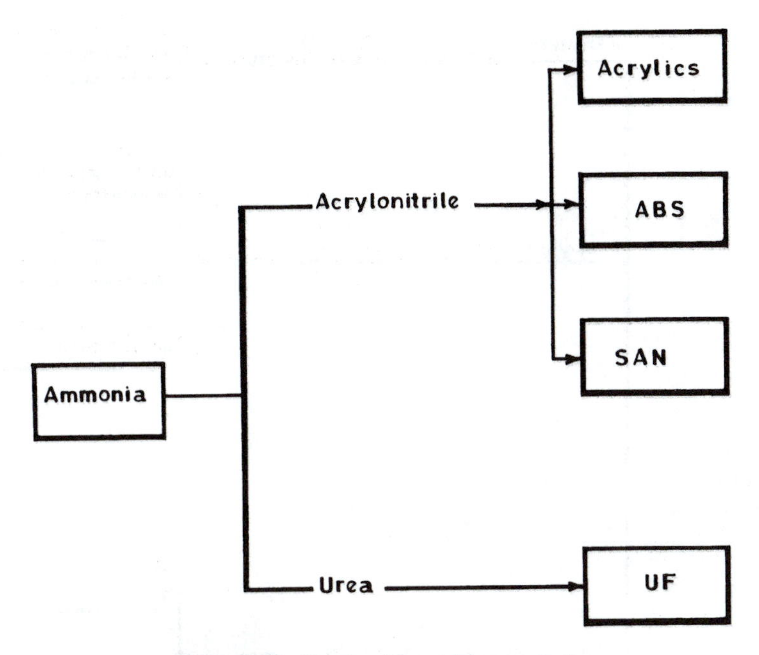

Figure 15.6 Some polymers from ammonia.

b. LDPE, on exposure to light and O_2, ages with a consequent loss of strength and a reduction in some other physical properties. It is therefore usual to protect LDPE from the effects of environmental degradation by the addition of carbon black and fillers/stabilizers. Cross-linking of the polyethylene insulating jacket either through chemical or radiation means can also prevent environmental aging. For example, incorporation of relatively stable peroxides provides a chemical means of cross-linking polyethylene. The peroxides, which are stable at normal processing temperatures, decompose and initiate cross-linking in post-processing reactions.

EVA and EEA copolymers as insulating materials show relatively easier acceptance of carbon black and other stabilizers than does conventional LDPE. They are also easier to cross-link presumably because of the greater preponderance of tertiary hydrogens (on the chain at branch points), which are regarded as the probable points of attack during cross-linking. In addition, the low melting points of the copolymers permit a greater ease of incorporation of peroxides by minimizing the risk of premature cross-linking during compounding.

B. POLYPROPYLENE (PP)

$$\left[CH_2 - \underset{\displaystyle CH_3}{\underset{|}{CH}} \right]_n \qquad \text{(Str. 2)}$$

Polypropylene is the third-largest volume polyolefin and one of the major plastics worldwide (Table 15.3). The commercial plastic was first introduced in 1957.

Polypropylene is made by polymerizing high-purity propylene gas recovered from cracked gas streams in olefin plants and oil refineries. The polymerization reaction is a low-pressure process that utilizes Ziegler–Natta catalysts (aluminum alkyls and titanium halides). The catalyst may be slurried in a hydrocarbon mixture to facilitate heat transfer. The reaction is carried out in batch or continuous reactors operating at temperatures between 50 and 80°C and pressure in the range of 5 to 25 atm.

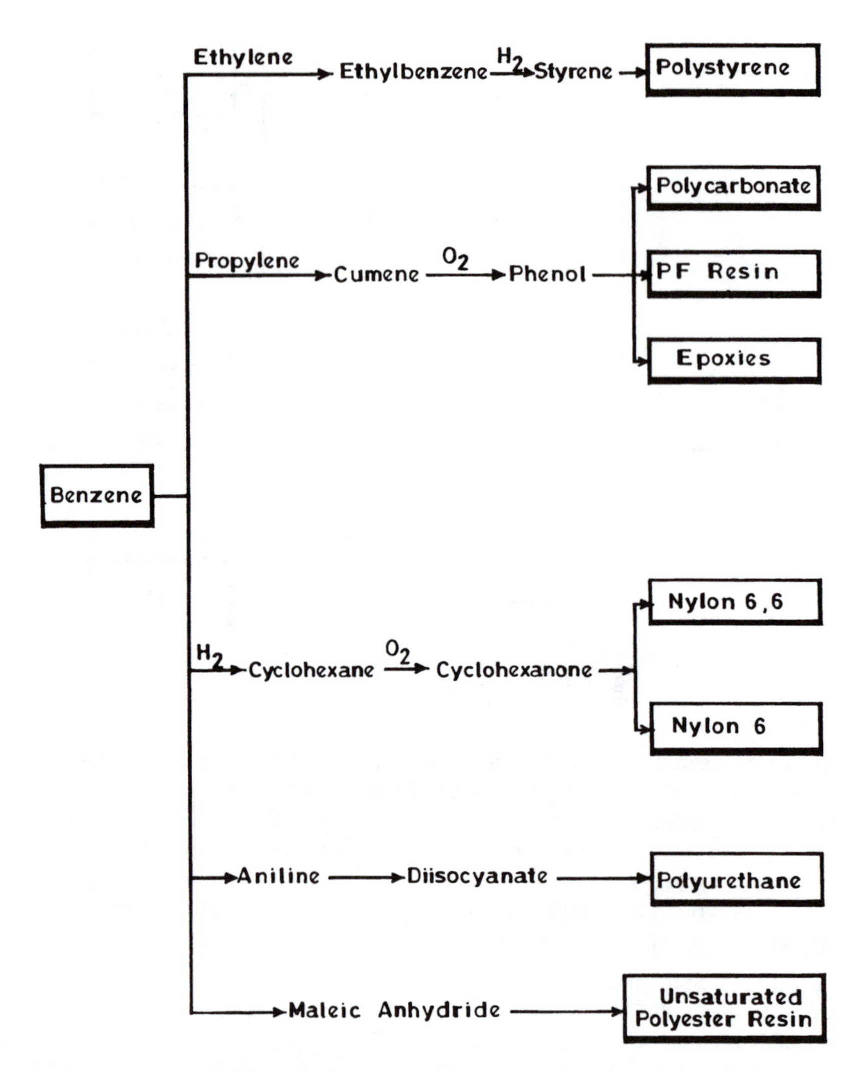

Figure 15.7 Some polymers from benzene.

Polypropylene can be made in isotactic (*i*-PP), syndiotactic (*s*-PP), and atactic (*a*-PP) forms. Ziegler–Natta-type catalysts are used to produce stereoregular polypropylene. Usually 90% or more of the polymer is in the isotactic form, which is the form with properties of commercial interest. Isotactic polypropylene is essentially linear, with an ordered arrangement of propylene molecules in the polymer chain. Unlike polyethylene, isotactic polyethylene does not crystallize in a planar zigzag conformation due to steric hindrance from the relatively bulky methyl groups. Instead *i*-PP crystallizes in a helical form with three monomer units per turn of the helix. Isotactic polyethylene is highly crystalline with a melting point of 165 to 171°C (Table 15.1). With a density in the range of 0.90 to 0.91 g/cc, polypropylene is one of the lightest of the widely used commercial thermoplastics.

Polypropylene has excellent electrical and insulating properties, chemical inertness, and moisture resistance typical of nonpolar hydrocarbon polymers. It is resistant to a variety of chemicals at relatively high temperatures and insoluble in practically all organic solvents at room temperature. Absorption of solvents by polypropylene increases with increasing temperature and decreasing polarity. The high crystallinity of polypropylene confers on the polymer high tensile strength, stiffness, and hardness. Polypropylene is practically free from environmental stress cracking. However, it is intrinsically less stable than polyethylene to thermal, light, and oxidative degradation. Consequently, for satisfactory

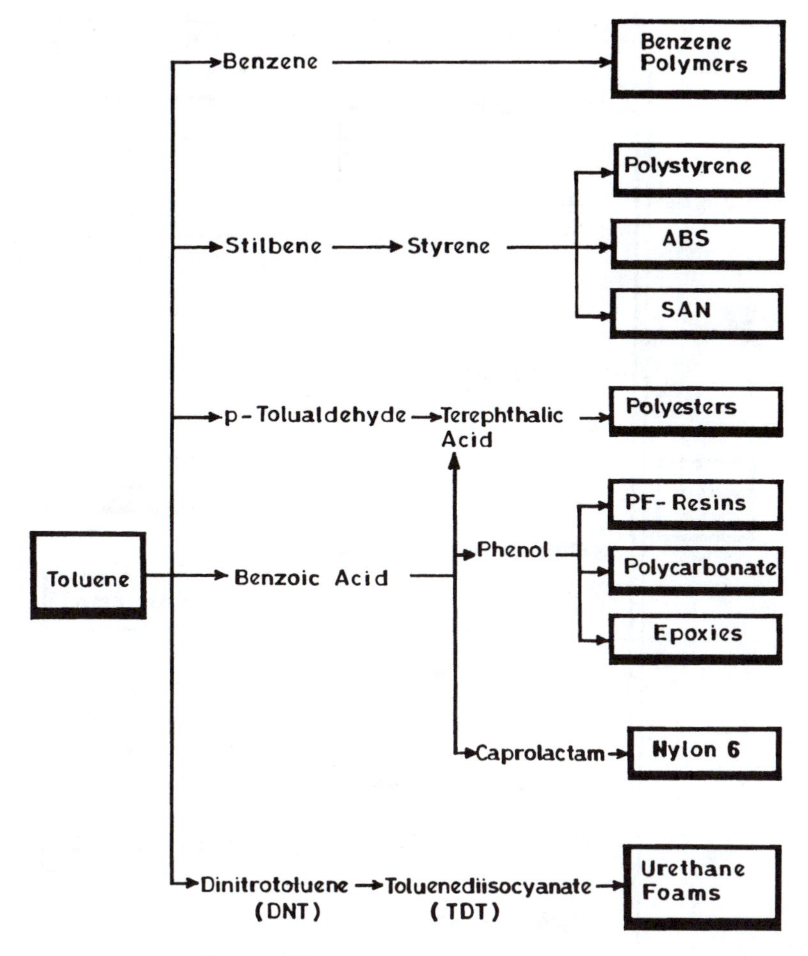

Figure 15.8 Some polymers from toluene.

processing and weathering, polypropylene must be stabilized by the incorporation of thermal stabilizers, UV absorbers, and antioxidants.

Polypropylene is used in applications ranging from injection-molded and blow-molded products and fibers and filaments to films and extrusion coatings. Injection molding uses, which account for about half of polypropylene produced, include applications in the automotive and appliance fields. Polypropylene can be designed with an integral hinge fabricated into products ranging from pillboxes to cabinet doors. Extruded polypropylene fibers are used in products such as yarn for carpets, woven and knitted fabrics, and upholstery fabrics. Nonwoven polypropylene fabrics are used in applications such as carpet backing, liners for disposable diapers, disposable hospital fabrics, reusable towels, and furniture dust covers. Polypropylene filaments are employed in rope and cordage applications. Nonwoven polypropylene soft film is suited for overwrap of such products as shirts and hose, while oriented polypropylene film is used as overwrap of such items as cigarettes, snacks, and phonograph records.

C. POLYSTYRENE

$$\left[CH_2 - CH(C_6H_5) \right]_n \qquad \text{(Str. 3)}$$

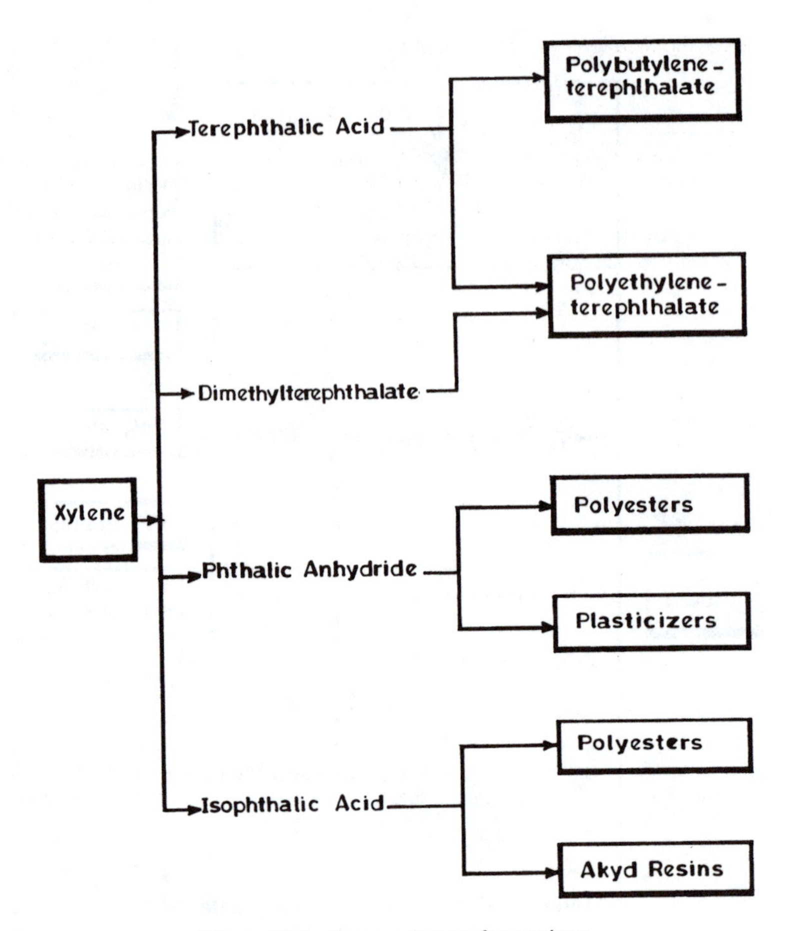

Figure 15.9 Some polymers from xylene.

Table 15.1 Properties of Commodity Polymer

Property	Polyethylene			Polystyrene			PVC	
	LDPE	HDPE	Polypropylene	GP-PS	HIPS	SAN	Rigid	Flexible
Specific gravity	0.912–0.94	0.941–0.965	0.902–905	1.04–1.065	1.03–1.06	1.07–1.08	1.30–158	1.16–135
Crystallinity (%)	50–70	80–95	40–68	—	—	—	—	—
Melting temperature (°C)	98–120	127–135	165–174	—	—	—	—	—
Tensile strength (MPa)	15.2–78.6	17.9–33.1	29.3–38.6	36.5–54.5	22.1–33.8	62.0–82.7	41.4–51.7	10.3–24.1
Tensile modulus (MPa)	55.1–172	413–1034	1032–1720	24.3–3378	1792–3240	2758–3861	2413–4136	—
Elongation at break (%)	150–600	20–130	500–900	1–2	13–50	1–4	40–80	200–450
Impact strength (Izod) (ft-lb/in. notch)	716	0.8–14	0.4–6.0	0.25–0.40	0.5–11	0.35–0.50	0.4–20	—
Heat deflection temperature (°C at 66 psi)	38–49	60–88	107–121	75–100	77–93	—	57–82	—

Table 15.2 Typical Applications of Polyethylene

Product Type	Typical Applications
Film extrusion	LDPE films used in packaging (bags and wrappings for frozen and perishable foods, produce and textile products); construction cover (moisture barriers and utility covering material); agriculture (greenhouse; ground cover; tank, pond, and canal liners); garment and stripping bags. HDPE film used for floral wrapping, grocery bags, snack and food packaging; EVA film used as produce bags, heavy-duty shipping bags, disposable protective gloves
Extrusion coating	Laminates of foil, paper and used in milk-type cartons (LDPE) for a variety of foods and drinks. Some HDPE used in extrusion coatings over flexible foil packages either as a glue layer between paper on plastics film and aluminum foil or as a heat-seal layer
Wire and cable insulation	LDPE used as insulation of high-frequency electrical cables, insulation materials for television, radar, and multicircuit long-distance telephones
Blow molding	Squeeze bottles, toys, housewares, lids, containers made from LDPE. Blow-molded HDPE used as containers for bleaches, liquid detergents, milk and other beverages, automobile and truck gas tanks, crates and pails
Injection molding	Packaging containers, pails and lids, houseware dishpans, and waste baskets, molded furniture seats, medical labware. HDPE is used in molded structural foam pallets, and crates, underground conduits and housings

Table 15.3 Typical Applications of Polypropylene

Product Type	Typical Applications
Injection molded	Automotive and appliance fields: distributor caps, radiator fans, accelerator pedals, battery casings, pillboxes, cabinet doors, bottle crates, jerry cans, cups, plates, file jackets, toys, food and drug containers
Extruded	Yarn for carpets, woven sacks and upholstery fabrics, hoses, drinking straws, hypodermic syringes, reusable towels, overwrap for cigarettes and phonograph records, liners for disposable diapers, furniture dust covers, geotextiles for road stabilization and erosion control

Polystyrene is one of the largest volume thermoplastics. It is a versatile polymer whose principal characteristics include transparency, ease of coloring and processing, and low cost. It is usually available in general-purpose or crystal (GP-PS), high impact (HIPS), and expanded grades. Some members of this family of styrene polymers are copolymers of styrene with other vinyl monomers.

Commercial atactic polystyrene is made by free-radical bulk or suspension polymerization of styrene with peroxide initiators. The reaction exotherm in bulk or mass polymerization of styrene is controlled by using a two-stage polymerization process. In the first stage, inhibitor-free styrene is subjected to low conversion in a stirred tank reactor (prepolymerizer). This is then followed by high conversion in a cylindrical tower (about 40 ft long by 15 ft diameter) with increasing temperature gradient (Chapter 9). The pure molten polymer that emerges from the reactor goes through spinnerets or an extruder to provide the desired finished product.

Polystyrene is a linear polymer that, in principle, can be produced in isotactic, syndiotactic, and atactic forms. The commercial product or general-purpose polystyrene is atactic and as such amorphous: isotactic polystyrene is more brittle and more difficult to process than atactic polystyrene. It is therefore not of commercial interest.

GP-PS is a clear, rigid polymer that is relatively chemically inert. Polystyrene, as produced, has outstanding flow characteristics and consequently is very easy to process. Its excellent optical properties, including high refractive index, make it useful in optical applications. However, GP-PS has a number of limitations, including its brittleness, low heat-deflection temperature (Table 15.1), poor UV resistance, and susceptibility to attack by a variety of solvents. Polystyrene is sensitive to foodstuffs with high fat or oil content; it crazes and turns yellow during outdoor exposure.

Many of the problems associated with GP-PS can be alleviated, or at least minimized, through copolymerization, blending, or proper formulation. For example, polystyrene with enhanced impact resistance and toughness is produced by the incorporation of butadiene rubber. High-impact polystyrene

Table 15.4 Some Applications of Different Grades of Polystyrene

Processing Method	Typical Products
Extrusion	Containers, tubs and trays for food packaging, mirror and picture frames, refrigerator breaker strips, room dividers, shower doors, glazings, lighting
Thermoforming	Refrigerator door liners, signs, horticultural trays, luggage; furniture panels, glove compartment boxes; meat, poultry, and egg containers; fast-food containers, blister packs, container lids; cookie, candy, pastry and other food packages
Injection molding	Packaging and nonpackaging disposables; air conditioner grilles; refrigerator and freezer components; small appliance housings
Others	Consumer electronics such as cassettes, reels, radio, television; and stereo dust covers; business machine housings, smoke detectors, display racks, toys, disposable tumblers, cutlery, bottles, combs, brush blocks

(HIPS) is produced commercially by the emulsion polymerization of styrene monomer containing dispersed particles of polybutadiene or styrene–butadiene (SBR) latex. The resulting product consists of a glassy polystyrene matrix in which small domains of polybutadiene are dispersed. The impact strength of HIPS depends on the size, concentration, and distribution of the polybutadiene particles. It is influenced by the stereochemistry of polybutadiene, with low vinyl contents and 36% *cis*-1,4-polybutadiene providing optimal properties. Copolymers of styrene and maleic anhydride exhibit improved heat distortion temperature, while its copolymer with acrylonitrile, SAN — typically 76% styrene, 24% acrylonitrile — shows enhanced strength and chemical resistance. The improvement in the properties of polystyrene in the form of acrylonitrile–butadiene–styrene terpolymer (ABS) is discussed in Section VII.A.

The expandable grade of styrene homopolymer is used to make foamed products that are beads generally foamed in place during application. Expandable polystyrene beads may be prepared by the suspension polymerization of styrene monomer in the presence of a volatile organic blowing or foaming agent. The foaming agent, such as pentane or hexane, is normally a liquid under polymerization conditions, but volatilizes during subsequent heating to soften the polymer thus forming a foamed product. Requirements for various types of products are satisfied by varying bead size, foaming agent level and composition, polymer molecular weight, and molecular weight distribution. The larger beads, which generally have the lowest density, find uses in thermal insulation, ceiling tiles, and loose-fill applications. On the other hand, the smaller beads, which provide better mechanical properties and surface finish, are employed in custom packaging, insulated drinking cups, and structural and semistructural applications.

The applications for all grades of polystyrene include packaging, housewares, toys and recreational products, electronics, appliances, furniture, and building and construction insulation (Table 15.4).

D. POLY(VINYL CHLORIDE) (PVC)

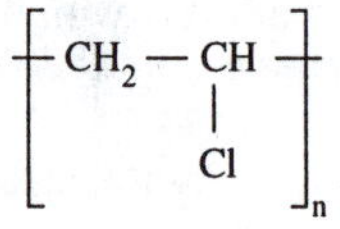

$$\left[CH_2 - \underset{\substack{|\\ Cl}}{CH} \right]_n \qquad \text{(Str. 4)}$$

Poly(vinyl chloride) is one of the largest volume thermoplastics in the world. It is chemically inert and versatile, ranging from soft to rigid products that are available at economic costs. PVC is available in essentially two grades — rigid and flexible.

Commercial grade PVC is produced primarily by free-radical-initiated suspension and emulsion polymerization of vinyl chloride. Suspension polymerization accounts for over 80% of PVC produced. Solution and bulk polymerization are also employed to some extent. However, there are difficulties with bulk polymerization because PVC is insoluble in its monomer and therefore precipitates. In suspension polymerization, vinyl chloride droplets are suspended in water by means of protective colloids such as poly(vinyl alcohol), gelatin, or methyl cellulose in pressure vessels equipped with agitators and heat

Table 15.5 Typical Applications of Poly(vinyl chloride)

Area	Typical Application
Piping systems	Pressure pipes — water supply and distribution, agricultural irrigation, chemical processing; nonpressure pipes — drain, waste and vent pipes, sewer systems, conduits for electrical and telephone cables
Building construction	Siding, window frames, gutters, interior molding and trim, flooring, wire and cable insulation, wall coverings, upholstery, shower curtains, refrigerator gaskets
Transportation	Upholstery, floor mats, auto tops, automotive wire, interior and exterior trim
Consumer products	Footwear, outerwear, phonograph records, sporting goods, toys

removal systems. Polymerization is conducted at temperatures of 40 to 70°C, typically around 60°C. Higher temperatures can result in minor branching and excessive formation of HCl through dehydrochlorination. Lower temperatures produce a high content of syndiotactic polymers. At a predetermined end point, unreacted vinyl chloride monomer is stripped from the slurry under vacuum. As a result of environmental and health concerns, stringent control of the escape of vinyl chloride monomer into the atmosphere is becoming an important issue.

Poly(vinyl chloride) is partially syndiotactic; it has a low degree of crystallinity due to the presence of structural irregularities. PVC is relatively unstable to heat and light. Unstabilized PVC undergoes dehydrochlorination when heated above its T_g (about 87°C) — for example, during melt processing. This leads to the production of hydrochloric acid, formation of intense color, and deterioration of polymer properties. Consequently, in practice, a number of ingredients must be added to PVC to enhance thermal stability and hence improve processing and product performance. Heat stabilizers are the most important additive. These are generally organometallic salts of tin, lead, barium–cadmium, calcium, and zinc. Other additives include lubricants, plasticizers, impact modifiers, fillers, and pigments.

The properties of PVC can be modified through chemical modification, copolymerization, and blending. PVC homopolymer contains about 57% chlorine. Chlorinated PVC with the chlorine content increased to 67% has a higher heat deflection temperature than the homopolymer. This extends the temperature range over which products can be used, allowing use in residential hot water pipes, for example. Over 90% of PVC produced is in the form of a homopolymer with the rest as copolymers and, to a small extent, terpolymers. Flexible film products are obtained by copolymerizing vinyl chloride with flexible chain monomers such as vinyl acetate and vinylidene chloride. Advantages of copolymers with small amounts of vinyl acetate over the homopolymer include lower softening point and higher solubility and, hence, improved processability, enhanced stability, and better color and clarity. The most commercially important copolymers contain about 13% vinyl acetate and are used for phonograph records and vinyl floor tiles. Copolymers with vinylidene chloride have better tensile properties than the homopolymer. They are used in coating applications because of their improved solubility. Copolymers of vinyl chloride and diethyl fumarate or diethyl maleate (10 to 20% content), while retaining the high softening temperature of poly(vinyl chloride) homopolymer, have enhanced workability and toughness. The toughness of PVC can also be improved by blending with high-impact resins like ABS.

As indicated earlier, PVC is available as rigid or flexible resins. Flexible PVC is obtained by incorporating internal or external plasticizers into PVC. Rigid PVC accounts for about 55% of PVC used while plasticized or flexible PVC accounts for the remainder. The largest single use of PVC is for piping systems. The major areas of use and typical applications of PVC are listed in Table 15.5.

Example 15.2: PVC has an advantage over other thermoplastic polyolefins in applications such as insulation for electrical circuitry in household electronic appliances. Explain.

Solution: Though polar, PVC-like nonpolar thermoplastics can be used as insulation for electrical wires in low-frequency applications. Prolonged use of household electronic appliances has a tendency to generate heat. Therefore, these appliances can be a potential source of fire hazard. Unlike other thermoplastic polyolefins, PVC has inherent (built-in) fire retardancy because of its 57% chlorine content. This reduces the susceptibility to fire outbreak arising from prolonged use of household electronic appliances.

V. OTHER VINYL POLYMERS

As discussed in Chapter 1, olefin polymers are represented by the generalized formula

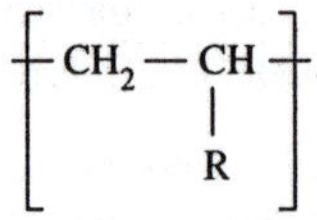

$$\left[CH_2 - \underset{\underset{R}{|}}{CH} \right] \qquad \text{(Str. 5)}$$

where R may be hydrogen alkyl or aryl, etc. In addition to poly(vinyl chloride) or other halogen-containing polymers, other vinyl polymers are essentially polyolefins in which the R substituents are bonded to the olefin monomers through an oxygen atom [poly(vinyl esters), poly(vinyl ethers)] or a nitrogen atom [poly(vinyl amides)]. In this discussion we focus attention on the commercially important poly(vinyl esters).

A. POLY(VINYL ACETATE) (PVAC)

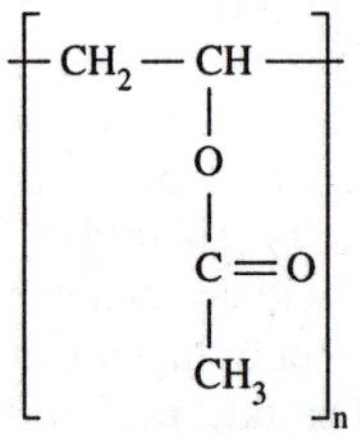

$$\left[CH_2 - \underset{\underset{\underset{\underset{\underset{CH_3}{|}}{C=O}}{|}}{O}}{|}}{CH} \right]_n \qquad \text{(Str. 6)}$$

Poly(vinyl acetate) is the most widely used vinyl ester polymer. It is also the precursor or starting material for the production of two other polymers that cannot be prepared by direct polymerization because the starting monomer is unstable. These are poly(vinyl alcohol) and poly(vinyl acetal). The most important of the latter are poly(vinyl butyral) and poly(vinyl formal).

As a result of its highly exothermic nature, bulk polymerization of vinyl acetate poses problems at high conversions. The properties of the resulting polymer are susceptible to deterioration due to chain branching. Therefore, bulk polymerization of vinyl acetate is usually stopped at 20 to 50% conversion. Thereafter, the unreacted monomer is either distilled off or the polymer precipitated with a suitable solvent (methanol, ethanol). Poly(vinyl acetate) is manufactured primarily by free-radical-initiated emulsion and, sometimes, solution polymerization.

Only atactic or amorphous poly(vinyl acetate) is currently commercially available. It has a glass transition temperature, T_g, of 29°C. Consequently, the polymer becomes sticky at temperatures slightly above ambient. The low-molecular polymers, which are normally brittle, become gumlike when masticated (used in chewing gums). Its adhesive strength is dictated by its water sensitivity.

Poly(vinyl acetate) latex is used in the production of water-based emulsion paints, adhesives, and textile and paper treatments. Emulsion paints are stable, dry quickly, and are relatively low cost. PVAC emulsion adhesives are used in labeling and packaging, and as the popular consumer white glue. Copolymers with dibutyl fumarate, vinyl stearate, 2-ethylhexyl acrylate, or ethyl acrylate are used to obtain compositions that are softer for emulsion use. As indicated above, a major use of poly(vinyl acetate) is in the production of poly(vinyl alcohol), which is itself the starting material for poly(vinyl butyral) and poly(vinyl formal).

B. POLY(VINYL ALCOHOL) (PVAL)

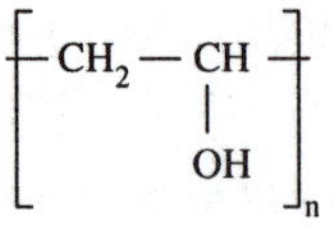

$$\left[CH_2 - \underset{\underset{OH}{|}}{CH} \right]_n \qquad \text{(Str. 7)}$$

Vinyl alcohol is unstable; it is isomeric with acetaldehyde. Therefore, poly(vinyl alcohol) is obtained indirectly by the alcoholysis of poly(vinyl acetate) in concentrated methanol or ethanol. The reaction is carried out in the presence of acid or base catalyst; base catalysis is usually faster:

$$\text{poly(vinyl acetate)}\ \ \sim\!CH_2{-}CH(OCOCH_3){-}CH_2{-}CH(OCOCH_3){-}CH_2{-}CH(OCOCH_3){-}CH_2\!\sim\ \xrightarrow[NaOCH_3,\ CH_3OH]{\text{alcoholysis}}\ \sim\!CH_2{-}CH(OH){-}CH_2{-}CH(OH){-}CH_2\!\sim\ \ \text{poly(vinyl alcohol)}\qquad \text{(Str. 8)}$$

$$\sim\!CH_2{-}CH(OH){-}CH_2{-}CH(OH){-}CH_2\!\sim\ \text{poly(vinyl alcohol)}\ \xrightarrow[EtOH,\ 80^{\circ}C\ H_2SO_4]{CH_3(CH_2)_2CHO,}\ \text{poly(vinyl butyral) (PVB)}$$

$$\sim\!CH_2{-}CH(OH){-}CH_2{-}CH(OH){-}CH_2\!\sim\ \text{poly(vinyl alcohol)}\ \xrightarrow[H_2SO_4,\ Na_2SO_4]{HCHO}\ \text{poly(vinyl formal) (PVF)}\qquad \text{(Str. 9)}$$

PVB repeat unit: $[-CH_2-CH-CH_2-CH-]_n$, where the two CH carbons are joined through $O-CH((CH_2)_2-CH_3)-O$ to form a six-membered ring.

PVF repeat unit: $[-CH_2-CH-CH_2-CH-]_n$, where the two CH carbons are joined through $O-CH_2-O$ to form a six-membered ring.

Poly(vinyl alcohol) has an atactic chain structure, and it is consequently amorphous. However, it can be stretched into a crystalline fiber. The small size of the OH groups permits them to fit into a crystal lattice. Poly(vinyl alcohol) is available in various grades defined by the molecular weight and by the degree of hydrolysis, which determines polymer water solubility.

The end uses of PVAL include textile and paper treatment and wet-strength adhesives. It is also used as a polymerization aid such as a thickening and stabilizing agent in emulsion polymerization in cosmetics and as packaging film requiring water solubility. With their much higher water absorption capacity and cottonlike feel, formaldehyde-modified poly(vinyl alcohol) fibers, vinal or vinylon fibers, can replace cotton in applications requiring body contact. These PVAL fibers have good dimensional stability and abrasion resistance, wash easily, and dry quickly.

Poly(vinyl alcohol) is also used in the manufacture of poly(vinyl butyral) (PVB) and poly (vinyl formal) (PVF). By far the largest single application of PVB is as an adhesive or plastic interlayer in the manufacture of laminated safety glass for automotive and aircraft uses. Compared with earlier cellulose acetate-based laminates, safety glass made from PVB has superior adhesion to glass; it is tough, stable on exposure to sunlight, clear, and insensitive to moisture. Poly(vinyl formal) is utilized in the manufacture of enamels for heat-resistant electrical wire insulation and in self-sealing gasoline tanks.

Example 15.3: Poly(vinyl butyral) is made by adding butyraldehyde and an acid catalyst, usually sulfuric acid, to an aqueous solution of poly(vinyl alcohol). For the poly(vinyl butyral) required for safety glass manufacture, the reaction is stopped at about 75% conversion of the hydroxyl groups. Explain.

Solution: For adequate performance of safety glass in end-use situations, a good bond between the components of the laminate is imperative. The residual or unreacted hydroxyl groups provide the required strength and adhesion to glass.

VI. ACRYLICS

Acrylics comprise a broad array of plastics derived from olefin monomers of the type $CH_2{=}CHR$ in which the R substituent is a cyamide, carboxylic acid, carboxylic acid ester, or carboxylic acid amide group. Two ester families, acrylates and methacrylates, constitute the major component of acrylic polymers. Used singly or as copolymers with monomers containing reactive functional groups, these monomers provide an array of products ranging from soft, flexible elastomers to hard, stiff plastics and thermosets, and from highly polar to oleophic resins. Acrylics generally exhibit crystal clarity and excellent chemical and environmental resistance.

A. POLY(METHYL METHACRYLATE) (PMMA)

$$\left[CH_2 - \underset{\underset{\underset{\underset{CH_3}{|}}{O}}{\overset{|}{C=O}}}{\overset{\overset{CH_3}{|}}{C}} - \right]_n \qquad \text{(Str. 10)}$$

The most important member of the acrylic polymers is poly(methyl methacrylate). It is a hard, clear, colorless, transparent plastic that is usually available as molding and extrusion pellets, reactive syrups, cast sheets, rods, and tubes.

Poly(methyl methacrylate) for molding or extrusion is produced commercially by free-radical-initiated suspension or bulk polymerization of methyl methacrylate. To minimize polymerization reaction exotherm and shrinkage, bulk polymerization, which is used in the production of sheets, rods and tubes, is carried out with a reactive syrup of partially polymerized methyl methacrylate, which has a viscosity convenient for handling.

Poly(methyl methacrylate) is an amorphous polymer composed of linear chains. The bulky nature of the pendant group (–O–CO–Me), and the absence of complete stereoregularity makes PMMA an amorphous polymer. Isotactic and syndiotactic PMMA may be produced by anionic polymerization of methyl methacrylate at low temperatures. However, these forms of PMMA are not available commercially. Modified PMMA can be obtained by copolymerizing methyl methacrylate with monomers such as acrylates, acrylonitrile, and butadiene.

Poly(methyl methacrylate) is characterized by crystal-clear light transparency, unexcelled weatherability, and good chemical resistance and electrical and thermal properties. It has a useful combination of stiffness, density, and moderate toughness. PMMA has a moderate T_g of 105°C, a heat deflection temperature in the range of 74 to 100°C, and a service temperature of about 93°C. However, on pyrolysis, it is almost completely depolymerized to its monomer. The outstanding optical properties of PMMA combined with its excellent environmental resistance recommend it for applications requiring light transmission and outdoor exposure. Poly(methyl methacrylate) is used for specialized applications such as hard contact lenses. The hydroxyethyl ester of methacrylic acid is the monomer of choice for the manufacture of soft contact lenses. Typical applications of poly(methyl methacrylate are shown in Table 15.6.

Table 15.6 Typical Applications of Poly(methyl methacrylate)

Area	Typical Applications
Construction	Enclosures for swimming pools, shopping malls and restaurants, tinted sunscreens to reduce air-conditioning and glare, domed skylights
Lighting	Lighted signs, luminous ceilings, diffusers, lenses and shields
Automotive	Lenses, instrument panels, signals and nameplates
Aviation	Windows, instrument panels, lighting fixture covers, radar plotting boards, canopies
Household	Housings, room dividers, decorating of appliances, furniture, vanities, tubs, counters
Others	Display cabinets and transparent demonstration models in museums, exhibits, and department stores

B. POLYACRYLATES

$$\left[CH_2 - \underset{\displaystyle COOR}{\underset{|}{CH}} \right]_n \qquad \text{(Str. 11)}$$

Polyacrylates are produced commercially by free-radical-initiated solution and emulsion polymerization of the appropriate monomer. Unlike for methacrylates, suspension and casting procedures are not feasible because of the rubber and adhesive nature of higher acrylates.

As shown in Table 15.7, the glass transition temperatures of acrylate polymers are generally below room temperature. This means that these polymers are usually soft and rubbery.

Solubility in oils and hydrocarbons increases with increasing length of the side group, while polymers become harder, tougher, and more rigid as the size of the ester group decreases. Polyacrylates have been used in finishes and textile sizing and in the production of pressure-sensitive adhesives. Poly(methyl acrylate) is used in fiber modification, poly(ethyl acrylate) in fiber modification and in coatings, and poly(butyl acrylate) and poly(2-ethylhexyl acrylate) are used in paints and adhesive formulation.

Table 15.7 Glass Transition Temperatures of Sample Polyacrylates

R	T_g (°C)
Methyl	3
Ethyl	–20
n-Propyl	–44
n-Butyl	–56

Quite frequently, copolymerization is used to optimize the properties of polyacrylates. For example, copolymers of ethyl acrylate with methyl acrylate provide the required hardness and strength, while small amounts of comonomers with hydroxyl, carboxyl, amine, and amide functionalities are used to produce high-quality latex paints for wood, wallboard, and masonry in homes. These functionalities provide the adhesion and thermosetting capabilities required in these applications. Monomers with the desired functional groups most often used in copolymerization with acrylates are shown in Table 15.8.

Table 15.8 Functional Comonomers Used with Acrylates

Monomer	Functional Group	Structure
Acrylic acid	$-COOH$	$CH_2{=}CH-COOH$
Methacrylic acid		$CH_2{=}\underset{\displaystyle CH_3}{\underset{|}{C}}-COOH$
Itaconic acid		$CH_2{=}\underset{\displaystyle CH_2COOH}{\underset{|}{C}}-COOH$
Dimethylaminoethyl methacrylate	$-NH_2$	$CH_2{=}\underset{\displaystyle CH_3}{\underset{|}{C}}-COO-CH_2-CH_2\overset{|}{N}-$
2-Hydroxyethyl acrylate	$-OH$	$CH_2{=}CH-COO-CH_2-CH_2-OH$
N-Hydroxyethyl acrylamide		$CH_2{=}CHCONH-CH_2-CH_2-OH$
Glycidyl methacrylate	CH_2-CH_2 (bridged by O, epoxide)	$CH_2{=}\underset{\displaystyle CH_3}{\underset{|}{C}}-COO-CH_2-CH_2-CH_2$ (last two CH_2 bridged by O, epoxide)

From Ulrich, H., *Introduction to Industrial Polymers,* Oxford Press, Oxford, 1982.

C. POLYACRYLONITRILE (PAN) — ACRYLIC FIBERS

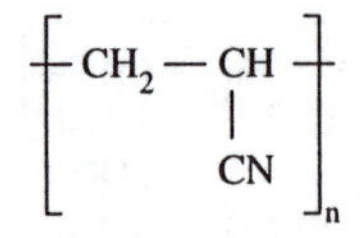

$$\left[CH_2 - \underset{\displaystyle CN}{\underset{|}{CH}} \right]_n \qquad \text{(Str. 12)}$$

Polyacrylonitrile, like PVC, is insoluble in its own monomer. Consequently, the polymer precipitates from the system during bulk polymerization. Acrylonitrile can be polymerized in solution in water or dimethyl formamide (DMF) with ammonium persulfate as the initiator (redox initiation). The polyacrylonitrile homopolymer can be dry spun from DMF directly from the polymerization reactor or wet spun from DMF into water.

Acrylic fibers are polymers with greater than 85% acrylonitrile content, while those containing 35 to 85% acrylonitrile are known as modacrylic. Acrylic fibers contain minor amounts of other comonomers, usually methyl acrylate, but also methyl methacrylate and vinyl acetate. These comonomers along with ionic monomers such as sodium styrene sulfonate are incorporated to enhance dyeability with conventional textile dyes. Modacrylics usually contain 20% or more vinyl chloride (or vinylidene chloride) to improve fire retardancy.

Polyacrylonitrile softens only slightly below its decomposition temperature. It cannot, therefore, be used alone for thermoplastic applications. In addition, it undergoes cyclization at processing temperatures.

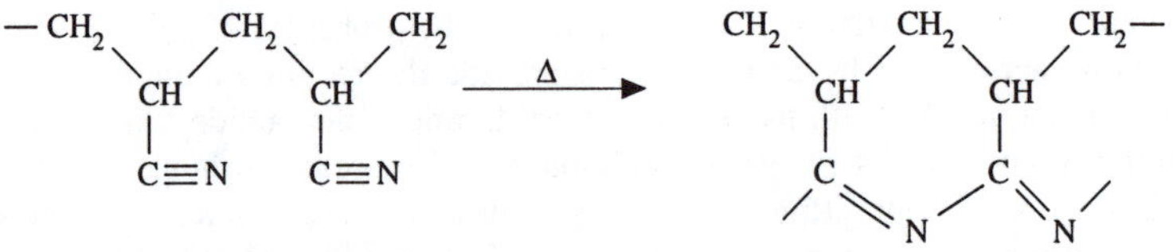

$$-CH_2-\underset{\displaystyle C\equiv N}{\underset{|}{CH}}-CH_2-\underset{\displaystyle C\equiv N}{\underset{|}{CH}}-CH_2 \xrightarrow{\Delta} \text{cyclized structure: } CH_2-CH(-C=N-)-CH_2-CH(-C=N-)-CH_2- \qquad \text{(Str. 13)}$$

The resulting polymer, on further heat treatment at elevated temperatures, is a source of graphite filaments. However, acrylonitrile copolymerized with other monomers finds extensive use in thermoplastic and elastomeric applications. Examples of such copolymers include styrene–acrylonitrile (SAN), acrylonitrile–butadiene–styrene terpolymer (ABS) and nitrile–butadiene rubber (NBR).

The presence of the highly polar nitrile group ($-C\equiv N$) in acrylic fibers results in strong intermolecular hydrogen bonding. This generates stiff, rodlike structures with high fiber strength from which acrylic fibers derive their properties. Acrylic fibers are used primarily in apparel and home furnishings. They are more durable than cotton and are suitable alternatives for wool. Typical applications of acrylic fibers include craft yarns, simulated fur, shirts, blouses, blankets, draperies, and carpets and rugs.

VII. ENGINEERING POLYMERS

Engineering plastics are high-performance polymers used in engineering applications because of their outstanding balance of properties. They generally attract a premium price due to their relatively low production volume and are replacing traditional materials in many engineering applications. For example, engineering plastics may be used as replacements for metals in automotive and home appliance applications where a high strength-to-weight ratio is an important requirement. As engineering polymers continue to replace traditional materials in many applications, they are being developed with more specialized properties so as to be distinctly superior to the displaced material in all significant respects.

Engineering polymers are strong, stiff, tough materials with high thermal stability, excellent chemical resistance, and good weatherability. They have relatively high tensile, flexural, and impact strengths and are capable of withstanding a wide range of temperatures. Engineering plastics derive their outstanding properties mainly from their inherently strong intermolecular forces. The superior properties of engineering polymers can be enhanced by the addition of various types of reinforcements, by blending and alloy formation, and through chemical modification such as cross-linking. Typical properties of some engineering polymers are shown in Table 15.9.

Table 15.9 Typical Properties of Some Engineering Polymers

Property	ABS[a]	Acetal	Nylon 6	Nylon 6,6	PC	PPO	PSF	PPS	Polyamide
Specific gravity	1.01–1.04	1.42	1.12–1.14	1.13–1.15	1.2	1.06	1.24	1.3	1.36–1.43
Tensile strength (MPa)	33.1–43.4	65.5–82.7	68.9	75.8	65.5	66.2	70.3	65.5	72.4–117.9
Tensile modulus (MPa)	1586–2275	3585	689	—	2378	2447	2482	3309	2068
Flexural strength (MPa)	55.2–75.8	96.5	34.5	42.1	93.1	93.1	106.2	96.5	131–199
Flexural modulus (MPa)	1724–2413	2620–2964	965	1275	2344	2482–2758	2689	3792	3102–3447
Impact strength, Izod (ft-lb/in notch)	3.0–12	1.3–2.3	3.9	2.1	16	1.8–5.0	1.2	1.8–5.0	1.5
Elongation at break	5–70	25–75	300	300	110	20–60	50–100	20–60	8–10
Heat deflection temperature (°C at 455 kPa)	102–107	124	150–185	180–240	138	137	181	135	—

[a] High impact grade

A. ACRYLONITRILE–BUTADIENE–STYRENE (ABS)

$$\left[CH_2-\underset{\displaystyle CN}{\underset{|}{CH}}-CH_2-CH{=}CH-CH_2-CH_2-\underset{\displaystyle C_6H_5}{\underset{|}{CH}} \right]_n \qquad \text{(Str. 14)}$$

Acrylonitrile–butadiene–styrene resins are terpolymers composed, as the name suggests, of acrylonitrile, butadiene, and styrene. Each component contributes special characteristics to the ultimate properties of products derived from the resins. Acrylonitrile provides heat and chemical resistance and high strength. Butadiene acts as the reinforcing agent providing impact strength and toughness even at low temperatures, while styrene contributes rigidity, easy processability, and gloss. By varying the ratio of the three components, the designer is provided with ABS resins with a wide range of properties to develop a variety of products with a well-balanced combination of properties. For example, general-purpose ABS includes both medium-impact-strength and high-impact-strength grades as well as the low-temperature, high-impact-strength variety. In addition, there are flame-retardant, structural foam, heat-resistant, low-gloss, and transparent grades.

ABS resins are produced primarily by grafting styrene and acrylonitrile onto polybutadiene latex in a batch or continuous polymerization process. They may also be made by blending emulsion latexes of styrene–acrylonitrile (SAN) and nitrile rubber (NBR).

ABS resins consist of two phases: a continuous glassy matrix of styrene–acrylonitrile copolymer and a dispersed phase of butadiene rubber or styrene–butadiene copolymer. The styrene–acrylonitrile matrix is ordinarily brittle; however, the reinforcing influence of the rubbery phase results in a product with greatly improved (high) load-bearing capacity. To optimize properties, it is usually necessary to graft the glassy and rubbery phases. A variety of ABS resins are produced by varying the ratio of components and the degree of bonding between the rubbery and glassy phases (graft level).

ABS resins have relatively good electrical insulating properties. They are resistant to weak acids and weak and strong bases. However, they have poor resistance to esters, ketones, aldehydes, and some chlorinated hydrocarbons. ABS resins are easily decorated by painting, vacuum metallizing, and electroplating. They are readily processed by all techniques commonly employed with thermoplastics and, like metals, can be cold-formed. ABS is hygroscopic and therefore requires drying prior to processing.

ABS resins are true engineering plastics particularly suited for high-abuse applications. Injection-molded ABS is used for housewares, small tools, telephones, and pipe fittings, which are applications requiring prolonged use under severe conditions. Extruded ABS sheet is used in one-piece camper tops and canoes. Applications of ABS in automobile and truck machinery include headliners, kick panel, wheel wells, fender extensions, wind deflectors, and engine covers. Profile-extruded ABS resins are used in pipes, sewer, well casing, and conduits. Specialty products from ABS include electroplating grades (automotive grilles and exterior decorative trim); high-temperature-resistant grades (automotive instrument panels, power tool housings); and structural foam grades, which are used in molded parts where high strength-to-weight ratio is required.

B. POLYACETAL (POLYOXYMETHYLENE — POM)

$$\left[CH_2 - O \right]_n \qquad \text{(Str. 15)}$$

Polyoxymethylene (polyacetal) — sometimes known as polyformaldehyde — is the polymer of formaldehyde. It is obtained either by anionic or cationic solution polymerization of formaldehyde or cationic ring-opening bulk polymerization of trioxane. Highly purified formaldehyde is polymerized in the presence of an inert solvent such as hexane at atmospheric pressure and a temperature usually in the range of –50 to 70°C. The cationic bulk polymerization of trioxane is the preferred method of production of polyoxymethylene.

Polyoxymethylene is susceptible to depolymerization, or unzipping, under molding conditions. To improve thermal stability, end capping is essential. The capping of the hydroxyl end groups is achieved by etherification or, preferably, by esterification using acetic anhydride:

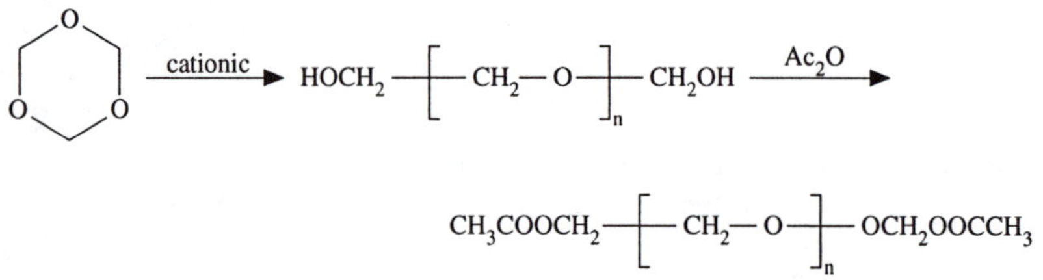

$$\text{trioxane} \xrightarrow{\text{cationic}} HOCH_2 \left[CH_2 - O \right]_n CH_2OH \xrightarrow{Ac_2O}$$

$$CH_3COOCH_2 \left[CH_2 - O \right]_n OCH_2OOCCH_3 \qquad \text{(Str. 16)}$$

Polyacetal can also be stabilized against degradative conditions by copolymerizing trioxane with small amounts of ethylene oxide. This introduces a random distribution of –C–C– bonds in the polymer chain. Hydrolysis of the copolymer with aqueous alkali gives a product with stable hydroxyethyl end groups. The presence of these stable end groups coupled with the randomly distributed C–C bonds prevents polymer depolymerization at high temperature.

Polyacetal is a linear, high-molecular-weight polymer with a highly ordered chain structure that permits an ordered arrangement of chain molecules in a crystalline structure. It is about 80% crystalline, with a melting point of 180°C. Polyacetal has excellent chemical resistance, has good dimensional stability due to negligible water absorption, and is insoluble in common solvents at room temperature. The stiffness, strength, toughness, and creep and fatigue resistance of polyacetals are higher than those of other unreinforced crystalline thermoplastics. It has good frictional and electrical properties. Products from polyacetal retain most of these engineering properties over a wide range of useful temperatures and other end-use conditions. Polyacetals can be processed by the usual molding and extrusion methods.

The relatively advantageous properties of polyacetals have led to applications in a variety of markets, particularly as a replacement for metals, where it provides enhanced properties and lower costs. Table 15.10 shows areas and typical applications of polyacetals.

C. POLYAMIDES (NYLONS)

The word *nylon* is a generic term used to describe a family of synthetic polyamides. Nylons are characterized by the amide group (–CONH–), which forms part of the polymer main chain (interunit

Table 15.10 Typical Applications of Polyacetals

Area	Typical Applications
Automotive	Filler necks for gasoline tanks, instrument panels, seat belts, steering columns, window support brackets, door handles, bearing and gear components, dashboard components, controls, wheel covers, gas caps
Plumbing	Shower heads, shower mixing valves, faucet cartridges, ball cocks
Consumer	Handles and other hardware items, the bodies of lighters, replaceable cartridges in showers, telephone components, lawn sprinklers, garden sprayers, stereo cassette cases, spools for video cassettes, zippers
Machinery	Machinery couplings, small engine starters, pump impellers and housings, fire extinguisher handles, gears

Table 15.11 Nomenclature of Nylons

Monomer(s)		Polymer
$H_2N-(CH_2)_6-NH_2$ Hexamethylenediamine	$HOOC-(CH_2)_4-COOH$ Adipic acid	$\left[-\overset{H}{\overset{\vert}{N}}-(CH_2)_6-\overset{H}{\overset{\vert}{N}}-\overset{O}{\overset{\Vert}{C}}-(CH_2)_4-\overset{O}{\overset{\Vert}{C}}-\right]_n$ Poly(hexamethylene adipamide), nylon 6,6
$H_2N-(CH_2)_6-NH_2$ Hexamethylenediamine	$HOOC-(CH_2)_6-COOH$ Sebacic acid	$\left[-\overset{H}{\overset{\vert}{N}}-(CH_2)_6-\overset{H}{\overset{\vert}{N}}-\overset{O}{\overset{\Vert}{C}}-(CH_2)_6-\overset{O}{\overset{\Vert}{C}}-\right]_n$ Poly(hexamethylene sebacamide), nylon 6,10
$H_2N-(CH_2)_5-COOH$ ω-amino caproic acid		$\left[-\overset{H}{\overset{\vert}{N}}-(CH_2)_5-\overset{O}{\overset{\Vert}{C}}-\right]_n$ Polycaprolactam, nylon 6

linkage). In terms of chemical structure, nylons may be divided into two basic types: those based on diamines and dibasic acids (A–A/B–B type); and those based on amino acids or lactams (A–B type). Nylons are described by a numbering system that reflects the number of carbon atoms in the structural units. A–B type nylons are designated by a single number. For example, nylon 6 represents polycaprolactam [poly(ω-amino caproic acid)]. A–A/B–B nylons are designated by two numbers, with the first representing the number of carbon atoms in the diamine and the second referring to the total number of carbon atoms in the acid (Table 15.11).

Among the nylons, nylon 6,6 and nylon 6 are of the greatest commercial importance and most widely used. Other commercially useful materials are the higher analogs such as nylon 6,9; 6,10; 6,12; 11; and 12. Nylon 6,6 and nylon 6 are widely used because they offer a good balance of properties at an economic price. Other nylons command relatively higher prices.

Nylon 6,6 is formed by the step-growth polymerization of hexamethylenediamine and adipic acid. The exact stoichiometric equivalence of functional groups needed to obtain a high-molecular-weight polymer is achieved by the tendency of hexamethylenediamine and adipic acid to form a 1:1 salt. This intermediate hexamethylene diammonium adipate is dissolved in water and then charged into an autoclave. Monofunctional acids such as aluric or acetic acid (0.5 to 1 mol%) may be added to the polymerization mixture for molecular-weight control. As the temperature is raised, the steam generated is purged by air. The temperature is raised initially to 220°C, and subsequently to 270 to 280°C when monomer conversion is about 80 to 90%, while maintaining the pressure of the steam generated at 250 psi. Pressure is then reduced to atmospheric, and heating is continued until polymerization is totally completed.

A–B type nylons are usually prepared by ring-opening polymerization of a cyclic lactam. Water is added in a catalytic amount to effect the ring-opening and then removed at higher temperature to encourage high polymer formation. High-molecular-weight nylon 6 is obtained from the anionic polymerization of ε-caprolactam with a strong base such as sodium hydride. Commercially, nylon 6 is produced by the hydrolytic polymerization of ε-caprolactam.

As a class, aliphatic polyamides exhibit excellent resistance to wear and abrasion, low coefficient of friction, good resilience, and high impact strength. Nylons are generally characterized by a good balance of high strength, elasticity, toughness, and abrasion resistance. They maintain good mechanical properties

Table 15.12 Melting Points and Moisture Absorption of Some Nylons

Nylon	Tm °C	Water Absorption (ASTM) D-570)
6,6	265	1.0–1.3
6,8	240	—
6,9	226	0.5
6,10	225	—
6,12	212	0.4
4	265	—
6	226	1.3–1.9
7	223	—
11	188	—
12	180	0.25–0.30

at elevated temperatures — sometimes as high as 150°C — while also retaining low-temperature flexibility and toughness.

Nylons are sensitive to water due to the hydrogen-bond-forming ability of the amide groups. Water essentially replaces amide–amide–hydrogen bond with amide–water–hydrogen bond. Consequently, water absorption decreases with decreasing concentration of amide groups in the polymer backbone (Table 15.12). Water acts as a plasticizer, which increases toughness and flexibility while reducing tensile strength and modulus. The absorption of moisture results in a deterioration of electrical properties and poor dimensional stability in environments of changing relative humidity. Therefore, care must be taken to reduce the water content of nylon resins to acceptable levels before melt processing to avoid surface imperfections and embrittlement due to hydrolytic degradation.

The markets for nylon products have been broadened considerably because of the apparent ease with which polyamides can be modified to produce improved properties for special applications. Modified nylons of various grades are produced by copolymerization and the incorporation of various additives (usually in small amounts) such as heat stabilizers; nucleating agents; mold-release agents; plasticizers; mineral reinforcements (glass fiber/beads, particulate minerals); and impact modifiers. For example, in applications involving long exposure to temperatures above 75 to 85°C such as automotive under-hood parts, polyamides have limited use due to their susceptibility to surface oxidation in air at elevated temperatures and the attendant loss of mechanical properties. Addition of less than 1% copper salt heat stabilizer permits use of nylons at elevated temperatures. The mechanical properties of nylons depend largely on their crystallization. Consequently, control of these properties can be achieved partly by controlling the degree of crystallinity and spherulite size by the use of nucleating agents.

Nylons are used in applications requiring durability, toughness, chemical inertness, electrical insulating properties, abrasion and low frictional resistance, and self-lubricating properties. Table 15.13 lists markets and typical applications of nylons. In many of these applications, nylons are used as small parts or elements in subassemblies of the finished commercial article.

The term *aramid* is used to describe aromatic polyamides, which were developed to improve the heat and flammability resistance of nylon. Nomex is a highly heat-resistance nylon introduced in 1961 by DuPont. It is produced by the solution or interfacial polymerization of isophthaloyl chloride and *m*-phenylenediamine:

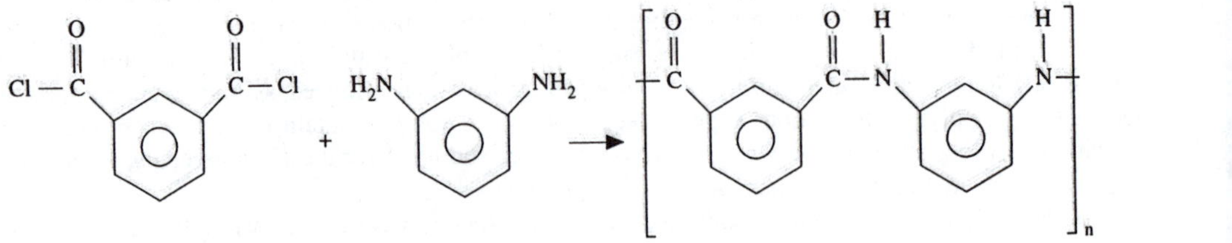

$$Cl-\overset{O}{\overset{\|}{C}}-C_6H_4-\overset{O}{\overset{\|}{C}}-Cl + H_2N-C_6H_4-NH_2 \longrightarrow \left[-\overset{O}{\overset{\|}{C}}-C_6H_4-\overset{O}{\overset{\|}{C}}-\overset{H}{\overset{|}{N}}-C_6H_4-\overset{H}{\overset{|}{N}}-\right]_n$$

poly(*m*-phenyleneisophthalamide) (Nomex)

(Str. 17)

Table 15.13 Markets and Typical Applications of Nylons

Market	Typical Applications
Automotive/truck	Speedometer gears, door lock wedges, distributor point blocks; automotive electrical system such as connectors, fuse blocks, generator parts, spark wire separators, wire insulation; monofilament thread for upholstery, license plate bolts and nuts, fuel vapor canisters, fender extensions, mirrors and grilles; pneumatic tubing and lubrication lines, fuel and fuel-vent lines
Electrical/electronics	Connector, tie straps, wire coil, bobbins, tuner gears
Industrial/machinery	Lawn mower carburetor components, window and furniture guides, pump parts, power tools, fans, housings, gears, pulleys, bearings, bushings, cams, sprockets, conveyor rollers, screws, nuts, bolts, washers
Monofilament/film coatings	Fishing lines, fish nets, brush bristles, food and medical packaging, coatings for wire and cable
Appliances	Refrigerators, dishwashers, ranges, hair dryers and curlers, corn poppers, smoke detectors
Consumer items	Combs, brushes, housewares, buttons, rollers, slides, racquetball racquets

Nomex has a melting point of about 365°C and is virtually nonflammable. It is used in many applications such as protective clothing and hot gas filtration equipment as a substitute for asbestos.

Another aramid, Kevlar, is the corresponding linear aromatic polyamide obtained from terephthaloyl chloride and phenylenediamine.

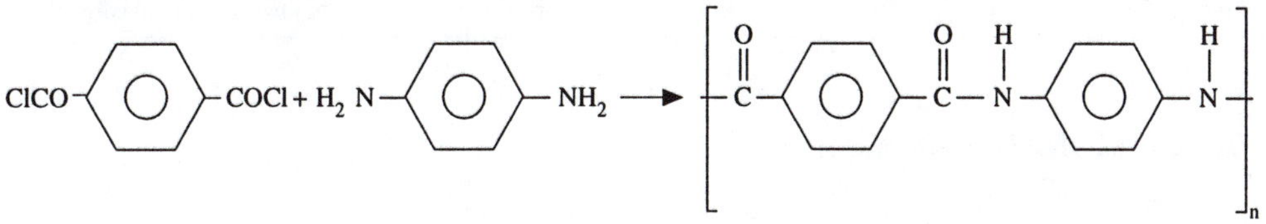

$$\text{ClCO}-\langle\bigcirc\rangle-\text{COCl} + \text{H}_2\text{N}-\langle\bigcirc\rangle-\text{NH}_2 \longrightarrow \left[\overset{\text{O}}{\overset{\|}{\text{C}}}-\langle\bigcirc\rangle-\overset{\text{O}}{\overset{\|}{\text{C}}}-\overset{\text{H}}{\overset{|}{\text{N}}}-\langle\bigcirc\rangle-\overset{\text{H}}{\overset{|}{\text{N}}}\right]_n$$

poly(*p*-phenylene terephthalamide) (Kevlar) (Str. 18)

Kevlar, which decomposes only above 500°C, provides a fiber material which is as strong as steel at one-fifth its weight. It is a good substitute for steel in belted radial tires and is used in the manufacture of mooring lines as well as bulletproof vests and other protective clothing. Fiber-reinforced plastic composites are also produced from Kevlar fiber. Typical applications of these composites include fishing rods, golf club shafts, tennis rackets, skis, and ship masts. Significant quantities of Kevlar composites are used in Boeing 757 and 767 planes.

Example 15.4: Explain the following observations.

a. Major uses of polyacetals are as direct replacements for metals such as brass, cast iron, aluminum, and zinc in many applications. For example, in the plumbing industry, shower heads, shower mixing valves, and faucet cartridges molded of acetal homopolymer are replacing brass and zinc parts.
b. Nylons have good resistance to solvents. However, good solvents for nylon 6,6 and nylon 6 are strong acids (such as H_2SO_4, HBr, trichloracetic acid) formic acid, phenols, cresols and perfluoro compounds.

Solution:

a. The use of polyacetals has a number of property advantages over the use of metals. These include a good balance of stiffness; light weight; resistance to corrosion, wear, and abrasion; and dimensional stability. In addition, cost savings result from elimination of metal assemblies, the reduced number of parts, and the use of low-cost plastic assembly techniques such as welding, snap fits, and self-threading screws. In other words, the provision of improved properties at reduced cost permits the use of polyacetals as replacements for metals.

b. The hydrogen bonding capability of nylons is the primary factor controlling their solvent resistance. The hydrogen bond must be broken or replaced during dissolution. Consequently, only strong acids, which can protonate the amide nitrogen atom and preclude formation of hydrogen bonds or other compounds (formic acid, phenols, etc.) that can form hydrogen bonds, are solvents for nylon 6,6 and nylon 6. These nylons are polyamides with a high content of amide groups.

Example 15.5: Indicate the polymer property/properties required for the following applications of nylons:

a. Gears, bearings, bushings, cams
b. Screws, nuts, bolts
c. Food packaging
d. Coatings for wire and cable insulation.

Solution:

Applications	Responsible Nylon Properties
a. Gears, bearings, bushings, cams	High tensile and impact strength; toughness; lubricity; low coefficient of friction
b. Screws, nuts, bolts	High strength; corrosion resistance
c. Food packaging	Low permeability to water and air to preserve freshness; high strength; puncture resistance
d. Coatings for wire and cable	Resistance to stress cracking, abrasion, and corrosion; low moisture absorption

D. POLYCARBONATE (PC)

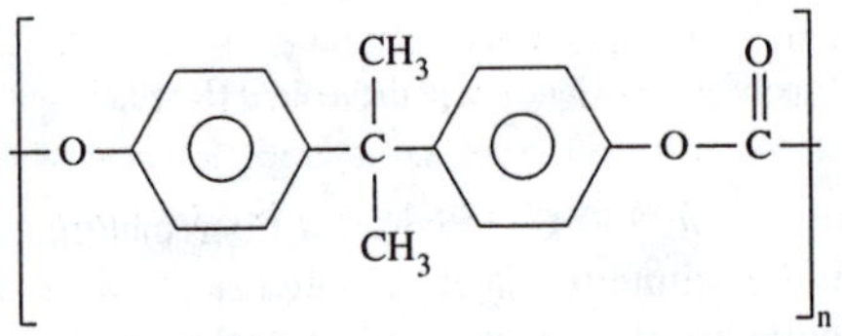

$$\left[-O-\langle\bigcirc\rangle-\underset{CH_3}{\overset{CH_3}{\overset{|}{\underset{|}{C}}}}-\langle\bigcirc\rangle-O-\overset{O}{\overset{\|}{C}}-\right]_n \qquad \text{(Str. 19)}$$

Polycarbonates are characterized by the carbonate (–O–COO–) interunit linkage. They may be prepared by interfacial polycondensation of bisphenol A and phosgene in methylene chloride–water mixture. The resulting hydrogen chloride is removed with sodium hydroxide or, in the case of solution polymerization, pyridine is used as the hydrogen chloride scavenger. Polycarbonate may also be made by ester interchange between bisphenol A and diphenyl carbonate.

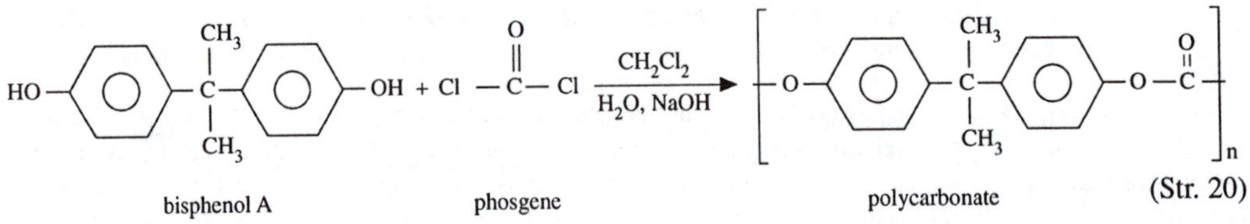

$$\underset{\text{bisphenol A}}{HO-\langle\bigcirc\rangle-\underset{CH_3}{\overset{CH_3}{\overset{|}{\underset{|}{C}}}}-\langle\bigcirc\rangle-OH} + \underset{\text{phosgene}}{Cl-\overset{O}{\overset{\|}{C}}-Cl} \xrightarrow[H_2O,\ NaOH]{CH_2Cl_2} \underset{\text{polycarbonate}}{\left[-O-\langle\bigcirc\rangle-\underset{CH_3}{\overset{CH_3}{\overset{|}{\underset{|}{C}}}}-\langle\bigcirc\rangle-O-\overset{O}{\overset{\|}{C}}-\right]_n} \qquad \text{(Str. 20)}$$

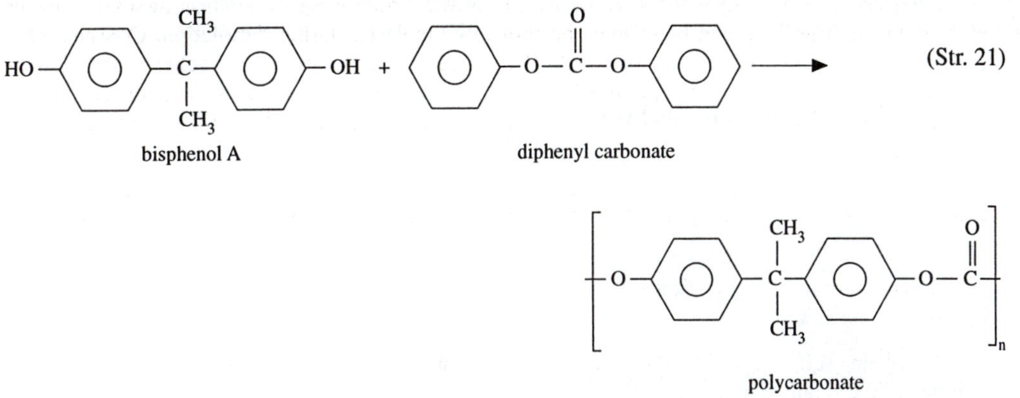

$$HO-C_6H_4-C(CH_3)_2-C_6H_4-OH + C_6H_5-O-C(=O)-O-C_6H_5 \longrightarrow \quad \text{(Str. 21)}$$

bisphenol A diphenyl carbonate

$$\left[-O-C_6H_4-C(CH_3)_2-C_6H_4-O-C(=O)-\right]_n$$

polycarbonate

Polycarbonate is an amorphous polymer with a unique combination of attractive engineering properties. These include exceptionally high-impact strength even at low temperatures, low moisture absorption, good heat resistance, good rigidity and electrical properties, and high light transmission. It possesses good dimensional stability (high creep resistance) over a broad temperature range. The transparency of polycarbonate has led to its use as an impact-resistant substitute for window glass. Polycarbonate, however, has a limited scratch and chemical resistance. It also has a tendency to yellow under long-term exposure to UV light. Copolymerization and/or incorporation of additives are used to modify the base resin for greater creep resistance, UV light performance, flame retardance, and thermal stability. For example, fire-retardant grades of polycarbonates are made by copolymerizing bisphenol A with tetrabromobisphenol A comonomer, while addition of glass fiber reinforcements greatly extends the level and range of creep resistance of polycarbonates.

Polycarbonates are processed by all the conventional techniques for processing thermoplastics. The balanced combination of properties permits polycarbonates to be used in a variety of applications. Markets for polycarbonates include automotive, construction, electronics, appliances, and lighting, while typical applications are automobile taillight lenses, lamp housings, bumpers, door and window components, drapery fixtures, furniture and plumbing, business machine housings, machinery housings, telephone parts, glazing signs, and returnable bottles.

E. POLY(PHENYLENE OXIDE) (PPO)

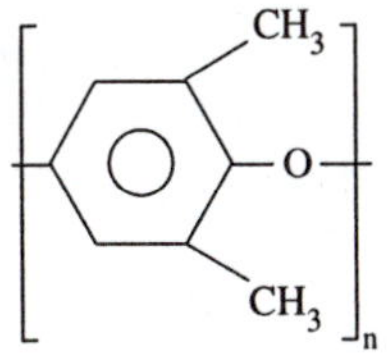

$$\left[-C_6H_2(CH_3)_2-O-\right]_n \quad \text{(Str. 22)}$$

Polyphenylene oxide is a large-volume engineering thermoplastic developed in 1956 by General Electric. It is made by free-radical step-growth oxidative coupling polymerization of 2,6-xylenol, with copper salts and pyridene as catalysts. Poly(phenylene oxide) homopolymer is difficult to process. Therefore, the commercial resins, marketed under the trade name Noryl, are modified poly(phenylene oxides) containing high-impact polystyrene (HIPS). The styrene component of HIPS forms homogeneous phase with PPO.

Poly(phenylene oxide) is an amorphous thermoplastic material with a low specific gravity, high impact strength, chemical resistance to mineral and organic acids, good electrical properties, and excellent dimensional stability at high temperatures. It has exceptionally low water absorption and complete hydrolytic stability.

Poly(phenylene oxide) resins are available in general-purpose, flame-retardant, glass-reinforced extrudable, foamable, and specialty grades. In addition to their applications as appliance, electrical, and business machine housings, PPO resins find use in a variety of pumps, showerheads, and components for underwater equipment. Platable grades are used for automotive grilles and wheel covers and plumbing fixtures.

F. POLY(PHENYLENE SULFIDE) (PPS)

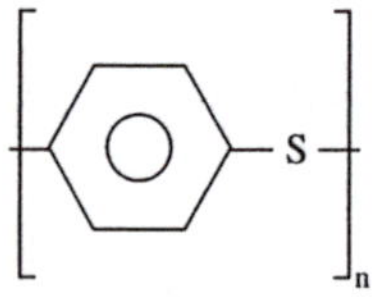

(Str. 23)

Poly(phenylene sulfide), sold by Phillips Chemical Company under the trade name Rylon, is a highly crystalline aromatic polymer (mp 285°C). It is obtained by the polycondensation of *p*-dichlorobenzene and sodium sulfide. The symmetrical arrangement of *p*-substituted benzene rings and sulfur atoms on the polymer backbone permits a high degree of crystallization. This, in addition to the extreme resistance of the benzene ring–sulfur bonds to thermal degradation, confers thermal stability, inherent nonflammability, and chemical resistance on PPS resin.

Even though concentrated oxidizing acids, some amines, and halogenated compounds can affect PPS, it has no known solvents below 200°C. Poly(phenylene sulfide) is characterized by high stiffness and good retention of mechanical properties at elevated temperatures. It has high tensile and flexural strengths, which can be increased substantially by addition of fillers such as glass fibers. PPS has good electrical properties, low coefficient of friction, and high transparency to microwave radiation. Poly(phenylene sulfide) is available in grades suitable for injection and compression molding and coating. Principal applications are in electrical and electronic components and industrial-mechanical uses such as parts for chemical processing equipment that require high temperature stability, mechanical strength, and chemical resistance.

G. POLYSULFONES

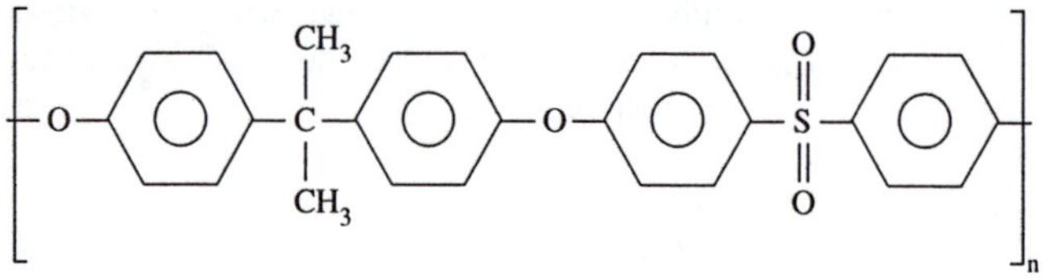

Polysulfone (Bisphenol A)

(Str. 24)

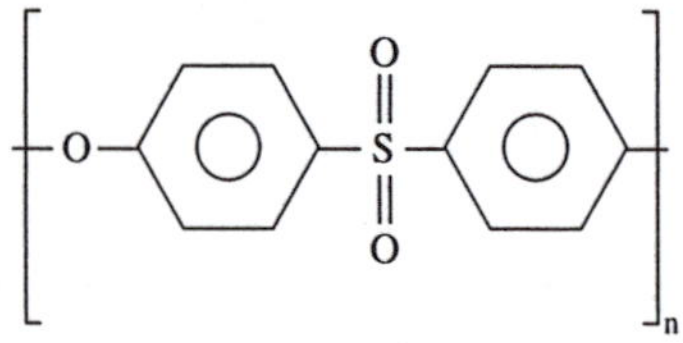

Polyethersulfone

(Str. 25)

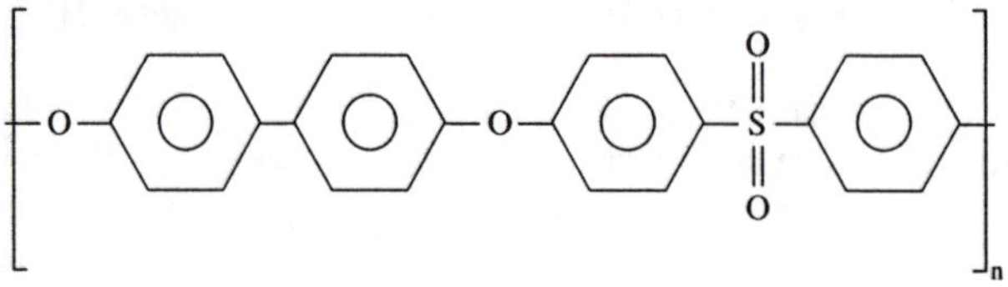

Polyphenylsulfone

(Str. 26)

Polysulfones constitute a family of high-performance transparent engineering thermoplastics with high oxidative and hydrolytic stability and excellent high-temperature properties. They may be prepared by condensation polymerization of 4,4′-dichlorophenyl sulfone with alkali salt of bisphenol A in polar solvents like dimethylsulfoxide (DMF) or sulfolane.

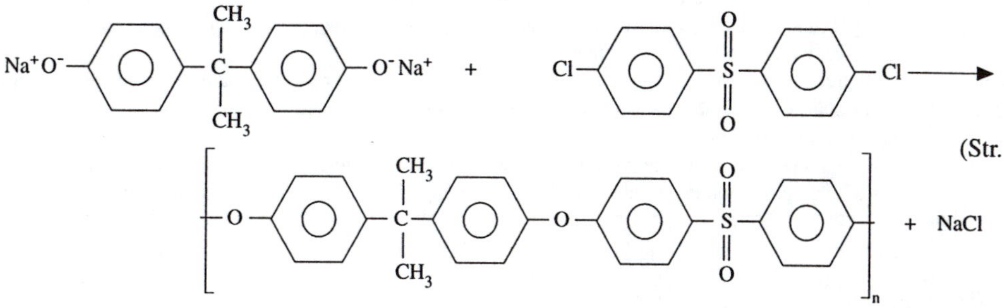

(Str. 27)

Alternatively, polysulfones may also be synthesized by a Friedel–Crafts reaction of aromatic sulfonyl chlorides using Lewis acid catalyst:

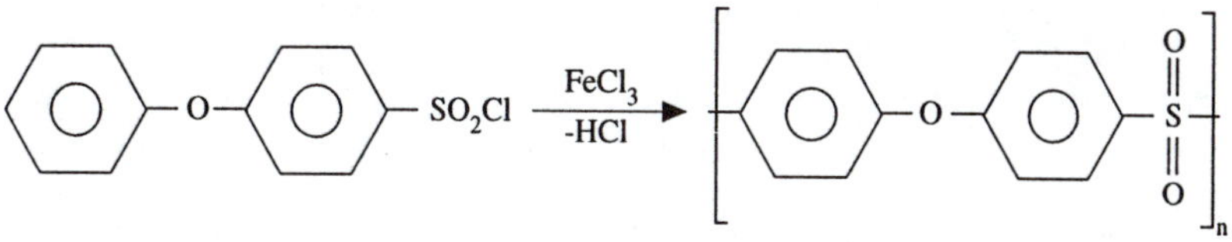

(Str. 28)

The presence of the diaryl sulfone group with a *para* oxygen atom confers oxidation resistance, good thermal stability, and rigidity at high temperatures. The ether linkages provide chain flexibility and, consequently, impart good impact strength. Polysulfones have good resistance to aqueous mineral acids, alkali, salt solutions, and oils and greases. They are strong, rigid, tough, amorphous polymers that can be extruded and injection molded on conventional equipment. Typical properties on some polysulfones are shown in Table 15.14.

Polysulfones have been used satisfactorily in a wide range of products, including consumer, medical, automotive, aircraft, aerospace, industrial, electrical, and electronic applications. Like other engineering thermoplastics, they are replacing metals in a variety of applications because they can be injection molded into complex shapes at reduced cost since costly machining and finishing operations can be avoided. Table 15.15 lists some of the applications of polysulfones.

Table 15.14 Typical Properties of Polysulfones

Property	Polysulfone (Bisphenol A)	Polyether Sulfone	Polyphenyl Sulfone
Specific gravity	1.24	1.37	1.29
Tensile strength (MPa)	70.3	84.1	71.7
Tensile modulus (MPa)	2482	2696	2137
Flexural strength (MPa)	106.2	128.6	85.5
Flexural modulus (MPa)	2689	2585	2275
Notched Izod impact strength (ft-lb/in)	1.2	1.6	12.

Table 15.15 Some Applications of Polysulfones

Polysulfone	Typical Applications
Polysulfone (bisphenol A)	Medical instrumentation; food processing and handling, including microwave ware, coffeemakers, beverage-dispensing tanks; electrical/electronic applications such as connectors, automotive fuses, switch housings, television components, structural circuit boards; chemical processing and other applications, including corrosion-resistant piping, tower packing, pumps, membranes, camera and watch cases, battery cell frames and housing
Polyether sulfone	High-temperature electrical parts such as connectors, motor components, lamp housings and alternator insulators; sterilizable medical components; oven windows; aerospace, aircraft and automobile composite structures
Polyphenyl sulfone	High-temperature coil bobbins, aircraft window reveals, automatic transmission housing, firemen's helmets, gas compressor valves, carbon fiber composites, flexible printed circuit boards

H. POLYIMIDES

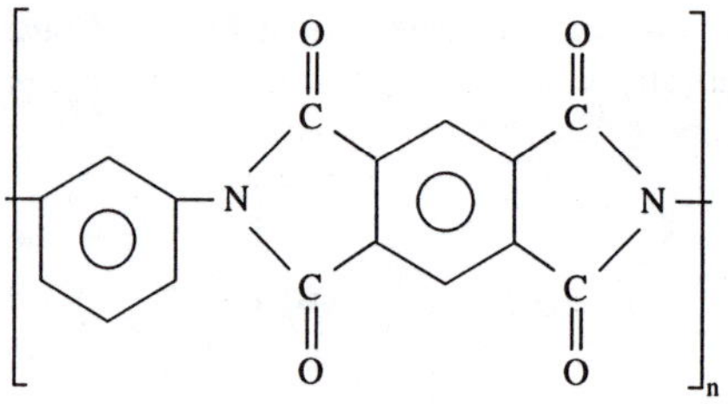

Polyimide

(Str. 29)

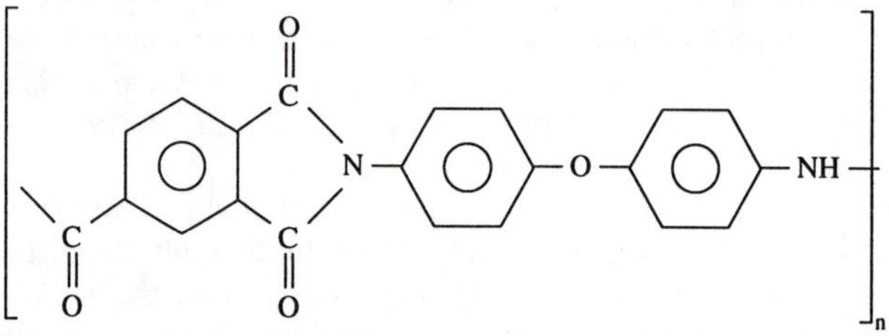

Polyamide-imide (Torlon)

(Str. 30)

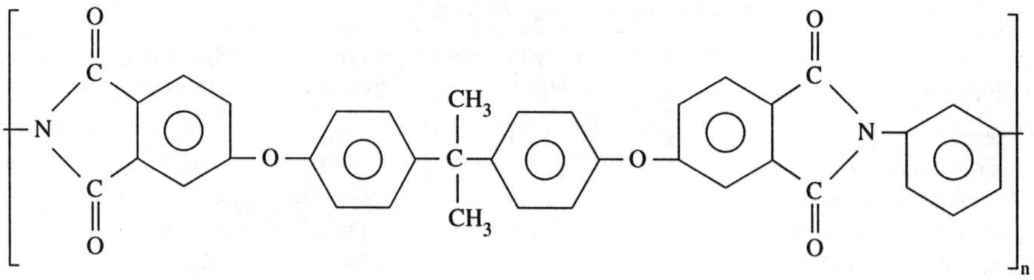

Polyetherimide (Ulterm)

(Str. 31)

Polyimides are made by a two-stage process involving an initial polycondensation of aromatic dianhydrides and aromatic diamines to produce a soluble intermediate, polyamic acid, which is then dehydrated at elevated temperatures to form the polyimide. While the intermediate, polyamic acid, is soluble, the cured or fully imidized polyimide is insoluble and infusible, with essentially the characteristics of a thermoset.

Polyimide molded parts and laminates display high heat, chemical, and wear resistance, with virtually no creep even at high temperatures. It has high oxidative stability, good electrical insulation properties, low coefficient of friction, and good cryogenic properties. However, polyimide has inherent structural weakness arising from imperfections and void formation due to water release in the curing process.

Typical uses of polyimide include electronic applications, sleeve bearings, valve seatings, and compressor vanes in jet engines. Other uses include aircraft and aerospace applications with high performance requirements. They are used for printed circuit boards in computers and electronic watches for both military and commercial uses. Polyimides are used in the insulation of automotive parts that require thermal and electrical insulation, such as wires used in electric motors, wheels, pistons, and bearings.

The thermoplastic variety of polyimides with enhanced melt processability is obtained by combining the basic imide structure with more flexible aromatic groups such as aromatic ethers or amides. Polyamide-imides are produced by condensing trimellitic anhydrides with aromatic diamines, while polyetherimides are made by the reaction between bisphenol A, 4,4′-methylene dianiline, and 3-nitrophthalic anhydride.

Both polyamide-imide and polyetherimide have high heat distortion temperature, tensile strength, and modulus. Polyamide-imide is useful from cryogenic temperatures up to 260°C. It is virtually unaffected by aliphatic and aromatic chlorinated and fluorinated hydrocarbons and by most acid and alkali solutions. These polymers are used in high-performance electrical and electronic parts, microwave appliances, and under-the-hood automotive parts. Typical automotive applications include timing gears, rocker arms, electrical connectors, switches, and insulators.

I. ENGINEERING POLYESTERS

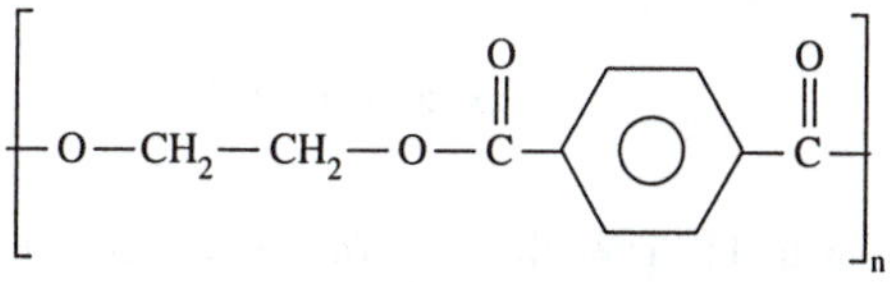

$$\left[-O-CH_2-CH_2-O-\overset{O}{\overset{\|}{C}}-C_6H_4-\overset{O}{\overset{\|}{C}}- \right]_n$$

Poly(ethylene terephthalate) (Mylar, Dacron, Torelene) (Str. 32)

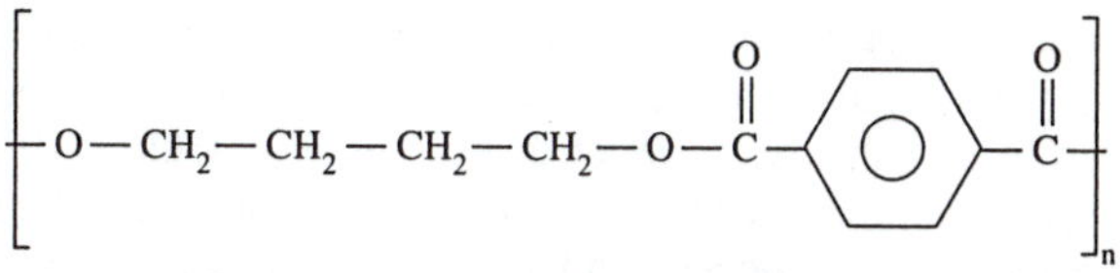

$$\left[-O-CH_2-CH_2-CH_2-CH_2-O-\overset{O}{\overset{\|}{C}}-C_6H_4-\overset{O}{\overset{\|}{C}}- \right]_n$$

Poly(butylene terephthalate) (PBT) (Str. 33)

Commercially important polyesters are based on polymers with the *p*-phenylene group in the polymer chain. In contrast to the low melting, linear aliphatic polyesters, the stiffening action of this group coupled with the high degree of symmetry results in a high melting point and other important engineering properties. For example, all commercial polyester fibers are based on terephthalic acid as the primary building block. Different products are obtained by varying the difunctional alcohols used in polycondensation reaction with this acid. However, the major engineering polyesters are poly(ethylene terephthalate) and poly(butylene terephthalate).

Polyesters are produced commercially by melt polymerization, ester interchanges, and interfacial polymerization. Commercial poly(ethylene terephthalate) is produced traditionally by two successive ester interchange reactions. In the first step, dimethyl terephthalate is heated with ethylene glycol at temperatures near 200°C. This yields an oligomeric dihydroxyethyl terephthalate (x = 1 to 4) and methanol, which is removed. In the second step, the temperature is increased, leading to polymer formation, while ethylene glycol is distilled off.

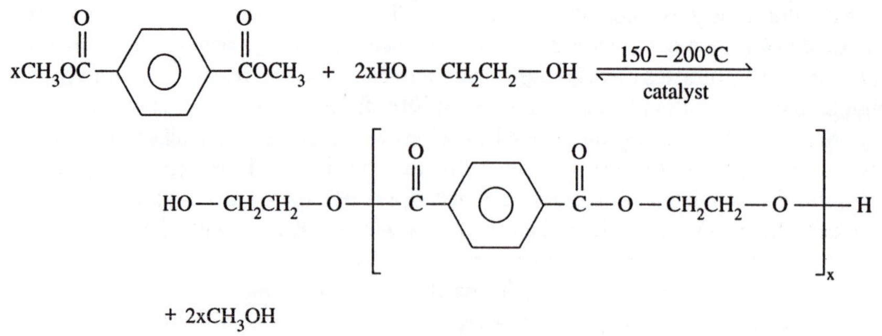

$$xCH_3O\overset{O}{\overset{\|}{C}}-C_6H_4-\overset{O}{\overset{\|}{C}}OCH_3 + 2xHO-CH_2CH_2-OH \xrightleftharpoons[\text{catalyst}]{150-200°C} HO-CH_2CH_2-O\left[\overset{O}{\overset{\|}{C}}-C_6H_4-\overset{O}{\overset{\|}{C}}-O-CH_2CH_2-O\right]_x H + 2xCH_3OH \quad \text{(Str. 34)}$$

$$HO-CH_2CH_2-O\left[\overset{O}{\overset{\|}{C}}-C_6H_4-\overset{O}{\overset{\|}{C}}-O-CH_2CH_2-O\right]_x H \xrightleftharpoons[\text{catalyst}]{260-300°C} \left[\overset{O}{\overset{\|}{C}}-C_6H_4-\overset{O}{\overset{\|}{C}}-O-CH_2CH_2-O\right]_{nx} + nx\,HO-CH_2CH_2-OH \quad \text{(Str. 35)}$$

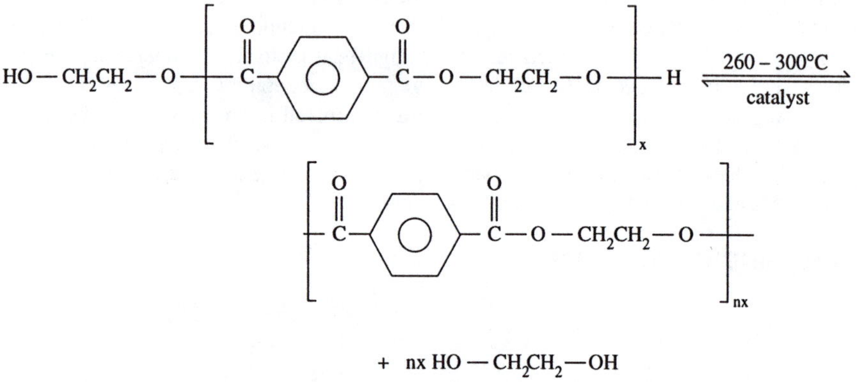

Poly(butylene terephthalate) is a highly crystalline thermoplastic polyester that is manufactured by condensation polymerization of 1,4-butanediol and dimethyl terephthalate in the presence of tetrabutyl titanate.

PET and PBT are characterized by high strength, rigidity, and toughness; low creep at elevated temperatures; excellent dimensional stability; low coefficient of friction; good chemical, grease, oil, and solvent resistance; minimal moisture absorption; and excellent electrical properties. Reinforcement of these polymers with glass fibers enhances many of their properties (Table 15.16).

PET has been used for a long time in fiber applications, including apparel, home furnishings, and tire cord. The fiber applications of PET depend on its outstanding crease resistance, work recovery, and low moisture absorption. Clothing made from PET fibers exhibit good wrinkle resistance. As a plastic, PET has been used in the production of films and, more recently, as blow-molded bottles for carbonated soft drinks. Biaxially oriented PET film is used industrially in magnetic tape, X-ray and photographic

Table 15.16 Properties of Engineering Polyesters

	Poly(ethylene terephthalate)		Poly(butylene terephthalate)	
Property	**Unfilled**	**30% Glass Fiber Reinforced**	**Unfilled**	**30% Glass Fiber Reinforced**
Specific gravity	1.34–1.39	1.27	1.31–1.38	1.52
Melting temperature (°C)	265	265	224	224
Tensile strength (MPa)	58.6–72.4	158.6	56.5	117.2–131.0
Tensile modulus (MPa)	2758–4136	9927	1930	8962
Flexural strength (MPa)	96.5–124.1	230.9	82.7–115.1	179.2–200.0
Flexural modulus (MPa)	2413–3102	8962	2275–2758	7583–8273
Impact strength (Izod) (ft-lb/in)	0.25–0.65	1.9	0.8–1.0	1.3–1.6
Water absorption (24 hr)	0.1–0.2	0.05	0.08–0.09	0.06–0.08

films, and electrical insulation. Both oriented and unoriented PET films are also used in food packaging applications such as boil-in-bag food pouches. Applications of 30% glass-fiber-reinforced PET resins include pump and power-tool housings, sporting goods, and automotive exterior components such as rearview mirror housings, hinges, and windshield-wiper components.

Automobile applications of PBT include window and door hardware, speedometer frames and gears, servo pistons, and automobile ignition system components (distributor caps, coil bobbins, and rotors). In addition, it is used as bases, handles, and housings for small appliances (toasters, cookers, fryers and irons) as well as small industrial pump housing, impellers, and support brackets, gears, showerhead and faucet components, and consumer products like buckles, clips, buttons, and zippers.

Example 15.6: Comment on the relative melting points as well as the impact strength and water absorption of unfilled PET and PBT.

Solution: PBT has four methylene groups compared with two in PET. This confers a greater chain flexibility on PBT. Consequently, PBT has a lower melting (due to greater susceptibility to thermal agitation) and a higher impact strength than PET. On the other hand, the hydrocarbon nature of PBT is greater than that of PET, making PBT more hydrophobic.

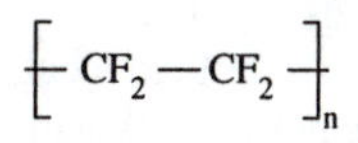

$$\left[CF_2 - CF_2 \right]_n$$

Polytetrafluorethylene (Teflon) (Str. 36)

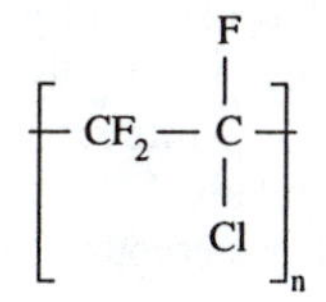

$$\left[CF_2 - \underset{\underset{Cl}{|}}{\overset{\overset{F}{|}}{C}} \right]_n$$

Polychlorotrifluoroethylene (CTFE) (Str. 37)

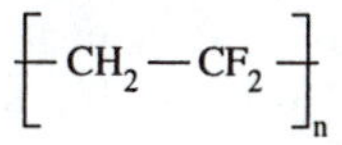

$$\left[CH_2 - CF_2 \right]_n$$

Poly (vinylidene fluoride) PVDF) (Str. 38)

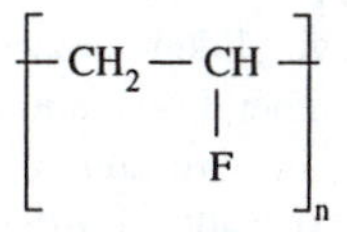

$$\left[CH_2 - \underset{\underset{F}{|}}{CH} \right]_n$$

Poly (vinyl fluoride) (PVF) (Str. 39)

J. FLUOROPOLYMERS

Fluoropolymers constitute a class of polyolefins in which some or all of the hydrogens are replaced by fluorine. The structures of some of these polymers are shown above. Fluoropolymers have a broad range of properties, offering unique performance characteristics. Within this family of polymers are those with high thermal stability and useful mechanical properties both at high temperatures and at cryogenic temperatures. Most fluoropolymers are chemically inert and totally insoluble in common organic solvents. The family of fluoropolymers has extremely low dielectric constants and high dielectric strength. Most fluoropolymers have unique nonadhesive and low friction properties.

Polytetrafluoroethylene, the most widely used fluoropolymer, is produced by emulsion free-radical polymerization of tetrafluoroethylene using redox initiators. As a result of its highly regular chain structure, PTFE is a highly crystalline polymer with high density and melting temperature (mp 327°C). It is a high-temperature-stable material characterized by low-temperature flexibility and extremely low coefficient of friction, dielectric constant, and dissipation factor. PTFE exhibits outstanding chemical inertness, is resistant to attack even by corrosive solvents, and is practically unaffected by water. While PTFE has a high impact strength, its tensile strength and wear and creep resistance are low relative to other engineering polymers.

Given its extremely high crystallinity and the associated high melting point, its melt viscosity, and its low melt flow rates, PTFE cannot be processed by conventional fabrication techniques used for polymers. Instead, unusual techniques have been developed for shaping polytetrafluoroethylene. Molding PTFE powders are processed by the two-staged press and sinter methods used in powder metallurgy. Granular PTFE is first pressed into the desired shape at room temperature and pressure (in the range 2000 to 10,000 psi). The resulting preform is then sintered at a temperature above the crystalline melting point (360 to 380°C) to obtain a dense, strong, homogeneous product. Better melt processing is achieved by reducing the crystallinity of PTFE through incorporation of a small concentration of a comonomer such as hexafluoropropylene. The resulting copolymer retains most of the desirable properties of polytetrafluoroethylene but has a reduced melt viscosity that permits processing by traditional techniques.

Polytetrafluoroethylene is used primarily in applications that require extreme toughness, outstanding chemical and heat resistance, good electrical properties, low friction, or a combination of these properties. Principal applications or PTFE are as components or linings for chemical process equipment, high-temperature cable insulation, molded electrical components, tape, and nonstick coatings. Chemical process equipment applications include linings for pipe, pipe fittings, valves, pumps, gaskets, and reaction vessels. PTFE is used as insulation for wire and cable, motors, generators, transformers, coils and capacitors, high-frequency electronic uses, and molded electrical components such as insulators and tube sockets. Nonstick low friction uses include home cookware, tools, and food-processing equipment.

In addition to polytetrafluoroethylene, several other partially fluorinated polymers are available commercially. These include poly(chlorotrifluoroethylene), which is also available as a copolymer with ethylene or vinylidene fluoride, poly(vinyl fluoride), and poly(vinylidene fluoride). Poly(chlorotrifluoroethylene) is a chemically inert and thermally stable polymer, soluble in a number of solvents above 100°C, tough at temperatures as low as –100°C while retaining its useful properties at temperatures as high as 150°C. Its melt viscosity, though relatively high, is sufficiently low to permit the use of conventional molding and extrusion processing methods. It is used for electrical insulators, gaskets and seals, and pump parts.

Poly(vinyl fluoride) is a highly crystalline polymer available commercially as a tough, flexible film sold under the trade name Tedlar by DuPont. It has excellent chemical resistance like other fluoropolymers, excellent outdoor weatherability, and good thermal stability, abrasion, and stain resistance. It maintains useful properties between –180°C to 150°C. It is used as protective coatings for materials like plywood, vinyl, hardboard, metals, and reinforced polyesters. These laminated materials find applications in aircraft interior panels, in wall covering, and in the building industry.

Poly(vinylidene chloride) is a crystalline polymer (mp 170°C), with significantly greater strength and creep and wear resistance than PTFE. Poly(vinylidene chloride), which is also available as a copolymer with hexafluoroethylene, has very good weatherability and chemical and solvent resistance. It is used primarily in coatings; as a gasket material; in wire and cable insulation; in piping, tanks, pumps and other chemical process equipment; and in extrusion of vinyl siding for houses.

K. IONOMERS

Ionomers are a family of polymers containing ionizable carboxyl groups, which can create ionic intermolecular cross-links. They are generally copolymers of α-olefins with carboxylic acid monomers that are partially neutralized by monovalent or divalent cations. A typical ionomer is DuPont Surlyn, which is a copolymer of ethylene and methacrylic acid partially neutralized with sodium or zinc cations.

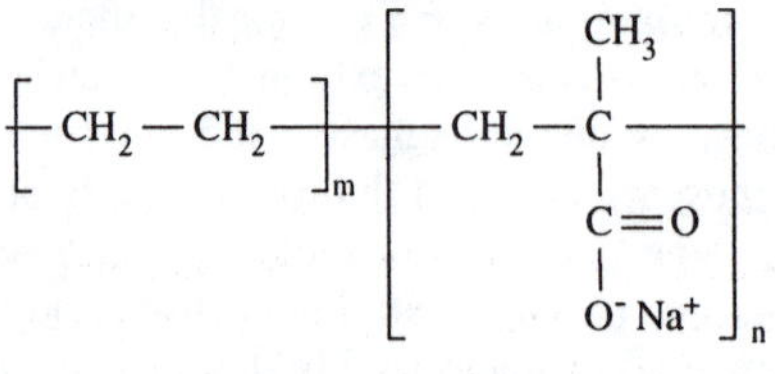

$$\left[-CH_2-CH_2- \right]_m \left[-CH_2-\underset{\underset{\underset{O^- Na^+}{|}}{\underset{C=O}{|}}}{\overset{\overset{CH_3}{|}}{C}}- \right]_n \qquad \text{(Str. 40)}$$

Table 15.17 Some Commercial Ionomers

Polymer System	Trade Name	Manufacturer	Uses
Poly(ethylene-*co*-methacrylic acid)	Surlyn	DuPont	Modified thermoplastic
Poly(butadiene-*co*-acrylic acid)	Hycar	BF Goodrich	High green strength elastomer
Perfluorosulfonate ionomers	Nafion	DuPont	Multiple membrane uses
Perfluorocarboxylate ionomer	Flemion	Asahi Glass	Chloralkali membrane
Telechelic polybutadiene	Hycar	BF Goodrich	Specialty Uses
Sulfonated ethylene–propylene–diene terpolymer	Ionic elastomer	Uniroyal	Thermoplastic elastomer

From Lundberg, R.D., in *Encyclopedia of Chemical Technology,* 3rd ed., Mark, H.F., Othmer, D.F., Overberger, C.G., and Seaborg, G.T., Eds., Interscience, New York, 1984. With permission.

For ionomers, the ratio n/m typically does not exceed 10 mol%. The metal ions act as the cross-links between chains. The ionic interchain forces confer solid-state properties normally associated with a cross-linked structure on ionomers, but the cross-links are labile at processing temperatures. Consequently, ionomers, like other thermoplastic materials, can be processed on conventional molding and extrusion equipment.

The preparation of ionomers involves either the copolymerization of a functionalized monomer with an olefinic unsaturated monomer or direct functionalization of a preformed polymer. Typically, free-radical copolymerization of ethylene, styrene, or other α-olefins with acrylic acid or methacrylic acid results in carboxyl-containing ionomers. The copolymer, available as a free acid, is then neutralized partially to a desired degree with metal hydroxides, acetates, or similar salts. The second route for the preparation of ionomers involves modification of a preformed polymer. For example, sulfonated polystyrene is obtained by direct sulfonation of polystyrene in a homogeneous solution followed by neutralization of the acid to the desired level. Some commercially available ionomers are listed in Table 15.17.

In contrast to homogeneous polymer systems, the pendant ionic groups in ionomers interact or associate, forming ion-rich aggregates immersed in the nonpolar matrix of polymer backbone (Figure 15.10). The extent of ionic interactions and, hence, the properties of ionomers, are dictated by the ionic content, degree of neutralization, type of polymer backbone, and cation.

Ionomers are characterized by outstanding abrasion and oil resistance, toughness, flexibility, good adhesion, and high transparency. These properties dictate the uses of these polymers. For example, the high melt viscosity of Surlyn provides good extrusion performance for paper and foil coatings of multiwall bags for food and drug packaging. Its toughness and abrasion resistance have resulted in extensive use for covering of golf balls (as a replacement of gutta-percha) and roller-skate wheels. The high impact strength coupled with its printability has led to the use of Surlyn in the manufacture of automotive bumper strips and guards. Nafion, with its selective permeability to ions, is used in the production of chlorine and caustic by electrolysis of salt solutions.

VIII. ELASTOMERS

As we saw in Chapter 1 elastomers are polymers that are amorphous in the unstretched state and are above their glass transition temperatures at normal ambient temperatures. They have low glass transition temperatures, usually in the range –50 to –70°C. Elastomers are composed of irregularly shaped chain molecules that are held together by a network of cross-links to prevent gross mobility of chains while permitting local mobility of chain segments. The network of cross-links may be formed by covalent bonds or may be due to physical links between chain molecules. The process of introducing covalent cross-links into an elastomer is referred to as vulcanization. Elastomers possess the unique ability to stretch usually to several times their initial dimensions without rupturing, but retract rapidly with full recovery on the release of the imposed stress. In the stretched state, elastomers exhibit high strength and modulus.

A. DIENE-BASED ELASTOMERS

Polymerization of conjugated dienes like butadiene, isoprene, and chloroprenes involves activation of either or both of the double bonds to give 1,2; 3,4; or 1,4 polymers (Figure 15.11). The residual unsaturation in the polymer chains provides convenient sites for the introduction of elastomeric network

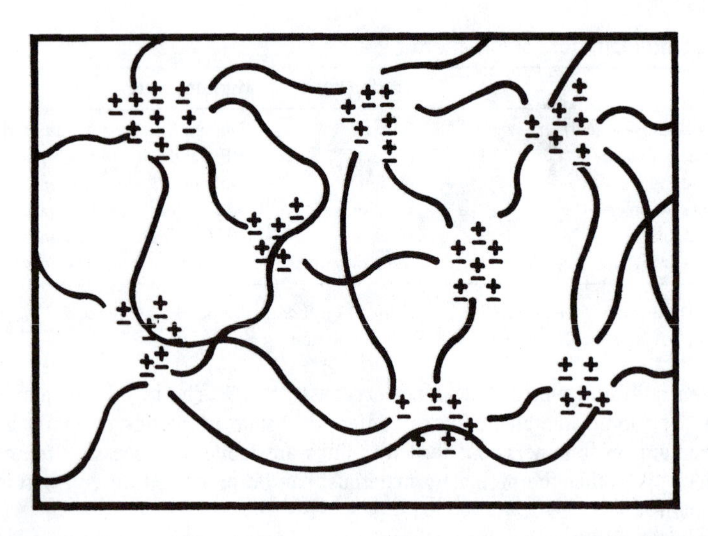

Figure 15.10 Schematic representation of ionomer. Lines represent nonpolar polymer backbone while + and − represent metal ions and anions, respectively.

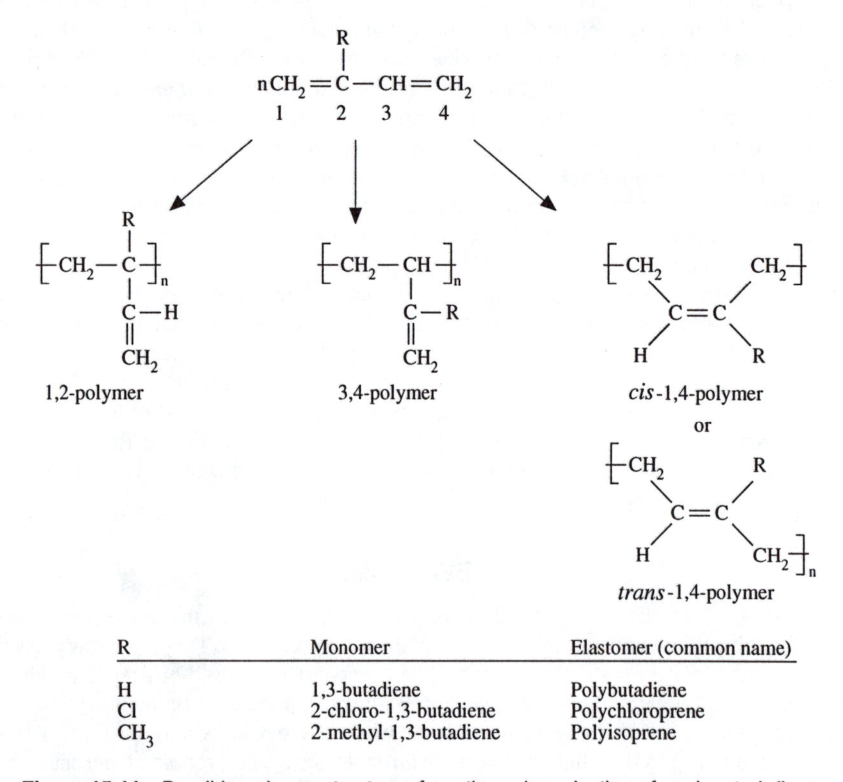

R	Monomer	Elastomer (common name)
H	1,3-butadiene	Polybutadiene
Cl	2-chloro-1,3-butadiene	Polychloroprene
CH_3	2-methyl-1,3-butadiene	Polyisoprene

Figure 15.11 Possible polymer structures from the polymerization of conjugated dienes.

of cross-links (vulcanization). Therefore, conjugated dienes are the source of some of the most important commercially available synthetic rubbers or elastomers. For isoprene and chloroprene, eight arrangements are theoretically possible: the 1,2 and 3,4 polymers (vinyl polymers) can be isotactic, syndiotactic, or atactic, while in the 1,4 polymer both *cis* and *trans* configurations are possible. In the case of 1,3 butadiene, the 1,2 and 3,4 structures are identical because of the absence of the asymmetrical substituent group. Both the thermal and physical properties of these polymers are influenced by the relative amounts of the various structures in the polymer chains. The proportion of each type of structure depends on the method and conditions of polymerization.

1. Polybutadiene (Butadiene Rubber, BR)

$$\left[CH_2 - CH = CH - CH_2 \right]_n \qquad \text{(Str. 41)}$$

Polybutadiene is second-largest-volume synthetic elastomer, next only to SBR. Polybutadiene can be produced by free-radical addition polymerization of butadiene. The resulting polymer has predominantly *trans*-1,4 units, with only about 20% 1,2 units. As the polymerization temperature is increased, the proportion of *cis*-1,4 units increases, while that of 1,2 structure remains essentially unchanged. Butadiene can also undergo anionic polymerization with lithium or organolithium initiators like *n*-butyllithium in nonpolar solvent, such as pentane or hexane. The resulting polymer has a high content of *cis*-1,4 structure, which decreases as either higher alkali-metal initiators or more polar solvents are used. High-molecular-weight polybutadiene with a high content of *trans*-1,4 polymers is prepared by solution polymerization of butadiene using stereo-selective coordination Ziegler–Natta catalysts. Slight changes in catalyst composition can produce drastic changes in polymer composition.

Like SBR, the principal use of BR is in the production of tires and tire products. BR exhibits good resilience and abrasion resistance and low heat buildup, which are important requirements for tire applications. However, in general, BR processes with more difficulty than SBR. Consequently, BR is blended with SBR and natural rubber in tire manufacturing for improved milling, traction, and wet skid resistance of tire treads, while BR contributes good resistance to wear and groove cracking, lowered rolling resistance, and low heat buildup.

2. Styrene–Butadiene Rubber (SBR)

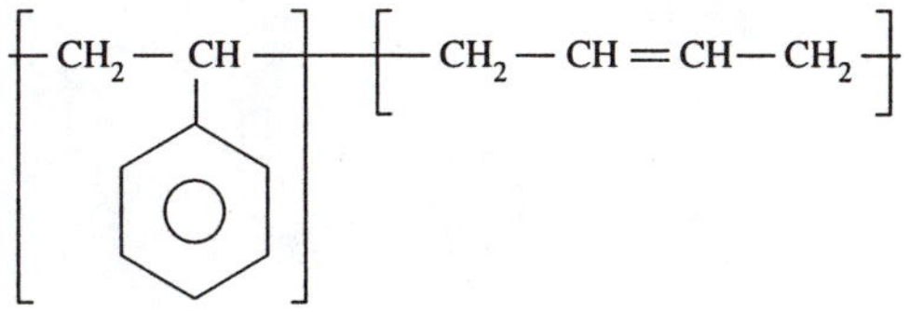

$$\left[CH_2 - \underset{C_6H_5}{CH} \right] \left[CH_2 - CH = CH - CH_2 \right] \qquad \text{(Str. 42)}$$

Styrene–butadiene rubber is the largest volume synthetic elastomer commercially available. It can be produced by free-radical emulsion polymerization of styrene and butadiene either at 50 to 60°C (hot emulsion SBR) or at about 5°C (cold emulsion SBR). The two kinds of SBR have significantly different properties. The hot emulsion SBR process, which was developed first, leads to a more branched polymer than the cold emulsion process. Cold SBR has a better abrasion resistance and, consequently, provides better tread wear and dynamic properties.

SBR may also be produced by anionic solution polymerization of styrene and butadiene with alkyllithium initiator (e.g., butyllithium) in a hydrocarbon solvent, usually hexane or cyclohexane. In contrast to emulsion SBR, which may have an emulsifier (soap) content of up to 5% and nonrubber materials sometimes in excess of 10%, solution SBR seldom has more than 2% nonrubber materials in its finished form. Solution SBR has a narrower molecular weight distribution, higher molecular weight, and higher *cis*-1,4-polybutadiene content than emulsion polymerization SBR.

SBR is a random copolymer with a styrene content in the range of 10 to 25%. The presence of styrene contributes to the good wearing and bonding characteristics of SBR and reduces its price. Also, compared with polybutadiene rubber alone, strength, abrasion resistance, and blend compatibility are improved. The butadiene units in SBR are composed approximately of 60 to 70% *trans*-1,4; 15 to 20% *cis*-1,4; and 15 to 20% 1,2 configuration for the polymer at 50°C. Polymers made at lower temperatures have a higher content of *trans*-1,4 polybutadiene units. In diene polymerization, high conversion of monomers or the absence of a chain transfer agent leads to branching due to chain transfer to polymer or reaction of both double bonds.

The major use of SBR is in the production of tires, particularly passenger-car and light-truck tires, and in other automotive applications where SBR is blended with other elastomers. These applications include belts, hoses, seals, and various extruded and molded items. Nontire and nonautomotive uses of SBR are in industries that require hoses, belts, gaskets, or seals. Others include footwear (shoe soles); various kinds of solid wheels; roll covers; coated fabrics; and electrical (wire and cable) insulation.

3. Acrylonitrile–Butadiene Rubber (Nitrile Rubber, NBR)

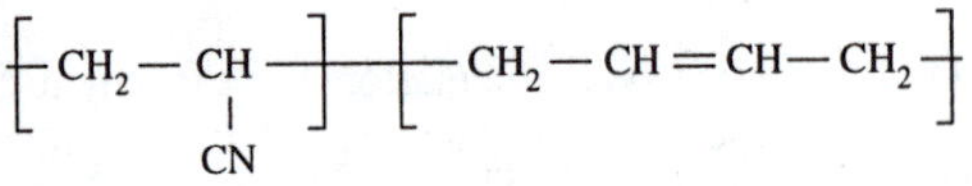

$$\left[CH_2 - \underset{CN}{\underset{|}{CH}} \right] \left[CH_2 - CH = CH - CH_2 \right] \quad \text{(Str. 43)}$$

Nitrile rubber is a unique elastomer that is a copolymer of butadiene and acrylonitrile. As in SBR production, butadiene can be copolymerized with 18 to 40% acrylonitrile in either cold or hot free-radical emulsion polymerization. Unlike SBR, NBR is not suitable for tire production, but it is unique for its excellent oil resistance, which increases with the proportion of acrylonitrile in the copolymer. Elastomers with high acrylonitrile content (40 to 50%) in addition to high hydrocarbon resistance generally are also more resistant to abrasion and have lower permeability to gases. Elastomers with about 20% acrylonitrile content exhibit enhanced resilience that is retained even at low temperatures. NBR retains its good tensile strength and abrasion resistance even after immersion in gasoline, water, alcohols, and aromatic solvents. While it has good heat resistance if properly protected with antioxidants, NBR has poor dielectric properties and ozone resistance.

As result of their excellent oil resistance, nitrile rubbers are used mainly to handle oils, fuels, and similar chemicals in the form of hoses, tubing, gaskets, seals, O-rings, and gasoline hose. These items are used in equipment for transportation of all kinds, food processing, and petroleum production. Nitrile rubbers are also used to enhance the impact strength of polymers such as PVC and ABS. In latex form, nitrile rubber is used to saturate paper for masking tapes, building papers, and labels.

4. Polyisoprene

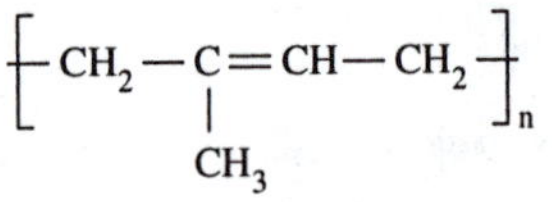

$$\left[CH_2 - \underset{CH_3}{\underset{|}{C}} = CH - CH_2 \right]_n \quad \text{(Str. 44)}$$

Polyisoprene is another widely used commercial synthetic rubber. It is produced by the polymerization of isoprene (2-methyl-1,3-butadiene) in a hydrocarbon solvent such as *n*-pentane using Ziegler–Natta catalyst systems. Out of the eight theoretically possible configurations, only three isomers: *cis*-1,4; *trans*-1,4; and atactic-3,4 forms have been isolated. Depending on the makeup of the catalyst employed in the polymerization reaction, a very high content of *cis*-1,4-polyisoprene can be obtained. The *cis*-1,4-polyisoprene is structurally identical to natural rubber. However, it is cleaner, lighter in color, more uniform and less expensive to process than natural rubber. The *trans*-1,4 isomer has limited commercial use in nonelastomeric applications. It was used originally for covering golf balls, but more recently as a material for orthopedic splints.

Cis-1,4-polyisoprene, like most diene-based elastomers, has poor resistance to attack by ozone, gasoline, oil, and organic solvents. It has, however, many of the good properties of natural rubber, including high resilience, strength, and abrasion resistance. Consequently, it is used mostly in tire making, usually as a replacement for natural rubber in blends with polybutadiene that are used for making heavy-duty truck and bus tires. Other uses of polyisoprene elastomer are in extruded and molded mechanical goods, footwear, sporting goods, and sealants.

5. Polychloroprene (Neoprene)

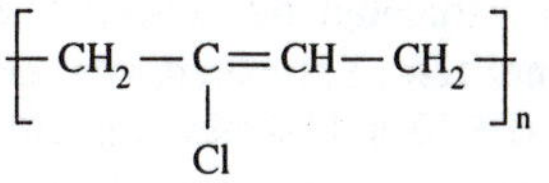

$$\left[CH_2 - \underset{\underset{Cl}{|}}{C} = CH - CH_2 \right]_n \qquad \text{(Str. 45)}$$

Polychloroprene, developed and sold under the trade name Neoprene by DuPont, was the first commercially successful synthetic elastomer. It is produced by free-radical emulsion polymerization of chloroprene (2-chloro-1,3-butadiene). The commercial material is mainly *trans*-1,4-polychloroprene, which is crystallizable.

Polychloroprenes are noted generally for their good resistance to abrasion, hydrocarbons, sunlight, oxygen, ozone, gas; weathering characteristics; and toughness. They are more difficult to process than most synthetic elastomers. Polychloroprene elastomer has a wide range of applications, ranging from adhesives to wire coverings. All the applications depend on its overall durability. The largest use of polychloroprene elastomer is in the fabrication of mechanical rubber goods for automotive products; petroleum production; and transportation, construction, and consumer products. The major uses include wire and cable coatings, industrial hoses, conveyor belts, diaphragms, seals, gaskets, O-rings, gasoline tubing, shoe heels, and solid tires. Neoprene latex is used in making gloves, adhesives, and binders.

6. Butyl Rubber

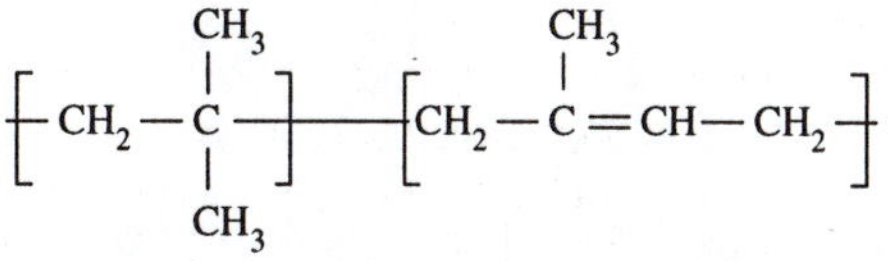

$$\left[CH_2 - \overset{\overset{CH_3}{|}}{\underset{\underset{CH_3}{|}}{C}} \right]\left[CH_2 - \overset{\overset{CH_3}{|}}{C} = CH - CH_2 \right] \qquad \text{(Str. 46)}$$

Butyl rubber is a copolymer of isobutylene with a small amount (0.5 to 2.5 mol%) of isoprene, which provides the unsaturation sites necessary for vulcanization. Butyl rubber is produced by cationic polymerization of isobutylene and chloroprene in methyl chloride in the presence of Friedel–Crafts catalysts such as aluminum chloride at about −100°C.

Butyl rubber exhibits unusually low permeability to gases and outstanding resistance to attack by oxygen and ozone. It has excellent chemical inertness, due to the very low residual unsaturation, and good electrical properties because of its nonpolar saturated nature. Butyl rubber has good tear resistance. Butyl elastomers can be tailored to have good thermal stability and vibrational damping characteristics.

As a result of its very low gas permeability, butyl rubber is used predominantly in the inner tubes of tires and inner liners for tubeless tires. Some of the other uses of butyl rubber include sealants, adhesives, hoses, gaskets, pads for truck cabs, bridge bearing mounts, and other places where vibration damping is important.

B. ETHYLENE–PROPYLENE RUBBERS

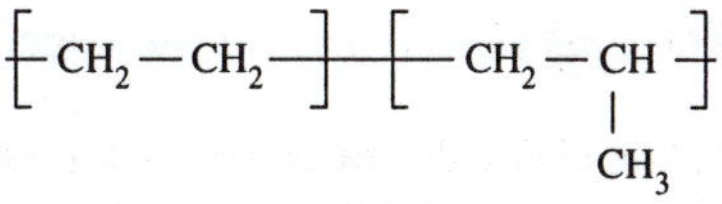

$$\left[CH_2 - CH_2 \right]\left[CH_2 - \underset{\underset{CH_3}{|}}{CH} \right] \qquad \text{(Str. 47)}$$

ethylene–propylene copolymer (EPM)

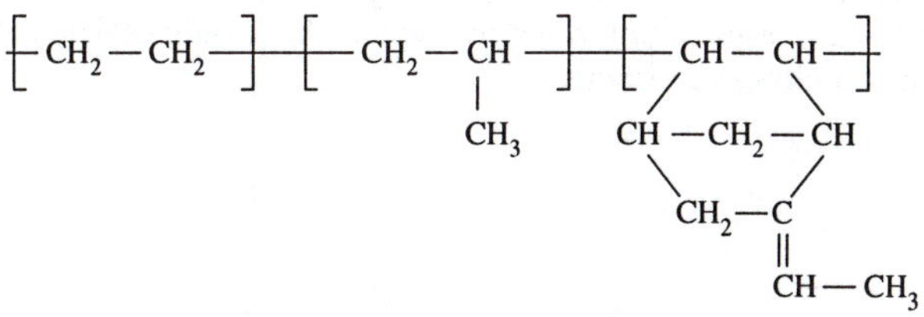

$$\left[CH_2 - CH_2 \right]\left[CH_2 - \underset{\underset{CH_3}{|}}{CH} \right]\left[CH - CH \right] \qquad \text{(Str. 48)}$$

(where the CH–CH unit is part of a ring: CH —CH₂— CH bridged, with CH₂—C(=CH—CH₃) closing the ring)

ethylene–propylene–diene terpolymer (EPDM)

Copolymerization of ethylene with propylene results in a random, noncrystalline copolymer that is a chemically inert and rubbery material. EPM is a saturated copolymer that can be cross-linked through the combination of the free radicals generated by peroxides or radiation. To incorporate sites for vulcanization, an unsaturated terpolymer can be prepared from ethylene, propylene, and a small amount (3 to 9%) of a nonconjugated diene (EPDM). The diene is either dicyclopentadiene, ethylidene norbornene, or 1,4-hexadiene. The resulting unsaturated terpolymer can be vulcanized by traditional techniques. Each of the termonomers confers different characteristics on the final elastomer.

Ethylene–propylene elastomers are made by solution polymerization of ethylene and propylene in a solvent such as hexane using Ziegler–Natta catalysts. EPDM terpolymers can be similarly made by adding 3 to 9% of any of the above dienes to the monoolefin mixture.

Even though tire use is small, automotive applications are still the largest market for ethylene–propylene elastomers. They are used predominantly in radiator and heater hoses, seals, gaskets, grommets, and weather stripping. Blends of polypropylene and EPDM are used as material in the manufacture of car bumpers, fender extensions, and rub strips. Other applications of ethylene–propylene elastomers include appliance parts, wire and cable insulation, and modification of polyolefins for improved impact and stress resistance.

C. POLYURETHANES

Polyurethane is the generic name of polymers with the urethane ($-\text{N}-\overset{\text{O}}{\overset{\|}{\text{C}}}-\text{O}-$) interunit linkage in the chain. There are two main synthetic routes for the preparation of linear urethane homopolymer. These are the condensation reaction between a bischloroformate and a diamine and the addition reaction of a diisocyanate with diol:

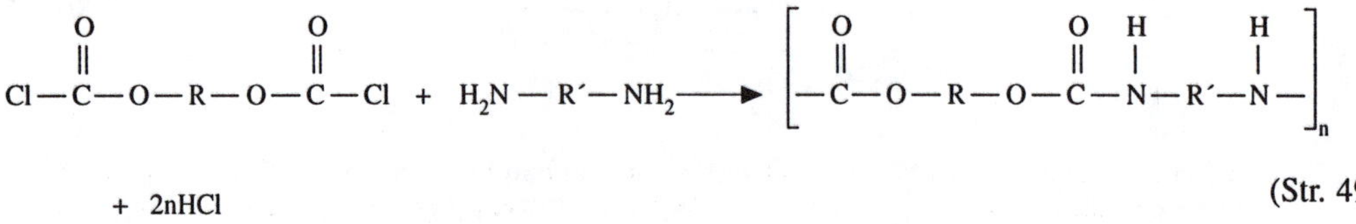

$$\text{Cl}-\overset{\overset{\text{O}}{\|}}{\text{C}}-\text{O}-\text{R}-\text{O}-\overset{\overset{\text{O}}{\|}}{\text{C}}-\text{Cl} + \text{H}_2\text{N}-\text{R}'-\text{NH}_2 \longrightarrow \left[-\overset{\overset{\text{O}}{\|}}{\text{C}}-\text{O}-\text{R}-\text{O}-\overset{\overset{\text{O}}{\|}}{\text{C}}-\overset{\overset{\text{H}}{|}}{\text{N}}-\text{R}'-\overset{\overset{\text{H}}{|}}{\text{N}}-\right]_n + 2n\text{HCl} \qquad \text{(Str. 49)}$$

$$\text{O}=\text{C}=\text{N}-\text{R}-\text{N}=\text{C}=\text{O} + \text{HO}-\text{R}'-\text{OH} \quad \left[-\overset{\overset{\text{O}}{\|}}{\text{C}}-\overset{\overset{\text{H}}{|}}{\text{N}}-\text{R}-\overset{\overset{\text{H}}{|}}{\text{N}}-\overset{\overset{\text{O}}{\|}}{\text{C}}-\text{O}-\text{R}'-\text{O}-\right]_n \qquad \text{(Str. 50)}$$

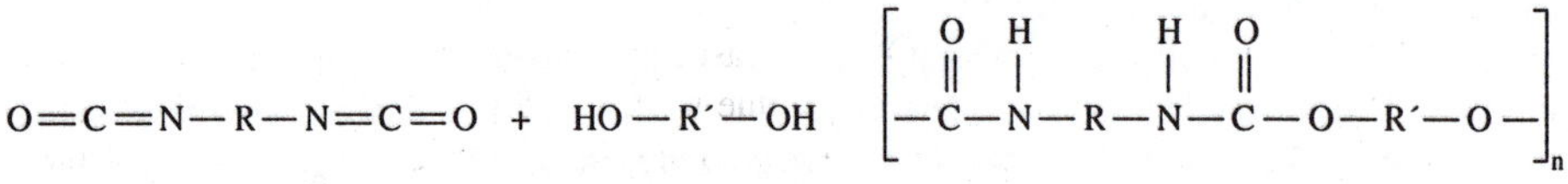

Typical diisocyanates include aromatic diisocyanates such as methylenediphenyl isocyanate (MDI) (or 4,4-diphenylmethane dissocyanate) and toluene diisocyanate (TDI). TDI is generally supplied as an 80:20 mixture of 2,4 and 2,6 isomers. Aliphatic diisocyanates include 1,6-hexamethylene diisocyanate (HMI). Dihydroxyl compounds employed are usually hydroxyl-terminated low-molecular-weight polyesters, polyethers, hydrocarbon polymers, and polydimethylsiloxanes. A typical example is polytetramethylene oxide (PTMO). Cross-linking or chain extension is achieved by using a number of di- and polyfunctional active hydrogen-containing compounds, the most significant of which are diol, diamines, and polydroxyl compounds.

Polyurethanes are used in four principal types of products: foams, elastomers, fibers, and coatings. The majority of polyurethane is used as rigid or flexible foams. However, about 15% is used for elastomer applications. Production of polyurethane elastomers involves a number of steps. As indicated above, an intermediate hydroxyl-terminated low-molecular-weight polyester or polyether is prepared. This intermediate is reacted with an isocyanate to form a prepolymer (macrodiisocyanate). The prepolymers are coupled or vulcanized by adding a diol or diamine:

$$2\,O{=}C{=}N{-}R{-}N{=}C{=}O + HO{-}(R')_n{-}OH \longrightarrow \underset{\text{intermediate}}{} \qquad \text{(Str. 51)}$$

$$O{=}C{=}N{-}R{-}\overset{H}{\overset{|}{N}}{-}\overset{O}{\overset{\|}{C}}{-}O{-}(R')_n{-}O{-}\overset{O}{\overset{\|}{C}}{-}\overset{H}{\overset{|}{N}}{-}R{-}N{=}C{=}O$$

prepolymer (macrodiisocyanate)

$$\underset{\text{prepolymer}}{O{=}C{=}N{-}R{-}\overset{H}{\overset{|}{N}}{-}\overset{O}{\overset{\|}{C}}{-}O{-}(R')_n{-}O{-}\overset{O}{\overset{\|}{C}}{-}\overset{H}{\overset{|}{N}}{-}R{-}N{=}C{=}O} + \underset{\text{diol}}{HO{-}R''{-}OH} \longrightarrow \qquad \text{(Str. 52)}$$

$$\left[{-}O{-}R''{-}O{-}\overset{O}{\overset{\|}{C}}{-}\overset{H}{\overset{|}{N}}{-}R{-}\overset{H}{\overset{|}{N}}{-}\overset{O}{\overset{\|}{C}}{-}O{-}(R')_n{-}O{-}\overset{O}{\overset{\|}{C}}{-}\overset{H}{\overset{|}{N}}{-}R{-}\overset{H}{\overset{|}{N}}{-}\overset{O}{\overset{\|}{C}}{-}O{-}R''{-}O{-}\right]_x$$

polyurethane

$$\underset{\text{prepolymer}}{O{=}C{=}N{-}R{-}\overset{H}{\overset{|}{N}}{-}\overset{O}{\overset{\|}{C}}{-}O{-}(R')_n{-}O{-}\overset{O}{\overset{\|}{C}}{-}\overset{H}{\overset{|}{N}}{-}R{-}N{=}C{=}O} + \underset{\text{diamine}}{H_2N{-}R''{-}NH_2} \longrightarrow \qquad \text{(Str. 53)}$$

$$\left[{-}\overset{H}{\overset{|}{N}}{-}R''{-}\overset{H}{\overset{|}{N}}{-}\overset{O}{\overset{\|}{C}}{-}\overset{H}{\overset{|}{N}}{-}R{-}\overset{H}{\overset{|}{N}}{-}\overset{O}{\overset{\|}{C}}{-}O{-}(R')_n{-}O{-}\overset{O}{\overset{\|}{C}}{-}\overset{H}{\overset{|}{N}}{-}R{-}\overset{H}{\overset{|}{N}}{-}\overset{O}{\overset{\|}{C}}{-}\overset{H}{\overset{|}{N}}{-}R''{-}\overset{H}{\overset{|}{N}}{-}\right]_x$$

polyurethane urea

Polyurethane elastomers have high strength; extremely good abrasion resistance; good resistance to gas, greases, oils, and hydrocarbons; and excellent resistance to oxygen and ozone. Applications include solid tires, shoe soles, gaskets, and impellers.

D. SILICONE ELASTOMERS

$$\left[{-}\underset{CH_3}{\underset{|}{\overset{CH_3}{\overset{|}{Si}}}}{-}O{-}\right]_n \qquad \text{(Str. 54)}$$

Polysiloxanes can be prepared by the hydrolysis of dichlorosilanes such as dimethydichlorosilanes:

$$n\,Cl{-}\underset{CH_3}{\underset{|}{\overset{CH_3}{\overset{|}{Si}}}}{-}Cl + n\,H_2O \longrightarrow \left[{-}\underset{CH_3}{\underset{|}{\overset{CH_3}{\overset{|}{Si}}}}{-}O{-}\right]_n + 2n\,HCl \qquad \text{(Str. 55)}$$

The process tends to result in the formation of cyclic products, typically trimers and tetramers. High-molecular-weight elastomers may be obtained by a subsequent base-catalyzed ring-opening polymerization of these cyclic siloxanes. Cross-linked siloxane elastomers are produced by partially cross-linking these linear polydimethylsiloxanes through the use of peroxide-based free-radical-initiated process. Alternatively, cross-linking may be effected by the incorporation of trifunctional monomers such as trichlorosiloxanes. To improve the efficiency of vulcanization, unsaturation is introduced into the polymer by copolymerizing a vinyl-group-containing siloxane. Typically about 10% of the methyl groups in polydimethylsiloxane are replaced by vinylmethylsilanol.

Silicone elastomers are noted for high temperature and oxidative stability, low temperature flexibility, good electrical properties, and high resistance to weathering and oil. They are used in wire and cable insulation and surgical implants and as material for gaskets and seals.

E. THERMOPLASTIC ELASTOMERS (TPE)

Thermoplastic elastomers are materials with functional properties of conventional thermoset rubbers and the processing characteristics of thermoplastics. They do not have the permanent cross-links present in vulcanized elastomers. Instead, elastomeric properties are the result of rigid-domain structures used to create a network structure. The domain structure is based on a block copolymer; one block consists of relatively long, flexible polymer chains (soft segment), while the other block is composed of stiff polymer molecules (hard segment). The rubber–plastic TPE is a two-phase mixture with a dispersion of the soft rubbery phase in a continuous glassy plastic matrix (Figure 15.12). Each polymer or major polymer segment or block has its softening temperature, T_s. The useful temperature range for TPE lies above the T_s of the elastomeric (soft) phase and below the T_s of the hard phase. Within this temperature range, the polymer molecules in the soft phase can undergo significant segmental motion. However, this motion is restricted by the bonds, such as hydrogen bonding, between chemical groups in the hard and soft phases. The reinforcing action of the hard phase disappears above its softening temperature, and the TPE behaves as a viscous liquid. Upon cooling, the hard phase resolidifies and the TPE becomes rubbery again. Similarly, cooling below the T_s of the soft phase changes the material from a rubbery to a hard, brittle solid. This process is also reversible. It is evident, therefore, that the physical nature of the domain structure of thermoplastic elastomers permits their reversibility. These elastomers are thermoplastic and can therefore be fabricated by conventional molding techniques by heating them above the softening temperature of the hard phase. The advantages and disadvantages of thermoplastic elastomers compared with conventional cross-linked rubbers are shown in Table 15.18.

A number of different classes of thermoplastic elastomers are currently commercially available. These include styrenics, polyurethanes, polyesters, polyolefin, and polyamides.

1. Styrene Block Copolymers (Styrenics)

The first and still most commercially important thermoplastic elastomers are those based on A–B–A block copolymers. A is a high-molecular-weight (50,000 to 100,000) polystyrene hard block, and B is low-molecular-weight (10,000 to 20,000) diene soft (elastomeric) block such as polybutadiene or polyisoprene or, sometimes, poly(ethylene–butylene).

These types of TPE, known as styrenics, e.g., Shell's kraton, have a useful temperature range of –70°C to 100°C. They are viscous melts above 130°C and can be molded and extruded into various articles. Typical applications include shoe soles, adhesives, and a number of molded products.

2. Thermoplastic Polyurethane Elastomers (TPUs)

Thermoplastic polyurethane elastomers are linear copolymers of the (AB)n type. They are composed of one block of polymer chain consisting of a relatively long, flexible "soft segment" derived from hydroxy-terminated polyester, polyether, or polyalkene. The second copolymer block is composed of a highly polar, rather stiff block (hard block) formed by the reaction of diisocyanates with low-molecular-weight diol or diamine chain extender. Intermolecular hydrogen bonding between the –NH group (proton donor) and the carbonyl group or ether oxygens (proton acceptor) results in a domain structure in which the hard blocks restrict the movement of the soft segments. This provides resistance to deformation when these elastomers are stretched.

Thermoplastic polyurethanes exhibit good abrasion and mar resistance. They are stable to attack by oxygen and hydrocarbons, but are sensitive to moisture and acidic and basic solutions, which can result in hydrolytic chain scission. They are used as coatings and adhesives and in footwear and automotive parts.

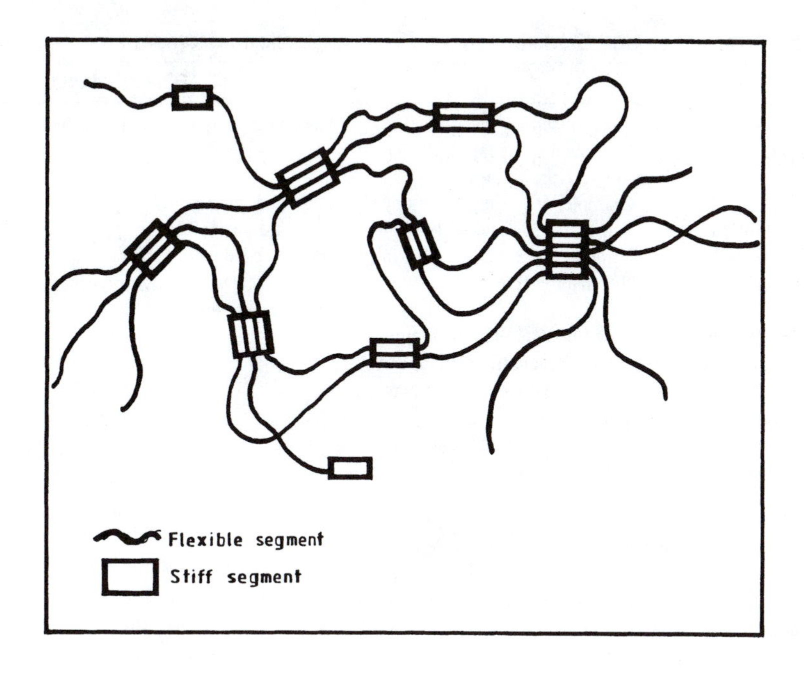

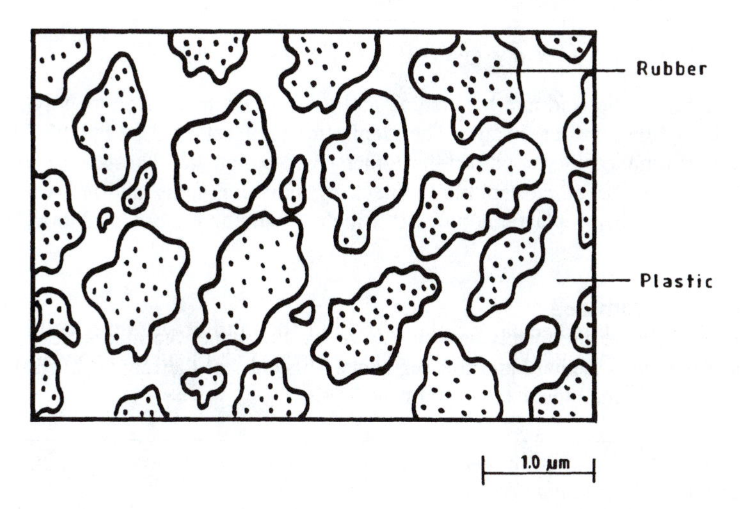

Figure 15.12 Thermoplastic elastomers are composed of a two-phase system either of a block copolymer (top) or a rubber–plastic mixture (bottom).

3. Polyolefin Blends

Olefinic thermoplastic elastomers are block copolymers or blends of polyolefins — commonly, polypropylene, which forms the hard crystalline block, and another olefin block, most commonly ethylene or EPDM. Some less common soft segments include natural rubber, nitrile rubber, and EVA. Olefinic thermoplastic elastomers exhibit better processability than neoprene and have excellent resistance to oils. Therefore, they offer attractive replacements for neoprene in oil-resistant wire and cable insulation.

4. Thermoplastic Copolyesters (COPE)

Thermoplastic copolyesters are condensation block copolymers based on a crystalline polyester hard segment and an amorphous long-chain polyether soft segment. The hard segment, for example, formed

Table 15.18 Advantages and Disadvantages of Thermoplastic Elastomers

Advantages	Disadvantages
TPEs require little or no compounding since they are fully formulated and are ready for use as received. Offer rapid, efficient, and economical means of fabricating rubber articles.	Most TPEs may require drying before processing
Processing of TPEs is cheaper and simpler requiring fewer steps than does a thermoset rubber. A single step is required to shape TPEs into final article, whereas thermoset rubbers require mixing and vulcanization in addition to the shaping step.	TPEs melt at elevated temperatures; consequently, they are unsuitable for applications requiring even brief exposures to temperatures above their melting points. Many thermoset rubbers can withstand such a short exposure.
Processing times for TPEs are shorter since the long cycle time required for vulcanization is eliminated. Cycle time is typically seconds instead of the minutes required for thermoset rubbers. This enhances productivity and reduces costs.	
Being thermoplastic, TPE scrap generated during processing can be recycled without any loss in functional properties. In contrast, thermoset elastomers are not recyclable.	
As a result of the fewer processing steps and shorter cycle time, lower energy is consumed in TPE processing than processing of thermoset rubbers.	
It is easier to obtain finished parts having closer tolerances with TPEs because of consistent material base and greater ease and control of processing.	
Elastomers are generally purchased on weight basis but used on volume basis. Consequently, TPEs offer cost advantages since they normally have lower specific gravity than thermoset rubbers.	

From Heineck, D.W. and Rader, C.P., *Plast. Eng.*, 45(3), 87, 1987. With permission.

by the reaction of terephthalic acid and butanediol, and the soft segment, such as poly(tetramethylene ether glycol), are joined by an ester linkage. Thermoplastic copolyesters, for example, DuPont Hytrel, are durable, high-performance thermoplastic elastomers that exhibit good resistance to abrasion and hydrocarbons. They are, however, susceptible to attack by moisture, acids, and bases at elevated temperatures. Applications of thermoplastic copolyesters include wire and cable insulation, gaskets, seals, hoses, and automotive parts.

5. Thermoplastic Polyamides

Thermoplastic polyamides are a relatively new class of block-copolymer thermoplastic elastomers. They have amide linkages between the hard and soft segments. This class of thermoplastic elastomers has the highest performance and is the most expensive of the TPEs.

Example 15.7: Currently available thermoplastic elastomers cannot be used in applications with use temperatures above 170°C. Comment on this statement.

Solution: The thermoplastic elastomers that are currently available are used in numerous applications requiring the elastomeric properties of conventional thermoset rubbers. These include nontire automotive applications (fascia, hoses, gaskets, seals, bushings, protective boots, weather stripping); mechanical rubber goods (components of modern appliances, including dishwashers, power drills, telephone, and electronic equipment); architectural and construction uses (window glazing systems, seals for windows and doors, primary electrical insulation and jacketing for cables, and foamed seals and tapes); and use in rubber articles in contact with food and in medical applications. However, the upper limit of the useful temperature range of thermoplastic elastomers is set by the softening or melting temperature of the hard segment. Above this temperature, the TPE loses its elastomeric properties and becomes a viscous liquid. Thermoplastic elastomers that are currently available have hard segments, with melting or softening temperatures much below 170°C.

IX. THERMOSETS

The principal feature that distinguishes thermosets and conventional elastomers from thermoplastics is the presence of a cross-linked network structure. As we have seen from the above discussion, in the case of elastomers the network structure may be formed by a limited number of covalent bonds (cross-linked rubbers) or may be due to physical links resulting in a domain structure (thermoplastic elastomers). For elastomers, the presence of these cross-links prevents gross mobility of molecules, but local molecular mobility is still possible. Thermosets, on the other hand, have a network structure formed exclusively by covalent bonds. Thermosets have a high density of cross-links and are consequently infusible, insoluble, thermally stable, and dimensionally stable under load. The major commercial thermosets include epoxies, polyesters, and polymers based on formaldehyde. Formaldehyde-based resins, which are the most widely used thermosets, consist essentially of two classes of thermosets. These are the condensation products of formaldehyde with phenol (or resorcinol) (phenoplasts or phenolic resins) or with urea or melamine (aminoplastics or amino resins).

A. PHENOLIC RESINS

Phenolic resins, introduced in 1908, are formed by either base- or acid-catalyzed addition of formaldehyde to phenol to give ortho- and para-substituted products. The nature of these products depends largely on the type of catalyst and the mole ratio of formaldehyde to phenol. In resole formation, excess formaldehyde is reacted with phenol under basic conditions. The initial reaction products are ortho- and para-substituted mono-, di-, and trimethylolphenols:

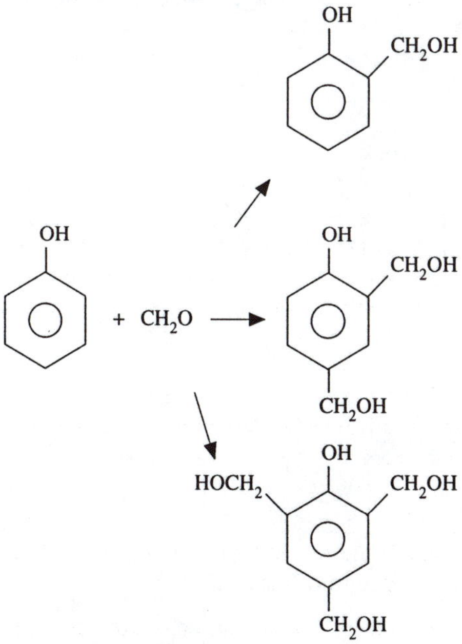

OH, CH_2OH

OH + CH_2O → OH, CH_2OH, CH_2OH (Str. 56)

$HOCH_2$, OH, CH_2OH, CH_2OH

When heated, methylol phenols, condense either through methylene or methylene oxide linkages to give a low-molecular-weight prepolymer called resole, which is soluble and fusible and contains alcohol groups:

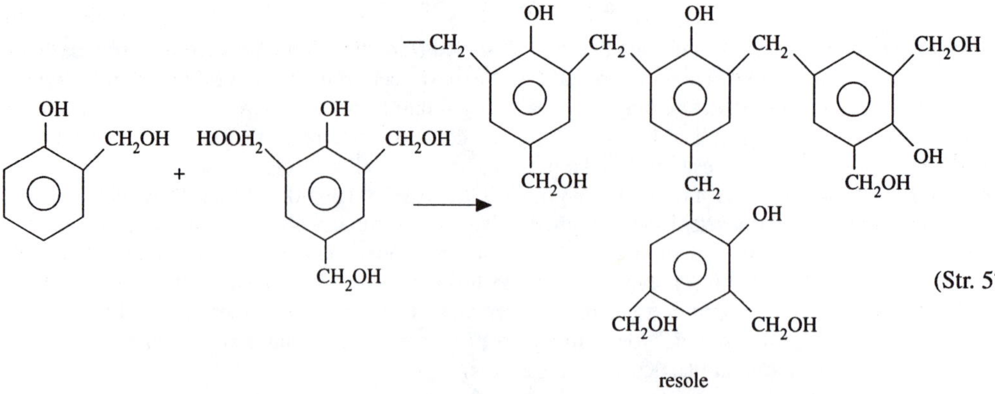

(Str. 57)

resole

When resoles are heated at elevated temperatures, under basic, neutral, or slightly acidic conditions, condensation of large numbers of phenolic nuclei takes place resulting in a high-molecular-weight cross-linked network structure.

Formation of novolak involves an acid-catalyzed reaction of formaldehyde with excess phenol (i.e., formaldehyde-to-phenol mole ratio less than 1). The initial methylol phenols condense with the excess phenol to form dihydroxydiphenyl methane, which undergoes further condensation yielding low-molecular-weight prepolymer or novolak. Unlike resoles, novolaks do not contain residual methylol groups. They are fusible and insoluble.

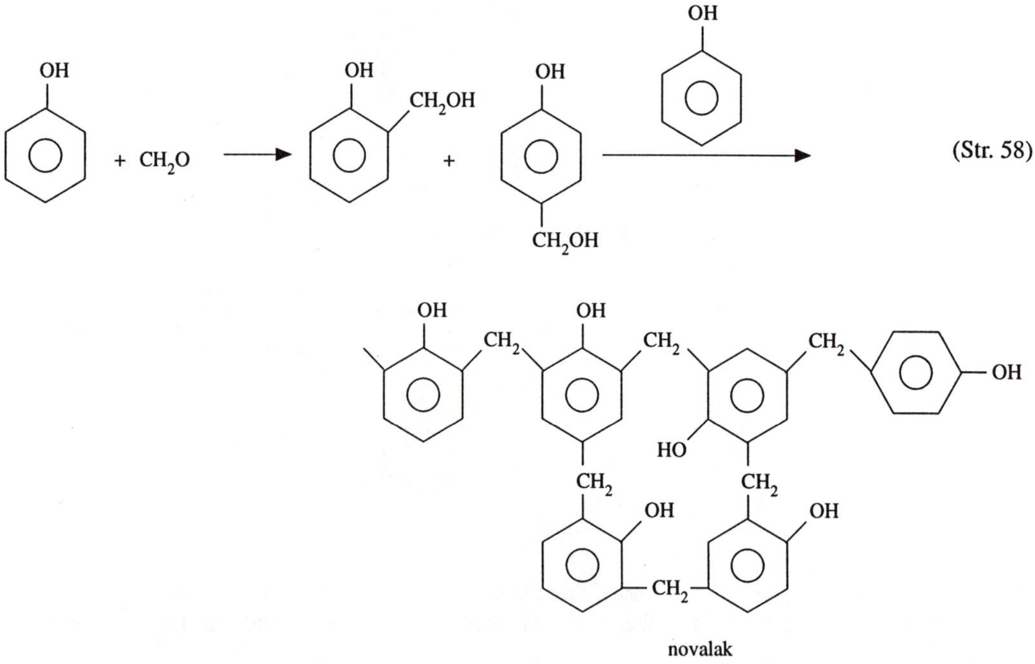

(Str. 58)

novalak

When novolaks are heated with additional paraformaldehyde or hexamethylene tetramine to raise the formaldehyde-to-phenol ratio above unity, high-molecular-weight cross-linked network structure is formed.

The largest use of phenol–formaldehyde resins is in plywood manufacture. Other applications include lacquers and varnishes, cutlery handles, and toilet seats. Molded parts are used in distributor caps, fuse boxes, and other electrical outlets because of the superior dimensional stability and electrical properties of PF resins. Decorative laminates from PF resins are used for countertops and wall coverings, while industrial laminates are used for electrical parts, including printed circuits. Other industrial applications of phenolics based on their excellent adhesive properties and bond strength include brake linings, abrasive wheels and sandpaper, and foundry molds.

B. AMINO RESINS

Urea–formaldehyde (UF) resins are obtained by a two-staged polymerization process. The first stage involves the reaction of formaldehyde with urea at a formaldehyde-to-urea (F/U) mole ratio equal to or greater than one under slightly alkaline conditions. This results in the formation of mono- and dimethylol urea and, possibly, trimethylol urea, depending on the F/U ratio (Figure 15.13A). The next stage involves condensation of these methylol ureas, usually under acidic conditions. Depending on the extent of reaction, the condensation reactions lead ultimately to a cross-linked product (Figure 15.13B). The production of melamine–formaldehyde resins, like UF resins, involves initial methoylation reaction followed by the formation of a rigid network structure (Figure 15.13C,D).

In contrast to phenolic resins, amino resins are clear and colorless. They are harder and have higher strength but lower moisture and heat resistance than phenolics. Melamine–formaldehyde resins have better moist-heat aging resistance than UF resins, but are more costly. Like phenolics, aminoplasts can be used to improve the shrink and crease properties, fire retardance, and water repellency of textiles and the wet and bursting strength of paper. UF resins are used in molding and in laminating applications. However, a greater part of UF resins are used for adhesive applications, particularly where the darker color of phenolics may be objectionable and the relatively inferior durability of UF resins does not pose

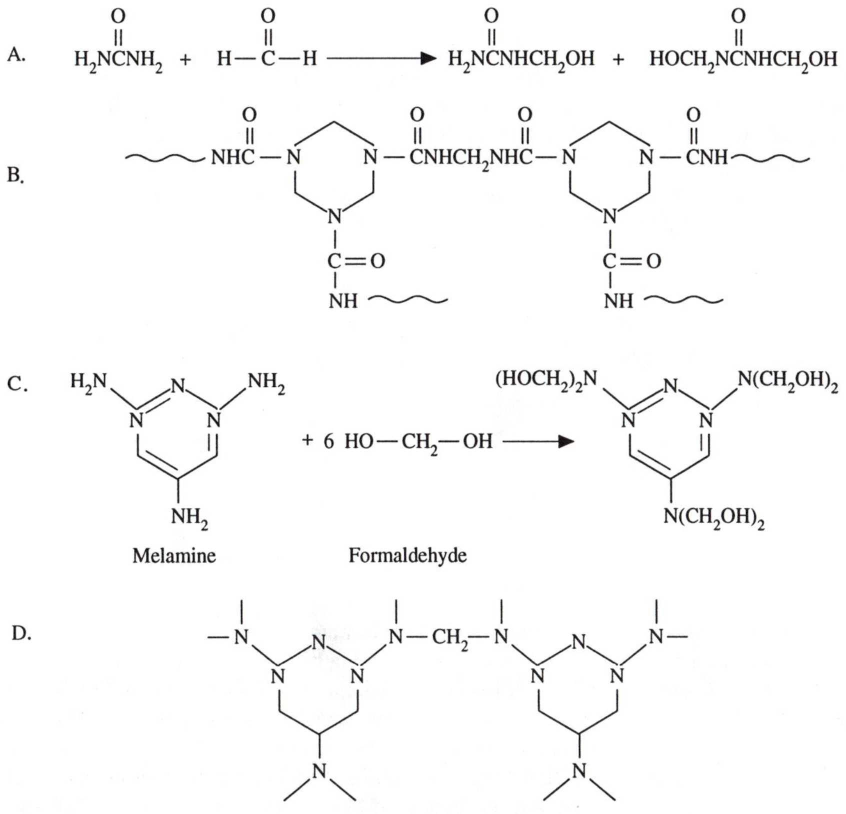

A. $H_2NC(=O)NH_2 + H{-}C(=O){-}H \longrightarrow H_2NC(=O)NHCH_2OH + HOCH_2NC(=O)NHCH_2OH$

B. ~NHC(=O)–N(ring)N–C(=O)NHCH₂NHC(=O)–N(ring)N–C(=O)NH~ (ring N–C=O–NH~)

C. Melamine + 6 HO—CH₂—OH (Formaldehyde) ⟶ $(HOCH_2)_2N$, $N(CH_2OH)_2$, $N(CH_2OH)_2$ substituted triazine

D. —N(triazine)N—CH_2—N(triazine)N— network

Figure 15.13 Two-staged process for the formation of network structure by UF resin (A and B) and MF resin (C and D).

a serious problem. A typical example is in the production of interior grade plywood. UF resins are also used in the manufacture of electrical switches and plugs. Typical applications of MF resins include high-quality decorative dinnerware, laminated counter-, cabinet, and tabletops; and electrical fittings. Formaldehyde-based resins may be used neat or compounded with various additives such as wood flour to improve strength properties and reduce cost.

C. EPOXY RESINS

Epoxy resins are complex network polyethers usually formed in a two-staged process. The first stage involves a base-catalyzed step-growth reaction of an excess epoxide, typically epichlorohydrin with a dihydroxy compound such as bisphenol A. This results in the formation of a low-molecular-weight prepolymer terminated on either side by an epoxide group.

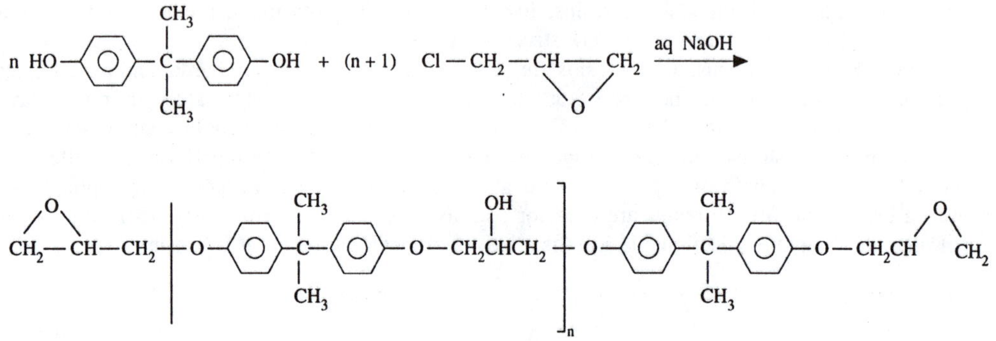

$$n\ HO-C_6H_4-C(CH_3)_2-C_6H_4-OH + (n+1)\ Cl-CH_2-\underset{\backslash O /}{CH-CH_2} \xrightarrow{\text{aq NaOH}}$$

$$\underset{\backslash O /}{CH_2-CH}-CH_2\left[O-C_6H_4-C(CH_3)_2-C_6H_4-O-CH_2\overset{OH}{\overset{|}{C}H}CH_2\right]_n O-C_6H_4-C(CH_3)_2-C_6H_4-O-CH_2\underset{\backslash O /}{CH-CH_2}$$

(Str. 59)

The prepolymer may be a viscous liquid, generally for commercial epoxies, or a hard and tough solid depending on the value of n. Other hydroxyl-containing compounds such as resorcinol, glycol, and glycerol can also be used but commercial epoxy resins are based on bisphenol A. In the second stage, a cross-linked network structure is formed by curing the prepolymer with active hydrogen-containing compounds. These curing agents include polyamines, polyacids and acid anhydrides, polyamides, and formaldehyde resins. Amines, preferably liquid amines like triethylene diamine, effect cure of the prepolymer by ring-opening of the terminal epoxide groups.

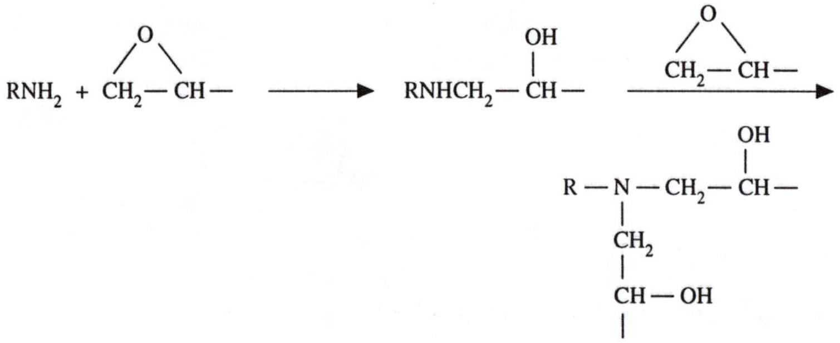

$$RNH_2 + \underset{\backslash O /}{CH_2-CH}- \longrightarrow RNHCH_2-\overset{OH}{\overset{|}{C}H}- \xrightarrow{\underset{\backslash O /}{CH_2-CH}-} R-\underset{\underset{\underset{|}{CH-OH}}{\overset{|}{CH_2}}}{\overset{|}{N}}-CH_2-\overset{OH}{\overset{|}{C}H}- \quad \text{(Str. 60)}$$

Epoxy resins are very versatile materials with a wide range of applications. They are tough and flexible and have high tensile, compressive, and flexural strengths. Epoxy resins have excellent chemical and corrosion resistance and good electrical insulation properties. They can be cured over a wide range of temperatures with very low shrinkage. Epoxy resins have outstanding adhesion to a variety of substrates, including metals and concrete. Consequently, the major uses of epoxy resins are in protective coating applications. Other uses of epoxies include laminates and composites. Potting, encapsulation, and casting with epoxy resins are common procedures in the electrical and tooling industries.

D. NETWORK POLYESTER RESINS

Network polyester resins may be categorized into saturated and unsaturated polyesters. Unlike linear saturated polyesters such as PET, which are made from difunctional monomers, saturated polyesters (glyptal) are formed by the reaction of polyols such as glycerol with dibasic acids such as phthalic anhydride.

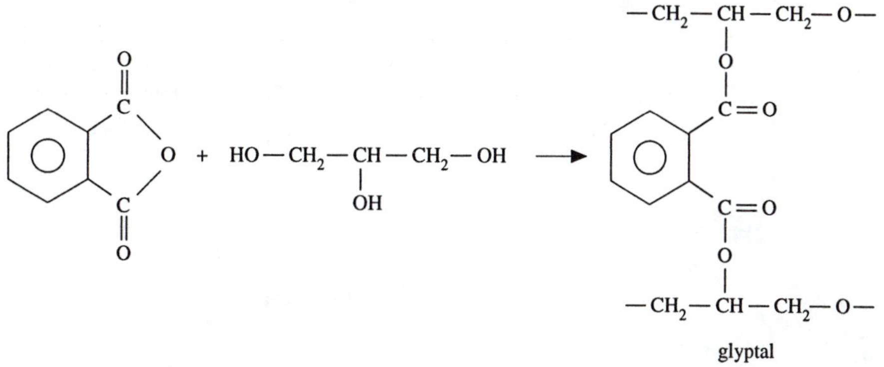

$$\text{phthalic anhydride} + HO{-}CH_2{-}\underset{\displaystyle OH}{\underset{|}{CH}}{-}CH_2{-}OH \longrightarrow \text{glyptal} \quad \text{(Str. 61)}$$

glyptal: $-CH_2-CH(-O-C(=O)-C_6H_4-C(=O)-O-)-CH_2-O-$ network, linking $-CH_2-CH-CH_2-O-$ units

The reaction involves an initial formation of a viscous liquid that is then transferred to a mold for hardening or network formation.

Unsaturated network polyesters may be produced by using an unsaturated dibasic acid or glycol or both. Typically the unsaturated dibasic acid, such as maleic anhydride or fumaric acid, is copolymerized with a saturated acid such as phthalic acid and a glycol (e.g., propylene glycol or diethylene glycol).

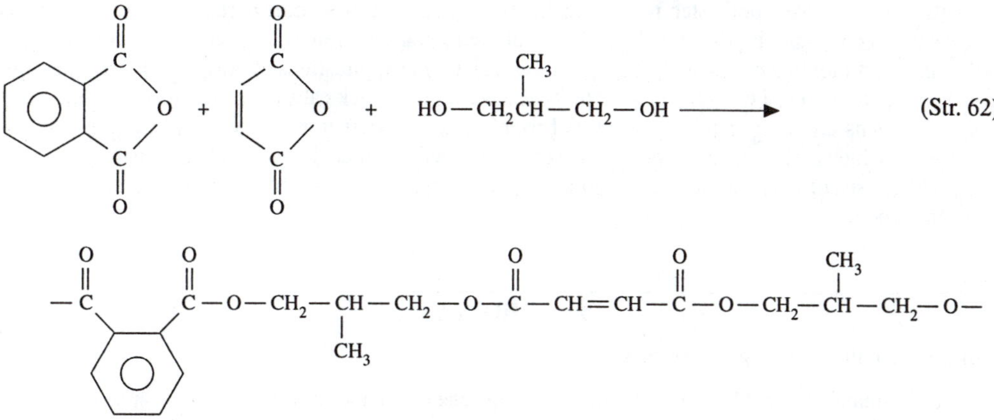

$$\text{phthalic anhydride} + \text{maleic anhydride} + HO{-}CH_2\overset{\displaystyle CH_3}{\overset{|}{CH}}{-}CH_2{-}OH \longrightarrow \quad \text{(Str. 62)}$$

$$-\overset{O}{\overset{\|}{C}}-C_6H_4-\overset{O}{\overset{\|}{C}}-O-CH_2-\underset{\displaystyle CH_3}{\underset{|}{CH}}-CH_2-O-\overset{O}{\overset{\|}{C}}-CH=CH-\overset{O}{\overset{\|}{C}}-O-CH_2-\overset{\displaystyle CH_3}{\overset{|}{CH}}-CH_2-O-$$

The acid and glycol components are allowed to react until a low-molecular-weight (1500 to 3000) viscous liquid (usually) is formed. This prepolymer is then dissolved in styrene monomer, which participates in cross-linking with the double bonds in the prepolymer by addition of peroxide or hydroperoxide to form the final network structure.

$$-O-\overset{\overset{\large O}{\|}}{C}-CH{=}CH-\overset{\overset{\large O}{\|}}{C}-O- \; + \; CH_2{=}CH(C_6H_5) \longrightarrow \qquad \text{(Str. 63)}$$

$$\begin{array}{c} -O-\overset{\overset{O}{\|}}{C}-\overset{|}{CH}-CH-\overset{\overset{O}{\|}}{C}-O- \\ \qquad\qquad | \\ \qquad\qquad CH_2 \\ \qquad\qquad | \\ C_6H_5-CH \\ \qquad\qquad | \\ -O-\overset{\overset{O}{\|}}{C}-\underset{|}{CH}-CH-\overset{\overset{O}{\|}}{C}-O- \end{array}$$

The saturated acid (phthalic anhydride) helps to reduce the cross-link density and, hence, the brittleness of the cured polyester resin. Resin composition can be varied so that product properties can be tailored to meet specific end-use requirements. For example, a resin with enhanced reactivity and improved stiffness at high temperatures is obtained by increasing the proportion of unsaturated acid. On the other hand, a less reactive resin with reduced stiffness is obtained with a higher proportion of the saturated acid.

Unsaturated polyester resins are light in color, easy to handle and cure rapidly without emission of volatiles. They are dimensionally stable and have good physical and electrical properties. The major markets for reinforced polyester resins are in marine applications, transportation, construction, and electrical and consumer products. Fabrication of pleasure boats as well as large commercial vessels from polyester sheet molding compounds (SMC) are typical marine applications of polyester. In transportation, polyester resins are used in passenger-car parts and bodies and truck cabs in an attempt to reduce weight. Polyester resins are used mainly as sheet and paneling and also as tubs, shower stalls, and pipes in the construction industry. Consumer products from polyesters include luggage, chairs, fishing rods, and trays. Other uses of polyester are electrical applications, appliances, business equipment, and corrosion-resistant products.

X. PROBLEMS

15.1. Explain the following observations.

a. Extremely broad MWD HDPE with polypropylene copolymer is used in the insulation of telephone cables.
b. Polypropylene is preferred to polyethylene and polystyrene in products that must be heat or steam sterilized.
c. Polypropylene is inert to ethanol and acetone, while solvents like benzene and carbon tetrachloride will cause swelling.
d. To retain transparency, random copolymers of propylene and ethylene are used for films applications, while polypropylene homopolymer is used almost exclusively for filaments.
e. Chlorinated PVC pipes are used in residential hot-water systems.
f. PVC currently has limited application in food packaging and in such applications it must meet stringent government standards.
g. General-purpose ABS finds extensive use as refrigerator door and food liners, while fire-retardant grades are used in appliance housing, business machines, and television cabinets and in aircraft for interior applications.
h. Nylon is used for low- and medium-voltage 60-Hz application, for example, as secondary insulation or jacket for primary electrical wire insulation.

i. Nylons 11, 12, 6,12 are preferred to nylons 6 and 6,6 for automotive wiring harness and pneumatic tubing.
j. Principal applications of poly(phenylene sulfide) include cookware, bearings, and pump parts for service in various corrosive environments.
k. Compared with PTFE, poly(chlorotrifluoroethylene) has a lower melting point (218°C vs. 327°C); has inferior electrical properties, particularly for high frequency applications; and can be quenched to quite clear sheets.
l. Butyl rubber has excellent resistance to ozone attack and is less sensitive to oxidative aging than most other elastomers — for example, natural rubber.

REFERENCES

1. Frados, J., Ed., *The Society of the Plastics Industry,* 13th ed., Society of the Plastics Industry, New York, 1977.
2. Ulrich, H., *Introduction to Industrial Polymers,* Macmillan, New York, 1982.
3. Waddams, A.L., *Chemicals from Petroleum,* Gulf Publishing Co., TX, 1980.
4. Chruma, J.L. and Chapman, R.D., *Chem. Eng. Prog.,* 81(1), 49, 1985.
5. Fried, J.R., *Plast. Eng.,* 38(12), 21, 1982.
6. Fried, J.R., *Plast. Eng.,* 39(5), 35, 1983.
7. Billmeyer, F.W., Jr., *Textbook of Polymer Science,* 3rd ed., Interscience, New York, 1984.
8. Mock, J.A., *Plast. Eng.,* 40(2), 25, 1984.
9. Lundberg, R.D., Ionomers, in *Encyclopedia of Chemical Technology,* 3rd ed., Mark, H.F., Othmer, D.F., Overberger, C.G., and Seaborg, G.T., Eds., Interscience, New York, 1984.
10. *1979/80 Modern Plastics Encyclopedia,* McGraw-Hill, New York, 1980.
11. MacKnight, W.J. and Earnest, T.R., Jr., *J. Polym. Sci. Macromol. Rev.,* 16, 41, 1981.
12. Fried, J.R., *Plast. Eng.,* 39(3), 67, 1983.
13. Greek, B.F., *Chem. Eng. News,* 25, March 21, 1988.
14. Gogolewski, S., *Colloid Polym. Sci.,* 267, 757, 1989.
15. Heineck, D.W. and Rader, C.P., *Plast. Eng.,* 45(3), 87, 1987.

Appendix I

Polymer Nomenclature

The IUPAC has specific guidelines for the nomenclature of polymers. However, these names are quite frequently discarded for common names and even principal trade names. Even though there is currently no completely systematic polymer nomenclature, there are some widely accepted guidelines that are used to identify individual polymers.

Simple vinyl polymers are named by attaching the prefix *poly* to the monomer name. For example, the polymer made from styrene becomes polystyrene. However, when the monomer name consists of more than one word or is preceded by a letter or a number, the monomer is enclosed in parenthesis with the prefix *poly.* Thus polymers derived from vinyl chloride or 4-chlorostyrene are designated poly(vinyl chloride) and poly(4-chlorostyrene), respectively. This helps to remove any possible ambiguity.

Diene polymerization may involve either or both of the double bonds. Geometric and structural isomers of butadiene, for example, are indicated by using appropriate prefixes — *cis* or *trans*; 1,2 or 1,4 — before *poly,* as in *cis*-1,2-poly(1,3-butadiene). Tacticity of the polymer may be indicated by using the prefix *i* (isotactic), *s* (syndiotactic), or *a* (atactic) before *poly,* such as *s*-polystyrene. Copolymers are identified by separating the monomers involved within parentheses by either *alt* (alternating), *b* (block), *g* (graft), or *co* (random), as in poly(styrene-*g*-butadiene).

When side groups are attached to the main chain, some ambiguity could result from naming the polymers. For example, poly(methylstyrene) is an appropriate designation for any of the following structures.

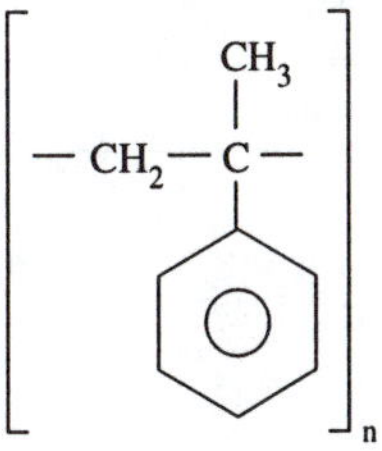

$$\left[-CH_2-C(CH_3)(C_6H_5)-\right]_n$$

(1)

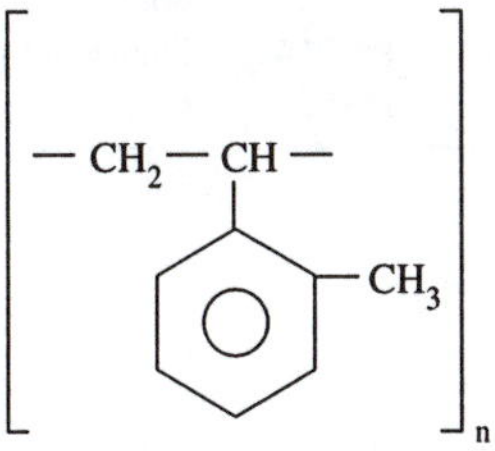

$$\left[-CH_2-CH(C_6H_4\text{-}o\text{-}CH_3)-\right]_n$$

(2)

$$\left[-CH_2-CH(C_6H_4\text{-}p\text{-}CH_3)-\right]_n$$

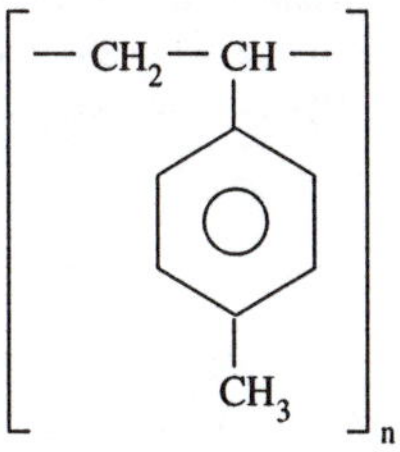

(3)

To avoid such ambiguity, these structures are designated poly(α-methylstyrene) (1), poly(*o*-methylstyrene) (2), and poly(*p*-methylstyrene) (3), respectively.

The nomenclature of step-reaction polymers is even more complicated than that of vinyl polymers and can be quite confusing. These polymers are usually named according to the source or initial monomer(s) and the type of reaction involved in the synthesis. For example, nylon 6,6 (4) is usually designated poly(hexamethylene adipamide), indicating an amidation reaction between hexamethylenediamine and adipic acid. Nylon 6 is called either poly(6-hexanoamide) or poly(ε-caprolactam). The former name indicates the structural and derivative method while the latter, which is more commonly used, is based on the source of the monomer.

$$\left[-NH-(CH_2)_6-NH-C(=O)-(CH_2)_4-C(=O)-\right]_n$$

(4)

$$\left[-NH-(CH_2)_5-C(=O)-\right]_n$$

(5)

Some polymers are referred to almost exclusively by their common names instead of the more appropriate chemical names. An example is polycarbonate in place of poly(2,2-bis(4-hydroxyphenyl) propane) (6).

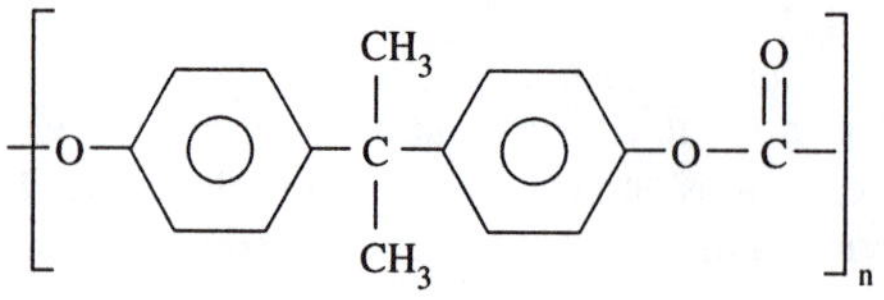

(6)

The following table lists the internationally accepted abbreviations for some common commercial polymers.

Name of Plastic	Abbrev.	Name of Plastic	Abbrev.
Cellulose acetate	CA	Polypropylene	PP
Chlorinated poly(vinyl chloride)	CPVC	Polystyrene	PS
Melamine–formaldehyde resins	MF	Polytetrafluoroethylene	PTFE
Poly(acrylonitrile-*co*-butadiene)	NBR	Polyurethane	PUR
Polyacrylonitrile	PAN	Poly(vinyl acetate)	PVAC
Bisphenol A polycarbonate	PC	Poly(vinyl alcohol)	PVAL
Polyethylene	PE	Poly(vinyl butyral)	PVB
Poly(ethylene terephthalate)	PETP	Poly(vinyl chloride)	PVC
Phenol–formaldehyde resins	PF	Poly(vinylidene fluoride)	PVDC
Poly(methyl methacrylate)	PMMA	Poly(vinyl pyrrolidone)	PVP
Polyoxymethylene	POM	Urea–formaldehyde resins	UF

Appendix II

Answers to selected problems

Chapter 1

1.5 (a) 1.13×10^5; (b) 1.92×10^5; (c) 1.18×10^5; (d) 2.54×10^5

Chapter 3

3.3 $\overline{M}_n = 3770$ g/mol; $\overline{M}_2 = 31{,}000$ g/mol; $\overline{M}_w/\overline{M}_n = 8.2$
3.4 $\overline{M}_n = 2.35 \times 10^5$ g/mol, $\overline{M}_w = 7.36 \times 10^5$
3.10 8.1

Chapter 4

4.2 $V_B = 32.4\%$
4.4 $\overline{X}_n = 50$
4.5 $T_m = 286.5°C$
4.6 $T_m = 194°C$
4.7 $f = 0.028$

Chapter 5

5.10 (a) 59.26 phr; (b) 296.30 phr

Chapter 6

6.1 (a) $\overline{M}_n = 5650$ g/gmol; (b) $\overline{X}_n = 30.45$; (c) $\overline{M}_n = 11{,}300$ g/gmol; (d) $\overline{M}_n = 7533$ g/gmol
6.2 (a) $\overline{M}_n = 6000$ g/gmol; $\overline{M}_w = 8667$ g/gmol; (b) melt viscosity will increase
6.3 (a) $\overline{M}_n = 2504$ g/gmol; (b) $\overline{M}_n = 4407$ g/gmol
6.4 (1) 2; 100; 00; (b) $\overline{X}_n = 199$; (c) $\overline{X}_n = 49$
6.5 (a) $\overline{M}_n = 9600$; (b) $\overline{M}_n = 19{,}104$; (c) 9.80×10^{-3}; (d) 2.94×10^{-6}
6.6 For n = 1 p = 0.1, $W_x = 0.98$
P = 0.9, $W_x = 0.01$
For n = 100 p = 0.1, $W_x = 8.1 \times 10^{-98}$
P = 0.9, $W_x = 2.95 \times 10^{-5}$
6.7 Nylon 6, $\overline{M}_n = 3260$
Nylon 12, $\overline{M}_n = 3240$
6.8 Fraction of monomers = 4.0×10^3

Chapter 7

7.2 (b) $R_i = 7.5 \times 10^{-11}$ mol/ml, $\overline{X}_n = 4.01 \times 10^3$; (c) 94.3%
7.3 (a) $R_p = 0.715[I]^{1/2}[M]$; (b) $R_p = 0.044$ mol/l.s.; (c) 15 ls
7.4 (a) $C_M = 0.6 \times 10^{-4}$; (b) $\overline{X}_n = 602$; (c) $\overline{X}_n = 833$; (d) V = 415; (e) G = 0.055; (f) f = 0.61
7.5 Cyclohexane $\overline{M}_n = 3.69 \times 10^5$
Carbon tetrachloride $\overline{M}_n = 2.87 \times 10^3$
7.6 Styrene [S]/[M] = 19.35
Methyl methacrylate [S]/[m] = 11.54
Vinyl Aceptate [S][M] = 0.07
7.10 (a) $\overline{M}_n = 1.04 \times 10^6$; (b) $\overline{M}_n = 2.08 \times 10^6$; (c) $\overline{M}_n = 1.49 \times 10^5$
7.11 (a) 250 mn; (b) E = 7.3 Kcal/mol

Chapter 9

9.2 $E_L = 2.88 \times 10^6$ psi

Chapter 10

10.2 $t = 11.1$ h
10.3 $t = 5.96 \times 10^t$ s
10.4 $[M]/[M_o] = 94.9\%$
10.5 $\Delta T = 617°C$
10.6 $(\overline{X}_n)_b/(\overline{X}_n)_s = 4.75$
10.8 Ratio of total surface area of micelles to droplets = 20×10^7
10.9 (a) $t = 0.40$ h; (b) $D = 4.44 \times 10^{-5}$ cm
10.10 $T_c = 30.1°C$
10.11 (a) $\tau = 2.53$ h; (b) flow rate = 19.73 m^3/h
10.12 (a) $p = 0.632$; (b) $L = 100$ m
10.16 Feed temperature = 50°F

Chapter 11

11.1 Power Requirement = 7–18 hp
11.3 (a) Polystyrene Q = 591/b/h; (b) Polyethylene Q = 259/b/h; (c) Nylon 6,6 Q = 329/b/h
11.4 (a) 176 hp; (b) 233 hp
11.5 (a) 1.3%; (b) 2.1%; (c) 3.6%

Chapter 12

12.1 Polymer Most Suitable Solvent
Natural Rubber Dichlorobenzene
Polyacrylonitrite Nitromethane
12.4 (a) 6.9; (b) 1.86; (c) 6.04; (d) 3.47×10^4 A°
12.6 $P_1 = 92$ mmHg
12.7 $M = 1.58 \times 10^6$
12.8 $(\overline{r_{of}^2})^{1/2} = 302$ A°
12.9 (a) 3.98×10^4 A°; (b) $r = 76$ A°; (c) $(\overline{r_{of}^2})^{1/2} = 750$ A°
12.11 $k = 0.561$; $\overline{M}_n = 2.22 \times 10^5$ g/mol
12.12 (a) Solution A: $\overline{M}_n = 1.89 \times 10^5$ g/mol
Solution B: $\overline{M}_n = 1.01 \times 10^5$ g/mol
(b) $\overline{M}_n = 1.14 \times 10^5$ g/mol
(c) $\overline{M}_w/\overline{M}_n = 2.16$

Chapter 13

13.1 79.78×10^4 J/m^3
13.4 7.07×10^5 N/m^2
13.5 2×10^2 N
13.6 67.55×10^6 N/m^2

Chapter 14

14.2 78.04×10^5 N/m^2
14.3 0.517
14.4 0.701
14.5 83°C
14.6 2.5×10^2 s
14.7 194°C
14.8 1.29×10^{-3}

Appendix III

Some Useful Conversion Factors

To Convert From	To	Multiply By
atmosphere (760mm Hg)	pascal (Pa)	1.013×10^5
Btu	joule (J)	1.055×10^3
calorie	joule (J)	4.187
centipoise	pascal-second (Pa·s)	1.00×10^{-3}
foot	meter (m)	3.048×10^{-1}
ft-lb_f	joule (J)	1.356
gallon (U.S. liquid)	cubic meter (m^3)	3.785×10^{-3}
horsepower	watt (W)	7.460×10^2
inch	meter (m)	2.540×10^{-2}
inch of mercury (60°F)	pascal (Pa)	3.377×10^3
inch of water (60°F)	pascal (Pa)	2.488×10^2
kilogram-force (K_{gf})	newton (N)	9.807
micron	meter (m)	1.000×10^{-6}
pound-force (lb_f)	newton (N)	4.448
$lb_f/in.^2$ (psi)	pascal (Pa)	6.895×10^3
watt-hour	joule (J)	3.600×10^3
yard	meter (m)	9.144×10^{-1}

Values of Some Useful Physical Constants

	cgs	SI
Avogadro's number, No.	6.02×10^{23} molecules/mol	6.02×10^{23} molecules/mol
Velocity of light, c	3.00×10^{10} cm/s	3.00×10^8 m/s
Boltzmann's constant, K	1.38×10^{-16} erg/K	1.38×10^{-23} J/K
Gas constant, R	8.31×10^7 erg/g mol·K (1.98 cal/mol·K)	8.31×10^3 J/kg mol·K

INDEX

C

D

E

F

G

H

I

N

O

P

Q

R

S

T

U

V

W

X

Y

Z